D1065245

Acta Numerica 1992

Acta
Numerica
1992

CAMBRIDGE
UNIVERSITY PRESS

Published by the Press Syndicate of the University of Cambridge
The Pitt Building, Trumpington Street, Cambridge CB2 1RP
40 West 20th Street, New York, NY 10011-4211, USA
10 Stamford Road, Oakleigh, Victoria 3166, Australia

First published 1992

Printed in the United States of America

A catalog record for this book is available from the British Library

ISBN 0-521-41026-6 hardback
ISSN 0962-4929

Contents

Preface

In these days and age, when the sheer number of publications in numerical mathematics increases so rapidly, it is surely necessary to provide valid justification to a new publication. The reason for *Acta Numerica* is, paradoxically, to counteract the information explosion by presenting selected and important developments in numerical mathematics and scientific computation on an annual basis.

Each year, the Editorial Board of *Acta Numerica* poses itself the question 'what are recent significant developments in our subject, developments that are important enough to merit interest by the numerical community as a whole'. Having selected a shortlist of topics, we ask named individuals to write survey papers. The purpose of the exercise being to disseminate knowledge outside restricted professional boundaries, the authors are expected to pitch their exposition so that it can be understood and appreciated by all practitioners of the numerical art, and not just by workers in a narrow sub-discipline. We are guided in our choice of authors both by their contribution to the underlying topic *and* by their track record as expositors.

Numerical analysts, like other professionals in a competitive world, are busy with their own research, academic and administrative duties. It is difficult (and sometimes impossible) to keep up with developments outside one's narrow experience. This, we believe, is an unhealthy and undesirable situation, not only because of broader cultural considerations but also since developments in different parts of numerical mathematics frequently impinge upon each other. We hope that *Acta Numerica* will play a role in bridging gaps and presenting many new and exciting ideas – algorithms and mathematical analysis alike – to a wider audience.

Acta Numerica (1991), pp. 1–56

Wavelets*

Ronald A. DeVore

Department of Mathematics
University of South Carolina,
Columbia, SC 29208 USA
E-mail: devore@math.scarolina.edu

Bradley J. Lucier

Department of Mathematics
Purdue University,
West Lafayette, IN 47907 USA
E-mail: lucier@math.purdue.edu

CONTENTS

1. Introduction

The subject of 'wavelets' is expanding at such a tremendous rate that it is impossible to give, within these few pages, a complete introduction to all aspects of its theory. We hope, however, to allow the reader to become sufficiently acquainted with the subject to understand, in part, the enthusiasm of its proponents toward its potential application to various numerical problems. Furthermore, we hope that our exposition can guide the reader who wishes to make more serious excursions into the subject. Our viewpoint is biased by our experience in approximation theory and data compression; we warn the reader that there are other viewpoints that are either not represented here or discussed only briefly. For example, orthogonal wavelets were developed primarily in the context of signal processing, an application upon

* This work was supported in part by the National Science Foundation (grants DMS-8922154 and DMS-9006219), the Air Force Office of Scientific Research (contract 89-0455-DEF), the Office of Naval Research (contracts N00014-90-1343, N00014-91-J-1152, and N00014-91-J-1076), the Defense Advanced Research Projects Agency (AFOSR contract 90-0323), and the Army High Performance Computing Research Center.

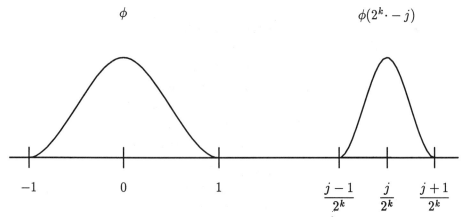

Fig. 1. An example of functions ϕ and $\phi(2^k \cdot - j)$.

which we touch only indirectly. However, there are several good expositions (e.g. Daubechies (1990) and Rioul and Vetterli (1991)) of this application. A discussion of wavelet decompositions in the context of Littlewood–Paley theory can be found in the monograph of Frazier *et al.* (1991). We shall also not attempt to give a complete discussion of the history of wavelets. Historical accounts can be found in the book of Meyer (1990) and the introduction of the article of Daubechies (1990). We shall try to give sufficient historical commentary in the course of our presentation to provide some feeling for the subject's development.

The term 'wavelet' (originally called wavelet of constant shape) was introduced by J. Morlet. It denotes a uni-variate function ψ (multi-variate wavelets exist as well and will be discussed subsequently), defined on $\mathbb{R}$, which, when subjected to the fundamental operations of shifts (i.e. translation by integers) and dyadic dilation, yields an orthogonal basis of $L_2(\mathbb{R})$. That is, the functions $\psi_{j,k} := 2^{k/2}\psi(2^k \cdot - j)$, $j, k \in \mathbb{Z}$, form a complete orthonormal system for $L_2(\mathbb{R})$. In this work, we shall call such a function an orthogonal wavelet, since there are many generalizations of wavelets that drop the requirement of orthogonality. The Haar function $H := \chi_{[0,1/2)} - \chi_{[1/2,1)}$, which will be discussed in more detail in the section that follows, is the simplest example of an orthogonal wavelet. Orthogonal wavelets with higher smoothness (and even compact support) can also be constructed. But before considering that and other questions, we wish first to motivate the desire for such wavelets.

We view a wavelet ψ as a 'bump' (and think of it as having compact support, though it need not). Dilation squeezes or expands the bump and translation shifts it (see Figure 1). Thus, $\psi_{j,k}$ is a scaled version of ψ centred at the dyadic integer $j2^{-k}$. If k is large positive, then $\psi_{j,k}$ is a bump with small support; if k is large negative, the support of $\psi_{j,k}$ is large.

The requirement that the set $\{\psi_{j,k}\}_{j,k\in\mathbb{Z}}$ forms an orthonormal system means that any function $f \in L_2(\mathbb{R})$ can be represented as a series

$$f = \sum_{j,k\in\mathbb{Z}} \langle f, \psi_{j,k}\rangle \, \psi_{j,k} \qquad (1.1)$$

with $\langle f, g\rangle := \int_{\mathbb{R}} f\bar{g}\,\mathrm{d}x$ the usual inner product of two $L_2(\mathbb{R})$ functions. We view (1.1) as building up the function f from the bumps $\psi_{j,k}$. Bumps corresponding to small values of k contribute to the broad resolution of f; those corresponding to large values of k give finer detail.

The decomposition (1.1) is analogous to the Fourier decomposition of a function $f \in L_2(\mathbb{T})$ in terms of the exponential functions $e_k := \mathrm{e}^{\mathrm{i}k\,\cdot}$, but there are important differences. The exponential functions e_k have global support. Thus, all terms in the Fourier decomposition contribute to the value of f at a point x. On the other hand, wavelets are usually either of compact support or fall off exponentially at infinity. Thus, only the terms in (1.1) corresponding to $\psi_{j,k}$ with $j2^{-k}$ near x make a large contribution at x. The representation (1.1) is in this sense local. Of course, exponential functions have other important properties; for example, they are eigenfunctions for differentiation. Many wavelets have a corresponding property captured in the 'refinement equation' for the function ϕ from which the wavelet ψ is derived, as discussed in Section 3.1.

Another important property of wavelet decompositions not present directly in the Fourier decomposition is that the coefficients in wavelet decompositions usually encode all information needed to tell whether f is in a smoothness space, such as the Sobolev and Besov spaces. For example, if ψ is smooth enough, then a function f is in the Lipschitz space $\mathrm{Lip}(\alpha, L_\infty(\mathbb{R}))$, $0 < \alpha < 1$, if and only if

$$\sup_{j,k} 2^{k(\alpha+\frac{1}{2})}|\langle f, \psi_{j,k}\rangle| \qquad (1.2)$$

is finite, and (1.2) is an equivalent semi-norm for $\mathrm{Lip}(\alpha, L_\infty(\mathbb{R}))$.

All this would be of little more than theoretical interest if it were not for the fact that one can efficiently compute wavelet coefficients and reconstruct functions from these coefficients. Such algorithms, known as 'fast wavelet transforms' are the analogue of the Fast Fourier Transform and follow simply from the refinement equation mentioned earlier.

In many numerical applications, the orthogonality of the translated dilates $\psi_{j,k}$ is not vital. There are many variants of wavelets, such as the pre-wavelets proposed by Battle (1987) and the ϕ-transform of Frazier and Jawerth (1990), that do not require orthogonality. Typically, for a given function ψ, one wants the translated dilates $\psi_{j,k}$, $j, k \in \mathbb{Z}$, to form a stable basis (also called a Riesz basis) for $L_2(\mathbb{R})$. This means that each $f \in L_2(\mathbb{R})$ has a unique series decomposition in terms of the $\psi_{j,k}$, and that the ℓ_2

norm of the coefficients in this series is equivalent to $\|f\|_{L_2(\mathbb{R})}$ (this will be discussed in more detail in Section 3.1). In other applications, when approximating in $L_1(\mathbb{R})$, for example, one must abandon the requirement that $\psi_{j,k}$, $j, k \in \mathbb{Z}$, form a stable basis of $L_1(\mathbb{R})$, because none exists. (The Haar system is a Schauder basis for $L_1([0,1])$, for example, but the representation is not $L_1([0,1])$-stable.) For such applications, one can use redundant representations of f, with ψ a box spline, for example.

We have, to this point, restricted our discussion to uni-variate wavelets. There are several constructions of multi-variate wavelets but the final form of this theory is yet to be decided. We shall discuss two methods for constructing multi-variate wavelets; one is based on tensor products while the other is truly multi-variate.

The plan of the paper is as follows. Section 2 is meant to introduce the topic of wavelets by studying the simplest orthogonal wavelets, which are the Haar functions. We discuss the decomposition of $L_p(\mathbb{R})$ using the Haar expansion, the characterization of certain smoothness spaces in terms of the coefficients in the Haar expansion, the fast Haar transform, and multi-variate Haar functions. Section 3 concerns itself with the construction of wavelets. It begins with a discussion of the properties of shift-invariant spaces, and then gives an overview of the construction of uni-variate wavelets and pre-wavelets within the framework of multi-resolution. Later, mention is made of Daubechies' specific construction of orthonormal wavelets of compact support. We finish with a discussion of wavelets in several dimensions.

Section 4 examines how to calculate the coefficients of wavelet expansions via the so-called Fast Wavelet Transform. Section 5 is concerned with the characterization of functions in certain smoothness classes called Besov spaces in terms of the size of wavelet coefficients. Section 6 turns to numerical applications. We briefly mention some uses of wavelets in nonlinear approximation, data compression (and, more specifically, image compression) and numerical methods for partial differential equations.

2. The Haar wavelets

2.1. Overview

The Haar functions are the most elementary wavelets. While they have many drawbacks, chiefly their lack of smoothness, they still illustrate in the most direct way some of the main features of wavelet decompositions. For this reason, we shall consider in some detail their properties which make them suitable for numerical applications. We hope that the detail we provide at this stage will render more convincing some of the later statements we make, without proof, about more general wavelets.

We consider first the uni-variate case. Let $H := \chi_{[0,1/2)} - \chi_{[1/2,1)}$ be the Haar function that takes the value 1 on the left half of $[0,1]$ and the value

-1 on the right half. By translation and dilation, we form the functions

$$H_{j,k} := 2^{k/2} H(2^k \cdot - j), \qquad j,k \in \mathbb{Z}. \qquad (2.1.1)$$

Then, $H_{j,k}$ is supported on the dyadic interval $I_{j,k} := [j2^{-k}, (j+1)2^{-k})$.

It is easy to see that these functions form an orthonormal system. In fact, given two of these functions $H_{j,k}$, $H_{j',k'}$, $k' \geq k$ and $(j,k) \neq (j',k')$, we have two possibilities. The first is that the dyadic intervals $I_{j,k}$ and $I_{j',k'}$ are disjoint, in which case $\int_{\mathbb{R}} H_{j,k} H_{j',k'} = 0$ (because the integrand is identically zero). The second possibility is that $k' > k$ and $I_{j',k'}$ is contained in one of the halves J of $I_{j,k}$. In this case $H_{j,k}$ is constant on J while $H_{j',k'}$ takes the values ± 1 equally often on its support. Hence, again $\int_{\mathbb{R}} H_{j,k} H_{j',k'} = 0$.

We want next to show that $\{H_{j,k} \mid j,k \in \mathbb{Z}\}$ is complete in $L_2(\mathbb{R})$. The following development gives us a chance to introduce the concept of multi-resolution, which is the main vehicle for constructing wavelets and which will be discussed in more detail in the section that follows. Let $\mathcal{S} := \mathcal{S}^0$ denote the subspace of $L_2(\mathbb{R})$ that consists of all piecewise-constant functions with integer breakpoints; i.e. functions in $\mathcal{S}$ are constant on each interval $[j, j+1)$, $j \in \mathbb{Z}$. Then $\mathcal{S}$ is a *shift-invariant space*: if $S \in \mathcal{S}$, each of its *shifts*, $S(\cdot + k)$, $k \in \mathbb{Z}$, is also in $\mathcal{S}$. A simple orthonormal basis for $\mathcal{S}$ is given by the shifts of the function $\phi := \chi_{[0,1]}$. Namely, each $S \in \mathcal{S}$ has a unique representation

$$S = \sum_{j \in \mathbb{Z}} c(j)\phi(\cdot - j), \qquad (c(j)) \in \ell_2(\mathbb{Z}). \qquad (2.1.2)$$

By dilation, we can form a *scale* of spaces

$$\mathcal{S}^k := \{S(2^k \cdot) \mid S \in \mathcal{S}\}, \qquad k \in \mathbb{Z}.$$

Thus, $\mathcal{S}^k$ is the space of piecewise-constant $L_2(\mathbb{R})$ functions with breakpoints at the dyadic integers $j2^{-k}$. The normalized dyadic shifts $\phi_{j,k} := 2^{k/2}\phi(2^k \cdot - j) = 2^{k/2}\phi(2^k(\cdot - j2^{-k}))$ with step $j2^{-k}$, $j \in \mathbb{Z}$, of the function $\phi(2^k \cdot)$, form an orthonormal basis for $\mathcal{S}^k$. However, to avoid possible confusion, we note that the totality of all such functions $\phi_{j,k}$ is not a basis for the space $L_2(\mathbb{R})$ because there is redundancy. For example, $\phi = (\phi_{0,1} + \phi_{1,1})/\sqrt{2}$.

Clearly, we have $\mathcal{S}^k \subset \mathcal{S}^{k+1}$, $k \in \mathbb{Z}$, so the spaces $\mathcal{S}^k$ get 'thicker' as k gets larger and 'thinner' as k gets smaller. We are interested in the limiting spaces

$$\mathcal{S}^\infty := \overline{\bigcup \mathcal{S}^k} \quad \text{and} \quad \mathcal{S}^{-\infty} := \bigcap \mathcal{S}^k, \qquad (2.1.3)$$

since these spaces hold the key to showing that the Haar basis is complete. We claim that

$$\mathcal{S}^\infty = L_2(\mathbb{R}) \quad \text{and} \quad \mathcal{S}^{-\infty} = \{0\}. \qquad (2.1.4)$$

The first of these claims is equivalent to the fact that any function in $L_2(\mathbb{R})$ can be approximated arbitrarily well (in the $L_2(\mathbb{R})$ norm) by the piecewise-constant functions from $\mathcal{S}^k$ provided k is large enough. For example, it is enough to approximate f by its best $L_2(\mathbb{R})$ approximation from $\mathcal{S}^k$. This best approximation is given by the orthogonal projector P_k from $L_2(\mathbb{R})$ onto $\mathcal{S}^k$. It is easy to see that

$$P_k f(x) = \frac{1}{|I_{j,k}|} \int_{I_{j,k}} f, \qquad x \in I_{j,k}, \ j \in \mathbb{Z}.$$

To verify the second claim in (2.1.4), we suppose that $f \in \bigcap \mathcal{S}^k$. Then, f is constant on each of $(-\infty, 0)$ and $[0, \infty)$, and since $f \in L_2(\mathbb{R})$, we must have $f = 0$ a.e. on each of these intervals.

Now, consider again the projector P_k from $L_2(\mathbb{R})$ onto $\mathcal{S}^k$. By (2.1.4), $P_k f \to f$, $k \to \infty$. We also claim that $P_k f \to 0$, $k \to -\infty$. Indeed, if this were false, then we could find a $C > 0$ and a subsequence $m_j \to -\infty$ for which $\|P_{m_j} f\|_{L_2(\mathbb{R})} \geq C$ for all m_j. By the weak-$*$ compactness of $L_2(\mathbb{R})$, we can also assume that $P_{m_j} f \to g$, weak-$*$, for some $g \in L_2(\mathbb{R})$. Now, for any $m \in \mathbb{Z}$, all $P_{m_j} f$ are in $\mathcal{S}^m$ for m_j sufficiently large and negative. Since $\mathcal{S}^m$ is weak-$*$ closed, $g \in \mathcal{S}^m$. Hence $g \in \bigcap \mathcal{S}^m$ implies $g = 0$ a.e. This gives a contradiction because by orthogonality

$$\int_{\mathbb{R}} |P_{m_j} f|^2 \, \mathrm{d}x = \int_{\mathbb{R}} \overline{f} P_{m_j} f \, \mathrm{d}x \to \int_{\mathbb{R}} \overline{f} g \, \mathrm{d}x = 0.$$

It follows that each $f \in L_2(\mathbb{R})$ can be represented by the series

$$f = \sum_{k \in \mathbb{Z}} (P_k f - P_{k-1} f) = \sum_{k \in \mathbb{Z}} Q_k f, \qquad Q_{k-1} := P_k - P_{k-1} \qquad (2.1.5)$$

because the partial sums, $P_n f - P_{-n} f$, of this series tend to f as $n \to \infty$.

To complete the construction of the Haar wavelets, we need the following simple remarks about projections. If $Y \subset X$ are two closed subspaces of $L_2(\mathbb{R})$ and P_X and P_Y are the orthogonal projectors onto these spaces, then $Q := P_X - P_Y$ is the orthogonal projector from $L_2(\mathbb{R})$ onto $X \ominus Y$, the orthogonal complement of Y in X (this follows from the identity $P_Y P_X = P_Y$). Thus, the operator $Q_{k-1} := P_k - P_{k-1}$ appearing in (2.1.5) is the orthogonal projector onto $W^{k-1} := \mathcal{S}^k \ominus \mathcal{S}^{k-1}$. The spaces W^k are the dilates of the *wavelet* space

$$W := \mathcal{S}^1 \ominus \mathcal{S}^0. \qquad (2.1.6)$$

Since the spaces W^k, $k \in \mathbb{Z}$, are mutually orthogonal, we have $W^k \perp W^j$, $j \neq k$, and (2.1.5) shows that $L_2(\mathbb{R})$ is the orthogonal direct sum of the W^k:

$$L_2(\mathbb{R}) = \bigoplus_{k \in \mathbb{Z}} W^k. \qquad (2.1.7)$$

How does the Haar function fit into all this? Well, the main point is that H and its translates $H(\cdot - k)$ form an orthonormal basis for W. Indeed, $H = 2\phi(2\cdot) - P_0(2\phi(2\cdot))$, which shows that H is in $\mathcal{S}^1 \ominus \mathcal{S}^0 = W$. On the other hand, the identities $H + \phi = 2\phi(2\cdot)$ and $\phi - H = 2\phi(2\cdot -1)$ show that the shifts of ϕ together with the shifts of H will generate all the half-shifts of $\eta := \phi(2\cdot)$. Since the half-shifts of $\sqrt{2}\,\eta$ form an orthonormal basis for S^1, the shifts of H must be complete in W.

By dilation, the functions $H_{j,k}$, $j \in \mathbb{Z}$, form a complete orthonormal system for W^k. Hence, we can represent the orthogonal projector Q_k onto W^k by

$$Q_k f = \sum_{j \in \mathbb{Z}} \langle f, H_{j,k} \rangle \, H_{j,k}.$$

Using this in (2.1.5), we have for any $f \in L_2(\mathbb{R})$ the decomposition

$$f = \sum_{k \in \mathbb{Z}} \sum_{j \in \mathbb{Z}} \langle f, H_{j,k} \rangle \, H_{j,k}. \tag{2.1.8}$$

In other words, the functions $H_{j,k}$, $j, k \in \mathbb{Z}$, form an orthonormal basis for $L_2(\mathbb{R})$.

2.2. The Haar decomposition in $L_p(\mathbb{R})$

While the Haar decomposition is initially defined only for functions in $L_2(\mathbb{R})$, it is worth noting that Haar decompositions also hold for other spaces of functions. In this section, we shall discuss the Haar representation for functions in $L_p(\mathbb{R})$, $1 \le p < \infty$. A similar analysis can be given when $p = \infty$ if $L_\infty(\mathbb{R})$ is replaced by the space of uniformly continuous functions that vanish at ∞, equipped with the $L_\infty(\mathbb{R})$ norm.

If $f \in L_p(\mathbb{R})$, the Haar coefficients $\langle f, H_{j,k} \rangle$ are well defined and we can ask whether the Haar series (2.1.8) converges in $L_p(\mathbb{R})$ to f. To answer this question, we fix a value of $1 \le p < \infty$ and a $k \in \mathbb{Z}$ and examine the projector P_k, which is initially defined only on $L_2(\mathbb{R})$. For any $f \in L_2(\mathbb{R})$, we have $P_k f = \sum_{I \in D_k} f_I \chi_I$ where D_k denotes the collection of dyadic intervals of length 2^{-k}, and where $f_I := (1/|I|) \int_I f \, dx$, $I \in D_k$, is the average of f over I. In this form, the projector P_k has a natural extension to $L_p(\mathbb{R})$ and takes values in the space $S^k(\chi, L_p(\mathbb{R}))$ of all functions in $L_p(\mathbb{R})$ that are piecewise-constant functions with breakpoints at the dyadic integers $j2^{-k}$, $j \in \mathbb{Z}$.

In representing P_k on $L_p(\mathbb{R})$, it is useful to change our normalization slightly. We fix a value of p and consider the $L_p(\mathbb{R})$-normalized characteristic functions $\phi_{j,k,p} := 2^{k/p}\phi(2^k\cdot -j)$, $\phi := \chi_{[0,1]}$, which satisfy $\int_{\mathbb{R}} |\phi_{j,k,p}|^p \, dx = 1$. Then,

$$P_k f = \sum_{j \in \mathbb{Z}} \langle f, \phi_{j,k,p'} \rangle \, \phi_{j,k,p}, \qquad \frac{1}{p} + \frac{1}{p'} = 1.$$

From Hölder's inequality, we find $|\langle f, \phi_{j,k,p'}\rangle|^p \le \int_{I_{j,k}} |f|^p \, dx$ and so

$$\|P_k f\|_{L_p(\mathbb{R})}^p = \sum_{j\in\mathbb{Z}} |\langle f, \phi_{j,k,p'}\rangle|^p \le \sum_{j\in\mathbb{Z}} \int_{I_{j,k}} |f|^p \, dx = \int_{\mathbb{R}} |f|^p \, dx.$$

Therefore, P_k is a bounded operator with norm 1 on the space $L_p(\mathbb{R})$.

If $f \in L_p(\mathbb{R})$, then since P_k is a projector of norm 1,

$$\|f - P_k f\|_{L_p(\mathbb{R})} = \inf_{S\in\mathcal{S}^k} \|(I - P_k)(f - S)\|_{L_p(\mathbb{R})} \le 2\mathrm{dist}(f, \mathcal{S}^k)_{L_p(\mathbb{R})}. \quad (2.2.1)$$

It follows that $P_k f \to f$ in $L_p(\mathbb{R})$ for each $f \in L_p(\mathbb{R})$.

On the other hand, consider $P_k f$ as $k \to -\infty$. If f is continuous and of compact support then at most two terms in $P_k f$ are nonzero for k large negative and each coefficient is $\le C2^{k/p'}$. Hence $\|P_k f\|_{L_p(\mathbb{R})} \to 0$ provided $p' < \infty$, i.e. $p > 1$. This shows that

$$f = \sum_{k\in\mathbb{Z}} (P_k f - P_{k-1} f) = \sum_{k\in\mathbb{Z}} \sum_{j\in\mathbb{Z}} \langle f, H_{j,k}\rangle H_{j,k}. \quad (2.2.2)$$

in the sense of $L_p(\mathbb{R})$ convergence. We see that the Haar representation holds for functions in $L_p(\mathbb{R})$ provided $p > 1$.

But what happens when $p = 1$? Well, as is typical for orthogonal decompositions, the expansion (2.2.2) cannot be valid. Indeed, each of the functions appearing on the right in (2.2.2) has mean value zero. If $g \in L_1(\mathbb{R})$ has mean value zero and f is an arbitrary function from $L_1(\mathbb{R})$, then

$$\int_{\mathbb{R}} |f - g| \ge \left| \int_{\mathbb{R}} f - \int_{\mathbb{R}} g \right| = \left| \int_{\mathbb{R}} f \right|.$$

This means that the sum in (2.2.2) cannot possibly converge in $L_1(\mathbb{R})$ to f unless f has mean value zero.

This phenomenon is typical of decompositions for orthogonal wavelets ψ: They cannot represent all functions in $L_1(\mathbb{R})$. However, if ψ is smooth enough, the representation (2.2.2) will hold for the Hardy space $H_1(\mathbb{R})$ used in place of $L_1(\mathbb{R})$, and in fact this representation will then hold for functions in the Hardy spaces $H_p(\mathbb{R})$ for a certain range of $0 < p < 1$ that depends on the smoothness of ψ. We shall not discuss further the behaviour of orthogonal wavelets in H_p spaces but the interested reader can consult Frazier and Jawerth (1990) for a corresponding theory in a slightly different setting.

2.3. Smoothness spaces

We noted earlier the important fact that wavelet decompositions provide a description of smoothness spaces in terms of the wavelet coefficients. We wish to illustrate this point with the Haar wavelets and the Lipschitz spaces in $L_p(\mathbb{R})$, $1 < p < \infty$.

The Lipschitz spaces $\text{Lip}(\alpha, L_p(\mathbb{R}))$ of $L_p(\mathbb{R})$, $0 < \alpha \leq 1$, $1 \leq p \leq \infty$, consist of all functions $f \in L_p(\mathbb{R})$ for which

$$\|f - f(\cdot + h)\|_{L_p(\mathbb{R})} = \mathcal{O}(h^\alpha), \qquad h \to 0.$$

A semi-norm for this space is provided by

$$|f|_{\text{Lip}(\alpha, L_p(\mathbb{R}))} := \sup_{0 < h < \infty} h^{-\alpha} \|f - f(\cdot + h)\|_{L_p(\mathbb{R})}.$$

The relationship between the smoothness of f and the size of its Haar coefficients rests on three fundamental inequalities. The first of these says that, for a fixed $k \in \mathbb{Z}$, the Haar functions $H_{j,k}$, $j \in \mathbb{Z}$, are $L_p(\mathbb{R})$-stable. Because of the disjoint support of the $H_{j,k}$, $j \in \mathbb{Z}$, stability takes the following particularly simple form: for any sequence $(c(j)) \in \ell_p(\mathbb{Z})$ and $S = \sum_{j \in \mathbb{Z}} c(j) H_{j,k}$, we have

$$\|S\|_{L_p(\mathbb{R})} = \left(\sum_{j \in \mathbb{Z}} |c(j)|^p 2^{kp(1/2 - 1/p)} \right)^{1/p}. \tag{2.3.1}$$

This follows by integrating the identity

$$|S|^p = \sum_{j \in \mathbb{Z}} |c(j)|^p |H_{j,k}|^p.$$

The other two inequalities are related to the approximation properties of $\mathcal{S}^k$ and the projectors P_k:

(J) $\|f - P_k f\|_{L_p(\mathbb{R})} \leq 2 \cdot 2^{-k\alpha} |f|_{\text{Lip}(\alpha, L_p(\mathbb{R}))}$, $0 < \alpha \leq 1$, $1 \leq p \leq \infty$.

(B) $|S|_{\text{Lip}(1/p, L_p(\mathbb{R}))} \leq 2 \cdot 2^{k/p} \|S\|_{L_p(\mathbb{R})}$, $S \in \mathcal{S}^k(\chi, L_p(\mathbb{R}))$, $1 \leq p \leq \infty$.

$$\tag{2.3.2}$$

The first of these, often called a Jackson inequality (after similar inequalities established by D. Jackson for polynomial approximation), tells how well functions from $\text{Lip}(\alpha, L_p(\mathbb{R}))$ can be approximated by the elements of $\mathcal{S}^k$. The second inequality is known as a Bernstein inequality because of its similarity with the classical Bernstein inequalities for polynomials, established by S. Bernstein.

We shall prove (2.3.2J) and (2.3.2B) for $1 \leq p < \infty$. If $I \in D_k$ and $h := |I| = 2^{-k}$, then, for all $x \in I$, we obtain

$$|f(x) - f_I| \leq \frac{1}{|I|} \int_I |f(x) - f(y)| \, dy$$

$$\leq \left(\frac{1}{|I|} \int_I |f(x) - f(y)|^p \, dy \right)^{1/p}$$

$$\leq \left(\frac{1}{|I|} \int_{-h}^{h} |f(x) - f(x+s)|^p \, ds \right)^{1/p}.$$

If we raise these last inequalities to the power p, integrate over I, and then sum over all $I \in D_k$, we obtain

$$\|f - P_k f\|^p_{L_p(\mathbb{R})} \le \frac{1}{h} \int_{-h}^{h} \int_{\mathbb{R}} |f(x) - f(x+s)|^p \, dx \, ds \le 2|h|^{\alpha p} |f|^p_{\mathrm{Lip}(\alpha, L_p(\mathbb{R}))},$$

which implies the Jackson inequality.

The Jackson inequality can also be proved from more general principles. Since the P_k have norm 1 on $L_p(\mathbb{R})$ and are projectors, we have

$$\|f - P_k f\|_{L_p(\mathbb{R})} \le (1 + \|P_k\|)\mathrm{dist}(f, \mathcal{S}^k)_{L_p(\mathbb{R})} \le 2\mathrm{dist}(f, \mathcal{S}^k)_{L_p(\mathbb{R})}. \quad (2.3.3)$$

Thus, the Jackson inequality follows from the fact that functions in

$$\mathrm{Lip}(\alpha, L_p(\mathbb{R}))$$

can be approximated by the elements of $\mathcal{S}^k$ with an error not exceeding the right-hand side of (2.3.2J).

To prove the Bernstein inequality, we note that any $S = \sum_{j \in \mathbb{Z}} c(j)\chi_{I_{j,k}}$ in $\mathcal{S}^k(\chi, L_p(\mathbb{R}))$ has norm:

$$\|S\|^p_{L_p(\mathbb{R})} = \sum_{j \in \mathbb{Z}} |c(j)|^p |I_{j,k}| = \sum_{j \in \mathbb{Z}} |c(j)|^p 2^{-k}. \quad (2.3.4)$$

We fix an $h > 0$. If $h \ge 2^{-k}$ (i.e. $h^{-1/p} \le 2^{k/p}$), then

$$h^{-1/p}\|S(\cdot + h) - S\|_{L_p(\mathbb{R})} \le 2^{k/p}(\|S(\cdot + h)\|_{L_p(\mathbb{R})} + \|S\|_{L_p(\mathbb{R})}) = 2 \cdot 2^{k/p}\|S\|_{L_p(\mathbb{R})}.$$

If $h < 2^{-k}$ then

$$|S(x+h) - S(x)| = \begin{cases} 0, & x \in [j2^{-k}, (j+1)2^{-k} - h), \\ |c(j+1) - c(j)|, & x \in [(j+1)2^{-k} - h, (j+1)2^{-k}). \end{cases}$$

Therefore

$$\begin{aligned}
h^{-1/p}\|S(\cdot + h) - S\|_{L_p(\mathbb{R})} &= h^{-1/p}\left(\sum_{j \in \mathbb{Z}} |c(j+1) - c(j)|^p h\right)^{1/p} \\
&= 2^{k/p}\left(\sum_{j \in \mathbb{Z}} |c(j+1) - c(j)|^p 2^{-k}\right)^{1/p} \\
&= 2^{k/p}\|S(\cdot + 2^{-k}) - S\|_{L_p(\mathbb{R})} \le 2 \cdot 2^{k/p}\|S\|_{L_p(\mathbb{R})}.
\end{aligned}$$

With the Jackson and Bernstein inequalities in hand, it is now easy to show that

$$|f|_{\mathrm{Lip}(\alpha, L_p(\mathbb{R}))} \approx \sup_{k \in \mathbb{Z}} 2^{k(\alpha + 1/2 - 1/p)}\left(\sum_{j \in \mathbb{Z}} |\langle f, H_{j,k}\rangle|^p\right)^{1/p}, \quad 0 < \alpha < 1/p. \tag{2.3.5}$$

(It will be convenient to use the notation $A \approx B$ to mean that the two ratios A/B and B/A of the functions A and B are bounded from above independently of the variables; in (2.3.5), independently of f.) First, from the Jackson inequality,

$$\|P_k f - P_{k-1} f\|_{L_p(\mathbb{R})} \leq \|f - P_k f\|_{L_p(\mathbb{R})} + \|f - P_{k-1} f\|_{L_p(\mathbb{R})}$$
$$\leq C 2^{-k\alpha} |f|_{\mathrm{Lip}(\alpha, L_p(\mathbb{R}))}.$$

If we write $P_k f - P_{k-1} f = \sum_{j \in \mathbb{Z}} \langle f, H_{j,k-1} \rangle H_{j,k-1}$ and replace $\|P_k f - P_{k-1} f\|_{L_p(\mathbb{R})}$ by the sum in (2.3.1) (with $c(j) = \langle f, H_{j,k-1} \rangle$), we obtain that the right-hand side of (2.3.5) does not exceed a multiple of the left.

To reverse this inequality, we fix a value of h and choose $n \in \mathbb{Z}$ so that $2^{-n} \leq h \leq 2^{-n+1}$. We write $f = \sum_{k \in \mathbb{Z}} w_k$ with $w_k := (P_{k+1} f - P_k f)$ and estimate

$$\|f(\cdot + h) - f\|_{L_p(\mathbb{R})}$$
$$\leq \sum_{k \geq n} \|w_k(\cdot + h) - w_k\|_{L_p(\mathbb{R})} + \sum_{k < n} \|w_k(\cdot + h) - w_k\|_{L_p(\mathbb{R})}.$$
$$(2.3.6)$$

The first sum does not exceed

$$\left(\sup_{k \geq n} 2^{k\alpha} \|w_k\|_{L_p(\mathbb{R})} \right) \left(\sum_{k \geq n} 2^{-k\alpha} \right) \leq C h^\alpha \sup_{k \geq n} 2^{k\alpha} \|w_k\|_{L_p(\mathbb{R})}.$$

Similarly, using the Bernstein inequality, the second sum does not exceed

$$h^{1/p} \sum_{k < n} |w_k|_{\mathrm{Lip}(1/p, L_p(\mathbb{R}))} \leq C h^{1/p} \sum_{k < n} 2^{k/p} \|w_k\|_{L_p(\mathbb{R})}$$
$$\leq C h^{1/p} \left(\sup_{k < n} 2^{k\alpha} \|w_k\|_{L_p(\mathbb{R})} \right) \left(\sum_{k < n} 2^{k(1/p - \alpha)} \right)$$
$$\leq C h^\alpha \sup_{k < n} 2^{k\alpha} \|w_k\|_{L_p(\mathbb{R})}.$$

If we write $w_k = \sum_{j \in \mathbb{Z}} \langle f, H_{j,k} \rangle H_{j,k}$ and use (2.3.1) to replace $\|w_k\|_{L_p(\mathbb{R})}$ by

$$2^{k(1/2 - 1/p)} \left(\sum_{j \in \mathbb{Z}} |\langle f, H_{j,k} \rangle|^p \right)^{1/p}$$

in each of these expressions, and then use the resulting expression in (2.3.6), we obtain

$$\|f(\cdot + h) - f\|_{L_p(\mathbb{R})} \leq C h^\alpha \sup_{k \in \mathbb{Z}} 2^{k(\alpha + 1/2 - 1/p)} \left(\sum_{j \in \mathbb{Z}} |\langle f, H_{j,k} \rangle|^p \right)^{1/p},$$

which shows that the left-hand side of (2.3.5) does not exceed a multiple of the right.

The restriction $\alpha < 1/p$ arises because the Haar function is not smooth; for smoother wavelets, the range of α can be increased.

2.4. The fast Haar transform

In numerical applications of the Haar decomposition, one must work with only a finite number of the functions $H_{j,k}$. The choice of which functions to use is often made as follows. Given a function $f \in L_2(\mathbb{R})$, we choose a large value of n, compatible with the accuracy we wish to achieve, and we replace f by $P_n f$ with P_n, as before, the $L_2(\mathbb{R})$ projector onto $\mathcal{S}^n$, the space of piecewise-constant functions in $L_2(\mathbb{R})$ with breakpoints at the dyadic integers $j2^{-n}$, $j \in \mathbb{Z}$. If f has compact support then $P_n f$ is a finite linear combination of the characteristic functions χ_I, $I \in D_n$. If f does not have compact support, it is necessary to truncate this sum (which is justified because $\int_{\mathbb{R} \setminus [-a,a]} |f|^2 \, dx \to 0$, $a \to \infty$).

We can now write

$$P_n f = (P_n f - P_{n-1} f) + \cdots + (P_1 f - P_0 f) + P_0 f = P_0 f + \sum_{k=0}^{n-1} Q_k f, \quad (2.4.1)$$

which is a finite Haar decomposition. We have started this decomposition with $P_0 f$ but we could have equally well started at any other dyadic level.

The fast Haar transform gives an efficient method for finding the coefficients in the expansions

$$Q_k f = \sum_{j \in \mathbb{Z}} d(j,k) H_{j,k}, \quad d(j,k) := \langle f, H_{j,k} \rangle, \quad (2.4.2)$$

and

$$P_k f = \sum_{j \in \mathbb{Z}} c(j,k) \phi_{j,k}, \quad c(j,k) := \langle f, \phi_{j,k} \rangle. \quad (2.4.3)$$

These coefficients are related to the integrals of f over the intervals $I_{j,k} := [j2^{-k}, (j+1)2^{-k})$:

$$c(j,k) = 2^{k/2} \int_{I_{j,k}} f \, dx,$$

$$d(j,k) = 2^{k/2} \left(\int_{I_{2j,k+1}} f \, dx - \int_{I_{2j+1,k+1}} f \, dx \right).$$

Therefore, if the coefficients $c(j, k+1)$, $j \in \mathbb{Z}$, are known, then

$$c(j,k) = \frac{1}{\sqrt{2}} (c(2j, k+1) + c(2j+1, k+1)),$$

$$d(j,k) = \frac{1}{\sqrt{2}} (c(2j, k+1) - c(2j+1, k+1)). \quad (2.4.4)$$

In other words, starting with the known values of $c(j, n)$ at level n, we can iteratively compute all values $d(j,k)$ and $c(j,k)$ needed for (2.4.1) from (2.4.4). The computation of the $c(j,k)$ at dyadic levels $k \neq 0$ is necessary for the recurrence even though we are, in the end, not interested in their values.

There is a similar formula for reconstructing a function from its Haar coefficients. Now, suppose that we know the coefficients appearing in (2.4.1), i.e. the values $c(j, 0)$, $j \in \mathbb{Z}$, and $d(j, k)$, $j \in \mathbb{Z}$, $k = 1, \ldots, n$, and we wish to find $c(j, n)$, i.e. to reconstruct f. For this we need only use the recursive formulae

$$c(2j, k+1) = \frac{1}{\sqrt{2}}(c(j, k) + d(j, k)),$$

$$c(2j+1, k+1) = \frac{1}{\sqrt{2}}(c(j, k) - d(j, k)).$$

More information on the structure of the fast Haar transform can be found in Section 5.

2.5. Multi-variate Haar functions

There is a simple method to construct multi-variate wavelets from a given uni-variate wavelet, which, for the Haar wavelets, takes the following form. Let $\phi_0 := \phi = \chi_{[0,1]}$ and $\phi_1 := \psi = H$ and let V denote the set of vertices of the cube $\Omega := [0, 1]^d$. For each $v = (v_1, \ldots, v_d)$ in V and $x = (x_1, \ldots, x_d)$ from $\mathbb{R}^d$, we let $\psi_v(x) := \prod_{i=1}^d \phi_{v_i}(x_i)$. The functions ψ_v are piecewise constant, taking the values ± 1 on the d-tants of Ω. The set

$$\Psi := \{\psi_v \mid v \in V, \ v \neq 0\}$$

is the set of multi-dimensional Haar functions; there are $2^d - 1$ of them. They generate by dilation and translation an orthonormal basis for $L_2(\mathbb{R}^d)$. That is, the collection of functions $2^{kd/2}\psi_v(2^k \cdot - j)$, $j \in \mathbb{Z}^d$, $k \in \mathbb{Z}$, $v \in V \setminus \{0\}$, forms a complete orthonormal basis for $L_2(\mathbb{R}^d)$.

Another way to view the multi-dimensional Haar functions is to consider the shift-invariant space $\mathcal{S}$ of piecewise-constant functions on the dyadic cubes of unit length in $\mathbb{R}^d$. A basis for $\mathcal{S}$ is provided by the shifts of $\chi_{[0,1]^d}$. Note that the space $\mathcal{S}$ is the tensor product of the uni-variate spaces of piecewise-constant functions with integer breakpoints. The collection of all shifts of the Haar functions $\psi_v \in \Psi$ forms an orthonormal basis for the space $W := \mathcal{S}^1 \ominus \mathcal{S}^0$. Properties of the multi-variate Haar wavelets follow from the uni-variate Haar function. For example, there is a fast Haar transform and a characterization of smoothness spaces in terms of Haar coefficients. We leave the formulation of these properties to the reader.

3. The construction of wavelets

3.1. Overview

We turn now to the construction of smoother orthogonal wavelets. Almost all constructions of orthogonal wavelets begin by using multi-resolution, which was introduced by Mallat (1989) (an interesting exception, presented by Strömberg (1981), apparently gave the first smooth orthogonal wavelets). We begin with a brief overview of multi-resolution that we will expand on in later sections.

Let $\phi \in L_2(\mathbb{R}^d)$ and let $\mathcal{S} := \mathcal{S}(\phi)$ be the shift-invariant subspace of $L_2(\mathbb{R}^d)$ generated by ϕ. That is, $\mathcal{S}(\phi)$ is the $L_2(\mathbb{R}^d)$ closure of finite linear combinations of ϕ and its shifts $\phi(\cdot + j)$, $j \in \mathbb{Z}^d$. By dilation, we form the scale of spaces

$$\mathcal{S}^k := \{ S(2^k \cdot) \mid S \in \mathcal{S} \}. \tag{3.1.1}$$

Then $\mathcal{S}^k$ is invariant under dyadic shifts $j2^{-k}$, $j \in \mathbb{Z}^d$. In the construction of Haar functions, we had $d = 1$, and $\mathcal{S}$ was the space of piecewise-constant functions with integer breakpoints. That is, $\mathcal{S} = \mathcal{S}(\phi)$ with $\phi := \chi := \chi_{[0,1]}$. Other examples for the reader to keep in mind, which result in smoother wavelets, are to take for $\mathcal{S}$ the space of cardinal spline functions of order r in $L_2(\mathbb{R})$. A cardinal spline is a piecewise polynomial function defined on $\mathbb{R}$, of local degree $< r$, that has breakpoints at the integers and has global smoothness C^{r-2}. Then $\mathcal{S} = \mathcal{S}(N_r)$ with N_r the (nonzero) cardinal B-spline that has knots at $0, 1, \ldots, r$. These B-splines are easiest to define recursively: $N_1 := \chi$ and $N_r := N_{r-1} * N_1$, with the usual operation of convolution

$$f * g(x) := \int_{\mathbb{R}} f(x - y)g(y)\, dy.$$

For example, N_2 is a hat function, N_3 a C^1 piecewise quadratic, and so on. In the multi-variate case, the primary examples to keep in mind are the tensor product of uni-variate B-splines: $N(x) := N(x_1, \ldots, x_d) := N(x_1) \cdots N(x_d)$, and the box splines, which will be introduced and discussed later.

Multi-resolution begins with certain assumptions on the scale of spaces $\mathcal{S}^k$ and shows under these assumptions how to construct an orthogonal wavelet ψ from the generating function ϕ. The usual assumptions are:

(i) $\mathcal{S}^k \subset \mathcal{S}^{k+1}, \quad k \in \mathbb{Z};$

(ii) $\overline{\bigcup \mathcal{S}^k} = L_2(\mathbb{R}^d);$

(iii) $\bigcap \mathcal{S}^k = \{0\};$

(iv) $\{\phi(\cdot - j)\}_{j \in \mathbb{Z}^d}$ forms an $L_2(\mathbb{R}^d)$-stable basis for $\mathcal{S}$. (3.1.2)

We have already seen the role of (ii) and (iii) in the context of Haar decompositions. The assumption (iv) means that there exist positive constants C_1 and C_2 such that each $S \in \mathcal{S}$ has a unique representation

(i) $S = \sum_{j \in \mathbb{Z}^d} c(j)\phi(\cdot - j), \quad$ and

(ii) $C_1\|S\|_{L_2(\mathbb{R}^d)} \leq \left(\sum_{j \in \mathbb{Z}^d} |c(j)|^2 \right)^{1/2} \leq C_2\|S\|_{L_2(\mathbb{R}^d)}.$ (3.1.3)

If ϕ has $L_2(\mathbb{R}^d)$-stable shifts then it follows by a change of variables that for each $k \in \mathbb{Z}$, the function $2^{kd/2}\phi(2^k \cdot)$ has $L_2(\mathbb{R}^d)$-stable $2^{-k}\mathbb{Z}^d$ shifts. We shall mention later how the assumption $((3.1.2)(iv))$ can be weakened.

Assumption $((3.1.2)(i))$ is a severe restriction on the underlying function ϕ. Because each space $\mathcal{S}^k$ is obtained from $\mathcal{S}$ by dilation, we see that $((3.1.2)(i))$ is satisfied if and only if $\mathcal{S} \subset \mathcal{S}^1$, or, equivalently, if ϕ is in the space $\mathcal{S}^1$. From the $L_2(\mathbb{R}^d)$-stability of the set $\{\phi(\cdot - j)\}_{j \in \mathbb{Z}^d}$, this is equivalent to

$$\phi(x) = \sum_{j \in \mathbb{Z}^d} a(j)\phi(2x - j) \tag{3.1.4}$$

for some sequence $(a(j)) \in \ell_2(\mathbb{Z}^d)$. Equation (3.1.4) is called the *refinement equation* for ϕ, since it says that ϕ can be expressed as a linear combination of the scaled functions $\phi(2 \cdot - j)$, which are at the finer dyadic level. We shall discuss the refinement equation in more detail later and for now only point out that this equation is well known for the B-spline of order r, for which it takes the form

$$N_r(x) = 2^{-r+1} \sum_{j=0}^{r} \binom{r}{j} N_r(2x - j). \tag{3.1.5}$$

Because of $((3.1.2)(i))$, the *wavelet space*

$$W := \mathcal{S}^1 \ominus \mathcal{S}^0$$

is a subspace of $\mathcal{S}^1$. By dilation, we obtain the scaled wavelet spaces W^k, $k \in \mathbb{Z}$. Then, W^k is orthogonal to $\mathcal{S}^k$ and

$$\mathcal{S}^{k+1} = \mathcal{S}^k \oplus W^k. \tag{3.1.6}$$

Since $W^j \subset \mathcal{S}^k$ for $j < k$, it follows that W_j and W_k are orthogonal. From this and $((3.1.2)(ii))$ and $((3.1.2)(iii))$, we obtain

$$L_2(\mathbb{R}^d) = \bigoplus_{k \in \mathbb{Z}} W^k. \tag{3.1.7}$$

We find wavelets by showing that W is shift invariant and finding its generators. For example, when $d = 1$, W is a principal shift-invariant space, that is it can be generated by one element ψ, i.e. $W = \mathcal{S}(\psi)$. Of course, there are many such generators ψ for W. In the multi-variate case, the space W will be generated by $2^d - 1$ such functions.

We find an orthogonal wavelet in one dimension by determining a ψ whose shifts form an orthonormal basis for W. Indeed, once such a function ψ is found, the scaled functions $\psi_{j,k} := 2^{k/2}\psi(2^k \cdot - j)$ will then form an orthonormal basis for $L_2(\mathbb{R})$.

Generators ψ for W whose shifts are not orthogonal are nonorthogonal wavelets. For example, if ψ has shifts that are $L_2(\mathbb{R})$-stable (but not orthonormal), the functions $\psi_{j,k} := 2^{k/2}\psi(2^k \cdot - j)$ form an $L_2(\mathbb{R})$-stable basis for $L_2(\mathbb{R})$. While they do not form an orthonormal system, they still possess orthogonality between levels,

$$\int_{\mathbb{R}} \psi_{j,k}\psi_{j',k'}\,\mathrm{d}x = 0, \quad k \neq k',$$

which is enough for most applications. After Battle (1987), we call such functions ψ *pre-wavelets*.

The construction of (uni-variate) orthogonal wavelets introduced by Mallat (1989), begins with a function ϕ that has orthonormal shifts (rather than just $L_2(\mathbb{R})$-stability). Mallat shows that the function

$$\psi := \sum_{j \in \mathbb{Z}}(-1)^j\overline{a(1-j)}\phi(2\cdot - j), \qquad (3.1.8)$$

with $a(j)$ the refinement coefficients of (3.1.4), is an orthogonal wavelet. (It is easy to check that ψ is orthogonal to the shifts of ϕ by using the refinement equation (3.1.4).) A construction similar to that of Mallat was used by Chui and Wang (1990) and Micchelli (1991) to produce pre-wavelets. In the construction of pre-wavelets, they begin with a function ϕ that has $L_2(\mathbb{R})$-stable shifts (but not necessarily orthonormal shifts). Then a formula similar to (3.1.8) gives a pre-wavelet ψ (see (3.4.15)).

To find generators for the wavelet space W, we shall follow the construction of de Boor *et al.* (1991b), which is somewhat different from that of Mallat. We simply take suitable functions η in the space $\mathcal{S}^1$ and consider their orthogonal projections $P\eta$ onto the space $\mathcal{S}$. The error function $w := \eta - P\eta$ is then an element of W. By choosing appropriate functions η, we obtain a set of generators for W. In one dimension, only one function is needed to generate W and any reasonable choice for η results in such a generator. The most obvious choices, $\eta := \phi(2\cdot)$ or $\eta := \phi(2\cdot - 1)$, lead to the wavelet (3.1.8) or its pre-wavelet analogue.

If we begin with the orthonormalized shifts of the B-spline $\phi = N_r$ as the basis for $\mathcal{S}$, the construction of Mallat gives the spline wavelets ψ of Battle–Lemarié (see, e.g., Battle (1987)), which have smoothness C^{r-2}. The support of ψ is all of $\mathbb{R}$, although ψ does decay exponentially at infinity. More details are given in Section 3.4. If we do not orthonormalize the shifts, we obtain the spline pre-wavelets of Chui and Wang (1991), which have compact support (in fact minimal support among all functions in W).

It is a more substantial problem to construct smooth orthogonal wavelets of compact support and this was an outstanding achievement of Daubechies (1988) (see Section 3.5). Daubechies' construction depends on finding a *compactly supported* function $\phi \in C^r$ that satisfies the assumptions of multi-resolution and has *orthonormal* shifts. In this way, she was able to apply Mallat's construction to obtain a compactly supported orthogonal wavelet ψ in C^r.

The construction of multi-variate wavelets by multi-resolution is based on similar ideas. We want now to find a set of generators $\Psi = \{\psi\}$ for the wavelet space W. There are typically $2^d - 1$ functions in Ψ. This is an orthogonal wavelet set if the totality of functions $\psi_{j,k}$, $j \in \mathbb{Z}^d$, $k \in \mathbb{Z}$, $\psi \in \Psi$, forms an orthonormal basis for $L_2(\mathbb{R}^d)$. For this to hold, it is sufficient to have orthogonality between $\psi(\cdot - j)$ and $\tilde{\psi}(\cdot - j')$, $(j, \psi) \neq (j', \tilde{\psi})$. If the shifts of the functions $\psi \in \Psi$ form an $L_2(\mathbb{R}^d)$-stable basis for W, we say this is a pre-wavelet set. In this case, we shall still have the orthogonality between levels: $\psi_{j,k} \perp \tilde{\psi}_{j',k'}$ if $k \neq k'$. Sometimes we also require orthogonality between $\tilde{\psi} \in \Psi$ and all of the $\psi(\cdot - j)$, $j \in \mathbb{Z}^d$, $\psi \neq \tilde{\psi}$. Because the construction of multi-variate wavelets is significantly more complicated and more poorly understood than the construction of wavelets of one variable, we shall postpone the discussion of multi-variate wavelets until Section 3.6.

In the following sections, we shall show how to construct wavelets and pre-wavelets in the setting of multi-resolution. These constructions depend on a good description of the space $\mathcal{S} := \mathcal{S}(\phi)$ in terms of Fourier transforms, which is the topic of the next section.

3.2. Shift-invariant spaces

Because multi-resolution is based on a family of shift-invariant spaces, it is useful to have in mind the structure of these spaces before proceeding with the construction of wavelets and pre-wavelets. The structure of shift-invariant spaces and their application to approximation and wavelet construction were developed in a series of papers by de Boor *et al.* (1991a,b,c); much of the material in our presentation is taken from these references.

We recall that a closed subspace $\mathcal{S}$ of $L_2(\mathbb{R}^d)$ is shift invariant if $S(\cdot + j)$, $j \in \mathbb{Z}^d$, is in $\mathcal{S}$ whenever $S \in \mathcal{S}$. We have already encountered the space $\mathcal{S}(\phi)$, which is the $L_2(\mathbb{R}^d)$-closure of finite linear combinations of the shifts of ϕ. We say that such a space is a *principal shift-invariant space* (in analogy with principal ideals). More generally, if Φ is a finite set of $L_2(\mathbb{R}^d)$ functions, then the space $\mathcal{S}(\Phi)$ is the $L_2(\mathbb{R}^d)$-closure of finite linear combinations of the shifts of the functions $\phi \in \Phi$. Of course, a general shift-invariant subspace of $L_2(\mathbb{R}^d)$ need not be finitely generated.

We are interested in describing the space $\mathcal{S}(\Phi)$ in terms of its Fourier transforms. We let

$$\hat{f}(x) := \int_{\mathbb{R}^d} f(y) e^{-ix \cdot y} \, dy$$

denote the Fourier transform of an $L_1(\mathbb{R}^d)$ function f. The Fourier transform has a natural extension from $L_1(\mathbb{R}^d) \cap L_2(\mathbb{R}^d)$ to $L_2(\mathbb{R}^d)$ and, more generally, to tempered distributions. We assume that the reader is familiar with the rudiments of Fourier transform theory.

The Fourier transform of $f(\cdot + t)$, $t \in \mathbb{R}^d$, is $e_t \hat{f}$; we shall use the abbreviated notation

$$e_t(x) := e^{ix \cdot t}$$

for the exponential functions. Now, suppose that the shifts of ϕ form an $L_2(\mathbb{R}^d)$-stable basis for $\mathcal{S}(\phi)$. Then from (3.1.3), each $S \in \mathcal{S}(\phi)$ can be written as $S = \sum_{j \in \mathbb{Z}^d} c(j)\phi(\cdot - j)$ with $(c(j)) \in \ell_2(\mathbb{Z}^d)$. Therefore,

$$\hat{S}(y) = \sum_{j \in \mathbb{Z}^d} c(j) e_{-j}(y)\hat{\phi}(y) = \tau(y)\hat{\phi}(y), \quad \tau(y) := \sum_{j \in \mathbb{Z}^d} c(j) e_{-j}(y). \quad (3.2.1)$$

Here τ is an $L_2(\mathbb{T}^d)$ function (i.e. of period 2π in each of the variables $y_1, \ldots, y_d$). The $L_2(\mathbb{R}^d)$-stability of the shifts of ϕ can easily be seen to be equivalent to the statement

$$\|\tau\|_{L_2(\mathbb{T}^d)} \approx \|S\|_{L_2(\mathbb{R}^d)}. \quad (3.2.2)$$

The characterization (3.2.1) allows one to readily decide when a function is in $\mathcal{S}(\phi)$. Even when the shifts of ϕ are not $L_2(\mathbb{R}^d)$-stable, one can characterize $\mathcal{S}(\phi)$ by (see de Boor et $al.$ (1991a))

$$\widehat{\mathcal{S}(\phi)} = \{\hat{S} = \tau\hat{\phi} \in L_2(\mathbb{R}^d) \mid \tau \text{ is } 2\pi\text{-periodic}\}. \quad (3.2.3)$$

By dilation, (3.2.3) gives a characterization of the scaled spaces $\mathcal{S}^k$, $\mathcal{S} = \mathcal{S}(\phi)$. For example, the functions in $\mathcal{S}^1$ are characterized by $\hat{S} = \tau\hat{\eta} \in L_2(\mathbb{R}^d)$, $\eta := \phi(2\cdot)$, with τ a 4π-periodic function.

A similar characterization holds for a finite set Φ of generators for a shift-invariant space $\mathcal{S}(\Phi)$. We say that this set provides $L_2(\mathbb{R}^d)$-stable shifts if the totality of all functions $\phi(\cdot - j)$, $j \in \mathbb{Z}^d$, $\phi \in \Phi$, forms an $L_2(\mathbb{R}^d)$-stable basis for $\mathcal{S}(\Phi)$. In this case, a function $S \in \mathcal{S}(\Phi)$ is described by its Fourier transform

$$\hat{S} = \sum_{\phi \in \Phi} \tau_\phi \hat{\phi},$$

where the functions τ_ϕ, $\phi \in \Phi$, are in $L_2(\mathbb{T}^d)$ and

$$\|S\|_{L_2(\mathbb{R}^d)} \approx \sum_{\phi \in \Phi} \|\tau_\phi\|_{L_2(\mathbb{T}^d)}.$$

It is clear that the values at points congruent modulo 2π of the Fourier transform of a function S in $\mathcal{S}(\phi)$ are related. If we know $\hat{\phi}(x)$ and $\hat{S}(x)$, then, because τ has period 2π, all other values $\hat{S}(x + \alpha)$, $\alpha \in 2\pi\mathbb{Z}^d$, are determined. It is natural to try to remove this redundancy. The following *bracket product* is useful for this purpose. If f and g are in $L_2(\mathbb{R}^d)$, we define

$$[f,g] := \sum_{\beta \in 2\pi \mathbb{Z}^d} f(\cdot + \beta)\overline{g(\cdot + \beta)}. \qquad (3.2.4)$$

Then $[f,g]$ is a function in $L_1(\mathbb{T}^d)$.

One particular use of the bracket product is to relate inner products on $\mathbb{R}^d$ to inner products on $\mathbb{T}^d$. For example, if $f, g \in L_2(\mathbb{R}^d)$ and $j \in \mathbb{Z}^d$, then

$$(2\pi)^d \int_{\mathbb{R}^d} f(x)\overline{g(x-j)}\,\mathrm{d}x = \int_{\mathbb{R}^d} e_j(y)\hat{f}(y)\overline{\hat{g}(y)}\,\mathrm{d}y = \int_{\mathbb{T}^d} e_j(\theta)[\hat{f},\hat{g}](\theta)\,\mathrm{d}\theta. \qquad (3.2.5)$$

Thus, these inner products are the Fourier coefficients of $[\hat{f},\hat{g}]$. In particular, a function f is orthogonal to all the shifts of g if and only if $[\hat{f},\hat{g}] = 0$ a.e., in which case one obtains that all shifts of f are orthogonal to all the shifts of g (which also follows directly by a simple change of variables).

Another application of the bracket product is to relate integrals over $\mathbb{R}^d$ to integrals over $\mathbb{T}^d$. For example, if $\hat{S} = \tau\hat{\phi}$ with τ of period 2π, then

$$(2\pi)^d\|S\|_{L_2(\mathbb{R}^d)} = \|\tau[\hat{\phi},\hat{\phi}]^{1/2}\|_{L_2(\mathbb{T}^d)}. \qquad (3.2.6)$$

Returning to $L_2(\mathbb{R}^d)$-stability for a moment, it follows from (3.2.6) and (3.2.2) that the shifts of ϕ are $L_2(\mathbb{R}^d)$-stable if and only if $C_1 \le [\hat{\phi},\hat{\phi}] \le C_2$, a.e., for constants $C_1, C_2 > 0$. Also, the shifts of ϕ are orthonormal if and only if $[\hat{\phi},\hat{\phi}] = 1$ a.e. For example, if we begin with a function ϕ with $L_2(\mathbb{R}^d)$-stable shifts, then the function ϕ_* with Fourier transform

$$\hat{\phi}_* := \frac{\hat{\phi}}{[\hat{\phi},\hat{\phi}]^{1/2}} \qquad (3.2.7)$$

has orthonormal shifts (this is the standard way to orthogonalize the shifts of ϕ). Incidentally, this orthogonalization procedure applies whenever $[\hat{\phi},\hat{\phi}]$ vanishes only on a set of measure zero in $\mathbb{T}^d$, in particular for any compactly supported ϕ. That is, it is not necessary to assume that ϕ has $L_2(\mathbb{R}^d)$-stable shifts in order to orthonormalize its shifts

The bracket product is useful in describing projections onto shift-invariant spaces. Let ϕ be an arbitrary $L_2(\mathbb{R}^d)$ function and let $P := P_\phi$ denote the $L_2(\mathbb{R}^d)$ projector onto the space $\mathcal{S}(\phi)$. Then for each $f \in L_2(\mathbb{R}^d)$, Pf is the best $L_2(\mathbb{R}^d)$ approximation to f from $\mathcal{S}(\phi)$. It was shown in de Boor *et al.* (1991a) that

$$\widehat{Pf} = \frac{[\hat{f},\hat{\phi}]}{[\hat{\phi},\hat{\phi}]}\hat{\phi}. \qquad (3.2.8)$$

Here and later, we use the convention that $0/0 = 0$. We note some properties of (3.2.8). First, $[\hat{f},\hat{\phi}]$ is 2π-periodic and therefore the form of $\widehat{Pf}$ matches that required by (3.2.3). If ϕ has orthonormal shifts, then $[\hat{\phi},\hat{\phi}] = 1$ a.e., and in view of (3.2.5), the formula (3.2.8) is the usual one for the $L_2(\mathbb{R}^d)$ projector. If $\hat{\phi}/[\hat{\phi}, \hat{\phi}]$ is the Fourier transform of an $L_2(\mathbb{R}^d)$ function γ (this holds, for example, if ϕ has $L_2(\mathbb{R}^d)$-stable shifts), then

$$Pf = \sum_{j \in \mathbb{Z}^d} \gamma_j(f)\phi(\cdot - j), \quad \gamma_j(f) := \int_{\mathbb{R}^d} f(x)\overline{\gamma(x-j)}\,\mathrm{d}x, \qquad (3.2.9)$$

as follows from (3.2.5). Whenever ϕ has compact support and $L_2(\mathbb{R}^d)$-stable shifts, the function γ decays exponentially.

The bracket product is also useful in the description of the properties of the shift-invariant spaces $\mathcal{S}(\Phi)$ that are generated by a finite set Φ of functions from $L_2(\mathbb{R}^d)$. The properties of the generating set Φ are contained in its Gramian

$$G(\Phi) := \left([\hat{\phi}, \hat{\psi}]\right)_{\phi, \psi \in \Phi}. \qquad (3.2.10)$$

This is a matrix of 2π-periodic functions from $L_1(\mathbb{T}^d)$. For example, the shifts of the functions in Φ form an orthonormal basis for $\mathcal{S}(\Phi)$ if and only if $G(\Phi)$ is the identity matrix a.e. on $\mathbb{T}^d$. The generating set Φ provides an $L_2(\mathbb{R}^d)$-stable basis for $\mathcal{S}(\Phi)$ if and only if $G(\Phi)$ and $G(\Phi)^{-1}$ exist and are a.e. bounded on $\mathbb{T}^d$ with respect to some (and then every) matrix norm. For proofs, see de Boor *et al.* (1991c).

3.3. The conditions of multi-resolution

The question arises as to when the conditions (3.1.2) of multi-resolution are satisfied for a function $\phi \in L_2(\mathbb{R}^d)$. We mention, without proof, two sufficient conditions on ϕ for ((3.1.2)(ii)) and (iii) to hold. Jia and Micchelli (1991) have shown that if the shifts of ϕ are $L_2(\mathbb{R}^d)$-stable, if ϕ satisfies the refinement equation (3.1.4) with coefficients $(a(j))$ in $\ell_1(\mathbb{Z}^d)$, and if $\sum_{j \in \mathbb{Z}^d} |\phi(x+j)|$ is in $L_2(\mathbb{T}^d)$, then ((3.1.2)(ii)) and ((3.1.2)(iii)) are satisfied.

On the other hand, in de Boor *et al.* (1991b) it is shown that ((3.1.2)(ii)) and ((3.1.2)(iii)) are satisfied whenever $\phi \in L_2(\mathbb{R}^d)$ satisfies the refinement condition ((3.1.2)(i)) and, in addition, $\mathrm{supp}[\hat{\phi}, \hat{\phi}] = \mathbb{T}^d$; by this we mean that $[\hat{\phi}, \hat{\phi}]$ vanishes only on a set of measure zero. In particular, these conditions are satisfied whenever ϕ has compact support and satisfies the refinement condition. Since these conditions are satisfied for all functions ϕ that we shall encounter (in fact for all functions ϕ that have been considered in wavelet construction by multi-resolution), it is not necessary to verify separately ((3.1.2)(ii)) and ((3.1.2)(iii))—they automatically hold. We also note that in the construction of wavelets and pre-wavelets in de Boor *et al.* (1991b) it is not necessary to assume that ϕ has $L_2(\mathbb{R}^d)$-stable shifts.

We now discuss the refinement condition ((3.1.2)(i)). In view of the characterization (3.2.3) of shift-invariant spaces, this condition is equivalent to

$$\hat{\phi} = A\hat{\eta}, \quad \eta := \phi(2 \cdot) \tag{3.3.1}$$

for some 4π-periodic function A. If ϕ has $L_2(\mathbb{R}^d)$-stable shifts then this condition becomes the refinement equation (3.1.4) and $A(y) = \sum_{j \in \mathbb{Z}^d} a(j)e_{-j/2}$ in the sense of $L_2(2\mathbb{T}^d)$ convergence.

It was shown in de Boor *et al.* (1991b) that one can construct generators for the wavelet space W even when ((3.1.2)(iv)) does not hold. For example, this condition can be replaced by the assumption that supp $\hat{\phi} = \mathbb{R}^d$. We also note that if $[\hat{\phi}, \hat{\phi}]$ is nonzero a.e., then we can always find a generator ϕ_* for $\mathcal{S}$ with orthonormal shifts, so the condition ((3.1.2)(ii)) is satisfied for this generator (and the other conditions of multi-resolution remain the same). However, the generator ϕ_* does not satisfy the same refinement equation as ϕ (for example, the refinement equation for ϕ_* may be an infinite sum even if the equation for ϕ is a finite sum) and ϕ_* may not have compact support even if ϕ has compact support, so the construction that gives ϕ_* is not completely satisfactory. Furthermore, we would like to describe the wavelets and prewavelets directly in terms of the original ϕ. This is especially the case when ϕ does not have $L_2(\mathbb{R}^d)$-stable shifts, since then we can say nothing about the decay of ϕ_* even when ϕ has compact support. In the remainder of this presentation, we shall assume that ϕ has $L_2(\mathbb{R}^d)$-stable shifts.

3.4. Constructions of uni-variate wavelets

In this section we restrict our attention to wavelets in one variable, because multi-resolution is simpler and better understood for a single variable than for several variables. We suppose that ϕ satisfies the assumptions (3.1.2) of multi-resolution and follow the ideas presented in de Boor *et al.* (1991b).

Fundamentally, the approach of de Boor *et al.* (1991b) is quite simple. We take a function $\eta \in \mathcal{S}^1$ and consider its error $w := \eta - P\eta$ of best $L_2(\mathbb{R})$ approximation by the elements of $\mathcal{S}^0 = \mathcal{S} = \mathcal{S}(\phi)$. Here P is the $L_2(\mathbb{R})$ projector onto $\mathcal{S}(\phi)$ given by (3.2.8). Clearly $w \in W$ and we shall show that with any reasonable choice for η, the function w is a generator of W, i.e. $W = \mathcal{S}(w)$. Thus, because of the characterization (3.2.3) of principal shift-invariant spaces, we can obtain all other generators for w by operations on the Fourier transform side. Here are the details.

We take $\eta := \phi(2 \cdot)$, which is clearly in $\mathcal{S}^1$. Then, $w := \eta - P\eta$ is in W and by virtue of (3.2.8) has Fourier transform

$$\hat{w} = \hat{\eta} - \frac{[\hat{\eta}, \hat{\phi}]}{[\hat{\phi}, \hat{\phi}]} \hat{\phi}. \tag{3.4.1}$$

It is convenient to introduce (for a function $f \in L_2(\mathbb{R}^d)$) the abbreviated notation

$$\tilde{f} := [\hat{f}, \hat{f}]^{1/2}, \qquad (3.4.2)$$

since this expression occurs often in wavelet constructions. Another description of $\tilde{f}$ is

$$\tilde{f} = \left(\sum_{j \in 2\pi\mathbb{Z}^d} |\hat{f}(\cdot + j)|^2 \right)^{1/2}.$$

We see that $\tilde{f}$ is a 2π-periodic function, and if f has compact support then $\tilde{f}^2$ is a trigonometric polynomial, because of (3.2.5). The analogue of this function for half-shifts is

$$\tilde{\tilde{f}} := \left(\sum_{j \in 4\pi\mathbb{Z}^d} |\hat{f}(\cdot + j)|^2 \right)^{1/2}, \qquad (3.4.3)$$

which is now a 4π-periodic function. In particular f has orthonormal half-shifts if and only if $\tilde{\tilde{f}} = 2^{-d/2}$ a.e., and $L_2(\mathbb{R}^d)$-stable half-shifts if and only if $C_1 \leq \tilde{\tilde{f}} \leq C_2$, a.e., for constants $C_1, C_2 > 0$.

We return now to the construction of wavelets. We can multiply $\hat{w}$ by any 2π-periodic function, and as long as the resulting function is in $L_2(\mathbb{R})$, it will be the Fourier transform of a function in W. We multiply (3.4.1) by $\tilde{\phi}^2$, which clears the denominator. The result is the function w_0 with Fourier transform

$$\hat{w}_0 := \tilde{\phi}^2 \hat{\eta} - [\hat{\eta}, \hat{\phi}]\hat{\phi}. \qquad (3.4.4)$$

We note that w_0 has compact support whenever ϕ does.

We can calculate the bracket products appearing in (3.4.4) by using the refinement relation $\hat{\phi} = A\hat{\eta}$ (see (3.3.1)) with A a 4π-periodic function. For example, to calculate $\tilde{\phi}$,

$$
\begin{aligned}
\tilde{\phi}^2 &= \sum_{j \in 2\pi\mathbb{Z}} |\hat{\phi}(\cdot + j)|^2 \\
&= \sum_{j \in 4\pi\mathbb{Z}} (|A(\cdot + j)|^2 |\hat{\eta}(\cdot + j)|^2 + |A(\cdot + j + 2\pi)|^2 |\hat{\eta}(\cdot + j + 2\pi)|^2) \\
&= |A|^2 \tilde{\tilde{\eta}}^2 + |A(\cdot + 2\pi)|^2 \tilde{\tilde{\eta}}^2(\cdot + 2\pi).
\end{aligned}
$$

Similarly, $[\hat{\eta}, \hat{\phi}] = \overline{A}\tilde{\tilde{\eta}}^2 + \overline{A(\cdot + 2\pi)}\tilde{\tilde{\eta}}^2(\cdot + 2\pi)$. Therefore,

$$
\begin{aligned}
\hat{w}_0 &= \{|A|^2 \tilde{\tilde{\eta}}^2 + |A(\cdot + 2\pi)|^2 \tilde{\tilde{\eta}}^2(\cdot + 2\pi) - \overline{A}A\tilde{\tilde{\eta}}^2 \\
&\quad - \overline{A(\cdot + 2\pi)}A\tilde{\tilde{\eta}}^2(\cdot + 2\pi)\}\hat{\eta} \\
&= \{A(\cdot + 2\pi) - A\}\overline{A(\cdot + 2\pi)}\tilde{\tilde{\eta}}^2(\cdot + 2\pi)\hat{\eta}.
\end{aligned}
$$

We can make one more simplification in the last representation for $\hat{w}_0$.

The function $\frac{1}{2}e_{1/2}\{A - A(\cdot + 2\pi)\}$ is 2π-periodic. Therefore, dividing by this function, we obtain the function

$$\hat{\psi} := 2e_{-1/2}\overline{A(\cdot + 2\pi)}\tilde{\tilde{\eta}}^2(\cdot + 2\pi)\hat{\eta}. \qquad (3.4.5)$$

It is easy to see (and is shown in (3.4.14)) that ψ is in $L_2(\mathbb{R})$. It follows, therefore, that ψ is in W and $\mathcal{S}(\psi) \subset W$. The following argument shows that we really have $\mathcal{S}(\psi) = W$.

If we replace η by $\eta_1 := \eta(\cdot - \frac{1}{2})$ (which is also in $\mathcal{S}^1$) and repeat the previous construction, in place of w_0 we obtain the function w_1 whose Fourier transform is

$$\hat{w}_1 = e_{-1/2}\{A(\cdot + 2\pi) + A\}\overline{A(\cdot + 2\pi)}\tilde{\tilde{\eta}}^2(\cdot + 2\pi)\hat{\eta}.$$

Hence, dividing by $A(\cdot + 2\pi) + A$ (which is 2π-periodic), we arrive at the same function ψ. The importance of this fact is that we can reverse these two processes. In other words we can multiply $\hat{\psi}$ by a 2π-periodic function and obtain $\hat{\eta} - \widehat{P\eta}$ (respectively $\hat{\eta}_1 - \widehat{P\eta_1}$). Hence, both of these functions are in $\mathcal{S}(\psi)$. Since $P\eta$ is in $\mathcal{S}(\phi)$, $\eta = P\eta + (\eta - P\eta)$ is in $\mathcal{S}(\phi) + \mathcal{S}(\psi)$. Similarly, η_1 is in this space. Since the full shifts of η and η_1 generate $\mathcal{S}^1(\phi)$, we must have $W = \mathcal{S}(\psi)$. This confirms our earlier remark that W is a principal shift-invariant space. Since we can obtain ψ from w and w_0 by multiplying by 2π-periodic functions, both w and w_0 are also generators of W.

We consider some examples that show that ψ is the (pre)wavelet constructed by various authors.

Orthogonal wavelets To obtain orthogonal wavelets, Mallat (1989) begins with a function ϕ that satisfies the assumptions (3.1.2) of multi-resolution and whose shifts are orthonormal. This is equivalent to $\tilde{\phi} = 1$ a.e., and (by a change of variables) to the half-shifts of $\sqrt{2}\,\eta$ being orthonormal, i.e. to $\tilde{\tilde{\eta}} = \frac{1}{2}$ a.e. When this is used in (3.4.5), we obtain

$$\hat{\psi} = e_{-1/2}\overline{A(\cdot + 2\pi)}\hat{\eta}, \qquad (3.4.6)$$

which is the orthogonal wavelet of Mallat. To see that the shifts of ψ are orthonormal, one simply computes

$$\tilde{\psi}^2 = |A(\cdot + 2\pi)|^2\tilde{\tilde{\eta}}^2 + |A|^2\tilde{\tilde{\eta}}^2(\cdot + 2\pi) = \tfrac{1}{4}\{|A(\cdot + 2\pi)|^2 + |A|^2\} = 1, \quad (3.4.7)$$

where the last equality follows from the identity

$$
\begin{aligned}
1 = \tilde{\phi}^2 = \tilde{\tilde{\phi}}^2 + \tilde{\tilde{\phi}}^2(\cdot + 2\pi) &= |A|^2\tilde{\tilde{\eta}}^2 + |A(\cdot + 2\pi)|^2\tilde{\tilde{\eta}}^2(\cdot + 2\pi) \\
&= \tfrac{1}{4}\{|A(\cdot + 2\pi)|^2 + |A|^2\}. \qquad (3.4.8)
\end{aligned}
$$

The Fourier transform identity (3.4.6) is equivalent to the identity (3.1.8).

We note that from the orthogonal wavelet ψ of (3.1.8) (respectively (3.4.6)), we obtain all other orthogonal wavelets in W by multiplying $\hat{\psi}$ by a 2π-periodic function τ of unit modulus. Indeed, we know that any element $w \in W$ satisfies $\hat{w} = \tau\hat{\psi}$ with $\tau \in L_2(\mathbb{T})$. To have $[\hat{w}, \hat{w}] = 1$ a.e., the function τ must satisfy $|\tau(y)| = 1$ a.e. in $\mathbb{T}$.

As an example, we consider the cardinal B-spline N_r of order r. To obtain orthogonal wavelets by Mallat's construction, one need only manipulate various Laurent series. First, one orthogonalizes the shifts of N_r. This gives the spline $\phi = \mathcal{N}_r$ whose Fourier transform is

$$\hat{\phi} = \hat{\mathcal{N}}_r := \frac{\hat{N}_r}{\tilde{N}_r}. \qquad (3.4.9)$$

It is easy to compute the coefficients in the expansion

$$\tilde{N}_r^2 = \sum_{j\in\mathbb{Z}} \alpha(j)e_{-j}. \qquad (3.4.10)$$

In fact, we know from (3.2.5) that this is a trigonometric polynomial whose coefficients are

$$\begin{aligned}
\alpha(j) &= \int_{\mathbb{R}} N_r(x-j)N_r(x)\,dx = \int_{\mathbb{R}} N_r(r+j-x)N_r(x)\,dx \\
&= [N_r * N_r](j+r) = N_{2r}(j+r), \quad j \in \mathbb{Z},
\end{aligned}$$

because N_r is symmetric about its midpoint.

The polynomial $\rho_{2r}(z) := z^r \sum_{j\in\mathbb{Z}} \alpha(j)z^{-j}$ is the Euler–Frobenius polynomial of order $2r$, which plays a prominent role in cardinal spline interpolation (see Schoenberg (1973)). It is well known that ρ_{2r} has no zeros on $|z| = 1$. Hence, the reciprocal $1/\rho_{2r}$ is analytic in a nontrivial annulus that contains the unit circle in its interior. One can easily find the coefficients of reciprocals and square roots of Laurent series inductively. By finding the coefficients of $\rho_{2r}^{-1/2}$, we obtain the coefficients $\beta(j)$ appearing in the expansion

$$\phi(x) = \mathcal{N}_r(x) = \sum_{j\in\mathbb{Z}} \beta(j)N_r(x-j). \qquad (3.4.11)$$

Because ρ_{2r} has no zeros on $|z| = 1$, we conclude that the coefficients $\beta(j)$ decrease exponentially. The spline $\mathcal{N}_r$ together with its shifts form an orthonormal basis for the cardinal spline space $\mathcal{S}(N_r)$. They are sometimes referred to as the Franklin basis for $\mathcal{S}(N_r)$.

Now that we have the spline $\phi := \mathcal{N}_r$ in hand, we can obtain the spline wavelet $\psi = \mathcal{N}_r^*$ of Battle–Lemarié (Battle, 1987) from formula (3.1.8). For this, we need to find the refinement equation for ϕ. We begin with the refinement equation (3.1.5) for the B-spline N_r, which we write in terms of celebrated construction of Daubechies (1988) leads to such wavelets, which are frequently used in numerical applications. Space prohibits us from giving all the details of Daubechies' construction, but the following discussion will outline the basic ideas.

To construct a compactly supported wavelet with a prescribed smoothness order r and compact support, one finds a special finite sequence $(a(j))$ such that the refinement equation (3.1.4) has a solution $\phi \in C^r$ with orthogonal shifts. The orthogonal wavelet ψ of (3.4.5) will then obviously have compact support and the same smoothness. Before we begin, it is necessary to understand which properties of the sequence $(a(j))$ guarantee the existence of a function ϕ with the desired properties, i.e. we need to understand the nature of solutions to the refinement equation (3.1.4). This has been studied in another context, namely in subdivision algorithms for computer aided geometric design (see, for example, the paper by Cavaretta $et\ al.$ (1991) for a discussion of subdivision). As was pointed out by Dahmen and Micchelli (1990), it is possible to derive part of Daubechies' construction from the subdivision approach. However, we shall describe Daubechies' original construction.

Let r be a nonnegative integer that corresponds to the desired order of smoothness, and let $(a(j))$ with $a(j) = 0$, $|j| > m$, and $a(m) \neq 0$, be the sequence of the refinement equation (3.1.4) for the function ϕ we want to construct. The sequence $(a(j))$ and the Fourier transform of ϕ are related by

$$\hat{\phi}(y) = \mathcal{A}(y/2)\hat{\phi}(y/2), \quad \mathcal{A}(y) := \frac{1}{2}\sum_{j=-m}^{m} a(j)e^{-ijy}. \tag{3.5.1}$$

Here we use a slightly different normalization for the refinement function $(\mathcal{A}(y) = \frac{1}{2}A(2y))$. If $\hat{\phi}$ is continuous at 0 and $\hat{\phi}(0) = 1$, we can, at least in a formal sense, write

$$\hat{\phi}(y) = \lim_{k \to \infty} A_k(y) \tag{3.5.2}$$

where

$$A_k(y) := \prod_{j=1}^{k} \mathcal{A}(y/2^j). \tag{3.5.3}$$

We note that $A_k^*(y) := A_k(2^k y)$ is a trigonometric polynomial of degree $(2^k - 1)m$. The key to Daubechies' construction is to impose conditions on $\mathcal{A}$ (which are therefore conditions on the sequence $(a(j))$) that not only make (3.5.2) rigorous but also guarantee that the function ϕ defined by (3.5.2) has the desired smoothness and has orthonormal shifts.

We first note that if the shifts of ϕ are orthonormal then, as was shown

Fourier transforms as $\hat{N}_r = A_0 \hat{\eta}_0$ with $\eta_0 := N_r(2\,\cdot\,)$ and

$$A_0 = 2^{-r+1}\sum_{j=0}^{r}\binom{r}{j}e_{-j/2}$$

a 4π-periodic trigonometric polynomial. It follows that

$$\hat{\phi} = A\hat{\eta}, \quad \eta := \mathcal{N}_r(2\,\cdot\,), \quad A(y) = \frac{\hat{\phi}(y)}{\frac{1}{2}\hat{\phi}(y/2)} = \tilde{N}_r(y/2)\tilde{N}_r^{-1}(y)A_0(y).$$
$$\tag{3.4.12}$$

In terms of the B-spline N_r, this gives

$$\hat{\psi}(y) = e_{-1/2}(y)\overline{A(y+2\pi)}\hat{\eta}(y) = \tfrac{1}{2}e_{-1/2}(y)\overline{A(y+2\pi)}\tilde{N}_r^{-1}(y/2)\hat{N}_r(y/2). \tag{3.4.13}$$

In other words, to find the orthogonal spline wavelet ψ of (3.4.13), we need to multiply out the various Laurent expansions making up $\overline{A(\cdot + 2\pi)}\tilde{N}_r^{-1}(\cdot/2)$. This gives the coefficients $\gamma(j)$, $j \in \mathbb{Z}$, in the representation

$$\psi(x) = \sum_{j\in\mathbb{Z}}\gamma(j)N_r(2x - j).$$

We emphasize that each of the Laurent series converges in an annulus containing the unit circle. This means that the coefficients $\gamma(j)$ converge exponentially to zero when $j \to \pm\infty$.

Pre-wavelets For the construction of pre-wavelets, we do not assume that the shifts of ϕ are orthonormal, but only that they are $L_2(\mathbb{R})$-stable, i.e. we assume ((3.1.2)(iv)). Then, it is easy to see that the function ψ defined by (3.4.5) is a pre-wavelet. Indeed, we already know that ψ is a generator for W and it is enough to check that it has $L_2(\mathbb{R})$-stable shifts. For this, we compute

$$\frac{\tilde{\psi}^2}{4} = |A(\cdot + 2\pi)|^2\tilde{\tilde{\eta}}^4(\cdot + 2\pi)\tilde{\tilde{\eta}}^2 + |A|^2\tilde{\tilde{\eta}}^4\tilde{\tilde{\eta}}^2(\cdot + 2\pi). \tag{3.4.14}$$

Since the shifts of ϕ are $L_2(\mathbb{R})$-stable, so are the half-shifts of η. This means that $C_1 \leq \tilde{\tilde{\eta}} \leq C_2$ for constants $C_1, C_2 > 0$. Moreover, the formula

$$\tilde{\phi}^2 = |A|^2\tilde{\tilde{\eta}}^2 + |A(\cdot + 2\pi)|^2\tilde{\tilde{\eta}}^2(\cdot + 2\pi)$$

shows that $C_1 \leq |A|^2 + |A(\cdot + 2\pi)|^2 \leq C_2$, again for positive constants C_1, C_2. Combining this information with (3.4.14) shows that $\tilde{\psi}$ is bounded above and below by positive constants, so that ψ has $L_2(\mathbb{R})$-stable shifts. This also shows that ψ is in $L_2(\mathbb{R})$. The pre-wavelet ψ was introduced by Chui and Wang (1991) and independently by Micchelli (1991).

We can also find a direct representation for ψ in terms of the shifts of $\phi(2\,\cdot\,)$. For this we need the Fourier coefficients $\mu(j)$ (of $e_{-j/2}$) for the 4π-

periodic function $2\overline{A}\tilde{\eta}$:

$$\begin{aligned}\mu(j) &:= \frac{1}{4\pi}\int_{-2\pi}^{2\pi}2\overline{A}\tilde{\eta}^2 e_{j/2} = \frac{1}{4\pi}\int_{\mathbb{R}}2\overline{\hat{\phi}}\hat{\eta}e_{j/2}\\ &= \int_{\mathbb{R}}\overline{\hat{\phi}}\eta(\,\cdot\,+j/2) = \int_{\mathbb{R}}\overline{\phi(x)}\phi(2x+j)\,\mathrm{d}x.\end{aligned}$$

Using this in (3.4.5), we find that

$$\psi = \sum_{j\in\mathbb{Z}}(-1)^{j+1}\mu(j-1)\phi(2\cdot-j), \quad \mu(j) := \int_{\mathbb{R}}\overline{\phi(x)}\phi(2x+j)\,\mathrm{d}x. \quad (3.4.15)$$

If ϕ has compact support, then clearly ψ also has compact support. Chui and Wang (1990) posed the interesting question as to whether ψ has the smallest support among all the elements in W, to which they gave the following answer. We assume that A is a polynomial, i.e. that ϕ satisfies a finite refinement equation. Next, we note that because $W \subset \mathcal{S}^1$, any $w \in W$ is represented as

$$\hat{w}(y) = e_{-1/2}(y)\overline{B(y+2\pi)}\hat{\eta}(y) \quad (3.4.16)$$

with B of period 4π. If $B = \sum_{j=m}^{M} b(j)e_{-j/2}$ is a Laurent polynomial with $b(m)b(M) \neq 0$, then w has compact support, and we define the length of B to be $M - m$. We know that there are nonzero polynomials B that satisfy (3.4.16) for some w because $\tilde{\eta}^2$ is a polynomial (since η has compact support) and (3.4.5) implies that for $B_0 := A\tilde{\eta}^2$, w is the pre-wavelet $\psi \in W$.

B_0 may not have minimal length among all such polynomials; however, because it may be possible to cancel certain symmetric factors from B_0. To see this, we write $B_0(y) = e_M(y/2)P(e^{-iy/2})$ with P an algebraic polynomial, and we let $Q(z^2) := \prod_\lambda(z-\lambda)$, with the product taken over all λ with λ and $-\lambda$ both zeros of P. Then, the factorization $P(z) = Q(z^2)P_*(z)$ gives the factorization $B_0(y) = \tau(y)B_*(y)$ with τ a trigonometric polynomial of period 2π that does not vanish. Therefore, the function ψ_* with Fourier transform

$$\hat{\psi}_*(y) = e_{-1/2}(y)\overline{B_*(y+2\pi)}\hat{\eta}(y), \quad B_*(y) := \tau^{-1}(y)B_0(y), \quad (3.4.17)$$

is in W and has smaller length than B_0. A simple argument (which we do not give) shows that B_* has smallest length. For most pre-wavelets of interest, $B_* = B_0$.

The problem of finding a wavelet w in the form (3.4.16) with B a polynomial of minimal length, which is solved by $w = \psi_*$, is not always equivalent to finding the wavelet with minimal support; here the word 'support' means the interval of smallest length outside of which w vanishes identically. In general, there may be wavelets w of compact support that can be represented by (3.4.16) with B not a polynomial. However, Ben-Artzi and Ron (1990) show that this is impossible whenever the following property holds:

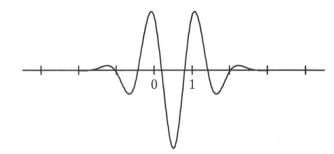

Fig. 2. The Chui–Wang spline pre-wavelet for $r = 4$, which has support $[-3, 4]$. The vertical scale is stretched by a factor of eight.

The linear combination $\sum_{j\in\mathbb{Z}}\gamma(j)\phi(\,\cdot\,-j)$ (which converges pointwise, since ϕ has compact support) is identically zero if and only if all the coefficients $\gamma(j)$ are 0. Under these assumptions, the wavelet ψ_* has minimal support (see de Boor et al. (1991b) for details).

For a pre-wavelet ψ, we have the wavelet decomposition

$$f = \sum_{k\in\mathbb{Z}}\sum_{j\in\mathbb{Z}^d}c_{j,k}(f)\psi_{j,k}, \quad c_{j,k}(f) := \int_{\mathbb{R}}f\overline{\gamma_{j,k}}, \quad (3.4.18)$$

where γ has Fourier transform $\hat{\gamma} = \hat{\psi}/[\hat{\psi},\hat{\psi}]$. This follows from the representation (3.2.9) for the projector P from $L_2(\mathbb{R})$ onto W. It is useful to note that when ψ has compact support, the function γ will generally not have compact support because of the division by the bracket product $[\hat{\psi},\hat{\psi}]$. Thus, there is in some sense a trade-off between the simplicity of the pre-wavelet and the complexity of the coefficient functional.

We consider the following important example. Let $\phi := N_r$ be the cardinal B-spline of order r, which is known to have linearly independent shifts. Then, the function ψ in (3.4.15) is a spline function with compact support. It is easy to see that $A\tilde{\eta}^2$ has no symmetric zeros so that ψ has minimal support. We note also that it is shown in de Boor et al. (1991b) that the shifts of ψ are themselves linearly independent. From formula (3.4.15), we see that ψ is supported on $[1 - r, r]$. Up to a shift, the spline ψ is the minimally supported spline pre-wavelet of Chui and Wang (1991); see Figure 2.

3.5. Daubechies' compactly supported wavelets

The orthogonal spline wavelets of Section 3.4, which decay exponentially at infinity, can be chosen to have any specified finite order of smoothness. It is natural to ask whether orthogonal wavelets can be constructed that have both any specified finite order of smoothness and compact support. A

in (3.4.8),
$$|\mathcal{A}(y)|^2 + |\mathcal{A}(y + \pi)|^2 = 1, \qquad y \in \mathbb{T}. \tag{3.5.4}$$

The converse to this is almost true. Namely, Daubechies' construction shows that (3.5.4) together with some mild assumptions (related to the convergence in (3.5.3)) imply the orthonormality of the shifts of ϕ. For this, the following identities, which follow from (3.5.4) by induction, are useful:

$$\sum_{j=0}^{2^k-1} |A_k^*(y + j2^{-k}2\pi)|^2 = 1, \quad k = 1, 2 \dots. \tag{3.5.5}$$

We next want to see what properties of $\mathcal{A}$ guarantee smoothness for ϕ. The starting point is the following observation. If $\int_{\mathbb{R}} \phi(x)\,dx \neq 0$, then integrating the refinement equation (3.1.4) gives $\sum_{j \in \mathbb{Z}} a(j) = 2$. Hence, $\mathcal{A}(0) = 1$ and $\mathcal{A}(\pi) = 0$. We can therefore write

$$\mathcal{A}(y) = (1 + e^{iy})^N \alpha(y), \quad \|\alpha\|_{L_\infty(\mathbb{T})} = 2^{-\theta}, \quad \alpha(0) = 2^{-N}, \tag{3.5.6}$$

for a suitable integer $N > 0$, a real number θ, and a function α.

By carefully estimating the partial products A_k, it can be shown that whenever $\mathcal{A}$ satisfies (3.5.6) for some $\theta > \frac{1}{2}$, the product (3.5.2) converges to a function in $L_2(\mathbb{R})$ that decays like $|x|^{-\theta}$ as $|x| \to \infty$. The limit function is the Fourier transform of the solution ϕ to the refinement equation (3.5.1). We see that the larger we can make θ in (3.5.6), the smoother ϕ is. For example, if $\theta > r + 1$, then ϕ is in C^r.

What is the role of the integer N in (3.5.6)? Practically, one must increase N to find a function $\alpha(y)$ that satisfies (3.5.6) for large θ. In addition, the local approximation properties of the spaces $S^k(\phi)$ are determined by N; see Section 5.

Once it is shown that there is a function ϕ that satisfies the refinement equation for the given sequence $a(j)$, it remains to show that ϕ has compact support and orthonormal shifts. Here the arguments have the same character as those used to analyse subdivision algorithms for the graphical display of curves and surfaces. Assume that $\mathcal{A}$ satisfies (3.5.4) and (3.5.6) for some $\theta > \frac{1}{2}$ and let χ denote the characteristic function of $[-\frac{1}{2}, \frac{1}{2}]$. Then $\hat{\chi}(y) = (\sin y/2)/(y/2)$. We define ϕ_k to be the function whose Fourier transform is $\hat{\phi}_k(y) := A_k(y)\hat{\chi}(2^{-k}y)$. It can then be shown that

$$\int_{\mathbb{R}} |\phi(y) - \phi_k(y)|^2 \, dy \to 0, \quad k \to \infty. \tag{3.5.7}$$

If $A_k^* = \sum_j a^*(j, k)e_{-j}$, then $\sum_j a^*(j, k)\hat{\chi}(x - j)$ has Fourier transform $A_k^*(y)\hat{\chi}(y)$ and $\hat{\phi}_k(y) = A_k^*(2^{-k}y)\hat{\chi}(2^{-k}y)$. Therefore,

$$\phi_k(x) = \sum_{j \in \mathbb{Z}} a^*(j, k)2^k \chi(2^k x - j). \tag{3.5.8}$$

Since the coefficients $a^*(j, k)$ of A_k^* are 0 for $|j| > (2^k - 1)m$, we obtain that ϕ_k is supported in $[-m, m]$. Letting $k \to \infty$, we obtain from (3.5.7) that ϕ is also supported on $[-m, m]$. From (3.5.8), (3.5.5), and the orthonormality of the functions $2^{k/2}\chi(2^k \cdot - j)$, $j \in \mathbb{Z}$, we have

$$\int_{\mathbb{R}} \phi_k(x)\phi_k(x - \ell)\,\mathrm{d}x = 2^k \sum_{\mu - \nu = 2^k \ell} a^*(\mu, k)a^*(\nu, k) = \delta(\ell), \quad \ell \in \mathbb{Z}.$$

Here the last equality follows by expanding the identity (3.5.5). Letting $k \to \infty$, we obtain that $\{\phi(\cdot - j)\}_{j \in \mathbb{Z}}$ is an orthonormal system.

This outline shows that a C^r, compactly supported function ϕ with orthonormal shifts exists if (3.5.4) and (3.5.6) hold for a sequence $(a(j))$ and two numbers N and $\theta > r + 1$. The following arguments show that such sequences exist.

We look for an $\mathcal{A}$ of the form (3.5.6) with α a trigonometric polynomial with real coefficients. Then, $|\alpha(y)|^2 = \alpha(y)\overline{\alpha(y)} = \alpha(y)\alpha(-y)$ is an even trigonometric polynomial, and

$$|\alpha(y)|^2 = T(\cos y) = T(1 - 2\sin^2 y/2) = R(\sin^2 y/2)$$

with R an algebraic polynomial. The identity (3.5.4) now becomes

$$(\cos^2 y/2)^N R(\sin^2 y/2) + (\sin^2 y/2)^N R(\cos^2 y/2) = 2^{-2N}.$$

With $t := \sin^2 y/2$, we have

$$(1 - t)^N R(t) + t^N R(1 - t) = 2^{-2N}. \tag{3.5.9}$$

Therefore, to find $\mathcal{A}$, we must find an algebraic polynomial R that satisfies (3.5.9). It is easy to see that the degree of R must be at least $N - 1$. We can find R of this degree by writing R in the Bernstein form

$$R(t) = \sum_{k=0}^{N-1} \lambda_k \binom{N-1}{k} t^k (1 - t)^{N-k-1}.$$

Then, (3.5.9) becomes

$$(1-t)^N \sum_{k=0}^{N-1} \lambda_k \binom{N-1}{k} t^k (1-t)^{N-k-1} + t^N \sum_{k=0}^{N-1} \lambda_k \binom{N-1}{k} (1-t)^k t^{N-k-1}$$

$$= 2^{-2N} = 2^{-2N} \sum_{k=0}^{2N-1} \binom{2N-1}{k} t^k (1-t)^{2N-1-k}. \tag{3.5.10}$$

We see that

$$\lambda_k := 2^{-2N} \frac{\binom{2N-1}{k}}{\binom{N-1}{k}}, \qquad k = 0, 1, \ldots, N-1,$$

satisfies (3.5.10), and we denote the polynomial with these coefficients by R_N.

It is important to observe that $R_N(t)$ is nonnegative for $0 \leq t \leq 1$, because we wish to show that there is a trigonometric polynomials $\alpha(y)$ such that $R_N(\sin^2 y/2) = |\alpha(y)|^2$, i.e. we somehow have to take a 'square root' of R_N. For this, we use the classical theorem of Fejér–Riesz (see, for example, Karlin and Studden (1966) p. 185) that says that if R is nonnegative on $[0, 1]$, then $R(\sin^2 y/2) = |\alpha(y)|^2$ for some trigonometric polynomial α with real coefficients and of the same degree as R. We let α_N be the trigonometric polynomial corresponding to R_N.

We now set $\mathcal{A}_N(y) := (1 + e^{iy})^N \alpha_N(y)$ and note that $\mathcal{A}_N$ satisfies (3.5.4) because R_N satisfies (3.5.9). Therefore, the function ϕ defined via the limit process (3.5.2) has compact support and orthonormal shifts. The corresponding orthogonal wavelet $\psi =: \mathcal{D}_{2N}$ defined by (3.1.8) for the refinement coefficients $(a(j))$ is an orthogonal wavelet with compact support. It is easy to show that $\mathcal{D}_{2N}$ is supported in $[-(N-1), N]$.

The question now is what is the smoothness of $\mathcal{D}_{2N}$. Here the matter can become somewhat technical (see Daubechies (1988) and Meyer (1987)). However, the following 'poor man's' argument based on Stirling's formula at least shows that given any integer r, if we choose N sufficiently large, the orthogonal wavelet $\mathcal{D}_{2N}$ will have smoothness C^r.

Because the Bernstein coefficients of R_N are monotonic, it follows that R_N is increasing on $[0, 1]$. Therefore, $\max_{0 \leq t \leq 1} R_N(t) = R_N(1) = \lambda_{N-1} = 2^{-2N}\binom{2N-1}{N}$. Therefore, $\|\alpha_N\|^2_{L_\infty(\mathbb{T})}$ is bounded by

$$2^{-2N}\binom{2N-1}{N} \leq \frac{2^{-2N}\sqrt{2\pi(2N-1)}(2N-1)^{2N-1}e^{-(2N-1)}}{\sqrt{2\pi N}N^N e^{-N}\sqrt{2\pi(N-1)}(N-1)^{N-1}e^{-(N-1)}}$$

$$\leq C_0 N^{-1/2},$$

by Stirling's formula. We see that given any value of $\theta > 0$, we can choose N large enough so that (3.5.6) is satisfied for that θ, and the function $\hat{\phi}$ satisfies $|\hat{\phi}(x)| \leq C(1 + |x|)^{-\theta}$. Hence, for any $r < \theta - 1$, ϕ, and hence $\mathcal{D}_{2N}$, is in C^r.

For $N = 1$, the Daubechies construction gives $\phi = \chi_{[0,1]}$ and $\mathcal{D}_2$ is the Haar function. For $N = 2$, the polynomial $R_2(t) = (1 + 2t)/16$ and

$$\mathcal{A}_2(y) = (1 + e^{iy})^2 \left(\frac{\sqrt{3}+1}{8} - \frac{\sqrt{3}-1}{8}e^{-iy}\right) = (1 + e^{iy})^2 \alpha_2(y).$$

Then $\alpha_2(y)$ satisfies $|\alpha_2(y)| \leq \sqrt{3}/4 < 2^{-1}$. Therefore, the function ϕ and the wavelet $\mathcal{D}_4 := \psi$ corresponding to this choice is continuous. (See Figure 3 for a graph of ϕ and ψ.) A finer argument shows that $\mathcal{D}_4$ is in $\mathrm{Lip}(.55, L_\infty(\mathbb{R}))$. The reader can consult Daubechies (1988) for a table of

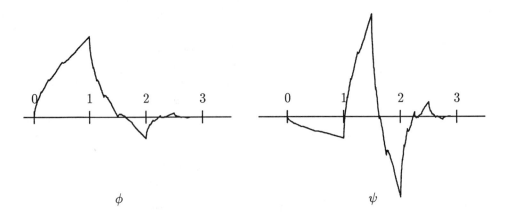

Fig. 3. The function ϕ and the Daubechies wavelet $\psi = \mathcal{D}_{2N}$ when $N = 2$.

the refinement coefficients of $\mathcal{D}_{2N}$ for other values of N and a more precise discussion of the smoothness of $\mathcal{D}_{2N}$ in $L_\infty(\mathbb{R})$.

3.6. Multi-variate wavelets

There are two approaches to the construction of multi-variate wavelets for $L_2(\mathbb{R}^d)$. The first, the tensor product approach, we now briefly describe. In this section, V will denote the set of vertices of the cube $[0, 1]^d$ and $V' := V \setminus \{0\}$. Let ϕ be a uni-variate function satisfying the conditions (3.1.2) of multi-resolution and let ψ be an orthogonal wavelet obtained from ϕ. For $\phi_0 := \phi$, $\phi_1 := \psi$, the collection Ψ of functions

$$\psi_v(x_1, \ldots, x_d) := \phi_{v_1}(x_1) \cdots \phi_{v_d}(x_d), \quad v \in V', \qquad (3.6.1)$$

generates, by dilation and translation, a complete orthonormal system for $L_2(\mathbb{R}^d)$. More precisely, the collection of functions $\psi_{j,k,v} := 2^{kd/2}\psi_v(2^k \cdot - j)$, $j \in \mathbb{Z}^d$, $k \in \mathbb{Z}$, $v \in V'$, forms a complete orthonormal system for $L_2(\mathbb{R}^d)$: each $f \in L_2(\mathbb{R}^d)$ has the series representation

$$f = \sum_{v \in V'} \sum_{k \in \mathbb{Z}} \sum_{j \in \mathbb{Z}^d} \langle f, \psi_{j,k,v} \rangle \, \psi_{j,k,v} \qquad (3.6.2)$$

in the sense of convergence in $L_2(\mathbb{R}^d)$. This construction also applies to pre-wavelets, thereby yielding a stable basis for $L_2(\mathbb{R}^d)$.

Another view of the tensor product wavelets is the following. We let $\mathcal{S}$ be the space generated by the shifts of the function $x \longmapsto \phi(x_1) \cdots \phi(x_d)$. Then, the wavelets ψ_v are generators for the wavelet space $W := \mathcal{S}^1 \ominus \mathcal{S}^0$.

The second way to construct multi-variate wavelets uses multi-resolution in several dimensions. We let ϕ be a function in $L_2(\mathbb{R}^d)$ that satisfies the conditions (3.1.2) of multi-resolution for $S := S(\phi)$, and we seek a set Ψ of generators for the wavelet space $W := S^1 \ominus S^0$. For example, if we want an orthonormal wavelet basis for $L_2(\mathbb{R}^d)$, we would seek Ψ such that the totality of functions $\psi(\cdot - j)$, $j \in \mathbb{Z}^d$, $\psi \in \Psi$, forms an orthonormal basis for W. By dilation and translation, we would obtain the collection of functions $\psi_{j,k} := 2^{kd/2}\psi(2^k \cdot - j)$, $\psi \in \Psi$, $j \in \mathbb{Z}^d$, $k \in \mathbb{Z}$, which together form an orthonormal basis for $L_2(\mathbb{R}^d)$. Each function f in $L_2(\mathbb{R}^d)$ has the representation

$$f = \sum_{\psi \in \Psi} \sum_{k \in \mathbb{Z}} \sum_{j \in \mathbb{Z}^d} \langle f, \psi_{j,k} \rangle \, \psi_{j,k}. \qquad (3.6.3)$$

For a pre-wavelet set Ψ, we would require $L_2(\mathbb{R}^d)$-stability in place of orthogonality. Sometimes, we might require additionally that the shifts of ψ and those of $\tilde{\psi}$ are orthogonal whenever ψ and $\tilde{\psi}$ are different functions in Ψ.

Constructing orthogonal wavelets and pre-wavelets by this second approach is complicated by the fact that there does not seem to be a straightforward way to choose a canonical orthogonal wavelet set from the many possible wavelet sets Ψ.

The book of Meyer (1990) contains first results on the construction of multi-variate wavelet sets by the second approach. This was expanded upon in the paper of Jia and Micchelli (1991). These treatments are not always constructive; for example, the latter paper employs in some contexts the Quillen–Suslin theorem from commutative algebra. Several papers (Riemenschneider and Shen, 1991, 1992; Chui *et al.*, 1991; Lorentz and Madych, 1991) treat special cases, such as the construction of orthogonal wavelet and pre-wavelet sets when ϕ is taken to be a box spline. The paper of Riemenschneider and Shen (1991) is particularly important, since it gives a constructive approach that applies in two and three dimensions to a wide class of functions ϕ.

We shall follow the approach of de Boor *et al.* (1991b), which is based on the structure of shift-invariant spaces. This approach immediately gives a generating set for W, which can then be exploited to find orthogonal wavelet and pre-wavelet sets. De Boor *et al.* start with a function ϕ that satisfies the refinement relation $((3.1.2)(\text{i}))$ and whose Fourier transform satisfies $\operatorname{supp} \hat{\phi} = \mathbb{R}^d$. It is not necessary in this approach to assume $((3.1.2)(\text{ii}))$ and $((3.1.2)(\text{iii}))$ – they follow automatically. It is also not necessary to assume $((3.1.2)(\text{iv}))$. In particular, this approach applies to any compactly supported ϕ. To simplify our discussion, we shall assume in addition to $((3.1.2)(\text{i}))$ that ϕ has compact support and that the shifts of ϕ are $L_2(\mathbb{R}^d)$-stable; we refer the reader to de Boor *et al.* (1991b) for a discussion of the more general theory.

The usual starting point for the construction of multi-variate wavelets is the fact that the dilated space S^1 of $S := S(\phi)$ is generated by the half-shifts of $\eta := \phi(2\cdot)$, and therefore also by the full shifts of the functions $\eta_v := \eta(\cdot - v/2)$, $v \in V$. The assumption that supp $\hat{\phi} = \mathbb{R}^d$ is important because it implies that the set $\Phi := \{\phi_v := \phi(x - v/2) \mid v \in V\}$ is also a generating set for S^1, i.e. $S^1 = S(\Phi)$. Φ is more useful than $\{\eta_v \mid v \in V\}$ as a generating set because Φ contains a function that is in S^0, namely ϕ. In analogy with the uni-variate construction, we see that with P the $L_2(\mathbb{R}^d)$ projector onto S, the functions $\phi_v - P\phi_v$, $v \in V'$, form a generating set for W. From (3.2.8), we calculate the Fourier transforms of these functions and multiply them by $[\hat{\phi}, \hat{\phi}]$ to obtain the functions w_v, with Fourier transform

$$\hat{w}_v := [\hat{\phi}, \hat{\phi}]\hat{\phi}_v - [\hat{\phi}_v, \hat{\phi}]\hat{\phi}, \quad v \in V'. \tag{3.6.4}$$

The set $W := \{w_v \mid v \in V'\}$ is another generating set for W. We note that because we assume ϕ has compact support, the two bracket products appearing in (3.6.4) are trigonometric polynomials, and hence the functions w_v also have compact support.

The set TW, with $T = (\tau_{v,v'})_{v,v' \in V'}$ a matrix of 2π-periodic functions, is another generating set for W if $\det(T) \neq 0$ a.e.

It is easy to find an orthogonal wavelet set by this approach. Because the functions in W have compact support, the Gramian matrix $([\hat{w}_v, \hat{w}_{v'}])_{v,v' \in V'}$ has trigonometric polynomials as its entries. Since this matrix is symmetric and positive semi-definite, its determinant is nonzero a.e. We can use Gauss elimination (Cholesky factorization) without division or pivoting to diagonalize $G(W)$. That is, we can find a (symmetric) matrix $T = (\tau_{v,v'})_{v,v' \in V'}$ of trigonometric polynomials such that $W^* := TW$ has Gramian $G(TW^*) = TG(W)T^*$ that is a diagonal matrix with trigonometric polynomial entries. If w_v^* are the functions in W^*, then the functions w_v^{**} with Fourier transforms $\hat{w}_v^{**} := \hat{w}_v^*/[\hat{w}_v^*, \hat{w}_v^*]^{1/2}$, $v \in V'$, have shifts that form an orthonormal basis for W. Indeed,

$$[\hat{w}_v^{**}, \hat{w}_{v'}^{**}] = \frac{[\hat{w}_v^*, \hat{w}_{v'}^*]}{[\hat{w}_v^*, \hat{w}_v^*]^{1/2}[\hat{w}_{v'}^*, \hat{w}_{v'}^*]^{1/2}},$$

which shows that the new set of generators W^{**} has the identity matrix as its Gramian.

The disadvantage of the orthogonal wavelet set W^{**} is that usually we can say nothing about the decay of the functions w_v^{**}, since this construction may involve division by trigonometric polynomials that have zeros. However, when ϕ has $L_2(\mathbb{R}^d)$-stable half-shifts, this construction can be modified to give an orthogonal wavelet set whose elements decay exponentially (see de Boor et al. (1991b)).

While the assumption that the half-shifts of ϕ are $L_2(\mathbb{R}^d)$-stable is often not realistic, we shall assume it a little longer in order to introduce some new ideas that can later be modified to drop the stability assumption. Under the half-shift stability assumption, we have that $\tilde{\tilde{\phi}}$ is a trigonometric polynomial of period 4π that has no zeros. Therefore, $\hat{\phi}_* := \hat{\phi}/\tilde{\tilde{\phi}}$ serves to define an $L_2(\mathbb{R}^d)$ function in $\mathcal{S}^1(\phi)$ that decays exponentially and has orthogonal half-shifts. Moreover, the function w with Fourier transform $\hat{w} := \hat{\phi}/\tilde{\tilde{\phi}}^2$ is also in $\mathcal{S}^1(\phi)$ and decays exponentially. Therefore, with $[\cdot, \cdot]_{1/2}$ the bracket product for half-shifts (which is defined as in (3.2.4) except that the sum is taken over $4\pi\mathbb{Z}^d$), we have

$$[\hat{\phi}, \hat{w}]_{1/2} = [\hat{\phi}_*, \hat{\phi}_*]_{1/2} = 1 \quad \text{a.e.} \tag{3.6.5}$$

The Fourier coefficients (with respect to the $e_{-j/2}$, $j \in \mathbb{Z}^d$) of $[\hat{\phi}, \hat{w}]_{1/2}$ are the inner products of ϕ with half-shifts of w. Hence, all nontrivial half-shifts of w are orthogonal to ϕ. This means that the functions in $\mathcal{W}_0 := \{w(\cdot + v/2) \mid v \in V'\}$ are all in W. It is easy to see that they generate W, that is $W = \mathcal{S}(\mathcal{W}_0)$.

Thus, in the special case we are considering, W is generated by the non-trivial half-shifts of a single function w. It is natural to ask whether this holds in general (i.e. when we do not assume stability of half-shifts). To see that this is indeed true, we modify the argument in (3.6.5). If we multiply $\hat{w}$ by the 2π-periodic function $\prod_{\lambda \in 2\pi V} \tilde{\tilde{\phi}}(\cdot + \lambda)^2$, the result is a compactly supported function $w_* \in L_2(\mathbb{R}^d)$, with Fourier transform

$$\hat{w}_* := \hat{\phi} \prod_{\lambda \in 2\pi V'} \tilde{\tilde{\phi}}(\cdot + \lambda)^2. \tag{3.6.6}$$

We find that

$$[\hat{\phi}, \hat{w}_*]_{1/2} = \prod_{\lambda \in 2\pi V} \tilde{\tilde{\phi}}(\cdot + \lambda)^2. \tag{3.6.7}$$

Because the right-hand side is 2π-periodic, we deduce that the inner product of ϕ with $w_*(\cdot - j/2)$ is zero whenever $j = v + 2k$ with $v \in V'$ and $k \in \mathbb{Z}^d$. Hence, the functions $w_*(\cdot + v/2)$, $v \in V'$, are all in W and it is easy to see that they are also a generating set for W.

While the nontrivial half-shifts of w_* are a generating set for W, they have the drawback that they usually do not provide an $L_2(\mathbb{R}^d)$-stable basis. The usual approach towards constructing an $L_2(\mathbb{R}^d)$-stable basis for W is to begin with the generating set $\{\eta_v \mid v \in V\}$, $\eta := \phi(2\cdot)$, for $\mathcal{S}^1(\phi)$. With this as a starting point, Meyer (1990) III, Section 6 and Jia and Micchelli (1991) have shown the existence of a set of generators for W consisting of compactly supported functions whose shifts provide an $L_2(\mathbb{R}^d)$-stable basis for W. However, their proofs are not constructive. In one, two or three dimensions, and with an additional assumption on the symmetry of ϕ, Riemenschneider and Shen (1991, 1992), have given a constructive proof for the existence of such a generating set, which we now describe.

We begin again with the function $\hat{w} := \hat{\phi}/\tilde{\hat{\phi}}^2$. If $\tau = \sum_{j\in\mathbb{Z}^d} c(j)e_{-j/2}$ is a 4π-periodic function whose Fourier coefficients $c(j) = 0$ whenever $j \in 2\mathbb{Z}^d$, then the function with Fourier transform $\tau\hat{w}$ is in W provided it is in $L_2(\mathbb{R}^d)$. The condition on these Fourier coefficients is equivalent to requiring that

$$\sum_{v\in 2\pi V} \tau(\cdot + v) = 0 \quad \text{a.e.} \tag{3.6.8}$$

In particular, we can use this method to produce functions in W as follows.

We assume that ϕ is real valued, has $L_2(\mathbb{R}^d)$-stable shifts, and satisfies the refinement equation (3.3.1) for a *real valued function* A. This assumption on A is a new ingredient; it will be fulfilled for example if $\phi(-x) = \phi(x)$. (More generally, one only needs symmetry about the centre of the support of ϕ.) If $\alpha \in V'$ and $v_\alpha \in 2\pi V$, then we claim that the function ψ_α with Fourier transform

$$\hat{\psi}_\alpha := 2e_{\alpha/2}\,(A\tilde{\hat{\eta}}^2)(\cdot + v_\alpha)\hat{\eta} \tag{3.6.9}$$

is in W, provided $e_{\alpha/2}(v_\alpha) = -1$. Indeed, the refinement equation says that $\hat{w} = \hat{\eta}/(A\tilde{\hat{\eta}}^2)$. We obtain $\hat{\psi}_\alpha$ from $\hat{w}$ by multiplying by $\tau_\alpha := 2e_{\alpha/2}A\tilde{\hat{\eta}}^2 A(\cdot + v_\alpha)\tilde{\hat{\eta}}^2(\cdot + v_\alpha)$. The vertices v and $v + v_\alpha$ (modulo 2π) contribute values in (3.6.8) that are negatives of one another. Hence, (3.6.8) is satisfied for τ_α and $\psi_\alpha \in W$. The functions ψ_α have compact support when A is a polynomial and ϕ has compact support.

We are allowed to make any assignment of $\alpha \longmapsto v_\alpha$ with $e_{\alpha/2}(v_\alpha) = -1$. To obtain an $L_2(\mathbb{R}^d)$-stable basis for W, we need a special assignment with the property that $\alpha - \beta$ (modulo 2) is assigned $v_\alpha - v_\beta$ (modulo 2π). If such a special assignment can be made, then a simple computation shows that with $\mu := 2A\tilde{\hat{\eta}}^2$,

$$[\hat{\psi}_\alpha, \hat{\psi}_\beta] = \sum_{v\in 2\pi V} e_{(\alpha-\beta)/2}(\cdot + v)\,\mu(\cdot + v + v_\alpha)\,\mu(\cdot + v + v_\beta)\,\tilde{\hat{\eta}}^2(\cdot + v). \tag{3.6.10}$$

For example, if the shifts of ϕ are orthonormal, then $\tilde{\hat{\eta}}^2 = \frac{1}{2}$. In this case, for $\alpha \neq \beta$, the terms of the sum in (3.6.10) are negatives of one another for the two values v and $v + (v_\alpha + v_\beta)$ (this is where we need to assume that a special assignment exists), and hence the sum in (3.6.10) is 0. When $\alpha = \beta$, this sum is

$$\sum_{v\in 2\pi V} A^2(\cdot + v)\tilde{\hat{\eta}}^2(\cdot + v) = \sum_{v\in 2\pi V} \tilde{\hat{\phi}}^2(\cdot + v) = \tilde{\hat{\phi}}^2 = 1 \quad \text{a.e.}$$

Hence, the Gramian of $\Psi := \{\psi_\alpha \mid \alpha \in V'\}$ is the identity matrix. We obtain an orthonormal basis for W in this way.

If we begin with a ϕ that has $L_2(\mathbb{R}^d)$-stable shifts then a slightly more complicated argument shows that the functions ψ_α, $\alpha \in V'$, are an $L_2(\mathbb{R}^d)$-stable basis for W.

This leaves the question of when we can make an assignment $\alpha \longmapsto v_\alpha$ of this special type. Such assignments are possible for $d = 1, 2, 3$ but not for $d > 3$. For example, when $d = 2$, we can make the assignments as follows:

$$(0,0) \longmapsto 2\pi(0,0), \qquad (0,1) \longmapsto 2\pi(0,1),$$
$$(1,0) \longmapsto 2\pi(1,1), \qquad (1,1) \longmapsto 2\pi(1,0).$$

This construction will give orthogonal wavelet and pre-wavelet sets for box splines. To illustrate this, we consider briefly the following special box splines on $\mathbb{R}^2$. Let Δ be the triangulation of $\mathbb{R}^2$ obtained from grid lines $x_1 = k$, $x_2 = k$, and $x_2 - x_1 = k$, $k \in \mathbb{Z}$. Let M be the Courant element for this partition. Thus, M is the piecewise linear function for this triangulation that takes the value 1 at $(0,0)$ and the value 0 at all other vertices. The Fourier transform of M is

$$\hat{M}(y_1, y_2) = \left(\frac{\sin(y_1/2)}{y_1/2}\right)\left(\frac{\sin(y_2/2)}{y_2/2}\right)\left(\frac{\sin((y_1 + y_2)/2)}{(y_1 + y_2)/2}\right).$$

By convolving M with itself, we obtain higher order box splines defined recursively by $M_1 := M$ and $M_r := M * M_{r-1}$. Then M_r is a compactly supported piecewise polynomial (with respect to Δ) of degree $3r - 2$ and smoothness C^{2r-2}. Since $\hat{M}$ is real, the box spline M_r satisfies the refinement identity (3.3.1) with A real. Therefore, if we take $\phi := M_r$ and $\mathcal{S} = \mathcal{S}(M_r)$, the construction of Riemenschneider and Shen applies to give a pre-wavelet set Ψ consisting of three compactly supported piecewise polynomials for the partition $\Delta/2$. The set Ψ provides an $L_2(\mathbb{R}^d)$-stable basis for the wavelet space W.

4. Fast wavelet transforms

It is easy to compute the coefficients in wavelet decompositions iteratively with a technique similar to the fast Haar transform. We shall limit our discussion to Daubechies' orthogonal wavelets with compact support in one dimension. However, the ideas presented here apply equally well to other orthogonal wavelets and to pre-wavelets. We let ϕ be a real-valued, compactly-supported function with orthonormal shifts that satisfies the conditions of multi-resolution and in particular the refinement equation (3.1.4). The function ϕ is real and the refinement coefficients are real and finite in number. The orthogonal wavelet ψ is then obtained from ϕ by (3.1.8).

A numerical application begins with a finite representation of a function f as a wavelet sum. This is accomplished by choosing a large value of n, commensurate with the numerical accuracy desired, and taking an approximation to f of the form

$$S_n = \sum_{j \in \mathbb{Z}} f_j \phi_{j,n}, \tag{4.1}$$

with only a finite number of nonzero coefficients f_j. The coefficients f_j are obtained from f in some suitable way. For many applications, it suffices to take $f_j := f(j2^{-n})$. The point $j2^{-n}$ corresponds to the support of $\phi_{j,n}$.

Since $S_n \in \mathcal{S}^n$, we have

$$S_n = P_n S_n = P_0 S_n + \sum_{k=1}^{n}(P_k S_n - P_{k-1}S_n) = P_0 S_n + \sum_{k=0}^{n-1}\sum_{j\in\mathbb{Z}} c(j,k)\psi(2^k x - j).$$
$$(4.2)$$

We will present an efficient algorithm for computing the coefficients $c(j,k)$ from the f_j and an efficient method to recover S_n from the coefficients $c(j,k)$.

The algorithm presented later has two main features. First, it computes the coefficients $c(j,k)$ using only f_j and the coefficients $a(j)$ of the refinement equation (3.1.4) for ϕ. In other words, one never needs to find a concrete realization of the functions ϕ and ψ. Second, the iterative computations are particularly simple to program and run very quickly—the complexity of the fast wavelet transform of 2^n coefficients is $\mathcal{O}(2^n)$; in contrast, the complexity of the Fast Fourier Transform is $\mathcal{O}(n2^n)$.

During one step of our algorithm, we wish to find the coefficients of $P_{k-1}S$ when $S = \sum_{j\in\mathbb{Z}} s(j)\phi_{j,k}$ is in $\mathcal{S}^k$. The coefficients of $P_{k-1}S = \sum_{i\in\mathbb{Z}} s'(i)\phi_{i,k-1}$ are the inner product of S with the $\phi_{i,k-1}$. We therefore compute, using (3.1.4),

$$\begin{aligned}
s'(i) &= \int_{\mathbb{R}} \left[\sum_{j\in\mathbb{Z}} s(j)\phi_{j,k}\right]\phi_{i,k-1} \\
&= \int_{\mathbb{R}} \left[\sum_{j\in\mathbb{Z}} s(j)\phi_{j,k}\right]\left[\frac{1}{\sqrt{2}}\sum_{\ell\in\mathbb{Z}} a(\ell)\phi_{2i+\ell,k}\right] = \frac{1}{\sqrt{2}}\sum_{j\in\mathbb{Z}} a(j-2i)s(j).
\end{aligned}$$

In other words, the sequence $s' := (s'(i))$ is obtained from $s := (s(i))$ by matrix multiplication:

$$s' = As, \quad A := (\alpha_{i,j}), \quad \alpha_{i,j} := \frac{1}{\sqrt{2}}a(j-2i), \ i,j \in \mathbb{Z}. \quad (4.3)$$

Let Q_k be the projector onto the wavelet space W^k. A similar calculation tells us how to compute the coefficients $t = (t(i))$ of the projection $Q_{k-1}S = \sum_{i\in\mathbb{Z}} t(i)\psi_{i,k-1}$ of $S \in \mathcal{S}^k$ onto W_{k-1}:

$$t = Bs, \quad B := (\beta_{i,j}), \quad \beta_{i,j} := \frac{1}{\sqrt{2}}b(j-2i), \ i,j \in \mathbb{Z}, \quad (4.4)$$

where $b_j := (-1)^j a(1-j)$ are the coefficients of the wavelet ψ given in (3.1.8).

This gives the following schematic diagram for computing the wavelet coefficients $(c(j,k))$:

$$
\begin{array}{ccccc}
& \boldsymbol{A} & & \boldsymbol{A} & \boldsymbol{A} \\
S_n & \to & P_{n-1}S_n & \to & \cdots \to \quad P_0 S_n \\
& \searrow \boldsymbol{B} & & \searrow \boldsymbol{B} & \searrow \boldsymbol{B} \\
& Q_{n-1}S_n & & \cdots & Q_0 S_n
\end{array}
\qquad (4.5)
$$

In other words, to compute the coefficients of $P_{k-1}S_n$ we apply the matrix $\boldsymbol{A}$ to the coefficients of $P_k S_n$, while to compute those of $Q_{k-1}S_n$ we apply the matrix $\boldsymbol{B}$ to the coefficients of $P_k S_n$. The coefficients $c(j,k)$, $j \in \mathbb{Z}$, are the coefficients of $Q_k S_n$. This is valid because $Q_{k-1}P_k S_n = Q_{k-1}S_n$ and $P_{k-1}P_k S_n = P_{k-1}S_n$.

Now suppose that we know the coefficients of $P_0 S_n$ and $Q_k S_n$, $k = 0, \ldots, n-1$. How do we reconstruct S_n? We need to rewrite an element $S \in \mathcal{S}^k$, $S = \sum_{j \in \mathbb{Z}} s(j)\phi_{j,k}$ as an element of $\mathcal{S}^{k+1}$, $S = \sum_{i \in \mathbb{Z}} s'(i)\phi_{i,k+1}$. From the refinement equation (3.1.4), we find

$$
S = \sum_{j \in \mathbb{Z}} s(j)\left[\frac{1}{\sqrt{2}}\sum_{\ell \in \mathbb{Z}} a(\ell)\phi_{\ell+2j,k+1}\right] = \sum_{i \in \mathbb{Z}}\left[\frac{1}{\sqrt{2}}\sum_{j \in \mathbb{Z}} a(i-2j)s(j)\right]\phi_{i,k+1}.
$$

Therefore, we can express the computation of $\boldsymbol{s}'$ from $\boldsymbol{s}$ as multiplication by the transpose $\boldsymbol{A}^*$ of $\boldsymbol{A}$:

$$
\boldsymbol{s}' = \boldsymbol{A}^*\boldsymbol{s}, \quad \boldsymbol{A}^* := (\alpha^*_{i,j}), \quad \alpha^*_{i,j} := \frac{1}{\sqrt{2}}a(i-2j), \ i,j \in \mathbb{Z}. \qquad (4.6)
$$

A similar calculation tells us how to rewrite a sum $S = \sum_{i \in \mathbb{Z}} t(i)\psi_{i,k}$ as a sum $S = \sum_{i \in \mathbb{Z}} t'(i)\phi_{i,k+1}$:

$$
\boldsymbol{t}' = \boldsymbol{B}^*\boldsymbol{t}, \quad \boldsymbol{B}^* := (\beta^*_{i,j}), \quad \beta^*_{i,j} := \frac{1}{\sqrt{2}}b(i-2j), \ i,j \in \mathbb{Z}. \qquad (4.7)
$$

The reconstruction of S_n from $Q_k S_n$, $k = 0, \ldots, n-1$, and $P_0 S_n$ is then given schematically by

$$
\begin{array}{ccccc}
& \boldsymbol{A}^* & & \boldsymbol{A}^* & \boldsymbol{A}^* \\
P_0 S_n & \to & P_1 S_n & \to & \cdots \to \quad S_n \\
& \nearrow \boldsymbol{B}^* & & & \nearrow \boldsymbol{B}^* \\
Q_0 S_n & & & \cdots & Q_{n-1}S_n
\end{array}
\qquad (4.8)
$$

The matrices $\boldsymbol{A}$, $\boldsymbol{B}$, $\boldsymbol{A}^*$, and $\boldsymbol{B}^*$ have a small finite number of nonzero elements in each row, so each of the operations in (4.5) and (4.6) has computational complexity proportional to the number of unknowns.

The reconstruction algorithm can be used to display graphically a finite wavelet sum S. We choose a large value of n, and use the reconstruction algorithm to write $S = \sum_{j \in \mathbb{Z}} s(j,n)\phi_{j,n}$. The piecewise linear function with values $s(j,n)$ at $j2^{-n}$ is an approximation to the graph of S. Such procedures for graphical displays are known as subdivision algorithms in computer aided geometric design.

The matrices A and B have many remarkable properties summarized by

$$BA^* = 0,$$
$$AA^* = I, \qquad BB^* = I,$$
$$A^*A + B^*B = I. \tag{4.9}$$

The first equation represents the orthogonality between W and S^0, the second the fact that the shifts of ϕ and ψ are orthonormal, and the third the orthogonal decomposition $S^1 = S^0 \oplus W$.

5. Smoothness spaces and wavelet coefficients

We have seen in Section 2 that one can determine when a function $f \in L_p(\mathbb{R})$ is in a Lipschitz space $\mathrm{Lip}(\alpha, L_p(\mathbb{R}))$, $0 < \alpha < 1/p$, by examining the coefficients of the Haar expansion of f. In fact, one can often characterize membership in general smoothness spaces in terms of the size of coefficients in general wavelet or pre-wavelet expansions. We do not have the space to explain in detail how such characterizations are proved, but we shall outline one approach, based on approximation, that parallels the arguments in Section 2 about Haar wavelets. A more complete presentation, much along the lines given here, can be found in the book of Meyer (1990). The article of Frazier and Jawerth (1990) gives a much more general and extensive treatment of wavelet-like decompositions from the viewpoint of Littlewood–Paley theory.

We shall suppose that ϕ satisfies the conditions (3.1.2) of multi-resolution. We also suppose that ϕ has compact support. This is not a necessary assumption for the characterizations given below (it can be replaced by suitable polynomial decay at infinity) but it will simplify our discussion. We shall also assume that $1 < p < \infty$. The arguments that follow can be modified simply to apply when $p = \infty$; the analysis for $p \leq 1$ can also be developed as shown later, but then it must be carried out in the setting of the Hardy spaces $H_p(\mathbb{R}^d)$.

We fix a value of p and let $S := S(\phi, L_p(\mathbb{R}^d))$ be the $L_p(\mathbb{R}^d)$ closure of the finite linear combination of shifts of ϕ.

We assume that the shifts of ϕ form an $L_p(\mathbb{R}^d)$-stable basis for S. For functions with compact support, this holds whenever the shifts of ϕ form an $L_2(\mathbb{R}^d)$-stable basis for $S(\phi, L_2(\mathbb{R}^d))$ (see Jia and Micchelli (1991)). It follows that the dilated functions $\phi_{j,k,p} := 2^{k/p} \phi(2^k \cdot - j)$, $j \in \mathbb{Z}^d$, form an $L_p(\mathbb{R}^d)$-stable basis of S^k, for each $k \in \mathbb{Z}$. That is, there are constants $C_1, C_2 > 0$ such that each $S \in S^k$ can be represented as $S = \sum_{j \in \mathbb{Z}^d} c(j, k, p)(S)\phi_{j,k,p}$

with

$$C_1\left(\sum_{j\in\mathbb{Z}^d}|c(j,k,p)|^p\right)^{1/p}\leq\|S\|_{L_p(\mathbb{R}^d)}\leq C_2\left(\sum_{j\in\mathbb{Z}^d}|c(j,k,p)|^p\right)^{1/p}. \qquad (5.1)$$

We will now show that the orthogonal projector P from $L_2(\mathbb{R}^d)$ onto $\mathcal{S}(\phi,L_2(\mathbb{R}^d))$ has a natural extension to a bounded operator from $L_p(\mathbb{R}^d)$ onto $\mathcal{S}$. We can represent P as in (3.2.9):

$$Pf=\sum_{j\in\mathbb{Z}}\gamma_j(f)\phi(\cdot-j),\quad \gamma_j(f):=\int_{\mathbb{R}^d}f(x)\overline{\gamma(x-j)}\,\mathrm{d}x. \qquad (5.2)$$

The function $\gamma\in L_\infty(\mathbb{R}^d)$ decays exponentially at infinity and hence is in $L_q(\mathbb{R}^d)$ for $1\leq q\leq\infty$. In particular, $\gamma\in L_{p'}(\mathbb{R}^d)$, and (5.2) serves to define P on $L_p(\mathbb{R}^d)$. The compact support of ϕ and the exponential decay of γ then combine to show that P is bounded on $L_p(\mathbb{R}^d)$. By dilation, we find that the projectors P_k (which map $L_p(\mathbb{R}^d)$ onto $\mathcal{S}^k$) are bounded independently of k.

The projector Q from $L_2(\mathbb{R}^d)$ onto the wavelet space W is also represented in the form (5.2) and has an extension to a bounded operator on $L_p(\mathbb{R}^d)$ for the same reasons as before. We can also derive the boundedness of Q from the formula $Q=P_1-P_0$.

Since the P_k are bounded projectors onto $\mathcal{S}^k$, their approximation properties are determined by the approximation properties of the spaces $\mathcal{S}^k$. Consequently, we want to bound the error of approximation by elements in $\mathcal{S}^k$ of functions in certain smoothness classes. In particular, we are interested in determining for which spaces $\mathcal{S}^k$ it is true that

$$\mathrm{dist}(f,\mathcal{S}^k)_{L_p(\mathbb{R}^d)}\leq C2^{-kr}|f|_{W^r(L_p(\mathbb{R}^d))}; \qquad (5.3)$$

here $W^r(L_p(\mathbb{R}^d))$ is the Sobolev space of functions with r (weak) derivatives in $L_p(\mathbb{R}^d)$ with its usual norm and semi-norm. This well-studied problem originated with the work of Schoenberg (1946), and was later developed by Strang and Fix (1973) for application to finite elements. Strang and Fix show that when ϕ has compact support, a sufficient condition for (5.3) to hold is that

$$\hat{\phi}(0)\neq 0\quad\text{and}\quad D^\nu\hat{\phi}(2\pi\alpha)=0,\ |\nu|<r,\ \alpha\in\mathbb{Z}^d,\ \alpha\neq 0. \qquad (5.4)$$

This condition is also necessary in a certain context; see de Boor *et al.* (1991a) and de Boor and Ron (1991) for a history of the Strang–Fix conditions.

Schoenberg (1946) showed that (5.4) guarantees that algebraic polynomials of (total) degree $<r$ are contained locally in the space $\mathcal{S}^k$. This means that for any compact set Ω and any polynomial R with $\deg(R)<r$, there is an $S\in\mathcal{S}$ that agrees with R on Ω.

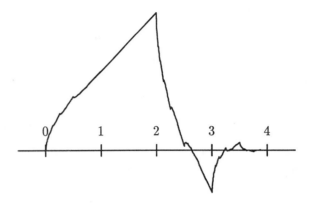

Fig. 4. The function $0 \cdot \phi(x+1) + 1 \cdot \phi(x) + 2 \cdot \phi(x-1)$, with ϕ given by Daubechies' formula (3.5.1) with $N = 2$; see Figure 3. Note the linear segment between $x = 1$ and $x = 2$.

In summary, the approximation properties of P_k are determined by the largest value of r for which (5.4) is valid. Because we usually know a lot about the Fourier transform of ϕ, the best value of r is easy to determine. For this r, we have (5.3). For example, for $\phi = N_r$, the B-spline of order r, $\hat{\phi}(y) = (1 - e^{-iy})^r / (iy)^r$ satisfies (5.4) for this value of r. Similarly, the Daubechies wavelets satisfy (5.4) for $r = N$, with N the integer appearing in the representation (3.5.6); see Figure 4.

Assume now that there are positive integers r and s such that the following Jackson and Bernstein inequalities hold:

(J) $\|f - P_k f\|_{L_p(\mathbb{R}^d)} \leq C \cdot 2^{-kr} |f|_{W^r(L_p(\mathbb{R}^d))}$,

(B) $|S|_{W^s(L_p(\mathbb{R}^d))} \leq C \cdot 2^{ks} \|S\|_{L_p(\mathbb{R}^d)}$, $S \in \mathcal{S}^k(\phi, L_p(\mathbb{R}^d))$, $1 \leq p \leq \infty$.

$$(5.5)$$

(Actually, s need not be an integer.) The Jackson inequality is just a reformulation of (5.3), and the largest value of r for which (J) holds is determined by (5.4). The Bernstein inequality holds if $\phi \in W^s(L_p(\mathbb{R}^d))$ and in particular (since ϕ has compact support) whenever ϕ is in C^s. It is enough to verify (B) for $k = 0$, since (B) would then follow for general k by rescaling. The left semi-norm in (B) for $S = \sum_{j \in \mathbb{Z}^d} c(j)\phi(\cdot - j)$ is bounded by the $\ell_p(\mathbb{Z}^d)$ norm of the coefficients $(c(j))_{j \in \mathbb{Z}^d}$, which is bounded in turn by the right-hand side of (B) by using the $L_p(\mathbb{R}^d)$-stability of the shifts of ϕ.

Once the Jackson and Bernstein inequalities have been established, we can invoke a general procedure to characterize smoothness spaces in terms of wavelet coefficients. To describe these results, we introduce the Besov

spaces, which are a family of smoothness spaces that depend on three parameters. We introduce the Besov spaces for only one reason: They are
the spaces that are needed to describe precisely the smoothness of functions
that can be approximated to a given order by wavelets. The following discussion is meant as a gentle introduction to Besov spaces for the reader who
instinctively dislikes any space that depends on so many parameters.

The Besov space $B_q^\alpha(L_p(\mathbb{R}^d))$, $\alpha > 0$ and $0 < q, p \leq \infty$, is a smoothness subspace of $L_p(\mathbb{R}^d)$. The parameter α gives the order of smoothness
in $L_p(\mathbb{R}^d)$. The second parameter q gives a finer scaling, which allows us
to make subtle distinctions in smoothness of fixed order α. This second
parameter is necessary in many embedding and approximation theorems.

To define these spaces, we introduce, for $h \in \mathbb{R}^d$, the rth difference in the
direction h:

$$\Delta_h^r(f, x) := \sum_{j=0}^{r} (-1)^{r+j} \binom{r}{j} f(x + jh).$$

Thus, $\Delta_h(f, x) := f(x + h) - f(x)$ is the first difference of f and the other
differences are obtained inductively by a repeated application of Δ_h. With
these differences, we can define the moduli of smoothness

$$\omega_r(f, t)_p := \sup_{|h| \leq t} \|\Delta_h^r(f, \cdot)\|_{L_p(\mathbb{R}^d)}, \quad t > 0,$$

for each $r = 1, 2, \ldots$. The rate at which $\omega_r(f, t)_p$ tends to zero gives information about the smoothness of f in $L_p(\mathbb{R}^d)$. For example, the spaces
$\text{Lip}(\alpha, L_p(\mathbb{R}^d))$, which we have discussed earlier, are characterized by the
condition $\omega_1(f, t)_p = \mathcal{O}(t^\alpha)$, $0 < \alpha \leq 1$.

The Besov spaces are defined for $0 < \alpha < r$ and $0 < p, q \leq \infty$ as the set
of all functions $f \in L_p(\mathbb{R}^d)$ for which

$$|f|_{B_q^\alpha(L_p(\mathbb{R}^d))} := \begin{cases} \left(\int_0^\infty [t^{-\alpha} \omega_r(f, t)_p]^q \, \dfrac{dt}{t} \right)^{1/q}, & 0 < q < \infty, \\ \sup_{t \geq 0} t^{-\alpha} \omega_r(f, t)_p, & q = \infty, \end{cases} \tag{5.6}$$

Because we allow values of p and q less than one, this 'norm' does not always
satisfy the triangle inequality, but it is always a quasi-norm: There exists a
constant C such that for all f and g in $B_q^\alpha(L_p(\mathbb{R}^d))$,

$$\|f + g\|_{B_q^\alpha(L_p(\mathbb{R}^d))} \leq C \left(\|f\|_{B_q^\alpha(L_p(\mathbb{R}^d))} + \|g\|_{B_q^\alpha(L_p(\mathbb{R}^d))} \right).$$

Even though the definition of the $B_q^\alpha(L_p(\mathbb{R}^d))$ norm depends on r through
the modulus of smoothness, we have not parametrized the spaces by r, for
two reasons. First, *no one* can stand spaces that are parametrized by more
than three parameters. Second, it can be shown that all values of r greater

than α give rise to equivalent norms, so the set of functions in $B_q^\alpha(L_p(\mathbb{R}^d))$ does not depend on r as long as $r > \alpha$.

We note that the family of Besov spaces contains both the Lipschitz spaces $\mathrm{Lip}(\alpha, L_p(\mathbb{R}^d)) = B_\infty^\alpha(L_p(\mathbb{R}^d))$, $0 < \alpha < 1$, and the Sobolev spaces $W^\alpha(L_2(\mathbb{R}^d)) = B_2^\alpha(L_2(\mathbb{R}^d))$, which are frequently denoted by $H^\alpha(\mathbb{R}^d)$.

We have mentioned that once Jackson and Bernstein inequalities have been established, there is a general theory for characterizing membership in Besov spaces by the decay of wavelet coefficients. This is based on results from approximation theory described (among other places) in the articles by DeVore *et al.* (1991d), DeVore and Popov (1988), and the forthcoming book by DeVore and Lorentz (1992). Among other things, this theory states that whenever (5.5) holds, we have that

$$|f|_{B_q^\alpha(L_p(\mathbb{R}^d))} \approx \left(\sum_{k \in \mathbb{Z}} [2^{k\alpha} \|Q_k(f)\|_{L_p(\mathbb{R}^d)}]^q \right)^{1/q} \tag{5.7}$$

for $0 < \alpha < \min(r, s)$, $1 < p < \infty$, and $0 < q < \infty$.

If Ψ is a wavelet set associated with ϕ, then

$$Q_k(f) = \sum_{\psi \in \Psi} \sum_{j \in \mathbb{Z}^d} \gamma_{j,k,\psi,p}(f) \psi_{j,k,p},$$

with $\psi_{j,k,p} := 2^{kd/p}\psi(2^k \cdot - j)$ the $L_p(\mathbb{R}^d)$-normalized (pre)wavelets, and the $\gamma_{j,k,\psi,p}$ the associated dual functionals. Using the $L_p(\mathbb{R}^d)$-stability of Ψ, we can replace $\|Q_k(f)\|_{L_p(\mathbb{R}^d)}^p$ by $\sum_{j \in \mathbb{Z}^d} |\gamma_{j,k,\psi,p}(f)|^p$ and obtain

$$|f|_{B_q^\alpha(L_p(\mathbb{R}^d))} \approx \left(\sum_{k \in \mathbb{Z}} [2^{k\alpha p} \sum_{\psi \in \Psi} \sum_{j \in \mathbb{Z}^d} |\gamma_{j,k,\psi,p}(f)|^p]^{q/p} \right)^{1/q}, \tag{5.8}$$

with the usual change to a supremum when $q = \infty$. This is the characterization of the Besov space in terms of wavelet coefficients. When $q = p$, (5.8) takes an especially simple form:

$$|f|_{B_p^\alpha(L_p(\mathbb{R}^d))}^p \approx \sum_{k \in \mathbb{Z}} [2^{k\alpha} \sum_{\psi \in \Psi} \sum_{j \in \mathbb{Z}^d} |\gamma_{j,k,\psi,p}(f)|^p]. \tag{5.9}$$

In particular, (5.9) gives an equivalent semi-norm for the Sobolev space $H^\alpha(\mathbb{R}^d)$ by taking $p = 2$.

6. Applications

6.1. Wavelet compression

We shall present a few examples that indicate how wavelets can be used in numerical applications. Wavelet techniques have had a particularly significant impact on data compression. We begin by discussing a problem in nonlinear approximation that is at the heart of compression algorithms.

Suppose that Ψ is a (pre)wavelet set and $f \in L_p(\mathbb{R}^d)$, $1 \leq p \leq \infty$, has the wavelet representation

$$f = \sum_{k \in \mathbb{Z}} \sum_{\psi \in \Psi} \sum_{j \in \mathbb{Z}^d} c_{j,k,\psi,p}(f)\psi_{j,k,p}. \tag{6.1.1}$$

with respect to the $L_p(\mathbb{R}^d)$-normalized functions $\psi_{j,k,p} := 2^{kd/p}\psi(2^k \cdot - j)$. In numerical applications, we must replace the sum in (6.1.1) by a finite sum, and the question arises as to the most efficient way to accomplish this. To make this into a well defined mathematical problem, we fix an integer n, which represents the number of terms we shall allow in the finite sum. Thus, we want to approximate f in the $L_p(\mathbb{R}^d)$ norm by an element from the set

$$\Sigma_n := \left\{ S = \sum_{(j,k,\psi) \in \Lambda} d_{j,k,\psi}\psi_{j,k,p} \mid |\Lambda| \leq n \right\}, \tag{6.1.2}$$

where $d_{j,k,\psi}$ are arbitrary complex numbers. We have the error of approximation

$$\sigma_n(f)_p := \inf_{S \in \Sigma_n} \|f - S\|_{L_p(\mathbb{R}^d)}. \tag{6.1.3}$$

In contrast to the usual problems in approximation, the set Σ_n is not a linear space since adding two elements of Σ_n results in an element of Σ_{2n}, but not generally an element of Σ_n.

The approximation problem (6.1.3) has a particularly simple solution when $p = 2$ and Ψ is an orthogonal wavelet set. We order the coefficients $c_{j,k,\psi,2}(f)$ by their absolute value. If Λ_n is the set of indices (j, k, ψ) corresponding to n largest values (this set is not necessarily unique), then $S_n = \sum_{(j,k,\psi) \in \Lambda_n} c_{j,k,\psi,2}(f)\psi_{j,k,2}$ attains the infimum in (6.1.3). For prewavelet sets, this selection is optimal within constants, as follows from the $L_2(\mathbb{R}^d)$-stability of the basis $\psi_{j,k,2}$.

It is somewhat surprising that the strategy of the preceding paragraph also is optimal (in a sense to be made clear later) for approximation in $L_p(\mathbb{R}^d)$, $p \neq 2$. To describe this, we fix a value of $1 < p < \infty$ (slightly weaker results than those stated later are known when $p = \infty$) and for $f \in L_p(\mathbb{R}^d)$, we let Λ_n denote a set of indices corresponding to n largest values of $|c_{j,k,\psi,p}(f)|$. We define $S_n := \sum_{(j,k,\psi) \in \Lambda_n} c_{j,k,\psi,p}(f)\psi_{j,k,p}$ and $\tilde{\sigma}_n(f)_p := \|f - S_n\|_{L_p(\mathbb{R}^d)}$. DeVore et al. (1991d) have established various results that relate $\sigma_n(f)_p$ with $\tilde{\sigma}_n(f)_p$ under certain conditions on ψ and the generating function ϕ. For example, it follows from their results that

$$\tilde{\sigma}_n(f)_p = \mathcal{O}(n^{-\alpha/d}) \Longleftrightarrow \sigma_n(f)_p = \mathcal{O}(n^{-\alpha/d}) \tag{6.1.4}$$

for $0 < \alpha < r$. Here, the integer r is related to properties of ϕ. Namely, the generating function ϕ should satisfy the Strang–Fix conditions (5.4) of this order and ϕ and ψ should have sufficient smoothness (for example, C^r

is enough). It is also necessary to assume decay for the functions ϕ and ψ; sufficiently fast polynomial decay is enough. We caution the reader that the results in DeVore et al. (1991d) are formulated for one wavelet ψ and not a wavelet set Ψ. However, the same proofs apply in the more general setting.

It is also of interest to characterize the functions f that satisfy (6.1.4). That is, we would like to know when we can expect the order of approximation (6.1.4). This has not been accomplished in exactly the form of (6.1.4), but the following variant has been shown in DeVore et al. (1991d). The following are equivalent for $\tau := \tau(\alpha, p) := (\alpha/d + 1/p)^{-1}$:

$$\text{(i)} \quad \sum_{n=1}^{\infty} [n^{\alpha/d} \sigma_n(f)_p]^\tau \frac{1}{n} < \infty,$$

$$\text{(ii)} \quad \sum_{n=1}^{\infty} [n^{\alpha/d} \tilde{\sigma}_n(f)_p]^\tau \frac{1}{n} < \infty,$$

$$\text{(iii)} \quad f \in B_\tau^\alpha(L_\tau(\mathbb{R}^d)). \tag{6.1.5}$$

Some explanation regarding (6.1.5) is in order. First, ((6.1.5)(i)) is very close to the condition in (6.1.4). For example, ((6.1.5)(i)) implies (6.1.4), and if (6.1.4) holds for some $\beta > \alpha$ then ((6.1.5)(i)) is valid. So, roughly speaking, it is the functions $f \in B_\tau^\alpha(L_\tau(\mathbb{R}^d))$ for which the order of approximation in (6.1.4) holds. Second, the characterization (6.1.5) says that it is those functions with smoothness of order α in $L_\tau(\mathbb{R}^d)$, $\tau = (\alpha/d + 1/p)^{-1}$, that are approximated with order $\mathcal{O}(n^{-\alpha/d})$ in $L_p(\mathbb{R}^d)$. This should be contrasted with the usual results for approximation from linear spaces (such as finite element methods), which characterize functions with this approximation order as having smoothness of order α in $L_p(\mathbb{R}^d)$. Since $\tau < p$, the nonlinear approximation problem (6.1.3) provides the approximation order (6.1.4) for functions with less smoothness than required by linear methods.

The fact that functions with less smoothness can be approximated well by (6.1.3) is at the essence of wavelet compression. This means that functions with singularities can be handled numerically. Intuitively this is accomplished by retaining in the sum for S_n terms corresponding to functions $\psi_{j,k,p}$ that make a large contribution to f near the singularity. Here the situation is similar to adaptive methods for piecewise-polynomial (finite element) approximation that have refined triangulations near a singularity. However, we want to stress that in wavelet compression, it is simple (almost trivial) to approximate optimally without encountering problems of triangulation. An overview of this approach to data compression using wavelets can be found in DeVore et al. (1991a).

6.2. Image compression

We explain next how a typical algorithm for compression is implemented from the theoretical results of the previous section. This has been accomplished for surface compression in DeVore *et al.* (1991b) and image compression in DeVore *et al.* (1991c). We shall discuss only image compression.

A digitized grey-scale image consists of an array of picture elements (pixels) represented by numbers that correspond to the brightness (grey scale) of each pixel, with $0 \equiv$ black, say, and $255 \equiv$ white. A grey-scale image has, say, 1024×1024 such numbers taking integer values between 0 and 255. Thus, the image is given by a matrix $(p_j)_{j \in \{0,\dots,1023\}^2}$ with $p_j \in \{0, \dots, 255\}$. It would take a data file of about one million bytes to encode such an image. For purposes of transmission or storage, it is desirable to compress this file.

To use wavelets for image compression, we proceed as follows. We think of the pixel values as associated with the points $j2^{-m}$, $j \in [0, 2^m)^2$, $m = 10$, of the unit square $[0, 1]^2$. In this way, we can think of the image as a discretization of a function f defined on this square.

We choose a function ϕ satisfying the assumptions of multi-resolution, and a corresponding wavelet set Ψ that provides a stable basis for $L_2(\mathbb{R}^2)$. Thus, Ψ would consist of three functions, which we shall assume are of compact support.

We would like to represent the image as a wavelet sum. For this purpose, we select coefficients γ_j and consider the function

$$f = \sum_{j \in \Omega} \gamma_j \phi_{j,m} \tag{6.2.1}$$

with Ω the set of indices for which $\phi_{j,m}$ does not vanish identically on $[0, 1]^2$. We think of f as the image and apply the results of the preceding section to compress f.

The coefficients γ_j are to be determined numerically from the pixel values; choosing good values of γ_j is a nontrivial problem, which we do not discuss. A typical choice is to take $\gamma_j = p_j$ for $j2^{-m} \in [0, 1]^2$ and some extension of these values for other j.

Once the coefficients (γ_j) have been determined, we use a fast wavelet transform to write f in its wavelet decomposition

$$f = P_0 f + \sum_{j,k,\psi} c_{j,k,\psi}(f)\psi_{j,k} \tag{6.2.2}$$

with respect to the $L_2(\mathbb{R}^2)$-normalized $\psi_{j,k} := 2^k \psi(2^k \cdot - j)$. We can find the coefficients of f with respect to the $L_p(\mathbb{R}^2)$-normalized ψ's by the relation $c_{j,k,\psi,p} = 2^{2k(1/p-1/2)} c_{j,k,\psi}$. The projection $P_0 f$ has very few terms, which we take intact into the compressed representation at little cost.

To apply the compression algorithm of the previous section, we need to decide on a suitable norm in which to measure the error. The $L_2(\mathbb{R}^2)$ norm is most commonly used, but we argue in DeVore *et al.* (1991c) that the $L_1(\mathbb{R}^2)$ is a better model for the human eye for error with high spatial frequency.

If one decides to use the $L_p(\mathbb{R}^2)$ norm to measure compression error, then the algorithm of the previous section orders the $L_p(\mathbb{R}^2)$-normalized wavelet coefficients $c_{j,k,\psi,p}$ and chooses the largest of these to keep. Optimally, one would send coefficients in decreasing order of size. Thus, we find a (small) set Λ of ordered triples $\{(j,k,\psi)\}$ that index the largest values of $|c_{j,k,\psi,p}|$ and use for our compressed image

$$g := \sum_{(j,k,\psi)\in\Lambda} c_{j,k,\psi,p}\psi_{j,k,p}.$$

This sum has $|\Lambda|$ terms.

This method of sending coefficients sequentially across a communications link to allow gradual reconstruction of an image by the receiver is known as *progressive transmission*. Our criterion provides a new ordering for the coefficients to be transmitted that depends on the value of p. However, sorting the coefficients requires $\mathcal{O}(m2^{2m})$ operations, while the fast wavelet transform itself takes but $\mathcal{O}(2^{2m})$ operations. Thus, a faster compression method that does not rely on sorting is to be preferred; we proceed to give one.

We discuss compression in $L_2(\mathbb{R}^2)$ for a moment. As noted earlier, the optimal algorithm is to keep the largest $L_2(\mathbb{R}^2)$ coefficients and to discard the other coefficients. The coefficients to keep can be determined by fixing any the following quantities:

(i) $N := |\Lambda|$,

(ii) $\|f - g\|_{L_2(\mathbb{R}^2)} = \left(\displaystyle\sum_{(j,k,\psi)\notin\Lambda} |c_{j,k,\psi}|^2\right)^{1/2}$,

(iii) $\epsilon := \displaystyle\inf_{(j,k,\psi)\in\Lambda} |c_{j,k,\psi}|$.

Setting any one of these quantities determines the other two, and by extension the set Λ, for any function f. In other words, we can prescribe either the number of nonzero coefficients N, the total error $\|f - g\|_{L_2(\mathbb{R}^2)}$, or ϵ, which we consider to be a measure of the local error. If we determine which triples (j,k,ψ) to include in Λ by the third criterion, then we do not need to sort the coefficients, for we can sequentially examine each coefficient and put (j,k,ψ) into Λ whenever $|c_{j,k,\psi}| \geq \epsilon$. This is known as *threshold coding* to the engineers, because one keeps only those coefficients that exceed a specified threshold.

Even more compression can be achieved by noting that we should keep only the most significant bits of the coefficients $c_{j,k,\phi,p}$. Thus, we choose a tolerance $\epsilon > 0$ and we take in the compressed approximation for each $\psi_{j,k,p}$ a coefficient $\tilde{c}_{j,k,\psi,p}$ such that

$$|\tilde{c}_{j,k,\psi,p} - c_{j,k,\psi,p}| < \epsilon \qquad (6.2.3)$$

with the proviso that $\tilde{c}_{j,k,\psi,p} = 0$ whenever $|c_{j,k,\psi,p}| < \epsilon$. Then $\tilde{S} := \sum_{j,k,\psi} \tilde{c}_{j,k,\psi,p} \psi_{j,k,p}$ represents our compressed image, and (6.1.5) holds for this approximation.

This process of keeping only the most significant bits of $c_{j,k,\psi,p}$ is known in the engineering literature as *scalar quantization*. The dependence on the dyadic level k and the space $L_p([0,1]^2)$ in which the error is measured is brought out more clearly when using $L_\infty(\mathbb{R}^2)$-normalized wavelets, i.e. when $\psi_{j,k} = \psi(2^k \cdot - j)$. For these wavelets, as the dyadic level increases, the number of bits of $c_{j,k,\psi}$ taken in $\tilde{c}_{j,k,\psi}$ decreases. For example, if $p = 2$, we would take one less bit at each increment of the dyadic level. On the other hand, if the compression is done in the $L_1([0,1]^2)$ norm, than we would take two fewer bits as we change dyadic levels. See DeVore *et al.* (1991c).

6.3. The numerical solution of partial differential equations

Wavelets are currently being investigated for the numerical solution of differential and integral equations (see, e.g., the papers of Beylkin *et al.* (1991) and Jaffard (1991)). While these applications are only now being developed, we shall consider a couple of simple examples to illustrate the potential of wavelets in this direction.

Elliptic equations The Galerkin method applied to elliptic partial differential equations gives rise to a matrix problem that involves a so-called stiffness matrix. We present a simple example that illustrates the perhaps surprising fact that the stiffness matrix derived from the wavelet basis can be preconditioned trivially to have a uniformly bounded condition number. In general, this property allows one to use iterative methods, such as the conjugate gradient method, to solve linear systems with great efficiency. The linear systems that arise by discretizing elliptic PDEs have a lot of structure and can in no way be considered general linear systems, and there are many very efficient numerical methods, such as multi-grid, that exploit the special structure of these linear systems to solve these systems to high accuracy with very low operation counts. We do not yet know of a complete analysis that shows that computations with wavelets can be more efficient than existing multi-grid methods when applied to the linear systems that arise by discretizing elliptic PDEs in the usual way.

Rather than consider Dirichlet and Neumann boundary value problems in several space dimensions, as discussed in Jaffard (1991), we shall present only a simple, periodic, second-order ODE that illustrates the main points. We shall consider functions defined on the one-dimensional torus $\mathbb{T}$, which is equivalent to $[0, 1]$ with the end-points identified, and search for $u = u(x)$, $x \in \mathbb{T}$, that satisfies the equation

$$-u''(x) + u(x) = f(x), \qquad x \in \mathbb{T}, \tag{6.3.1}$$

with $f \in L_2(\mathbb{T})$. In variational form, the solution $u \in W^1(L_2(\mathbb{T}))$ of (6.3.1) satisfies

$$\int_{\mathbb{T}} (u'v' + uv) = \int_{\mathbb{T}} fv, \tag{6.3.2}$$

for all $v \in W^1(L_2(\mathbb{T}))$. We remark that we can take

$$\|u\|^2_{W^1(L_2(\mathbb{T}))} := \int_{\mathbb{T}} ([u']^2 + u^2).$$

To approximate u by Galerkin's method, we must choose a finite-dimensional subspace of $W^1(L_2(\mathbb{T}))$, which we shall choose to be a space spanned by wavelets defined on the circle $\mathbb{T}$. We indicate briefly how to construct periodic wavelets on $\mathbb{T}$ from wavelets on $\mathbb{R}$.

For any f with compact support defined on $\mathbb{R}$, the function

$$f^\circ := \sum_{j \in \mathbb{Z}} f(\cdot + j)$$

is a function of period 1, which we call the periodization of f. To obtain wavelets on $\mathbb{T}$, we apply this periodization to wavelets on $\mathbb{R}$. To be specific, we consider only the Daubechies wavelets $\psi := \mathcal{D}_{2N}$ with $N > 2$ because they are contained in $W^1(L_2(\mathbb{T}))$. (The first nontrivial Daubechies wavelet, $\mathcal{D}_4$, is in $W^\alpha(L_2(\mathbb{T}))$ for all $\alpha < 1$ (see Eirola (1992)), but we do not know if it is in $W^1(L_2(\mathbb{T}))$.) Let ϕ be the function of Section 3.5 that gives rise to ψ. For each $k \in \mathbb{Z}$, we let $\phi^\circ_{j,k}$ and $\psi^\circ_{j,k}$ be the periodization of the functions $\phi_{j,k}$ and $\psi_{j,k}$ respectively. We define $\tilde{S}^k$ to be the linear span of the functions $\phi^\circ_{j,k}$, $j = 0, \ldots, 2^k - 1$. The functions in this space are clearly of period one. We also note that the $\phi^\circ_{j,k}$ are orthogonal. Indeed, because they are periodic,

$$\int_0^1 \phi^\circ_{j,k} \phi^\circ_{j',k} = \sum_{\ell \in \mathbb{Z}} \int_0^1 \phi_{j,k}(\cdot + \ell) \phi^\circ_{j',k} = \sum_{\ell \in \mathbb{Z}} \int_0^1 \phi_{j,k}(\cdot + \ell) \phi^\circ_{j',k}(\cdot + \ell)$$

$$= \int_{\mathbb{R}} \phi_{j,k} \phi^\circ_{j',k} = \sum_{\ell \in \mathbb{Z}} \int_{\mathbb{R}} \phi_{j,k} \phi_{j',k}(\cdot + \ell) = \sum_{\ell \in \mathbb{Z}} \int_{\mathbb{R}} \phi_{j,k} \phi_{j'+\ell 2^k, k}.$$

If $j \neq j'$, each integral in the last sum is zero, since we never have $j = j' + 2^k \ell$. If $j = j'$, then exactly one integral in the last sum is nonzero and its value is one. Similarly, we find that $\psi^\circ_{j,k}$, $j = 0, \ldots, 2^k - 1$, $k \geq 0$, is an orthonormal system for $L_2(\mathbb{T})$. It is easy to check that by adjoining $\phi^\circ_{0,0}$, which is identically one on $\mathbb{T}$, this orthonormal system is complete.

Returning to our construction of periodic wavelet spaces, we define $\tilde{W}^k$

to be the linear span of the functions $\psi^{\circ}_{j,k}$, $j = 0, \ldots, 2^k - 1$. Then $\tilde{S}^{k+1} = \tilde{S}^k \oplus \tilde{W}^k$. To simplify our notation in this section, in which we refer to periodic wavelets only, we drop the superscripts and tildes, and denote by $\phi_{j,k}$, $\psi_{j,k}$, S^k, and W^k, the periodic wavelet bases and spaces.

Returning to the numerical solution of (6.3.1), we choose a positive value of m and approximate u by an element $u_m \in S^m$ that satisfies:

$$\int_{\mathbb{T}} (u'_m v' + u_m v) = \int_{\mathbb{T}} fv, \quad v \in S^m. \tag{6.3.3}$$

We can write u_m and v as in (6.2.2). For example,

$$u_m = P_0 u_m + \sum_{k=0}^{m-1} Q_k u_m = \gamma \phi_{0,0} + \sum_{k=0}^{m-1} \sum_{j=0}^{2^k-1} \beta(j,k)\psi_{j,k}. \tag{6.3.4}$$

Because $\phi_{0,0}$ is identically one on the circle $\mathbb{T}$, γ is just the average of u_m on $\mathbb{T}$. From (6.3.3) we see that with $v \equiv 1$, $\gamma = \int_{\mathbb{T}} f$. Thus, we need to determine $\beta(j,k)$.

If we replace u_m in (6.3.3) by its representation (6.3.4), we arrive at a system of equations

$$\sum_{k=0}^{m-1} \sum_{j=0}^{2^k-1} \beta(j,k) \int_{\mathbb{T}} (\psi'_{j,k}\psi'_{j',k'} + \psi_{j,k}\psi_{j',k'}) = \int_{\mathbb{T}} f\psi_{j',k'},$$

for $j' = 0, \ldots, 2^{k'} - 1$ and $k' = 0, \ldots, m - 1$, or, more succinctly,

$$T\beta = f \tag{6.3.5}$$

where the typical entry in T is

$$\int_{\mathbb{T}} (\psi'_{j,k}\psi'_{j',k'} + \psi_{j,k}\psi_{j',k'})$$

and f a vector with components

$$\int_{\mathbb{T}} f\psi_{j',k'}; \quad \beta = (\beta(j,k))_{j=0,\ldots,2^k-1, \ k=0,\ldots,m-1}$$

is the coefficient vector of the unknown function.

The convergence rate of the conjugate gradient method, a popular iterative method for the solution of systems like (6.3.5), depends on the condition number, $\kappa(T) = \|T\|\|T^{-1}\|$, of the symmetric, positive definite, stiffness matrix T. Now,

$$\|T\| = \sup_{\|\alpha\|_{\ell^2}=1} \alpha^* T\alpha \quad \text{and} \quad \|T^{-1}\|^{-1} = \inf_{\|\alpha\|_{\ell^2}=1} \alpha^* T\alpha,$$

where $\alpha := (\alpha(j,k))_{j=0,\ldots,2^k-1, \ k=0,\ldots,m-1}$. For any vector α, we form the function $S := S(\boldsymbol{\alpha}) \in \mathcal{S}^m$ by

$$S = \sum_{k=0}^{m-1} \sum_{j=0}^{2^k-1} \alpha(j,k)\psi_{j,k}.$$

For example, $u_m = \gamma\phi_{0,0} + S(\boldsymbol{\beta})$. It follows easily that

$$|\boldsymbol{\alpha}^* T \boldsymbol{\alpha}| = \int_0^1 ([S']^2 + S^2) = \|S\|^2_{W^1(L_2(\mathbb{R}^d))} \approx \sum_{k=0}^{m-1} \sum_{j=0}^{2^k-1} |2^k \alpha(j,k)|^2. \quad (6.3.6)$$

The last equivalence in (6.3.6) is a variant of (5.9) (for $q = p = 2$). In (5.9), we considered wavelet representations beginning at the dyadic level $k = -\infty$; we could have just as easily begin at the level $k = 0$ (by including P_0) and arrived at the equivalence in (6.3.6).

Equation (6.3.6) shows that the matrix DTD, where D has entries

$$2^{-k} \delta^j_{j'} \delta^k_{k'},$$

satisfies

$$|\boldsymbol{\alpha}^* DTD \boldsymbol{\alpha}| \approx \sum_{k=0}^{m-1} \sum_{j=0}^{2^k-1} |\alpha(j,k)|^2;$$

i.e. the stiffness matrix T can be trivially preconditioned to have a condition number $\kappa(DTD) := \|DTD\| \|D^{-1} T^{-1} D^{-1}\|$ that is bounded independently of m.

Because the subspace $\mathcal{S}^m$ of $L_2(\mathbb{T})$ generated by the Daubechies wavelet D_{2N} locally contains all polynomials of degree $< N$, we have that

$$\inf_{v \in \mathcal{S}^m} \|u - v\|_{W^1(L_2(\mathbb{T}))} \leq C 2^{-m(N-1)} |u|_{W^N(L_2(\mathbb{T}))}.$$

Because of the special form of (6.3.1), u_m is in fact the $W^1(L_2(\mathbb{T}))$ projection of u onto $\mathcal{S}^m$, so we have

$$\|u - u_m\|_{W^1(L_2(\mathbb{T}))} \leq C 2^{-m(N-1)} |u|_{W^N(L_2(\mathbb{T}))}.$$

One can treat in an almost identical way the equation (6.3.1) defined on the torus $\mathbb{T}^d$ with the left-hand side replaced by $-\nabla \cdot (a\nabla u) + bu$ with a and b bounded, smooth, positive functions on $\mathbb{T}^d$, and obtain the same uniform bound on the condition number. Of course, ultimately, one would like to handle elliptic boundary value problems for a domain Ω. First results in this direction have been obtained by Jaffard (1991) and in a slightly different (nonwavelet) setting by Oswald (see for example Oswald (1991)). For example, Jaffard's approach to elliptic equations with Dirichlet boundary conditions is to transform the equation to one with zero boundary conditions by extending the boundary function into the interior of the domain Ω. He then employs in a Galerkin method wavelets whose support is contained strictly inside Ω. However, there has not yet been an analysis of the desired relationship between the extension and the wavelets employed. Another approach is to develop wavelets for the given domain (which do not vanish on the boundary) (see Jaffard and Meyer (1989)).

Employing wavelets for elliptic problems as outlined earlier is similar to the use of hierarchical bases in the context of multi-grid. However, two points suggest that wavelet bases may be more useful than hierarchical bases. First, the wavelet bases are $L_2(\mathbb{R}^d)$-stable, while the hierarchical basis are not. Second, one can choose from a much greater variety of wavelet bases with various approximation properties (values of N).

Finally, we mention the great potential for compression to be used in conjunction with wavelets for elliptic problems in a similar way to adaptive finite elements (see also Jaffard (1991)).

6.4. Time-dependent problems

Wavelets have a potential application for the numerical solution of time-dependent parabolic and hyperbolic problems. We mention one particular application where the potential of wavelets has at least a theoretical foundation.

We consider the solution $u(x, t)$ of the scalar hyperbolic conservation law

$$
\begin{aligned}
u_t + f(u)_x &= 0, & x \in \mathbb{R}, \quad t > 0, \\
u(x, 0) &= u_0(x), & x \in \mathbb{R},
\end{aligned}
\tag{6.4.1}
$$

in one space dimension. It is well known that the solution to (6.4.1) develops discontinuities (called 'shocks') even when the initial condition is smooth. This makes the numerical treatment of (6.4.1), and even more so its analogue in several space dimensions, somewhat subtle. The appearance of shocks calls for adaptive or nonlinear methods.

Considering the appearance of discontinuities in the solution u to (6.4.1), the following regularity result of DeVore and Lucier (1990, 1988), is quite surprising. If the flux f in (6.4.1) is strictly convex and suitably smooth, and u_0 has bounded variation, then it has been shown that, for any $\alpha > 0$ and $\tau := \tau(\alpha) := (\alpha + 1)^{-1}$,

$$
u_0 \in B_\tau^\alpha(L_\tau(\mathbb{R})) \implies u(\cdot, t) \in B_\tau^\alpha(L_\tau(\mathbb{R})),
\tag{6.4.2}
$$

for all later time $t > 0$. That is, if u_0 has smoothness of order α in $L_\tau(\mathbb{R})$ then so will $u(\cdot, t)$ for all later time $t > 0$.

The regularity result (6.4.2) has an interpretation in terms of wavelet decompositions. We cannot use orthogonal wavelets in these decompositions because, as we pointed out earlier, they do not provide stable representations of functions in $L_1(\mathbb{R})$ because they have mean value zero. However, the characterization of Besov spaces can be carried out using other, nonorthogonal, wavelets, such as B-splines. With this caveat, (6.4.2) says that whenever u_0 has a wavelet decomposition $\sum_{j,k \in \mathbb{Z}} \gamma_{j,k}(u_0) \psi_{j,k,1}$, with certain $L_1(\mathbb{R})$-normalized wavelets ψ, whose coefficients satisfy

$$\sum_{k \in \mathbb{Z}} \sum_{j \in \mathbb{Z}} |\gamma_{j,k}(f)|^\tau < \infty, \qquad (6.4.3)$$

then $u(\cdot, t)$ has a similar wavelet decomposition with the same control (6.4.3) on the coefficients. We want to stress that the results in DeVore and Lucier (1990,1988) do not describe directly how to determine wavelet coefficients at a later time $t > 0$ from those of the initial function u_0. That is, there is no direct, theoretically correct, numerical method known to us that describes how to update coefficients with time so that (6.4.3) holds. The regularity result (6.4.2) is proved by showing that whenever u_0 can be approximated well in $L_1(\mathbb{R})$ by piecewise polynomials with free (variable) knots, then $u(\cdot, t)$ can be approximated in the same norm by piecewise algebraic functions (of a certain type) with free knots.

Finally, we mention that the authors have also shown that the analogue of regularity result (6.4.3) does not hold in more than one space dimension. From the point of view of wavelet decompositions, our results seem to indicate that the wavelets described in this presentation are too symmetric to effectively handle the diverse types of singularities that arise in the solution of conservation laws in several space dimensions.

REFERENCES

G. Battle (1987), 'A block spin construction of ondelettes, Part I: Lemarié functions', *Comm. Math. Phys.* **110**, 601–615.

A. Ben Artzi and A. Ron (1990), 'On the integer translates of a compactly supported function: Dual bases and linear projectors', *SIAM J. Math. Anal.* **21**, 1550–1562.

G. Beylkin, R. Coifman and V. Rokhlin (1991), 'Fast wavelet transforms and numerical algorithms I', *Comm. Pure Appl. Math.* **XLIV**, 141–183.

C. de Boor, R. DeVore and A. Ron (1991a), 'Approximation from shift invariant spaces', Preprint.

C. de Boor, R. DeVore and A. Ron (1991b), 'Wavelets and pre-wavelets', Preprint.

C. de Boor, R. DeVore and A. Ron (1991c), 'The structure of finitely generated shift invariant spaces in $L_2(\mathbb{R}^d)$', Preprint.

C. de Boor and A. Ron (1991), 'Fourier analysis of approximation orders from principal shift invariant spaces', *Constructive Approx.*, to appear.

A. Cavaretta, W. Dahmen and C. Micchelli (1991), 'Subdivision for computer aided geometric design', *Memoirs Amer. Math. Soc.* **93**.

C. K. Chui, J. Stöckler and J. D. Ward (1991), 'Compactly supported box-spline wavelets', *CAT Report, Texas A&M University* **230**

C. K. Chui and J. Z. Wang (1990), 'A general framework for compactly supported splines and wavelets', *CAT Report, Texas A&M University* **219**.

C. K. Chui and J. Z. Wang (1991), 'On compactly supported spline wavelets and a duality principle', *Trans. Amer. Math. Soc.*, to appear.

W. Dahmen and C. Micchelli (1990), 'On stationary subdivision and the construction of compactly supported wavelets', *Multivariate Approximation and Interpolation* Vol. 94 (W. Haussmann and K. Jetter, eds) ISNM, Birkhauser Verlag (Boston) pp. 69–89.

I. Daubechies (1988), 'Orthonormal bases of compactly supported wavelets', *Comm. Pure Appl. Math.* **XLI**, 909–996.

I. Daubechies (1990), 'The wavelet transform, time-frequency localization and signal analysis', *IEEE Trans. Information Theory* **36**, 961–1005.

R. DeVore, B. Jawerth and B. Lucier (1991a), 'Data compression using wavelets: Error, smoothness, and quantization', *DCC-91, Data Compression Conference* (A. Storer and J. H. Reif, eds), IEEE Computer Society Press (Los Alamitos, CA) pp. 186–195.

R. DeVore, B. Jawerth and B. Lucier (1991b), 'Surface compression', *Computer Aided Geometric Design*, to appear.

R. DeVore, B. Jawerth, and B. Lucier (1991c), 'Image compression through wavelet transform coding', *IEEE Trans. Information Theory* to appear.

R. DeVore, B. Jawerth and V. Popov (1991d), 'Compression of wavelet decompositions', *Amer. J. Math.*, to appear.

R. DeVore and G. Lorentz (1992), *Constructive Approximation* Springer *Grundlehren* (Heidelberg).

R. DeVore and B. Lucier (1989), 'High order regularity for solutions of the inviscid Burgers equation', *Nonlinear Hyperbolic Problems (Proc. Advanced Research Workshop, Bordeaux, France, June 1988, Springer Lecture Notes in Mathematics, 1402)* (C. Carasso, P. Charrier, B. Hanouzet and J.-L. Joly, eds) Springer-Verlag (New York) pp. 147–154.

R. DeVore and B. Lucier (1990), 'High order regularity for conservation laws', *Indiana Univ. Math. J.* **39**, 413–430.

R. DeVore and V. Popov (1988), 'Interpolation spaces and non-linear approximation', *Function Spaces and Applications (Springer Lecture Notes in Mathematics, 1302)* (M. Cwikel, J. Peetre, Y. Sagher and H. Wallin, eds) Springer (New York) pp. 191–205.

T. Eirola (1992) 'Sobolev characterization of solutions of dilation equations', *SIAM J. Math. Anal.*, to appear.

M. Frazier and B. Jawerth (1990), 'A discrete transform and decompositions of distribution spaces', *J. Funct. Anal.* **93**, 34–170.

M. Frazier, B. Jawerth and G. Weiss (1991), *Littlewood–Paley Theory and the Study of Function Spaces* Conference Board of the Mathematical Sciences, Regional Conference Series in Mathematics, Number 79, Amer. Math. Soc. (Providence, RI)

S. Jaffard (1991) 'Wavelet methods for fast resolution of elliptic problems', Preprint

S. Jaffard and Y. Meyer (1989), 'Bases ondelettes dans des ouverts de $\mathbb{R}^n$', *J. Math. Pures Appl.* **68**, 95–108.

R-Q. Jia and C. Micchelli (1991), 'Using the refinement equation for the construction of pre-wavelets II: power of two', *Curves and Surfaces*, (P. J. Laurent, A. LeMéhauté, and L. Schumaker, eds) Academic Press (New York) pp. 209–246.

S. Karlin and W. Studden (1966), *Tchebycheff Systems with Application to Analysis and Statistics* Wiley Interscience (New York).

R. A. H. Lorentz and W. R. Madych (1991) Wavelets and generalized box splines, preprint.

S. G. Mallat (1989), 'Multi-resolution approximations and wavelet orthonormal bases of $L^2(\mathbb{R})$', *Trans. Amer. Math. Soc.* **315**, 69–87.

Y. Meyer (1987), 'Wavelets with compact support', *Zygmund Lectures* (University of Chicago).

Y. Meyer (1990), *Ondelettes et Opérateurs I: Ondelettes* Hermann (Paris).

C. Micchelli (1991), 'Using the refinement equation for the construction of pre-wavelets', *Numer. Alg.* **1**, 75–116.

P. Oswald (1991) 'On discrete norm estimates related to multi-level preconditioners in the finite elment method', Preprint.

S. D. Riemenschneider and Z. W. Shen (1991), 'Box splines, cardinal splines, and wavelets', *Approximation Theory and Functional Analysis* (C. K. Chui, ed.) Academic Press (New York) pp. 133–149.

S. D. Riemenschneider and Z. W. Shen (1992) 'Wavelets and pre-wavelets in low dimensions', *J. Approx. Theory* to appear.

O. Rioul and M. Vetterli (1991), 'Wavelets and signal processing', *IEEE Signal Processing Magazine* **8**, (4) 14–38.

I. Schoenberg (1946), 'Contributions to the problem of approximation of equidistant data by analytic functions, Parts A & B', *Quart. Appl. Math.* **IV**, 45–99, 112–141.

I. Schoenberg (1973), *Cardinal Spline Interpolation (Regional Conference Series in Applied Mathematics 12)* SIAM (Philadelphia).

G. Strang and G. Fix (1973), 'A Fourier analysis of the finite element variational method', *C.I.M.E. II Ciclo 1971, Constructive Aspects of Functional Analysis* (G. Geymonat, ed.) pp. 793–840.

J. O. Strömberg (1981) 'A modified Franklin system and higher-order spline systems on $\mathbb{R}^n$ as unconditional basis for Hardy spaces', *Conference in Harmonic Analysis in Honor of Antoni Zygmund* Vol II (W. Beckner *et al.*, eds) Wadsworth pp. 475–493.

Acta Numerica (1991), *pp.* 57–100

Iterative solution of linear systems

Roland W. Freund *

RIACS, Mail Stop Ellis Street

NASA Ames Research Center

Moffett Field, CA 94035, USA

E-mail: na.freund@na-net.ornl.gov

Gene H. Golub †

Computer Science Department

Stanford University

Stanford, CA 94305, USA

E-mail: golub@sccm.stanford.edu

Noël M. Nachtigal ‡

RIACS, Mail Stop Ellis Street

NASA Ames Research Center

Moffett Field, CA 94035, USA

E-mail: na.nachtigal@na-net.ornl.gov

Recent advances in the field of iterative methods for solving large linear systems are reviewed. The main focus is on developments in the area of conjugate gradient-type algorithms and Krylov subspace methods for nonHermitian matrices.

CONTENTS

* The work of this author was supported by DARPA via Cooperative Agreement NCC 2-387 between NASA and the Universities Space Research Association (USRA).

† The work of this author was supported in part by the National Science Foundation under Grant NSF CCR-8821078.

‡ The work of this author was supported by Cooperative Agreement NCC 2-387 between NASA and the Universities Space Research Association (USRA).

1. Introduction

One of the fundamental building blocks of numerical computing is the ability to solve linear systems

$$A\mathbf{x} = \mathbf{b}. \tag{1.1}$$

These systems arise very frequently in scientific computing, for example from finite difference or finite element approximations to partial differential equations, as intermediate steps in computing the solution of nonlinear problems or as subproblems in linear and nonlinear programming.

For linear systems of small size, the standard approach is to use direct methods, such as Gaussian elimination. These algorithms obtain the solution of (1.1) based on a factorization of the coefficient matrix A. However, in practice linear systems arise that can be arbitrarily large; this is particularly true when solving partial differential equations. Fortunately, the resulting systems usually have some special structure; sparsity, i.e. matrices with only a few nonzero entries, is the most common case. Often, direct methods can be adapted to exploit the special structure of the matrix and then remain useful even for large linear systems. However, in many cases, especially for systems arising from three-dimensional partial differential equations, direct approaches are prohibitive both in terms of storage requirements and computing time, and then the only alternative is to use iterative algorithms.

Especially attractive are iterative methods that involve the coefficient matrix only in the form of matrix–vector products with A or A^H. Such schemes naturally exploit the special structure of large sparse linear systems. They are also well suited for the solution of certain dense large systems for which matrix–vector products can be obtained cheaply. The most powerful iterative scheme of this type is the conjugate gradient method (CG) due to Hestenes and Stiefel (1952), which is an algorithm for solving Hermitian positive definite linear systems. Although CG was introduced as early as 1952, its true potential was not appreciated until the work of Reid (1971) and Concus *et al.* (1976) in the 1970s. Since then, a considerable part of the research in numerical linear algebra has been devoted to generalizations of CG to indefinite and nonHermitian linear systems.

A straightforward extension to general nonHermitian matrices is to apply CG to either one of the Hermitian positive definite linear systems

$$A^H A\mathbf{x} = A^H \mathbf{b}, \tag{1.2}$$

or

$$A A^H \mathbf{y} = \mathbf{b}, \quad \mathbf{x} = A^H \mathbf{y}. \tag{1.3}$$

Solving (1.2) by CG was mentioned already by Hestenes and Stiefel (1952); we will refer to this approach as CGNR. Applying CG to (1.3) was proposed by Craig (1955); we will refer to this second approach as CGNE. Although there are special situations where CGNR or CGNE are the optimal extensions of CG, both algorithms generally converge very slowly and hence they are not usually satisfactory generalizations of CG to arbitrary nonHermitian matrices.

Consequently, CG-type algorithms were sought that are applied to the original system (1.1), rather than (1.2) or (1.3). A number of such methods have been proposed since the mid-1970s, the most widely used of which is the generalized minimum residual algorithm (GMRES) due to Saad and Schultz (1986). While GMRES and related schemes generate at each iteration optimal approximate solutions of (1.1), their work and storage requirements per iteration grow linearly. Therefore, it becomes prohibitive to run the full version of these algorithms and restarts are necessary, which often leads to very slow convergence.

For this reason, since the late 1980s, research in nonHermitian matrix iterations has focused mainly on schemes that can be implemented with low and roughly constant work and storage requirements per iteration. A number of new algorithms with this feature have been proposed, all of which are related to the nonsymmetric Lanczos process. It is these recent developments in CG-type methods for nonHermitian linear systems that we will emphasize in this survey.

The outline of this paper is as follows. In Section 2, we present some background material on general Krylov subspace methods, of which CG-type algorithms are a special case. We recall the outstanding properties of CG and discuss the issue of optimal extensions of CG to nonHermitian matrices. We also review GMRES and related methods, as well as CG-like algorithms for the special case of Hermitian indefinite linear systems. Finally, we briefly discuss the basic idea of preconditioning. In Section 3, we turn to Lanczos-based iterative methods for general nonHermitian linear systems. First, we consider the nonsymmetric Lanczos process, with particular emphasis on the possible breakdowns and potential instabilities in the classical algorithm. Then we describe recent advances in understanding these problems and overcoming them by using look-ahead techniques. Moreover, we describe the quasi-minimal residual algorithm (QMR) proposed by Freund and Nachtigal (1990), which uses the look-ahead Lanczos process to obtain quasi-optimal approximate solutions. Next, a survey of transpose-free Lanczos-based methods is given. We conclude this section with comments on other related work and some historical remarks. In Section 4, we elaborate on CGNR and CGNE and we point out situations where these approaches are optimal. The general class of Krylov subspace methods also contains parameter-dependent algorithms that, unlike CG-type schemes, require explicit information on the spectrum of the coefficient matrix. In Section 5, we discuss recent insights into obtaining appropriate spectral information for parameter-dependent Krylov subspace methods. After that, we turn to special classes of linear systems. First, in Section 6, we consider CG-type algorithms for complex symmetric and shifted Hermitian matrices. In Section 7, we review cases of dense large linear systems for which iterative algorithms are a viable alternative to direct methods. Finally, in Section 8, we make some concluding remarks.

Today, the field of iterative methods is a rich and extremely active research area, and it has become impossible to cover in a survey paper all recent advances. For example, we have not included any recent developments in preconditioning of linear systems, nor any discussion of the efficient use of iterative schemes on advanced architectures. Also, we would like to point the reader to the following earlier survey papers. Stoer (1983) reviews the state of CG-like algorithms up to the early 1980s. In the paper by Axelsson (1985), the focus is on preconditioning of iterative methods. More modern iterative schemes, such as GMRES, and issues related to the implementation of Krylov subspace methods on supercomputers are treated in the survey by Saad (1989). An annotated bibliography on CG and CG-like methods covering the period up to 1976 was compiled by Golub and O'Leary (1989). Finally, readers interested in direct methods for sparse linear systems are referred to the book by Duff *et al.* (1986) and, for the efficient use of these techniques on parallel machines, to Heath *et al.* (1991).

Throughout the article, all vectors and matrices are in general assumed to be complex. As usual, $i = \sqrt{-1}$. For any matrix $M = [m_{jk}]$, we use the following notation:

$$\overline{M} = [\overline{m_{jk}}] = \text{the complex conjugate of } M,$$
$$M^T = [m_{kj}] = \text{the transpose of } M,$$
$$M^H = \overline{M}^T = \text{the Hermitian of } M,$$
$$\text{Re}\, M = (M + \overline{M})/2 = \text{the real part of } M,$$
$$\text{Im}\, M = (M - \overline{M})/(2i) = \text{the imaginary part of } M,$$
$$\sigma(M) = \text{the set of singular values of } M,$$
$$\sigma_{\max}(M) = \text{the largest singular value of } M,$$
$$\sigma_{\min}(M) = \text{the smallest singular value of } M,$$
$$\|M\|_2 = \sigma_{\max}(M) = \text{the 2-norm of } M,$$
$$\kappa_2(M) = \sigma_{\max}(M)/\sigma_{\min}(M)$$
$$= \text{the 2-condition number of } M, \text{ if } M \text{ has full rank.}$$

For any vector $\mathbf{c} \in \mathbb{C}^m$ and any matrix $B \in \mathbb{C}^{m \times m}$, we denote by

$$\mathcal{K}_n(\mathbf{c}, B) = \text{span}\{\mathbf{c}, B\mathbf{c}, \ldots, B^{n-1}\mathbf{c}\}$$

the nth Krylov subspace of $\mathbb{C}^m$, generated by $\mathbf{c}$ and B. Furthermore, we use the following notation:

$$\|\mathbf{c}\|_2 \;=\; \sqrt{\mathbf{c}^H \mathbf{c}} \;=\; \text{Euclidean norm of } \mathbf{c},$$

$$\|\mathbf{c}\|_B \;=\; \sqrt{\mathbf{c}^H B \mathbf{c}}$$

$$= \; B\text{-norm of } \mathbf{c}, \text{ if } B \text{ is Hermitian positive definite},$$

$$\lambda(B) \;=\; \text{the set of eigenvalues of } B,$$

$$\lambda_{\max}(B) \;=\; \text{the largest eigenvalue of } B, \text{ if } B \text{ is Hermitian},$$

$$\lambda_{\min}(B) \;=\; \text{the smallest eigenvalue of } B, \text{ if } B \text{ is Hermitian}.$$

Moreover, we denote by I_n the $n \times n$ identity matrix; if the dimension n is evident from the context, we will simply write I. The symbol 0 will be used both for the number 0 and for the zero matrix; in the latter case, the dimension will always be apparent. We denote by

$$\mathcal{P}_n = \{\phi(\lambda) \equiv \sigma_0 + \sigma_1 \lambda + \cdots + \sigma_n \lambda^n \;\big|\; \sigma_0, \sigma_1, \ldots, \sigma_n \in \mathbb{C}\}$$

the set of complex polynomials of degree at most n.

Throughout this paper, N denotes the dimension of the coefficient matrix A of (1.1) and $A \in \mathbb{C}^{N \times N}$ is in general nonHermitian. In addition, unless otherwise stated, A is always assumed to be nonsingular. Moreover, we use the following notation:

$$\mathbf{x}_0 \;=\; \text{initial guess for the solution of } (1.1),$$

$$\mathbf{x}_n \;=\; n\text{th iterate},$$

$$\mathbf{r}_n \;=\; \mathbf{b} - A\mathbf{x}_n \;=\; n\text{th residual vector}.$$

If it is not apparent from the context which iterative method we are considering, quantities from different algorithms will be distinguished by superscripts, e.g. $\mathbf{x}_n^{CG}$ or $\mathbf{x}_n^{GMRES}$.

2. Background

In this section, we present some background material on general Krylov subspace methods.

2.1. Krylov subspace methods

Many iterative schemes for solving the linear system (1.1) belong to the class of Krylov subspace methods: they produce approximations $\mathbf{x}_n$ to $A^{-1}\mathbf{b}$ of the form

$$\mathbf{x}_n \in \mathbf{x}_0 + \mathcal{K}_n(\mathbf{r}_0, A), \quad n = 1, 2, \ldots . \tag{2.1}$$

Here, $\mathbf{x}_0 \in \mathbb{C}^N$ is any initial guess for the solution of (1.1), $\mathbf{r}_0 = \mathbf{b} - A\mathbf{x}_0$ is the corresponding residual vector, and $\mathcal{K}_n(\mathbf{r}_0, A)$ is the nth Krylov subspace generated by $\mathbf{r}_0$ and A. In view of

$$\mathcal{K}_n(\mathbf{r}_0, A) = \{\phi(A)\mathbf{r}_0 \;\big|\; \phi \in \mathcal{P}_{n-1}\}, \tag{2.2}$$

schemes with iterates (2.1) are also referred to as polynomial-based iterative methods. In particular, the residual vector corresponding to the nth iterate $\mathbf{x}_n$ can be expressed in terms of polynomials:

$$\mathbf{r}_n = \mathbf{b} - A\mathbf{x}_n = \psi_n(A)\mathbf{r}_0, \tag{2.3}$$

where

$$\psi_n \in \mathcal{P}_n, \quad \text{with} \quad \psi_n(0) = 1. \tag{2.4}$$

Generally, any polynomial satisfying (2.4) is called an nth residual polynomial.

As (2.3) shows, the goal in designing a Krylov subspace method is to choose at each step the polynomial ψ_n such that $\mathbf{r}_n \approx 0$ in some sense. One option is to actually minimize some norm of the residual $\mathbf{r}_n$:

$$
\begin{aligned}
\|\mathbf{r}_n\| &= \min_{\mathbf{x} \in \mathbf{x}_0 + \mathcal{K}_n(\mathbf{r}_0, A)} \|\mathbf{b} - A\mathbf{x}\| \\
&= \min_{\psi \in \mathcal{P}_n : \psi(0) = 1} \|\psi(A)\mathbf{r}_0\|.
\end{aligned}
\tag{2.5}
$$

Here $\| \cdot \|$ is a vector norm on $\mathbb{C}^N$, which may even depend on the iteration number n (see Section 3.3). Another option is to require that the residual satisfies a Galerkin-type condition:

$$\mathbf{s}^H \mathbf{r}_n = 0 \quad \text{for all} \quad \mathbf{s} \in \mathcal{S}_n, \tag{2.6}$$

where $\mathcal{S}_n \subset \mathbb{C}^N$ is a subspace of dimension n. Note that an iterate satisfying (2.6) need not exist for each n; in contrast, the existence of iterates with (2.5) is always guaranteed. The point is that iterates with (2.5) or (2.6) can be obtained from a basis for $\mathcal{K}_n(r_0, A)$ (and a basis for $\mathcal{S}_n$ in the case of (2.6)), without requiring any *a priori* choice of other iteration parameters.

In contrast to parameter-free schemes based on (2.5) or (2.6), parameter-dependent Krylov subspace methods require some advance information on the spectral properties of A for the construction of ψ_n. Usually, knowledge of some compact set $\mathcal{G}$ with

$$\lambda(A) \subseteq \mathcal{G} \subset \mathbb{C}, \quad 0 \notin \mathcal{G}, \tag{2.7}$$

is needed. For example, assume that A is diagonalizable, and let U be any matrix of eigenvectors of A. For the case of the Euclidean norm, it then follows from (2.3) and (2.7) that

$$
\begin{aligned}
\frac{\|\mathbf{r}_n\|_2}{\|\mathbf{r}_0\|_2} &\leq \kappa_2(U) \max_{\lambda \in \lambda(A)} |\psi_n(\lambda)| \\
&\leq \kappa_2(U) \max_{\lambda \in \mathcal{G}} |\psi_n(\lambda)|.
\end{aligned}
\tag{2.8}
$$

Ideally, one would like to choose the residual polynomial ψ_n such that the right-hand side in (2.8) is minimal, i.e.

$$\max_{\lambda \in \mathcal{G}} |\psi_n(\lambda)| = \min_{\psi \in \mathcal{P}_n : \psi(0)=1} \max_{\lambda \in \mathcal{G}} |\psi(\lambda)|. \tag{2.9}$$

Unfortunately, the exact solution of the approximation problem (2.9) is known only for a few special cases. For example, if $\mathcal{G}$ is a real interval, then shifted and scaled Chebyshev polynomials are optimal in (2.9); the resulting algorithm is the well-known Chebyshev semi-iterative method for Hermitian positive definite matrices (see Golub and Varga, 1961). Later, Manteuffel (1977) extended the Chebyshev iteration to the class of non-Hermitian matrices for which $\mathcal{G}$ in (2.7) can be chosen as an ellipse. We remark that, in this case, Chebyshev polynomials are always nearly optimal for (2.9), but – contrary to popular belief – in general they are not the exact solutions of (2.9), as was recently shown by Fischer and Freund (1990, 1991). The solution of (2.9) is also known explicitly for complex line segments $\mathcal{G}$ that are parallel to the imaginary axis and symmetric about the real line (see Freund and Ruscheweyh, 1986); this case corresponds to shifted skew-symmetric matrices A of the form (2.14). In the general case, however, the exact solution of (2.9) is not available and is expensive to compute numerically. Instead, one chooses polynomials that are only asymptotically optimal for (2.9). An elegant theory for semi-iterative methods of this type was developed by Eiermann *et al.* (1985).

In this survey, we will focus mainly on parameter-free algorithms with iterates characterized by (2.5) or (2.6). Parameter-dependent Krylov subspace methods will be only briefly discussed in Section 5.

2.2. CG and optimal extensions

Classical CG is a Krylov subspace method for Hermitian positive definite matrices A with two outstanding features. First, its iterates $\mathbf{x}_n$ satisfy a minimization property, namely (2.5) in the A^{-1}-norm:

$$\|\mathbf{b} - A\mathbf{x}_n\|_{A^{-1}} = \min_{\mathbf{x} \in \mathbf{x}_0 + \mathcal{K}_n(\mathbf{r}_0, A)} \|\mathbf{b} - A\mathbf{x}\|_{A^{-1}}. \tag{2.10}$$

Secondly, $\mathbf{x}_n$ can be computed efficiently, based on simple three-term recurrences.

An ideal extension of CG to nonHermitian matrices A would have similar features. However, since in general $\| \cdot \|_{A^{-1}}$ is no longer a norm, one usually replaces (2.10) with either the minimization property

$$\|\mathbf{b} - A\mathbf{x}_n\|_2 = \min_{\mathbf{x} \in \mathbf{x}_0 + \mathcal{K}_n(\mathbf{r}_0, A)} \|\mathbf{b} - A\mathbf{x}\|_2, \tag{2.11}$$

or the Galerkin condition

$$\mathbf{s}^H (\mathbf{b} - A\mathbf{x}_n) = 0 \quad \text{for all} \quad \mathbf{s} \in \mathcal{K}_n(\mathbf{r}_0, A). \tag{2.12}$$

In the sequel, a Krylov subspace algorithm with iterates (2.1) defined by

(2.11) or (2.12) will be called a minimal residual (MR) method or an orthogonal residual (OR) method, respectively. We remark that (2.10) and (2.12) are equivalent for Hermitian positive definite A, and hence (2.12) is an immediate extension of (2.10). Unfortunately, for nonHermitian A and even for Hermitian indefinite A, an iterate x_n with (2.12) need not exist at each step n. In contrast, there is always a unique iterate $x_n \in x_0 + \mathcal{K}_n(r_0, A)$ satisfying (2.11). We note that the conjugate residual algorithm (CR) due to Stiefel (1955) is a variant of CG that generates iterates characterized by (2.11) for the special case of Hermitian positive definite A.

An ideal CG-like scheme for solving nonHermitian linear systems would then have the following features:

1   its iterates would be characterized by the MR or OR property; and
2   it could be implemented based on short vector recursions, so that work and storage requirements per iteration would be low and roughly constant.

Unfortunately, it turns out that, for general matrices, the conditions 1 and 2 cannot be fulfilled simultaneously. This result is due to Faber and Manteuffel (1984, 1987) who proved the following theorem (see also Voevodin (1983) and Joubert and Young (1987)).

Theorem 2.1   (Faber and Manteuffel, 1984 and 1987.) Except for a few anomalies, ideal CG-like methods that satisfy both requirements 1 and 2 exist only for matrices of the special form

$$A = e^{i\theta}(T + \sigma I), \quad \text{where} \quad T = T^H, \quad \theta \in \mathbb{R}, \quad \sigma \in \mathbb{C}. \tag{2.13}$$

The class (2.13) consists of just the shifted and rotated Hermitian matrices. Note that the important subclass of real nonsymmetric matrices

$$A = I - S, \quad \text{where} \quad S = -S^T \quad \text{is real}, \tag{2.14}$$

is contained in (2.13), with $e^{i\theta} = i$, $\sigma = -i$, and $T = iS$. Concus and Golub (1976) and Widlund (1978) were the first to devise an implementation of an OR method for the family (2.14). The first MR algorithm for (2.14) was proposed by Rapoport (1978), and different implementations were given by Eisenstat *et al.* (1983) and Freund (1983). For a brief discussion of actual CG-type algorithms for the general class of complex nonHermitian matrices (2.14), we refer the reader to Section 6.2.

Finally, we remark that ideal CG-like methods also exist for the more general family of shifted and rotated B-Hermitian matrices

$$A = e^{i\theta}(T + \sigma I), \quad \text{where} \quad TB = (TB)^H, \quad \theta \in \mathbb{R}, \quad \sigma \in \mathbb{C}. \tag{2.15}$$

Here B is a fixed given Hermitian positive definite $N \times N$ matrix (see Ashby *et al.*, 1990). However, since for any matrix A of the form (2.15),

$$A' = B^{1/2}AB^{-1/2}$$

is of the type (2.13), without loss of generality the case (2.15) can always be reduced to (2.13).

2.3. CG-type algorithms for Hermitian indefinite linear systems

The family (2.13) also contains Hermitian indefinite matrices; next we review CG-type methods for this special case.

Luenberger (1969) was the first to propose a modification of standard CG for Hermitian indefinite matrices; however, his algorithm encountered some unresolved computational difficulties. The first numerically stable schemes for Hermitian indefinite linear systems were derived by Paige and Saunders (1975). Their SYMMLQ algorithm is an implementation of the OR approach and hence the immediate generalization of classical CG. As pointed out earlier, an OR iterate $\mathbf{x}_n$ satisfying (2.12) need not exist for each n, and, in fact, SYMMLQ generates $\mathbf{x}_n$ only indirectly. Instead of the OR iterates, a second sequence of well-defined iterates $\mathbf{x}_n^L$ is updated, from which existing $\mathbf{x}_n$ can then be obtained cheaply. Paige and Saunders also proposed the MINRES algorithm, which produces iterates defined by the MR property (2.11) and thus can be viewed as an extension of CR to Hermitian indefinite matrices. SYMMLQ and MINRES both use the Hermitian Lanczos recursion to generate an orthonormal basis for the Krylov subspaces $\mathcal{K}_n(\mathbf{r}_0, A)$, and, like the latter, they can be implemented based on simple three-term recurrences. We would like to stress that the work of Paige and Saunders was truly pioneering, in that they were the first to extend CG beyond the class of Hermitian positive definite matrices in a numerically stable manner.

SYMMLQ is also closely connected with an earlier algorithm due to Fridman (1963), which generates iterates $\mathbf{x}_n^{ME} \in \mathbf{x}_0 + \mathcal{K}_n(A\mathbf{r}_0, A)$ defined by the minimal error (ME) property

$$\|A^{-1}\mathbf{b} - \mathbf{x}_n^{ME}\|_2 = \min_{\mathbf{x} \in \mathbf{x}_0 + \mathcal{K}_n(A\mathbf{r}_0, A)} \|A^{-1}\mathbf{b} - \mathbf{x}\|_2. \qquad (2.16)$$

Unfortunately, Fridman's original implementation of the ME approach is unstable. Fridman's algorithm was later rediscovered by Fletcher (1976) who showed that, in exact arithmetic, the ME iterate $\mathbf{x}_n^{ME}$ coincides with the auxiliary quantity $\mathbf{x}_n^L$ in SYMMLQ. Hence, as a by-product, SYMMLQ also provides a stable implementation of the ME method. Another direct stabilization of Fridman's algorithm was proposed by Stoer and Freund (1982).

Finally, we remark that Chandra (1978) proposed the SYMMBK algorithm, which is a slightly less expensive variant of SYMMLQ, and derived another stable implementation of the MR method, different from MINRES.

2.4. GMRES and related algorithms

We now return to Krylov subspace methods for general nonHermitian matrices. Numerous algorithms for computing the iterates characterized by the MR or OR property (2.11) or (2.12), respectively, have been proposed; see Vinsome (1976), Axelsson (1980, 1987), Young and Jea (1980), Saad (1981, 1982, 1984), Elman (1982), Saad and Schultz (1985, 1986). Interestingly, a simple implementation of the MR approach was already described in a paper by Khabaza (1963), which is not referenced at all in the recent literature. In view of Theorem 2.1, all these algorithms generally involve long vector recursions, and typically work and storage requirements per iteration grow linearly with the iteration index n. Consequently, in practice, one cannot afford to run the full algorithms, and it becomes necessary to use restarts or to truncate the vector recursions.

The most elegant and most widely used scheme of this type is GMRES, due to Saad and Schultz (1986), and here we sketch only this particular algorithm. GMRES is modelled after MINRES, where now a generalization of the Hermitian Lanczos process, namely the Arnoldi process (see Arnoldi (1951) and Saad (1980)), is used to generate orthonormal basis vectors for the Krylov subpaces $\mathcal{K}_n(\mathbf{r}_0, A)$.

Algorithm 2.2   (Arnoldi process)

0) Choose $\mathbf{v}_1 \in \mathbb{C}^N$ with $\|\mathbf{v}_1\|_2 = 1$.

For $n = 1, 2, \ldots$, do :

1 For $k = 1, 2, \ldots, n$, compute

$$h_{kn} = \mathbf{v}_k^H A \mathbf{v}_n.$$

2 Set

$$\tilde{\mathbf{v}}_{n+1} = A\mathbf{v}_n - \sum_{k=1}^{n} h_{kn} \mathbf{v}_k.$$

3 Compute

$$h_{n+1,n} = \|\tilde{\mathbf{v}}_{n+1}\|_2.$$

4 If $h_{n+1,n} = 0$, stop.

Otherwise, set

$$\mathbf{v}_{n+1} = \tilde{\mathbf{v}}_{n+1}/h_{n+1,n}.$$

The vector recurrences in Step 2 of Algorithm 2.2 can be rewritten compactly in matrix form as follows:

$$AV_n = V_{n+1} H_n^{(e)}, \tag{2.17}$$

where

$$V_n = [\mathbf{v}_1 \quad \mathbf{v}_2 \quad \cdots \quad \mathbf{v}_n] \tag{2.18}$$

has orthonormal columns, and

$$H_n^{(e)} = \begin{bmatrix} h_{11} & h_{12} & \cdots & & h_{1n} \\ h_{21} & \ddots & & & \vdots \\ 0 & \ddots & \ddots & & \vdots \\ \vdots & \ddots & & h_{n,n-1} & h_{nn} \\ 0 & \cdots & & 0 & h_{n+1,n} \end{bmatrix} \tag{2.19}$$

is an $(n+1) \times n$ upper Hessenberg matrix of full rank n.

If one chooses the starting vector $\mathbf{v}_1 = \mathbf{r}_0/\|\mathbf{r}_0\|_2$ in Algorithm 2.2, then all possible iterates (2.1) can be parametrized as follows:

$$\mathbf{x}_n = \mathbf{x}_0 + V_n \mathbf{z}_n, \quad \text{where} \quad \mathbf{z}_n \in \mathbb{C}^n. \tag{2.20}$$

Moreover, with (2.20) and (2.17), the minimal residual property (2.11) reduces to the $(n+1) \times n$ least squares problem

$$\|\mathbf{d}_n - H_n^{(e)} \mathbf{z}_n\|_2 = \min_{\mathbf{z} \in \mathbb{C}^n} \|\mathbf{d}_n - H_n^{(e)} \mathbf{z}\|_2, \tag{2.21}$$

where

$$\mathbf{d}_n = [\,\|\mathbf{r}_0\|_2 \quad 0 \quad \cdots \quad 0\,]^T \in \mathbb{R}^{n+1}. \tag{2.22}$$

GMRES is an implementation of the minimal residual approach (2.11) that obtains the nth MR iterate $\mathbf{x}_n$ by first running n steps of the Arnoldi process and then solving the $(n+1) \times n$ least squares problem (2.21). Note that (2.21) always has a unique solution, since $H_n^{(e)}$ is of full column rank. For a detailed description of the algorithm, we refer the reader to Saad and Schultz (1986).

The Arnoldi Algorithm 2.2 can also be used to compute the nth OR iterate characterized by (2.12). Indeed, as Saad (1981) has shown, $\mathbf{x}_n$ is again of the form (2.20) where $\mathbf{z}_n$ is now the solution of the $n \times n$ linear system

$$H_n \mathbf{z}_n = \mathbf{d}_{n-1}. \tag{2.23}$$

Here

$$H_n = [\,I_n \quad 0\,] H_n^{(e)} \tag{2.24}$$

is the matrix obtained from $H_n^{(e)}$ by deleting the last row in (2.19). The problem with this approach is that H_n can be singular, and then the linear system (2.23) is inconsistent. In fact, H_n is singular if, and only if, no OR iterate satisfying (2.12) exists.

An interesting alternative is to use quasi-Newton techniques, such as Broyden's method (Broyden, 1965). Although designed for general nonlinear equations, these schemes can be applied to nonHermitian linear systems as a special case. In addition to the iterates $\mathbf{x}_n$, these algorithms also produce approximations to A^{-1}, updated from step to step by a simple rank-1 correction. While these schemes look different at a first glance, they also belong to the class of Krylov subspace methods, as was first observed by Elman (1982). Furthermore, Deuflhard *et al.* (1990) have demonstrated that Broyden's rank-1 update combined with a suitable line search strategy leads to an iterative algorithm that is competitive with GMRES. Eirola and Nevanlinna (1989) have proposed two methods based on a different rank-1 update and shown that one of the resulting schemes is mathematically equivalent to GMRES. These algorithms were studied further by Vuik (1990).

2.5. Preconditioned Krylov subspace methods

For the solution of realistic problems, it is crucial to combine Krylov subspace methods with an efficient preconditioning technique. The basic idea here is as follows. Let M be a given nonsingular $N \times N$ matrix, which approximates in some sense the coefficient matrix A of the original linear system (1.1). Moreover, assume that M is decomposed in the form

$$M = M_1 M_2. \tag{2.25}$$

The Krylov subspace method is then used to solve the preconditioned linear system

$$A'\mathbf{x}' = \mathbf{b}', \tag{2.26}$$

where

$$A' = M_1^{-1} A M_2^{-1}, \quad \mathbf{b}' = M_1^{-1}\mathbf{b}, \quad \mathbf{x}' = M_2\mathbf{x}.$$

Clearly, (2.26) is equivalent to (1.1). This process generates approximate solutions of (2.26) of the form

$$\mathbf{x}'_n \in \mathbf{x}'_0 + \mathcal{K}_n\left(\mathbf{r}'_0, A'\right). \tag{2.27}$$

Usually, one avoids the explicit calculation of primed quantities, and instead one rewrites the resulting algorithm in terms of the corresponding quantities for the original system. For example, iterates and residual vectors for (1.1) and (2.26) are connected by

$$\mathbf{x}_n = M_2^{-1}\mathbf{x}'_n \quad \text{and} \quad \mathbf{r}_n = M_1\mathbf{r}'_n. \tag{2.28}$$

In particular, note that, by (2.27) and (2.28), the resulting approximations to $A^{-1}\mathbf{b}$ are of the form

$$\mathbf{x}_n \in \mathbf{x}_0 + \mathcal{K}_n\left(M^{-1}\mathbf{r}_0, M^{-1}A\right).$$

We remark that the special cases $M_1 = I$ or $M_2 = I$ in (2.25) are referred to as right or left preconditioning, respectively. For right preconditioning, by

(2.28), the preconditioned residual vectors coincide with their counterparts for the original system. For this reason, right preconditioning is usually preferred for MR-type Krylov subspace methods based on (2.11). Moreover, if A has some special structure, the decomposition (2.25) can often be chosen such that the structure is preserved for A'. For example, for Hermitian positive definite A this is the case if one sets $M_2 = M_1^H$ in (2.25).

Obviously, there are two (in general conflicting) requirements for the choice of the preconditioning matrix M for a given Krylov subspace method. First, M^{-1} should approximate A^{-1} well enough so that the algorithm applied to (2.26) will converge faster than for the original system (1.1). On the other hand, preconditioned Krylov subspace methods require at each iteration the solution of one linear system of the type

$$M\mathbf{p} = \mathbf{q}. \tag{2.29}$$

Moreover, for algorithms that involve matrix–vector products with A^T (see Section 3), one has to solve an additional linear system of the form

$$M^T\mathbf{p} = \mathbf{q}. \tag{2.30}$$

Therefore, the preconditioner M needs to be such that linear systems (2.29) respectively (2.30) can be solved cheaply. In this article, the problem of how to actually construct such preconditioners is not addressed at all. Instead, we refer the reader to the papers by Axelsson (1985) and Saad (1989) for an overview of common preconditioning techniques.

Finally, one more note. In Section 3, explicit descriptions of some Krylov subspace algorithms are given. For simplicity, we have stated these algorithms without preconditioning. It is straightforward to incorporate preconditioning by using the transition rules (2.28).

3. Lanczos-based Krylov subspace methods

In this section, we discuss Krylov subspace methods that are based on the nonsymmetric Lanczos process.

3.1. The classical Lanczos algorithm and BCG

The nonsymmetric Lanczos method was proposed by Lanczos (1950) as a means to reduce an arbitrary matrix $A \in \mathbb{C}^{N \times N}$ to tridiagonal form. One starts the process with two nonzero vectors $\mathbf{v}_1 \in \mathbb{C}^N$ and $\mathbf{w}_1 \in \mathbb{C}^N$ and then generates basis vectors $\{\mathbf{v}_j\}$ for $\mathcal{K}_n(\mathbf{v}_1, A)$ and $\{\mathbf{w}_j\}$ for $\mathcal{K}_n(\mathbf{w}_1, A^T)$ such that the bi-orthogonality condition

$$\mathbf{w}_j^T \mathbf{v}_k = \begin{cases} 1 & k = j, \\ 0 & \text{otherwise,} \end{cases} \tag{3.1}$$

holds. The point is that the two bases can be built with just three-term recurrences, thus requiring minimal amounts of work and storage per step. The complete algorithm is straightforward and can be stated as follows.

Algorithm 3.1   (Classical Lanczos method)

0 Choose $\tilde{\mathbf{v}}_1$, $\tilde{\mathbf{w}}_1 \in \mathbb{C}^N$ with $\tilde{\mathbf{v}}_1$, $\tilde{\mathbf{w}}_1 \neq 0$.

   Set $\mathbf{v}_0 = \mathbf{w}_0 = 0$.

   For $n = 1, 2, \ldots$, do :

1 Compute $\delta_n = \tilde{\mathbf{w}}_n^T \tilde{\mathbf{v}}_n$.

   If $\delta_n = 0$, set $L = n - 1$ and stop.

2 Otherwise, choose $\beta_n, \gamma_n \in \mathbb{C}$ with $\beta_n \gamma_n = \delta_n$.

   Set $\mathbf{v}_n = \tilde{\mathbf{v}}_n / \gamma_n$ and $\mathbf{w}_n = \tilde{\mathbf{w}}_n / \beta_n$.

3 Compute

$$\alpha_n = \mathbf{w}_n^T A \mathbf{v}_n,$$
$$\tilde{\mathbf{v}}_{n+1} = A\mathbf{v}_n - \alpha_n \mathbf{v}_n - \beta_n \mathbf{v}_{n-1},$$
$$\tilde{\mathbf{w}}_{n+1} = A^T \mathbf{w}_n - \alpha_n \mathbf{w}_n - \gamma_n \mathbf{w}_{n-1}.$$

   If $\tilde{\mathbf{v}}_{n+1} = 0$ or $\tilde{\mathbf{w}}_{n+1} = 0$, set $L = n$ and stop.

The particular choice of the coefficients α_n, β_n, and γ_n ensures that the bi-orthogonality relation (3.1) is satisfied.

Similar to (2.18), (2.19) and (2.24), let

$$V_n = [v_1 \quad v_2 \quad \cdots \quad v_n], \quad W_n = [w_1 \quad w_2 \quad \cdots \quad w_n],$$

$$H_n^{(e)} = \begin{bmatrix} \alpha_1 & \beta_2 & 0 & \cdots & & 0 \\ \gamma_2 & \alpha_2 & \ddots & \ddots & & \vdots \\ 0 & \ddots & \ddots & \ddots & & 0 \\ \vdots & \ddots & \ddots & \ddots & & \beta_n \\ \vdots & & \ddots & \gamma_n & & \alpha_n \\ 0 & \cdots & & \cdots & 0 & \gamma_{n+1} \end{bmatrix} \in \mathbb{C}^{(n+1) \times n}, \qquad (3.2)$$

and

$$H_n = [I_n \quad 0]\, H_n^{(e)}.$$

Then the recurrences in the Lanczos process can be written compactly as

$$AV_n = V_n H_n + [0 \quad \cdots \quad 0 \quad \tilde{\mathbf{v}}_{n+1}],$$
$$A^T W_n = W_n H_n^T + [0 \quad \cdots \quad 0 \quad \tilde{\mathbf{w}}_{n+1}], \qquad (3.3)$$

while the bi-orthogonality relation (3.1) can be written as

$$W_n^T V_n = I_n. \qquad (3.4)$$

We note that the Lanczos method is invariant under shifts of the form

$$A \mapsto A + \sigma I, \quad \text{where} \quad \sigma \in \mathbb{C},$$

in that the process generates the same vectors $\{\mathbf{v}_j\}$ and $\{\mathbf{w}_j\}$ and only the tridiagonal matrix (3.2) is shifted:

$$H_n \mapsto H_n + \sigma I_n.$$

In particular, for the Lanczos algorithm it is irrelevant whether the matrix A is singular or not.

Moreover, we would like to stress that the Lanczos process can also be formulated with A^H instead of A^T, by simply conjugating the three-term recurrences for the vectors $\{\mathbf{w}_j\}$. We chose the transpose because one can then avoid complex conjugated recurrence coefficients. Finally, we remark that the Lanczos process reduces to only one recursion in two important special cases, namely $A = A^H$ (with starting vectors $\tilde{\mathbf{w}}_1 = \overline{\tilde{\mathbf{v}}_1}$) and complex symmetric matrices $A = A^T$ (with starting vectors $\tilde{\mathbf{w}}_1 = \tilde{\mathbf{v}}_1$). In both cases, one must also choose $\beta_n = \gamma_n$. In the first case, the resulting algorithm is the well-known Hermitian Lanczos method, which has been studied extensively (see, e.g., Golub and Van Loan (1989) and the references therein). In the second case, the resulting algorithm is the complex symmetric Lanczos process.

In exact arithmetic, the classical Lanczos method terminates after a finite number of steps. As indicated in Algorithm 3.1, there are two different situations in which the process can stop. The first one, referred to as regular termination, occurs when $\mathbf{v}_{L+1} = 0$ or $\mathbf{w}_{L+1} = 0$. In this case, the Lanczos algorithm has found an invariant subspace of $\mathbb{C}^N$: if $\mathbf{v}_{L+1} = 0$, then the right Lanczos vectors $\mathbf{v}_1, \ldots, \mathbf{v}_L$ span an A-invariant subspace, while if $\mathbf{w}_{L+1} = 0$, then the left Lanczos vectors $\mathbf{w}_1, \ldots, \mathbf{w}_L$ span an A^T-invariant subspace. The second case, referred to as a serious breakdown by Wilkinson (1965), occurs when $\mathbf{w}_L^T \mathbf{v}_L = 0$ with neither $\mathbf{v}_L = 0$ nor $\mathbf{w}_L = 0$. In this case, the Lanczos vectors span neither an A-invariant subspace nor an A^T-invariant subspace of $\mathbb{C}^N$. We will discuss in Section 3.2 techniques for handling the serious breakdowns. We remark that, in the special case of the Hermitian Lanczos process, breakdowns are excluded. In contrast, breakdowns can occur in the complex symmetric Lanczos algorithm (see Cullum and Willoughby (1985) and Freund (1989b, 1992)).

The Lanczos algorithm was originally introduced to compute eigenvalues, as – in view of (3.3) – the eigenvalues of H_n can be used as approximations for eigenvalues of A. However, Lanczos (1952) also proposed a closely related method, the biconjugate gradient algorithm (BCG), for solving general nonsingular nonHermitian linear systems (1.1). By and large, BCG was ignored until the mid-1970s, when Fletcher (1976) revived the method.

The BCG algorithm is a Krylov subspace approach that generates iterates defined by a Galerkin condition (2.6) with the special choice of subspaces

$$\mathcal{S}_n = \{\mathbf{s} = \overline{\mathbf{w}} \mid \mathbf{w} \in \mathcal{K}_n(\tilde{\mathbf{r}}_0, A^T)\}.$$

Here, $\tilde{\mathbf{r}}_0$ is a nonzero starting vector discussed later. The standard implementation of the BCG method is as follows.

Algorithm 3.2   (BCG method)

0 Choose $\mathbf{x}_0 \in \mathbb{C}^N$ and set $\mathbf{q}_0 = \mathbf{r}_0 = \mathbf{b} - A\mathbf{x}_0$.

   Choose $\tilde{\mathbf{r}}_0 \in \mathbb{C}^N$, $\tilde{\mathbf{r}}_0 \neq 0$, and set $\tilde{\mathbf{q}}_0 = \tilde{\mathbf{r}}_0$, $\rho_0 = \tilde{\mathbf{r}}_0^T \mathbf{r}_0$.

For $n = 1, 2, \ldots$, do :

1 Compute

$$\sigma_{n-1} = \tilde{\mathbf{q}}_{n-1}^T A\mathbf{q}_{n-1},$$
$$\alpha_{n-1} = \rho_{n-1}/\sigma_{n-1},$$
$$\mathbf{x}_n = \mathbf{x}_{n-1} + \alpha_{n-1}\mathbf{q}_{n-1},$$
$$\mathbf{r}_n = \mathbf{r}_{n-1} - \alpha_{n-1}A\mathbf{q}_{n-1},$$
$$\tilde{\mathbf{r}}_n = \tilde{\mathbf{r}}_{n-1} - \alpha_{n-1}A^T\tilde{\mathbf{q}}_{n-1}.$$

2 Compute

$$\rho_n = \tilde{\mathbf{r}}_n^T \mathbf{r}_n,$$
$$\beta_n = \rho_n/\rho_{n-1},$$
$$\mathbf{q}_n = \mathbf{r}_n + \beta_n\mathbf{q}_{n-1},$$
$$\tilde{\mathbf{q}}_n = \tilde{\mathbf{r}}_n + \beta_n\tilde{\mathbf{q}}_{n-1}.$$

3 If $\mathbf{r}_n = 0$ or $\tilde{\mathbf{r}}_n = 0$, stop.

Note that BCG requires a second nonzero starting vector $\tilde{\mathbf{r}}_0 \in \mathbb{C}^N$, which can be chosen freely. Usually, one sets $\tilde{\mathbf{r}}_0 = \mathbf{r}_0$ or $\tilde{\mathbf{r}}_0 = \overline{\mathbf{r}_0}$, or one chooses $\tilde{\mathbf{r}}_0$ as a vector with random entries.

The BCG algorithm is the archetype of an entire class of Lanczos-based Krylov subspace methods for nonHermitian matrices, some of which we will discuss later. Unfortunately, BCG typically exhibits an erratic convergence behaviour with wild oscillations in the residual norm $\|\mathbf{r}_n\|_2$. Even worse, the BCG process can break down completely. More precisely, the BCG Algorithm 3.2 cannot be continued if

$$\tilde{\mathbf{q}}_{n-1}^T A\mathbf{q}_{n-1} = 0, \quad \tilde{\mathbf{r}}_{n-1} \neq 0, \quad \mathbf{r}_{n-1} \neq 0, \tag{3.5}$$

or if

$$\tilde{\mathbf{r}}_{n-1}^T \mathbf{r}_{n-1} = 0, \quad \tilde{\mathbf{r}}_{n-1} \neq 0, \quad \mathbf{r}_{n-1} \neq 0. \tag{3.6}$$

The source of the breakdown (3.5) is the Galerkin condition (2.6) used to define the iterates. As was pointed out in Section 2.2, the existence of an iterate satisfying (2.6) is not guaranteed at every step, and in fact (3.5) occurs if, and only if, no BCG iterate exists. Furthermore, it can be shown that (3.5) is equivalent to the Lanczos matrix H_n being singular. The source of the second breakdown (3.6) is the underlying nonsymmetric Lanczos process, which can have a serious breakdown. It turns out that the vectors r_{n-1} and $\tilde{r}_{n-1}$ in the BCG Algorithm 3.2 are scalar multiples of the vectors v_n and w_n, respectively, that are generated by the classical Lanczos Algorithm 3.1 started with

$$\tilde{v}_1 = r_0 \quad \text{and} \quad \tilde{w}_1 = \tilde{r}_0.$$

Hence, a breakdown in the Lanczos process will be parallelled by a breakdown (3.6) in the BCG algorithm.

As the earlier discussion shows, BCG, while requiring little work and storage per step, is susceptible to breakdowns and numerical instabilities. In addition, another possible disadvantage of the classical BCG algorithm is its use of the transpose of A, which may not be readily available in some situations. As a result, variants of BCG were sought which would preserve the low work and storage requirements, while curing the possible breakdowns and avoiding the use of the transpose. In the next section, we will discuss the look-ahead Lanczos algorithm, an extension of the Lanczos method which handles in almost all cases the serious breakdowns in the Lanczos process. In Section 3.4 we present the quasi-minimal residual approach, based on the look-ahead Lanczos algorithm and using a quasi-minimization property to avoid the breakdowns caused by the Galerkin condition. Finally, in Section 3.5 we survey some of the so-called transpose-free algorithms, which typically replace the multiplication by A^T in the BCG algorithm by one or more multiplications by A.

3.2. A look-ahead Lanczos algorithm

One of the possible terminations of the Lanczos algorithm is a serious breakdown, when $\delta_n = 0$ in Algorithm 3.1, with neither $\tilde{v}_n = 0$ nor $\tilde{w}_n = 0$. As a result, the vectors $\tilde{v}_n$ and $\tilde{w}_n$ cannot be scaled to obtain the Lanczos vectors v_n and w_n corresponding to the basis vectors $A^n v_1$ and $(A^T)^n w_1$. Furthermore, it turns out that even if v_n and w_n were computed using a different scaling method, the next pair of vectors $\tilde{v}_{n+1}$ and $\tilde{w}_{n+1}$ could not be computed so as to fulfil (3.1). The problem here is not just one of scaling, but also that the bi-orthogonality required of the vectors $\tilde{v}_{n+1}$ and $\tilde{w}_{n+1}$ cannot be satisfied. However, it could happen that the bi-orthogonality condition (3.1) can once again be fulfilled for a pair of vectors corresponding to some higher power of A and A^T. A procedure which somehow advances to this next pair of Lanczos vectors will be called a look-ahead Lanczos procedure.

The main idea behind the look-ahead Lanczos algorithms is to relax the bi-orthogonality relation (3.1) when a breakdown is encountered. For each fixed $n = 1, 2, \ldots$, the vectors $\mathbf{v}_1, \ldots, \mathbf{v}_n$ and $\mathbf{w}_1, \ldots, \mathbf{w}_n$ generated by the look-ahead procedure can be grouped into $l = l(n)$ blocks

$$V^{(k)} = [\,\mathbf{v}_{n_k} \quad \mathbf{v}_{n_k+1} \quad \cdots \quad \mathbf{v}_{n_{k+1}-1}\,],$$
$$W^{(k)} = [\,\mathbf{w}_{n_k} \quad \mathbf{w}_{n_k+1} \quad \cdots \quad \mathbf{w}_{n_{k+1}-1}\,], \qquad k = 1, \ldots, l-1,$$

and

$$V^{(l)} = [\,\mathbf{v}_{n_l} \quad \mathbf{v}_{n_l+1} \quad \cdots \quad \mathbf{v}_n\,],$$
$$W^{(l)} = [\,\mathbf{w}_{n_l} \quad \mathbf{w}_{n_l+1} \quad \cdots \quad \mathbf{w}_n\,],$$

where

$$1 = n_1 < n_2 < \cdots < n_k < \cdots < n_l \leq n < n_{l+1}.$$

The blocks are constructed so that (3.1) is relaxed to

$$\left(W^{(j)}\right)^T \left(V^{(k)}\right) = \begin{cases} D^{(k)} & \text{if } j = k, \\ 0 & \text{if } j \neq k, \end{cases} \qquad j, k = 1, \ldots, l,$$

where $D^{(k)}$ is nonsingular for $k = 1, \ldots, l-1$, and $D^{(l)}$ is nonsingular if $n = n_{l+1} - 1$. The first vectors $\mathbf{v}_{n_k}$ and $\mathbf{w}_{n_k}$ in each block are called regular, and the remaining vectors are called inner. We note that, in practice, for reasons of stability, the computed vectors are usually scaled to have unit length (see Taylor, 1982).

Two such look-ahead procedures have been proposed, one by Parlett *et al.* (1985), and a second one by Freund *et al.* (1991b). The Freund *et al.* implementation requires the same number of inner products per step as the classical Lanczos algorithm, and reduces to the classical Lanczos procedure in the absence of look-ahead steps. In contrast, the Parlett *et al.* implementation always requires more work per step and does not reduce to the classical Lanczos algorithm in the absence of look-ahead steps. It also does not generalize easily to blocks of more than two vectors. Therefore, we will focus on the implementation proposed by Freund *et al.*. The basic structure of this look-ahead Lanczos algorithm is as follows.

Algorithm 3.3   (Sketch of the look-ahead Lanczos process)

   0 Choose $\mathbf{v}_1, \mathbf{w}_1 \in \mathbb{C}^N$ with $\|\mathbf{v}_1\|_2 = \|\mathbf{w}_1\|_2 = 1$.

   Set $V^{(1)} = \mathbf{v}_1$, $W^{(1)} = \mathbf{w}_1$, $D^{(1)} = (W^{(1)})^T V^{(1)}$.

   Set $n_1 = 1$, $l = 1$, $\mathbf{v}_0 = \mathbf{w}_0 = 0$, $V_0 = W_0 = \emptyset$, $\rho_1 = \xi_1 = 1$.

   For $n = 1, 2, \ldots$, do :

   1 Decide whether to construct $\mathbf{v}_{n+1}$ and $\mathbf{w}_{n+1}$ as regular or inner vectors and go to Steps 2 or 3, respectively.

2 (Regular step.) Compute

$$
\begin{aligned}
\tilde{\mathbf{v}}_{n+1} &= A\mathbf{v}_n - V^{(l)}(D^{(l)})^{-1}(W^{(l)})^T A\mathbf{v}_n \\
&\quad - V^{(l-1)}(D^{(l-1)})^{-1}(W^{(l-1)})^T A\mathbf{v}_n, \\
\tilde{\mathbf{w}}_{n+1} &= A^T\mathbf{w}_n - W^{(l)}(D^{(l)})^{-T}(V^{(l)})^T A^T\mathbf{w}_n \\
&\quad - W^{(l-1)}(D^{(l-1)})^{-T}(V^{(l-1)})^T A^T\mathbf{w}_n.
\end{aligned}
\tag{3.7}
$$

Set $n_{l+1} = n + 1$, $l = l + 1$, $V^{(l)} = W^{(l)} = \emptyset$, and go to Step 4.

3 (Inner step.) Compute

$$
\begin{aligned}
\tilde{\mathbf{v}}_{n+1} &= A\mathbf{v}_n - \zeta_n\mathbf{v}_n - (\eta_n/\rho_n)\,\mathbf{v}_{n-1} \\
&\quad - V^{(l-1)}(D^{(l-1)})^{-1}(W^{(l-1)})^T A\mathbf{v}_n, \\
\tilde{\mathbf{w}}_{n+1} &= A^T\mathbf{w}_n - \zeta_n\mathbf{w}_n - (\eta_n/\xi_n)\,\mathbf{w}_{n-1} \\
&\quad - W^{(l-1)}(D^{(l-1)})^{-T}(V^{(l-1)})^T A^T\mathbf{w}_n.
\end{aligned}
\tag{3.8}
$$

4 Compute $\rho_{n+1} = \|\tilde{\mathbf{v}}_{n+1}\|_2$ and $\xi_{n+1} = \|\tilde{\mathbf{w}}_{n+1}\|_2$.

If $\rho_{n+1} = 0$ or $\xi_{n+1} = 0$, set $L = n$, and stop.

Otherwise, set

$$
\mathbf{v}_{n+1} = \tilde{\mathbf{v}}_{n+1}/\rho_{n+1}, \quad \mathbf{w}_{n+1} = \tilde{\mathbf{w}}_{n+1}/\xi_{n+1},
$$

$$
V^{(l)} = \left[\begin{array}{cc} V^{(l)} & \mathbf{v}_{n+1} \end{array} \right], \quad W^{(l)} = \left[\begin{array}{cc} W^{(l)} & \mathbf{w}_{n+1} \end{array} \right],
$$

$$
D^{(l)} = (W^{(l)})^T V^{(l)}.
$$

Step 2 of the algorithm is a block version of the classical Lanczos recurrences of Step 3 in Algorithm 3.1. Step 3 builds inner vectors spanning the gap between two regular vectors. In this implementation, the inner vectors are generated by a three-term inner recurrence, and then bi-orthogonalized against the last block. The vectors generated by the look-ahead Lanczos algorithm still obey (3.3), but now H_n, instead of having simply the tridiagonal structure (3.2), is an upper Hessenberg block tridiagonal matrix with small blocks of size $(n_k - n_{k-1}) \times (n_k - n_{k-1})$ on the diagonal. Furthermore, the bi-orthogonality relation (3.4) now reads

$$
W_n^T V_n = D_n = \mathrm{diag}\left(D^{(1)}, D^{(2)}, \ldots, D^{(l)} \right).
$$

Note that D_n is guaranteed to be nonsingular if $n = n_{l+1} - 1$.

If only regular steps are performed, then Algorithm 3.3 reduces to the classical Lanczos process. Thus, the strategy used in Step 1 for deciding when to construct inner vectors should perform regular steps whenever possible. In addition, in practice the look-ahead algorithm must also be able to handle near-breakdowns, that is situations when

$$\tilde{\mathbf{w}}_n^T \tilde{\mathbf{v}}_n \approx 0, \quad \tilde{\mathbf{v}}_n \not\approx 0, \quad \tilde{\mathbf{w}}_n \not\approx 0.$$

Freund *et al.* (1991b) proposed a practical procedure for the decision in Step 1 based on three different checks. For a regular step, it is necessary that $D^{(l)}$ be nonsingular. Therefore, one of the checks monitors the size of $\sigma_{\min}(D^{(l)})$. The other two checks attempt to ensure the linear independence of the Lanczos vectors. The algorithm monitors the size of the components along the two previous blocks $V^{(l)}$ and $V^{(l-1)}$, respectively $W^{(l)}$ and $W^{(l-1)}$, in (3.7), and performs a regular step only if these terms do not dominate the components $A\mathbf{v}_n$ and $A^T\mathbf{w}_n$ in the new Krylov spaces. For details, see Freund *et al.* (1991b).

The look-ahead algorithm outlined here will handle serious breakdowns and near-breakdowns in the classical Lanczos algorithm, except for the special event of an incurable breakdown (Taylor, 1982). These are situations where the look-ahead procedure would build an infinite block, without ever finding a nonsingular $D^{(l)}$. Taylor (1982) has shown in his Mismatch Theorem that, in case of an incurable breakdown, one can still recover eigenvalue information, as the eigenvalues of the H_{n_l} are also eigenvalues of A. For linear systems, an incurable breakdown would require restarting the procedure with a different choice of starting vectors. Fortunately, in practice round-off errors will make an incurable breakdown highly unlikely.

Finally, we remark that, for the important class of p-cyclic matrices A, serious breakdowns in the Lanczos process occur in a regular pattern. In this case, look-ahead steps are absolutely necessary if one wants to exploit the p-cyclic structure. For details of a look-ahead Lanczos algorithm for p-cyclic matrices, we refer the reader to Freund *et al.* (1991a).

3.3. The QMR algorithm

We now turn to the quasi-minimal residual approach. The procedure was first proposed by Freund (1989b) for the case of complex symmetric linear systems, and then extended by Freund and Nachtigal (1991) for the case of general nonHermitian matrices.

Recall from (2.2) that the nth iterate of any Krylov subspace method is of the form

$$\mathbf{x}_n \in \mathbf{x}_0 + \mathcal{K}_n(\mathbf{r}_0, A).$$

If now we choose

$$\mathbf{v}_1 = \mathbf{r}_0 / \|\mathbf{r}_0\|_2 \tag{3.9}$$

in Algorithm 3.3, then the right Lanczos vectors $\mathbf{v}_1, \dots, \mathbf{v}_n$ span the Krylov space $\mathcal{K}_n(\mathbf{r}_0, A)$, hence we can write

$$\mathbf{x}_n = \mathbf{x}_0 + V_n \mathbf{z}_n,$$

for some $\mathbf{z}_n \in \mathbb{C}^n$. Together with (3.9) and the first relationship in (3.3), this gives for the residual

$$\mathbf{r}_n = \mathbf{r}_0 - AV_n\mathbf{z}_n = V_{n+1}\left(\mathbf{d}_n - H_n^{(e)}\mathbf{z}_n\right), \qquad (3.10)$$

where $\mathbf{d}_n$ is defined as in (2.22). As V_{n+1} is not unitary, it is not possible to minimize the Euclidean norm of the residual without expending $\mathcal{O}(Nn^2)$ work and $\mathcal{O}(Nn)$ storage. Instead, one minimizes just the Euclidean norm of the coefficient vector in (3.10), that is $\mathbf{z}_n \in \mathbb{C}^n$ is chosen as the solution of the least-squares problem

$$\|\mathbf{d}_n - H_n^{(e)}\mathbf{z}_n\|_2 = \min_{\mathbf{z}\in\mathbb{C}^n} \|\mathbf{d}_n - H_n^{(e)}\mathbf{z}\|_2. \qquad (3.11)$$

As was pointed out by Manteuffel (1991), solving the minimization problem (3.11) is equivalent to minimizing the residual in a norm that changes with the step number:

$$\min_{\mathbf{x}\in\mathbf{x}_0+\mathcal{K}_n(\mathbf{r}_0,A)} \|D_{n+1}^{-1}W_{n+1}^T(\mathbf{b} - A\mathbf{x})\|_2, \quad n = n_k - 2.$$

Thus, the QMR does not contradict the Faber and Manteuffel Theorem 2.1, which excludes only methods that minimize in a *fixed* norm.

To solve the least-squares problem (3.11), one uses a QR factorization of $H_n^{(e)}$. As $H_n^{(e)}$ is upper Hessenberg, its QR factorization can be easily computed and updated using Givens rotations; the approach is a standard one (see, e.g., Golub and Van Loan (1989)). One computes a unitary matrix $Q_n \in \mathbb{C}^{(n+1)\times(n+1)}$ and an upper triangular matrix $R_n \in \mathbb{C}^n$ such that

$$Q_n H_n^{(e)} = \begin{bmatrix} R_n \\ 0 \end{bmatrix}, \qquad (3.12)$$

and then obtains $\mathbf{z}_n$ from

$$\mathbf{z}_n = R_n^{-1}\mathbf{t}_n, \quad \mathbf{t}_n = [\,I_n \;\; 0\,]\,Q_n\mathbf{d}_n, \qquad (3.13)$$

which gives

$$\mathbf{x}_n = \mathbf{x}_0 + V_n R_n^{-1}\mathbf{t}_n. \qquad (3.14)$$

This gives the following QMR algorithm.

Algorithm 3.4 (QMR algorithm)

0 Choose $\mathbf{x}_0 \in \mathbb{C}^N$ and set $\mathbf{r}_0 = \mathbf{b} - A\mathbf{x}_0$, $\rho_0 = \|\mathbf{r}_0\|_2$, $\mathbf{v}_1 = \mathbf{r}_0/\rho_0$.

   Choose $\mathbf{w}_1 \in \mathbb{C}^N$ with $\|\mathbf{w}_1\|_2 = 1$.

 For $n = 1, 2, \ldots$, do :

 1 Perform the nth iteration of the look-ahead Lanczos Algorithm 3.3.

This yields matrices V_n, V_{n+1}, $H_n^{(e)}$ which satisfy

$$AV_n = V_{n+1}H_n^{(e)}.$$

2 Update the QR factorization (3.12) of $H_n^{(e)}$ and the vector $\mathbf{t}_n$ in (3.13).

3 Compute $\mathbf{x}_n$ from (3.14).

4 If $\mathbf{x}_n$ has converged, stop.

We note that since Q_n is a product of Givens rotations, the vector $\mathbf{t}_n$ is easily updated in Step 2. Also, as H_n is block tridiagonal, R_n also has a block structure that is used in Step 3 to update $\mathbf{x}_n$ using only short recurrences. For complete details, see Freund and Nachtigal (1991).

The point of the quasi-minimal residual approach is that the least-squares problem (3.11) always has a unique solution. From Step 4 of Algorithm 3.3, ρ_k, the subdiagonal entries of $H_n^{(e)}$, are all nonzero, hence $H_n^{(e)}$ has full column rank, R_n is nonsingular, and so (3.11) always defines a unique iterate $\mathbf{x}_n$. This then avoids the Galerkin breakdown in the BCG algorithm. But more importantly, the quasi-minimization (3.11) is strong enough to enable us to prove a convergence theorem for QMR. This is in contrast to BCG and methods derived from BCG, for which no convergence results are known. Indeed, one can prove two theorems for QMR, both relating the QMR convergence behaviour to the convergence of GMRES.

Theorem 3.5 (Freund and Nachtigal, 1991)
Suppose that the $L \times L$ matrix H_L generated by the look-ahead Lanczos algorithm is diagonalizable, and let $X \in \mathbb{C}^{L \times L}$ be a matrix of eigenvectors of H_L. Then for $n = 1, \ldots, L - 1$,

$$\frac{\|\mathbf{r}_n^{QMR}\|_2}{\|\mathbf{r}_0\|_2} \le \|V_{n+1}\|_2 \, \kappa_2(X) \, \epsilon_n,$$

where

$$\epsilon_n = \max_{\psi \in \mathcal{P}_n : \psi(0)=1} \min_{\lambda \in \lambda(A)} |\psi(\lambda)|.$$

By comparison, the convergence result for GMRES reads as follows.

Theorem 3.6 (Saad and Schultz, 1986)
Suppose that A is diagonalizable, and let U be a matrix of eigenvectors of A. Then, for $n = 1, 2, \ldots,$

$$\frac{\|\mathbf{r}_n^{GMRES}\|_2}{\|\mathbf{r}_0\|_2} \le \kappa_2(U) \, \epsilon_n,$$

where ϵ_n is as described previously.

Thus, Theorem 3.5 shows that GMRES and QMR solve the same approxima-tion problem. The second convergence result gives a quantitative description of the departure from optimality due to the quasi-optimal approach.

Theorem 3.7 (Nachtigal, 1991)

$$\|\mathbf{r}_n^{\mathrm{QMR}}\|_2 \le \kappa_2(V_{n+1})\,\|\mathbf{r}_n^{\mathrm{GMRES}}\|_2.$$

For both Theorems 3.5 and 3.7, we note that the right Lanczos vectors $\{\mathbf{v}_j\}$ obtained from Algorithm 3.4 are unit vectors, and hence the condition number of V_{n+1} can be bounded by a slowly growing function,

$$\|V_{n+1}\|_2 \le \sqrt{n+1}.$$

Next, we note that it is possible to recover BCG iterates, when they exist, from the corresponding QMR iterates. We have

Theorem 3.8 (Freund and Nachtigal, 1991)
Let $n = n_k - 1$, $k = 1, \dots$. Then,

$$\mathbf{x}_n^{\mathrm{BCG}} = \mathbf{x}_{n-1}^{\mathrm{QMR}} + \frac{\tau_n}{c_n}\mathbf{p}_n,$$

$$\|\mathbf{r}_n^{\mathrm{BCG}}\|_2 = \|\mathbf{r}_0\|_2 |s_1 \cdots s_n| \frac{1}{c_n}.$$

Here, $\mathbf{p}_n$ is the nth column of $V_n R_n^{-1}$ and is computed anyway as part of Step 3 of the QMR Algorithm 3.4, s_n and c_n are the sine and cosine of the nth Givens rotation involved in the QR decomposition of $H_n^{(e)}$, and τ_n is the $(n+1)$st component of $Q_n \mathbf{d}_n$. The point is that the BCG iterate and the norm of the BCG residual are both by-products of the QMR algorithm, available at little extra cost, and the existence of the BCG iterate can be checked by monitoring the size of the Givens rotation cosine c_n. Thus, QMR can also be viewed as a stable implementation of BCG.

Finally, we remark that, for the special case of Hermitian matrices, the QMR Algorithm 3.4 (with $\mathbf{w}_1 = \overline{\mathbf{v}_1}$) is mathematically equivalent to MIN-RES. Hence, QMR can also be viewed as an extension of MINRES to non-Hermitian matrices.

3.4. Transpose-free methods

In contrast to Krylov subspace methods based on the Arnoldi Algorithm 2.2, which only require matrix–vector multiplications with A, algorithms such as BCG and QMR, which are based directly on the Lanczos process, involve matrix–vector products with A and A^T. This is a disadvantage for certain applications, where A^T is not readily available. It is possible to devise Lanczos-based Krylov subspace methods that do not involve the transpose of A. In this section, we give an overview of such transpose-free schemes.

First, we consider the QMR algorithm. As pointed out by Freund and

Zha (1991), in principle it is always possible to eliminate A^T altogether, by choosing the starting vector $\mathbf{w}_1$ suitably. This observation is based on the fact that any square matrix is similar to its transpose. In particular, there always exists a nonsingular matrix P such that

$$A^T P = PA. \qquad (3.15)$$

Now suppose that in the QMR Algorithm 3.4 we choose the special starting vector $\mathbf{w}_1 = P\mathbf{v}_1/\|P\mathbf{v}_1\|_2$. Then, with (3.15), one readily verifies that the vectors generated by look-ahead Lanczos Algorithm 3.3 satisfy

$$\mathbf{w}_n = P\mathbf{v}_n/\|P\mathbf{v}_n\|_2 \quad \text{for all } n. \qquad (3.16)$$

Hence, instead of updating the left Lanczos vectors $\{w_n\}$ by means of the recursions in (3.7) or (3.8), they can be computed directly from (3.16). The resulting QMR algorithm no longer involves the transpose of A; in exchange, it requires one matrix–vector multiplication with P in each iteration step. Therefore, this approach is only viable for special classes of matrices A, for which one can find a matrix P satisfying (3.15) easily, and for which, at the same time, matrix–vector products with P can be computed cheaply. The most trivial case are complex symmetric matrices (see Section 6), which fulfil (3.15) with $P = I$. Another simple case are matrices A that are symmetric with respect to the antidiagonal. These so-called centrosymmetric matrices, by their very definition, satisfy (3.15) with $P = J$, where

$$J = \begin{bmatrix} 0 & \cdots & 0 & 1 \\ \vdots & & 1 & 0 \\ 0 & & & \vdots \\ 1 & 0 & \cdots & 0 \end{bmatrix}$$

is the $N \times N$ antidiagonal identity matrix. Note that Toeplitz matrices (see Section 7) are a special case of centrosymmetric matrices. Finally, the condition (3.15) is also fulfilled for matrices of the form

$$A = TM^{-1}, \quad P = M^{-1},$$

where T and M are real symmetric matrices and M is nonsingular. Matrices A of this type arise when real symmetric linear systems $Tx = b$ are preconditioned by M. The resulting QMR algorithm for the solution of preconditioned symmetric linear system has the same work and storage requirements as preconditioned SYMMLQ or MINRES. However, the QMR approach is more general, in that it can be combined with any nonsingular symmetric preconditioner M, while SYMMLQ and MINRES require M to be positive definite M (see, e.g., Gill *et al.*, 1990). For strongly indefinite matrices T, the use of indefinite preconditioners M typically leads to considerably faster convergence; see Freund and Zha (1991) for numerical examples.

Next, we turn to transpose-free variants of the BCG method. Sonneveld (1989) with his conjugate gradients squared algorithm (CGS) was the first to devise a transpose-free BCG-type scheme. Note that, in the BCG Algorithm 3.2, the matrix A^T appears merely in the update formulae for the vectors $\tilde{\mathbf{r}}_n$ and $\tilde{\mathbf{q}}_n$. On the other hand, these vectors are then used only for the computation of the vector products $\rho_n = \tilde{\mathbf{r}}_n^T \mathbf{r}_n$ and $\sigma_n = \tilde{\mathbf{q}}_n^T A \mathbf{q}_n$. Sonneveld observed that, by rewriting these products, the transpose can be eliminated from the formulae, while at the same time one obtains iterates

$$\mathbf{x}_{2n} \in \mathbf{x}_0 + \mathcal{K}_{2n}(\mathbf{r}_0, A), \quad n = 1, 2, \ldots, \tag{3.17}$$

that are contained in a Krylov subspace of twice the dimension, as compared to BCG. First, we consider ρ_n. From Algorithm 3.2 it is obvious that

$$\mathbf{r}_n = \psi_n(A)\mathbf{r}_0 \quad \text{and} \quad \tilde{\mathbf{r}}_n = \psi_n(A^T)\tilde{\mathbf{r}}_0, \tag{3.18}$$

where ψ_n is the nth residual polynomials (recall (2.3) and (2.4)) of the BCG process. With (3.18), one obtains the identity

$$\rho_n = \tilde{\mathbf{r}}_0^T \left(\psi_n(A)\right)^2 \mathbf{r}_0, \tag{3.19}$$

which shows that ρ_n can be computed without using A^T. Similarly,

$$\mathbf{q}_n = \varphi_n(A)\mathbf{r}_0 \quad \text{and} \quad \tilde{\mathbf{q}}_n = \varphi_n(A^T)\tilde{\mathbf{r}}_0,$$

for some polynomial $\varphi_n \in \mathcal{P}_n$, and hence

$$\sigma_n = \tilde{\mathbf{r}}_0^T A \left(\varphi_n(A)\right)^2 \mathbf{r}_0. \tag{3.20}$$

By rewriting the vector recursions in Algorithm 3.2 in terms of ψ_n and φ_n and by squaring the resulting polynomial relations, Sonneveld showed that the vectors in (3.19) and (3.20) can be updated by means of short recursions. Furthermore, the actual iterates (3.17) generated by CGS are characterized by

$$\mathbf{r}_{2n}^{\mathrm{CGS}} = \mathbf{b} - A\mathbf{x}_{2n} = \left(\psi_n^{\mathrm{BCG}}(A)\right)^2 \mathbf{r}_0. \tag{3.21}$$

Hence the CGS residual polynomials $\psi_{2n}^{\mathrm{CGS}} = \left(\psi_n^{\mathrm{BCG}}\right)^2$ are just the squared BCG polynomials. As pointed out earlier, BCG typically exhibits erratic convergence behaviour. As is clear from (3.21), these effects are magnified in CGS, and CGS typically accelerates convergence as well as divergence of BCG. Moreover, there are cases for which CGS diverges, while BCG still converges.

For this reason, more smoothly converging variants of CGS have been sought. Van der Vorst (1990) was the first to propose such a method. His Bi-CGSTAB again produces iterates of the form (3.17), but instead of squaring the BCG polynomials as in (3.21), the residual vector is now of the form

$$\mathbf{r}_{2n} = \psi_n^{\mathrm{BCG}}(A)\chi_n(A)\mathbf{r}_0.$$

Here $\chi_n \in \mathcal{P}_n$, with $\chi_n(0) = 1$, is a polynomial that is updated from step to step by adding a new linear factor:

$$\chi_n(\lambda) \equiv (1 - \eta_n\lambda)\chi_{n-1}(\lambda). \qquad (3.22)$$

The free parameter η_n in (3.22) is determined by a local steepest descent step, i.e. η_n is the optimal solution of

$$\min_{\eta \in \mathbb{C}} \|(I - \eta A)\chi_{n-1}(A)\psi_n^{\text{BCG}}(A)\mathbf{r}_0\|_2.$$

Due to the steepest descent steps, Bi-CGSTAB typically has much smoother convergence behaviour than BCG or CGS. However, the norms of the Bi-CGSTAB residuals may still oscillate considerably for difficult problems. Finally, Gutknecht (1991) has noted that, for real A, the polynomials χ_n will always have real roots only, even if A has complex eigenvalues. He proposed a variant of Bi-CGSTAB with polynomials (3.22) that are updated by quadratic factors in each step and thus can have complex roots in general.

In the CGS algorithm, the iterates (3.17) are updated by means of a formula of the form

$$\mathbf{x}_{2n}^{\text{CGS}} = \mathbf{x}_{2(n-1)}^{\text{CGS}} + \alpha_{n-1}(\mathbf{y}_{2n-1} + \mathbf{y}_{2n}). \qquad (3.23)$$

Here the vectors $\mathbf{y}_1, \mathbf{y}_2, \ldots, \mathbf{y}_{2n}$ satisfy

$$\text{span}\{\mathbf{y}_1, \mathbf{y}_2, \ldots, \mathbf{y}_m\} = \mathcal{K}_m(\mathbf{r}_0, A), \quad m = 1, 2, \ldots, 2n.$$

In other words, in each iteration of the CGS algorithm two search directions $\mathbf{y}_{2n-1}$ and $\mathbf{y}_{2n}$ are available, while the actual iterate is updated by the one-dimensional step (3.23) only. Based on this observation, Freund (1991b) has proposed a variant of CGS that makes use of all available search directions. More precisely, instead of one iterate $\mathbf{x}_{2n}^{\text{CGS}}$ per step it produces two iterates $\mathbf{x}_{2n-1}$ and $\mathbf{x}_{2n}$ of the form

$$\mathbf{x}_m = \mathbf{x}_0 + [\mathbf{y}_1 \quad \mathbf{y}_2 \quad \cdots \quad \mathbf{y}_m]\mathbf{z}_m, \quad \mathbf{z}_m \in \mathbb{C}^m. \qquad (3.24)$$

Furthermore, the free parameter vector $\mathbf{z}_m$ in (3.24) can be chosen such that the iterates satisfy a quasi-minimal residual condition, similar to the quasi-minimization property of the QMR Algorithm 3.4. For this reason, the resulting scheme is called transpose-free quasi-minimal residual algorithm (TFQMR). For details, we refer the reader to Freund (1991b), where the following implementation of TFQMR is derived.

Algorithm 3.9 (TFQMR algorithm)

0 Choose $\mathbf{x}_0 \in \mathbb{C}^N$.

Set $\mathbf{w}_1 = \mathbf{y}_1 = \mathbf{r}_0 = \mathbf{b} - A\mathbf{x}_0$, $\mathbf{v}_0 = A\mathbf{y}_1$, $\mathbf{d}_0 = 0$.

Set $\tau_0 = \|\mathbf{r}_0\|_2$, $\vartheta_0 = 0$, $\eta_0 = 0$.

Choose $\tilde{\mathbf{r}}_0 \in \mathbb{C}^N$, $\tilde{\mathbf{r}}_0 \neq 0$, and set $\rho_0 = \tilde{\mathbf{r}}_0^T\mathbf{r}_0$.

For $n = 1, 2, \ldots$, do :

1 Compute

$$\sigma_{n-1} = \tilde{\mathbf{r}}_0^T \mathbf{v}_{n-1},$$
$$\alpha_{n-1} = \rho_{n-1}/\sigma_{n-1},$$
$$\mathbf{y}_{2n} = \mathbf{y}_{2n-1} - \alpha_{n-1}\mathbf{v}_{n-1}.$$

2 For $m = 2n - 1, 2n$ do :

Compute

$$\mathbf{w}_{m+1} = \mathbf{w}_m - \alpha_{n-1}A\mathbf{y}_m,$$
$$\vartheta_m = \|\mathbf{w}_{m+1}\|_2/\tau_{m-1}, \quad c_m = 1/\sqrt{1 + \vartheta_m^2},$$
$$\tau_m = \tau_{m-1}\vartheta_m c_m, \quad \eta_m = c_m^2 \alpha_{n-1},$$
$$\mathbf{d}_m = \mathbf{y}_m + (\vartheta_{m-1}^2 \eta_{m-1}/\alpha_{n-1})\mathbf{d}_{m-1},$$
$$\mathbf{x}_m = \mathbf{x}_{m-1} + \eta_m \mathbf{d}_m.$$

If $\mathbf{x}_m$ has converged, stop.

3 Compute

$$\rho_n = \tilde{\mathbf{r}}_0^T \mathbf{w}_{2n+1},$$
$$\beta_n = \rho_n/\rho_{n-1},$$
$$\mathbf{y}_{2n+1} = \mathbf{w}_{2n+1} + \beta_n \mathbf{y}_{2n},$$
$$\mathbf{v}_n = A\mathbf{y}_{2n+1} + \beta_n(A\mathbf{y}_{2n} + \beta_n \mathbf{v}_{n-1}).$$

We would like to point out that the iterates generated by the QMR Algorithm 3.4 and the TFQMR Algorithm 3.9 are different in general.

Another transpose-free QMR method was proposed by Chan *et al.* (1991). Their scheme is mathematically equivalent to the QMR Algorithm 3.4, where the latter is based on the classical Lanczos process without look-ahead. The method first uses a transpose-free squared version of the Lanczos algorithm (see Gutknecht, 1990a) to generate the tridiagonal matrix (3.2). The right Lanczos vectors $\mathbf{v}_n$ are then computed by running the corresponding recursion in Step 3 of Algorithm 3.1, and finally the QMR iterates are obtained as in Algorithm 3.4.

Freund and Szeto (1991) have derived yet another transpose-free QMR scheme, which is modelled after CGS and is based on squaring the residual polynomials of the standard QMR Algorithm 3.4.

However, the algorithm by Chan *et al.* and the squared QMR approach both require per iteration three matrix–vector products with A and hence they are more expensive than CGS, Bi-CGSTAB or TFQMR, which involve only two such products per step.

Finally, we remark that none of the transpose-free methods considered in this section, except for Freund and Zha's simplified QMR algorithm based on (3.15), addresses the problem of breakdowns. Indeed, in exact arithmetic, all these schemes break down every time a breakdown (3.5) or (3.6) occurs in the BCG Algorithm 3.2. Practical look-ahead techniques for avoiding exact and near-breakdowns in these transpose-free methods still have to be developed.

3.5 Related work and historical remarks

The problem of breakdowns in the classical Lanczos algorithm has been known from the beginning. Although a rare event in practice, the possibility of breakdowns certainly has brought the method into discredit and has prevented many people from actually using the algorithm. On the other hand, it was also demonstrated (see Cullum and Willoughby (1986)) that the Lanczos process – even without look-ahead – is a powerful tool for sparse matrix computation.

The Lanczos method has intimate connections with many other areas of Mathematics, such as formally orthogonal polynomials (FOPs), Padé approximation, Hankel matrices and control theory. The problem of breakdowns has a corresponding formulation in all of these areas, and remedies for breakdowns in these different settings have been known for quite some time. For example, the breakdown in the Lanczos process is equivalent to a breakdown of the generic three-term recurrence relationship for FOPs, and it is well known how to overcome such breakdowns by modifying the recursions for FOPs (see Gragg (1974), Draux (1983), Gutknecht (1990b) and the references given there). Kung (1977) and Gragg and Lindquist (1983) presented remedies for breakdowns in the context of the partial realization problem in control theory. The Lanczos process is also closely related to fast algorithms for the factorization of Hankel matrices, and again it is well known how to overcome possible breakdowns of these algorithms (see Heinig and Rost (1984)). However, in all these cases, only the problem of exact breakdowns has been addressed. Taylor (1982) and Parlett *et al.* (1985) were the first to propose a modification of the classical Lanczos process that remedies both exact and near-breakdowns.

In recent years, there has been a revival of the nonsymmetric Lanczos algorithm, and since 1990, in addition to the papers we have already cited in this section, there are several others dealing with various aspects of the Lanczos process. We refer the reader to the papers by Boley *et al.* (1991), Boley and Golub (1991), Brezinski *et al.* (1991), Gutknecht (1990c), Joubert (1990), Parlett (1990), and the references given therein.

Note that Algorithm 3.2 is only one of several possible implementations of the BCG approach; see Joubert (1990) and Gutknecht (1990a) for an overview of the different BCG variants. As for the nonsymmetric Lanczos process, exact and near-breakdowns in the BCG methods can be avoided by incorporating look-ahead procedures. Such look-ahead BCG algorithms have been proposed by Joubert (1990) and Gutknecht (1990c). Particularly attractive in this context is the algorithm called Lanczos/Orthodir in Joubert (1990). Instead of generating the search directions $\mathbf{q}_n$ and $\tilde{\mathbf{q}}_n$ by coupled two-term recursions as in Algorithm 3.2, in Lanczos/Orthodir they are computed by three-term recurrences. This eliminates the vectors $\mathbf{r}_n$ and $\tilde{\mathbf{r}}_n$, and hence the second of the two possible breakdowns (3.5) and (3.6). We note that Brezinski et al. (1991) have proposed an implementation of the BCG approach that is mathematically equivalent to Lanczos/Orthodir.

Finally, recall that the algorithms QMR, Bi-CGSTAB, and TFQMR are designed to generate iterates that converge more smoothly than BCG and CGS. A different remedy for the erratic convergence behaviour of BCG or CGS was proposed by Schönauer (see Weiss (1990)). The approach used is to run plain BCG or CGS and then apply a smoothing procedure to the sequence of BCG or CGS iterates, resulting in iterates with monotonically decreasing residual norms. However, since the process is based directly on BCG or CGS, this approach inherits the numerical problems of BCG and CGS.

4. Solving the normal equations is not always bad

In this section, we consider CGNR and CGNE in more detail. Recall from Section 1 that CGNR and CGNE are the algorithms that result when (1.1) is solved by applying standard CG to either one of the normal equations (1.2) or (1.3), respectively. Clearly, for nonsingular A, both systems (1.2) and (1.3) are equivalent to the original system (1.1). From the minimization property (2.10) of CG, it follows that CGNR produces iterates

$$\mathbf{x}_n \in \mathbf{x}_0 + \mathcal{K}_n\left(A^H \mathbf{r}_0, A^H A\right), \quad n = 1, 2, \ldots, \tag{4.1}$$

that are characterized by the minimal residual condition

$$\|\mathbf{b} - A\mathbf{x}_n\|_2 = \min_{\mathbf{x} \in \mathbf{x}_0 + \mathcal{K}_n(A^H \mathbf{r}_0, A^H A)} \|\mathbf{b} - A\mathbf{x}\|_2.$$

Similarly, CGNE generates iterates (4.1) that satisfy the minimal error property

$$\|A^{-1}\mathbf{b} - \mathbf{x}_n\|_2 = \min_{\mathbf{x} \in \mathbf{x}_0 + \mathcal{K}_n(A^H \mathbf{r}_0, A^H A)} \|A^{-1}\mathbf{b} - \mathbf{x}\|_2.$$

Note that the letters 'R' and 'E' in CGNR and CGNE indicate the minimization of the residual or of the error, respectively. We also remark that the LSQR algorithm of Paige and Saunders (1982) is mathematically equivalent to CGNR, but has better numerical properties.

Since the convergence of CG depends on the spectrum of the coefficient matrix, the convergence behaviour of CGNR and CGNE depends on

$$\lambda(A^H A) = \{\sigma^2 \mid \sigma \in \sigma(A)\},$$

i.e. on the squares of the singular values of A. In particular, the worst-case convergence behaviour of CGNR and CGNE is governed by

$$\kappa_2\left(A^H A\right) = (\kappa_2(A))^2,$$

which suggests that convergence can be very slow even for matrices A with moderate condition numbers. This is indeed true in many cases, and generally it is preferable to use CG-type methods that are applied directly to (1.1) rather than CGNR or CGNE.

Nevertheless there are special cases for which solving the normal equations is optimal. A simple case are unitary matrices A for which CGNR and CGNE find the exact solution after only one step, while Krylov subspace methods with iterates (2.1) tend to converge very slowly (see Nachtigal *et al.*, 1990a). More interesting are cases for which CGNR and CGNE are optimal, in that they are mathematically equivalent to ideal CG-type methods based on the MR or OR Conditions (2.11) or (2.12). Typically, these situations arise when the spectrum of A has certain symmetries. Since these equivalences are not widely known, we collect here a few of these results. In the following, $\mathbf{x}_n^{\mathrm{MR}}$ and $\mathbf{x}_n^{\mathrm{OR}}$ denote iterates defined by (2.11) and (2.12), respectively.

First, we consider the case of real skew-symmetric matrices.

Theorem 4.1 (Freund, 1983)
Let $A = -A^T$ be a real $N \times N$ matrix, and let $\mathbf{b} \in \mathbb{R}^N$ and $\mathbf{x}_0 \in \mathbb{R}^N$. Then:

$$\mathbf{x}_{2n+1}^{\mathrm{MR}} = \mathbf{x}_{2n}^{\mathrm{MR}} = \mathbf{x}_n^{\mathrm{CGNR}}, \quad n = 0, 1, \ldots,$$

$$\mathbf{x}_{2n}^{\mathrm{OR}} = \mathbf{x}_n^{\mathrm{CGNE}}, \quad n = 1, 2, \ldots.$$

Moreover, no odd OR iterate $\mathbf{x}_{2n-1}^{\mathrm{OR}}$ exists.

Next, we turn to shifted skew-symmetric matrices of the form (2.13). For this class of matrices Eisenstat (1983a, b) has obtained the following result (see also Szyld and Widlund (1989)).

Theorem 4.2 Let $A = I - S$ where $S = -S^T$ is a real $N \times N$ matrix, and let $\mathbf{b} \in \mathbb{R}^N$ and $\mathbf{x}_0 \in \mathbb{R}^N$. Let $\mathbf{x}_n^{\mathrm{CGNE}}$ and $\tilde{\mathbf{x}}_n^{\mathrm{CGNE}}$ denote the iterates generated by CGNE started with initial guess $\mathbf{x}_0$ and $\tilde{\mathbf{x}}_0 = \mathbf{b} + S\mathbf{x}_0$, respectively. Then, for $n = 0, 1, \ldots$, it holds:

$$\mathbf{x}_{2n}^{\mathrm{OR}} = \mathbf{x}_n^{\mathrm{CGNE}},$$

$$\mathbf{x}_{2n+1}^{\mathrm{OR}} = \tilde{\mathbf{x}}_n^{\mathrm{CGNE}}.$$

We remark that for the MR and CGNR approaches a result corresponding to Theorem 4.2 does not hold (see Freund, 1983).

Finally, we consider linear systems (1.1) with Hermitian coefficient matrices A. Note that A has a complete set of orthonormal eigenvectors, and hence one can always expand the initial residual in the form:

$$\mathbf{r}_0 = \sum_{j=1}^{m} \rho_j \mathbf{z}_j, \tag{4.2}$$

where

$$\rho_j \in \mathbb{C}, \quad A\mathbf{z}_j = \lambda_j \mathbf{z}_j, \quad \lambda_1 < \lambda_2 < \cdots < \lambda_m, \quad \mathbf{z}_j^H \mathbf{z}_k = \delta_{jk}.$$

In the following theorem, $\mathbf{x}_n^{\mathrm{ME}}$ denotes the iterate defined by (2.16).

Theorem 4.3 (Freund, 1983)
Assume that the expansion (4.2) is 'symmetric', i.e. $m = 2l$ is even and

$$\lambda_j = -\lambda_{m+1-j}, \quad |\rho_j| = |\rho_{m+1-j}|, \quad j = 1, 2, \ldots, l.$$

Then, for $n = 0, 1, \ldots$, it holds:

$$\mathbf{x}_{2n+1}^{\mathrm{MR}} = \mathbf{x}_{2n}^{\mathrm{MR}} = \mathbf{x}_n^{\mathrm{CGNR}}, \quad n = 0, 1, \ldots,$$

$$\mathbf{x}_{2n}^{\mathrm{ME}} = \mathbf{x}_{2n-1}^{\mathrm{ME}} = \mathbf{x}_{2n}^{\mathrm{OR}} = \mathbf{x}_n^{\mathrm{CGNE}}, \quad n = 1, 2, \ldots .$$

Moreover, no odd OR iterate $\mathbf{x}_{2n-1}^{\mathrm{OR}}$ exists.

5. Estimating spectra for hybrid methods

We now turn to parameter-dependent schemes. As mentioned in Section 2.1, methods in this class require *a priori* knowledge of some spectral properties of the matrix. Typically, it is assumed that some compact set $\mathcal{G}$ is known, which in a certain sense approximates the spectrum of A and, in particular, satisfies (2.7). Given such a set $\mathcal{G}$, one then constructs residual polynomials ψ_n as some approximations to the optimal polynomials in (2.9). We would like to stress that, in view of (2.4) and (2.9), it is crucial that $\mathcal{G}$ excludes the origin. Since in general one does not know in advance a suitable set $\mathcal{G}$, parameter-dependent methods combine an approach for estimating the set $\mathcal{G}$ with an approach for solving the approximation problem (2.9), usually cycling between the two parts in order to improve the estimate for $\mathcal{G}$. For this reason, the algorithms in this class are often called hybrid methods. In this section, we review some recent insights in the problem of how to estimate the set $\mathcal{G}$.

The standard approach for obtaining $\mathcal{G}$ is to run a few steps of the Arnoldi Algorithm 2.2 to generate the upper Hessenberg matrix H_n (2.24) and then compute its spectrum $\lambda(H_n)$. Saad (1980) showed that the eigenvalues of H_n are Ritz values for A and can be used to approximate the spectrum

$\lambda(A)$, hence one takes the convex hull of $\lambda(H_n)$ as the set $\mathcal{G}$. Once the set is obtained, a hybrid method turns to solving the complex approximation problem (2.9), and the possibilities here are numerous.

However, there are problems with the approach outlined here, originating from the use of the Arnoldi Ritz values. In general, there is no guarantee that the convex hull of $\lambda(H_n)$ does not include the origin or, indeed, that one of the Ritz values is not at or near the origin. For matrices with spectrum in both the left and right half-planes, the convex hull of the Ritz values might naturally enclose the origin. Nachtigal et $al.$ (1990b) give an example of a matrix whose spectrum is symmetric with respect to the origin, so that the convex hull of $\lambda(H_n)$ will generally contain the origin, and on every other step, one of the Ritz values will be close to the origin. A second problem is that the approach aims to estimate the spectrum of A, which may be highly sensitive to perturbations, especially for nonnormal matrices. In these cases, a more natural concept is the pseudospectrum $\lambda_\epsilon(A)$, introduced by Trefethen (1991):

Definition 5.1 For $\epsilon \geq 0$ and $A \in \mathbb{C}^{N \times N}$, the ϵ-pseudospectrum $\lambda_\epsilon(A)$ is given by

$$\lambda_\epsilon(A) = \{\lambda \in \mathbb{C} \mid \lambda \in \lambda(A + \Delta), \ \|\Delta\|_2 \leq \epsilon\}. \tag{5.1}$$

As is apparent from (5.1), the pseudospectrum, in general, can be expected to be insensitive to perturbations of A, even for highly nonnormal matrices. For practical purposes, the sets $\lambda_\epsilon(A)$ of interest correspond to values of the parameter ϵ that are small relative to the norm of A but larger than round-off.

It is easy to construct examples where a hybrid algorithm using the exact spectrum $\lambda(A)$ of A will in fact diverge, while the same hybrid algorithm using the pseudospectrum $\lambda_\epsilon(A)$ will converge (see Nachtigal et $al.$ (1990b) and Trefethen (1991)). Unfortunately, in general the pseudospectrum $\lambda_\epsilon(A)$ cannot be easily computed directly. Fortunately, it turns out that one can compute approximations to the pseudospectrum; this is the approach taken in the hybrid introduced by Nachtigal et $al.$ (1990b). They observed that the level curves – or lemniscates – of the GMRES residual polynomial bound regions that approximate the pseudospectrum $\lambda_\epsilon(A)$. Let

$$\mathcal{C}_n(\eta) = \{\lambda \in \mathbb{C} \mid |\psi_n(\lambda)| = \eta\}, \quad \eta \geq 0,$$

be any lemniscate of the residual polynomial ψ_n. Due to the normalization (2.4), the region bounded by $\mathcal{C}_n(\eta)$ with $\eta < 1$ will automatically exclude the origin; in particular, 0 cannot be a root of a residual polynomial. In addition, since GMRES is solving the minimization problem (2.11), ψ_n will naturally be small on an appropriate set $\mathcal{G}$. Motivated by these considerations, Nachtigal et $al.$ observed that the region bounded by the lemniscate

$\mathcal{C}_n(\eta_n)$ with

$$\eta_n = \frac{\|\mathbf{r}_n\|_2}{\|\mathbf{r}_0\|_2} < 1$$

usually yields a suitable set $\mathcal{G}$.

The GMRES residual polynomials are kernel polynomials, and their roots are a type of Ritz values called pseudo-Ritz values by Freund (1989a). In fact, one is not restricted to using the GMRES residual polynomial, but could use the residual polynomials from other minimization methods, such as QMR or its transpose-free versions. In these cases, the residual polynomials are no longer kernel polynomials, but rather, they are quasi-kernel polynomials (Freund, 1991c). Nevertheless, their lemniscates still yield suitable sets $\mathcal{G}$, and their roots are also pseudo-Ritz values of A.

Freund has also proposed an algorithm to compute pseudo-Ritz values, using the upper Hessenberg matrix $H_n^{(e)}$ (2.19) appearing in the recurrence (2.17). One uses the fact that kernel and quasi-kernel polynomials can be defined by

$$\psi_n(\lambda) \equiv \frac{\det\left(\left(H_n^{(e)}\right)^H \left(H_n^{(e)}\right) - \lambda H_n^H\right)}{\det\left(\left(H_n^{(e)}\right)^H \left(H_n^{(e)}\right)\right)},$$

which makes it clear that the roots of ψ_n can be obtained from the generalized eigenvalue problem

$$\left(H_n^{(e)}\right)^H \left(H_n^{(e)}\right) \mathbf{z} = \lambda H_n^H \mathbf{z}, \quad \mathbf{z} \neq 0 \in \mathbb{C}^n.$$

Finally, we should point out another set that has been proposed as a candidate for $\mathcal{G}$ in (2.7), namely the field of values. Defined as

$$\mathcal{F} = \{\mathbf{z}^H A \mathbf{z} \mid \mathbf{z} \in \mathbb{C}^N, \ \|\mathbf{z}\|_2 = 1\},$$

the field of values has the advantage that it is easier to compute (see, e.g., Ruhe (1987)) than the pseudospectrum. However, the field of values is convex and always at least as large as the convex hull of $\lambda(A)$, and hence may once again enclose the origin. For more details on the field of values in iterative methods, we refer the reader to Eiermann (1991).

The literature on hybrid methods for nonHermitian linear systems starts with an algorithm proposed by Manteuffel (1978), which combines a modified power method with the Chebyshev iteration. Since then, literally dozens of hybrid methods have been proposed, most of which use Arnoldi in the first phase and differ in the way they compute the residual polynomial ψ_n in (2.9). For an overview of some of these algorithms, see Nachtigal et al. (1990b). The hybrid by Nachtigal et al. was the first to avoid the explicit computation of an estimate of $\lambda(A)$. Instead, it explicitly obtains the GMRES residual polynomial and applies it using Richardson's method until convergence. Finally, hybrids recently introduced include an algorithm by Saylor and Smolarski (1991), which combines Arnoldi with Richardson's method, and an algorithm proposed by Starke and Varga (1991), which combines Arnoldi with Faber polynomials.

6. CG-type methods for complex linear systems

While most linear systems arising in practice are real, there are important applications that lead to linear systems with complex[†] coefficient matrices A. Partial differential equations that model dissipative processes usually involve complex coefficient functions or complex boundary conditions, and discretizing them yields complex linear systems. An important example for this category is the complex Helmholtz equations. Other applications that lead to complex linear systems include the numerical solution of Schrödinger's equation, underwater acoustics, frequency response computations in control theory, semiconductor device simulation, and numerical computations in quantum chromodynamics; for details and further references, see Bayliss and Goldstein (1983), Barbour *et al.* (1987), Freund (1989b, 1991a), and Laux (1985). In all these applications, the resulting linear systems are usually nonHermitian. In this section, we review some recent advances in understanding the issues related to solving complex nonHermitian systems.

Until recently, the prevailing approaches used when solving complex linear systems have consisted of either solving the normal equations or rewriting the complex system as a real system of dimension $2N$. However, as indicated already in Section 4, the normal equations often lead to systems with very poor convergence rates, and this has indeed been observed for many of the complex systems of interest. The other option is to split the original matrix A into its real and imaginary parts and combine them into a real system of dimension $2N$. There are essentially only two different possibilities for doing this, namely

$$A_\star \begin{bmatrix} \mathrm{Re}\,\mathbf{x} \\ \mathrm{Im}\,\mathbf{x} \end{bmatrix} = \begin{bmatrix} \mathrm{Re}\,\mathbf{b} \\ \mathrm{Im}\,\mathbf{b} \end{bmatrix}, \quad A_\star := \begin{bmatrix} \mathrm{Re}\,A & -\mathrm{Im}\,A \\ \mathrm{Im}\,A & \mathrm{Re}\,A \end{bmatrix}, \tag{6.1}$$

and

$$A_{\star\star} \begin{bmatrix} \mathrm{Re}\,\mathbf{x} \\ -\mathrm{Im}\,\mathbf{x} \end{bmatrix} = \begin{bmatrix} \mathrm{Re}\,\mathbf{b} \\ \mathrm{Im}\,\mathbf{b} \end{bmatrix}, \quad A_{\star\star} := \begin{bmatrix} \mathrm{Re}\,A & \mathrm{Im}\,A \\ \mathrm{Im}\,A & -\mathrm{Re}\,A \end{bmatrix}. \tag{6.2}$$

Unfortunately, it turns out that this option is not viable either. As Freund (1989b, 1992) pointed out, both $A_\star$ and $A_{\star\star}$ have spectral properties that make them far more unsuitable for Krylov space iterations than the original

---

[†] In this section, 'complex' will imply the presence of imaginary components.

system was. The spectrum of $A_\star$ in (6.1) is given by

$$\lambda(A_\star) = \lambda(A) \cup \overline{\lambda(A)}, \qquad (6.3)$$

while the spectrum of $A_{\star\star}$ in (6.2) is given by

$$\lambda(A_{\star\star}) = \{\lambda \in \mathbb{C} \mid \lambda^2 \in \lambda(\overline{A}A)\}. \qquad (6.4)$$

In particular, note that the spectrum (6.3) is symmetric with respect to the real axis, while the spectrum (6.4) is symmetric with respect to both the real and imaginary axes. The point is that in both cases, the spectrum of the real system is either very likely to or is guaranteed to contain the origin, thus presenting a Krylov subspace iteration with the worst possible eigenvalue distribution. As a result, both approaches have had very poor results in practice, and have often led to the conclusion that complex systems are ill suited for iterative methods.

Therefore, instead of solving the normal equations or either one of the equivalent real systems (6.1) or (6.2), it is generally preferable to solve the original linear system by a CG-like Krylov subspace methods. In particular, if A is a general nonHermitian matrix, then we recommend using the Lanczos-based algorithms discussed in Section 3 or GMRES. However, in many applications, the resulting complex linear systems have additional structure that can be exploited. For example, often complex matrices of the form (2.13) arise. Recall from Theorem 2.1 that ideal CG-type algorithms based on the MR or OR property (2.11) or (2.12) exist for such shifted Hermitian matrices. Freund (1990) has derived practical implementations of the MR and OR approaches for the class of matrices (2.13). These algorithms can be viewed as extensions of Paige and Saunders' MINRES and SYMMLQ for Hermitian matrices (see Section 2.3). As in the case of SYMMLQ, the OR implementation for shifted Hermitian matrices also generates auxiliary iterates $\mathbf{x}_n^{\mathrm{ME}} \in \mathbf{x}_0 + \mathcal{K}_n(A^H \mathbf{r}_0, A)$ that are characterized by the minimal error property

$$\|A^{-1}\mathbf{b} - \mathbf{x}_n^{\mathrm{ME}}\|_2 = \min_{\mathbf{x} \in \mathbf{x}_0 + \mathcal{K}_n(A^H \mathbf{r}_0, A)} \|A^{-1}\mathbf{b} - \mathbf{x}\|_2.$$

Hence the OR algorithm proposed by Freund also generalizes Fridman's method to shifted Hermitian matrices. Unfortunately, when matrices A of the form (2.13) are preconditioned by standard techniques, the special structure of A is destroyed. In Freund (1989b) it is shown that the shift structure can be preserved when polynomial preconditioning is used, and results on the optimal choice of the polynomial preconditioner are given.

Another special case that arises frequently in applications are complex symmetric matrices $A = A^T$. For example, the complex Helmholtz equations leads to complex symmetric systems. As pointed out in Section 3.4, the QMR Algorithm 3.4 can take advantage of this special structure, and work and storage is roughly halved. We remark that the complex symmetry structure is preserved when preconditioning is used, if the preconditioner M is again symmetric, as is the case for all standard techniques. For an overview of other CG-type methods and further results for complex symmetric linear systems, we refer the reader to Freund (1991a, 1992).

7. Large dense linear systems

As mentioned in Section 1, there are certain classes of large dense linear systems for which it is possible to compute matrix–vector products cheaply; for these systems, iterative methods remain an attractive option. Typically, the matrix–vector product takes advantage of either some special structure of the matrix or of some special property of the underlying operator. We will briefly discuss one situation from each class.

The first case is the solution of integral equations involving a decaying potential, such as a gravitational or Coulombic potential. Some typical applications are the solution of N-body problems, vortex methods, potential theory, and others. In these problems, the effort required by a naive computation of a matrix–vector product is generally $\mathcal{O}(N^2)$, as it involves computing the influence of each of N points on all the other $N - 1$ points. However, Carrier *et al.* (1988) noticed that it is possible to approximate the result of the matrix–vector product $A\mathbf{x}$ in only $\mathcal{O}(N)$ time, rather than $\mathcal{O}(N^2)$.

The main idea behind their algorithm is to group points in clusters and evaluate the influence of entire clusters on faraway points. The cumulative potential generated by m charges in a cluster can be expressed as a power series. If the power series is truncated after p terms, then the work required to compute it turns out to be $\mathcal{O}(mp)$. Applying the cumulative influence to n points in a cluster well separated from the first requires an additional $\mathcal{O}(np)$ work, for a total of $\mathcal{O}(mp + np)$ work, as compared to the $\mathcal{O}(mn)$ work required to compute the individual interactions. Finally, the point is that the number p of terms in the series is determined by the preassigned precision to which the series is computed, and once chosen, p becomes a constant, giving an $\mathcal{O}(m + n)$ algorithm. For a complete description of the algorithm, see Carrier *et al.* (1988).

While the Carrier *et al.* algorithm is the first in the class of fast multipole methods, it falls in the bigger class of hierarchical algorithms. Several other algorithms have been proposed in this class; see for example Hockney and Eastwood (1981), Appel (1985), Rokhlin (1985). More recently, Hanrahan *et al.* (1991) have proposed a hierarchical algorithm for radiosity problems in computer graphics, and Greenbaum *et al.* (1991) have introduced an algorithm that uses the Carrier *et al.* algorithm combined with GMRES to

solve Laplace's equation in multiply connected domains. We refer the reader
to these articles for details of the algorithms, as well as further references.

Another important class of dense linear systems where the matrix–vector
product can be computed cheaply are Toeplitz systems, where the coefficient
matrix A has the form

$$
A = \begin{bmatrix}
t_0 & t_1 & \cdots & t_{N-1} \\
t_{-1} & t_0 & \ddots & \vdots \\
\vdots & \ddots & \ddots & t_1 \\
t_{-(N-1)} & \cdots & t_{-1} & t_0
\end{bmatrix}.
$$

A matrix–vector product with an $N \times N$ Toeplitz matrix can be computed
with $\mathcal{O}(N \log N)$ operations by means of the fast Fourier transform. Furthermore, as Chan and Strang (1989) observed, Toeplitz systems can be
preconditioned efficiently using circulant matrices

$$
M = \begin{bmatrix}
c_0 & c_1 & \cdots & \cdots & c_{N-1} \\
c_{N-1} & c_0 & \ddots & & c_{N-2} \\
\vdots & \ddots & \ddots & \ddots & \vdots \\
\vdots & & \ddots & \ddots & c_1 \\
c_1 & \cdots & \cdots & c_{N-1} & c_0
\end{bmatrix}.
$$

Recall from Section 2.5 that preconditioned iterative methods require the
solution of linear systems with the preconditioner M in each iteration. In
the case of circulant M, these systems can also be solved by means of fast
Fourier transform with only $\mathcal{O}(N \log N)$ operations per system.

8. Concluding remarks

In conclusion, we have covered some of the developments in iterative methods, especially for nonHermitian matrices. The introduction of CGS in the
late 1980s spurred renewed interest in the nonsymmetric Lanczos algorithm,
with most of the effort directed towards obtaining a method with better convergence properties than BCG or CGS. Several BCG-based algorithms were
proposed, such as Bi-CGSTAB, introduced by Van der Vorst (1990). The
quasi-minimal residual technique was introduced by Freund (1989b) in the
context of complex symmetric systems, then later coupled with a new variant of the look-ahead Lanczos approach to obtain a general nonHermitian
QMR algorithm. Finally, several transpose-free algorithms based on QMR
have been introduced recently, which trade the multiplication by A^T for
one or more multiplications by A. However, their convergence properties
are not well understood, and none of these algorithms have been combined
with look-ahead techniques yet. In general, it seems that the transpose-free

methods have more numerical problems than the corresponding methods that use A^T, and more research is needed into studying their behaviour.

With the advent of Lanczos-based methods that require little work and storage per step, the importance of preconditioners has decreased. With GMRES or similar algorithms, the high cost of the method makes restarts necessary in practice, which generally results in much slower convergence. As a result, preconditioned GMRES requires a very effective preconditioner, so that the preconditioned system requires few iterations to converge. The problem is that effective preconditioners are often too expensive, especially on parallel machines, where the inherently serial nature of many preconditioners makes their use unappealing. In contrast, restarts are not necessary with Lanczos-based methods, and hence a wider array of preconditioners – in particular, cheaper or more parallelizable preconditioners – becomes usable.

Finally, even though the field of iterative methods has made great progress in the last few years, it is still in its infancy, especially with regard to the packaged software available. Whereas there are well-established robust general-purpose solvers based on direct methods, the same cannot be said about solvers based on iterative methods. There are no established iterative packages of the same robustness and wide acceptance as, for example, the LINPACK library, and as a result many of the scientists who use iterative methods write their own specialized solvers. We feel that this situation needs to change, and we would like to encourage researchers to provide code for their methods.

Acknowledgments

We would like to thank Susanne Freund for her help with LaTeX and Marlis Hochbruck for her careful reading of this paper. The first and third authors are grateful to the San Jose radio station KSJO for keeping them awake late at night. The first author would like to thank the Printers Inc Book Store in Mountain View for providing countless mega espressos.

REFERENCES

A.W. Appel (1985), 'An efficient program for many-body simulation', *SIAM J. Sci. Statist. Comput.* **6**, 85–103.

W.E. Arnoldi (1951), 'The principle of minimized iterations in the solution of the matrix eigenvalue problem', *Quart. Appl. Math.* **9**, 17–29.

S.F. Ashby, T.A. Manteuffel and P.E. Saylor (1990), 'A taxonomy for conjugate gradient methods', *SIAM J. Numer. Anal.* **27**, 1542–1568.

O. Axelsson (1980), 'Conjugate gradient type methods for unsymmetric and inconsistent systems of linear equations', *Lin. Alg. Appl.* **29**, 1–16.

O. Axelsson (1985), 'A survey of preconditioned iterative methods for linear systems of algebraic equations', *BIT* **25**, 166–187.

O. Axelsson (1987), 'A generalized conjugate gradient, least square method', *Numer. Math.* **51**, 209–227.

I.M. Barbour, N.-E. Behilil, P.E. Gibbs, M. Rafiq, K.J.M. Moriarty and G. Schierholz (1987), 'Updating fermions with the Lanczos method', *J. Comput. Phys.* **68**, 227–236.

A. Bayliss and C.I. Goldstein (1983), 'An iterative method for the Helmholtz equation', *J. Comput. Phys.* **49**, 443–457.

D.L. Boley and G.H. Golub (1991), 'The nonsymmetric Lanczos algorithm and controllability', *Systems Control Lett.* **16**, 97–105

D.L. Boley, S. Elhay, G.H. Golub and M.H. Gutknecht (1991), 'Nonsymmetric Lanczos and finding orthogonal polynomials associated with indefinite weights', *Numer. Algorithms* **1**, 21–43.

C. Brezinski, M. Redivo Zaglia and H. Sadok (1991), 'A breakdown-free Lanczos type algorithm for solving linear systems', Technical Report ANO–239, Université des Sciences et Techniques de Lille Flandres-Artois, Villeneuve d'Ascq.

C.G. Broyden (1965), 'A class of methods for solving nonlinear simultaneous equations', *Math. Comput.* **19**, 577–593.

J. Carrier, L. Greengard and V. Rokhlin (1988), 'A fast adaptive multipole algorithm for particle simulations', *SIAM J. Sci. Stat. Comput.* **9**, 669–686.

R.H. Chan and G. Strang (1989), 'Toeplitz equations by conjugate gradients with circulant preconditioner', *SIAM J. Sci. Stat. Comput.* **10**, 104–119.

T.F. Chan, L. de Pillis and H.A. Van der Vorst (1991), 'A transpose-free squared Lanczos algorithm and application to solving nonsymmetric linear systems', Technical Report, University of California at Los Angeles.

R. Chandra (1978), 'Conjugate gradient methods for partial differential equations', Ph.D. Dissertation, Yale University, New Haven.

P. Concus and G.H. Golub (1976), 'A generalized conjugate gradient method for nonsymmetric systems of linear equations', in *Computing Methods in Applied Sciences and Engineering (Lecture Notes in Economics and Mathematical Systems 134)* (R. Glowinski and J.L. Lions, eds), Springer (Berlin) 56–65.

P. Concus, G.H. Golub and D.P. O'Leary (1976), 'A generalized conjugate gradient method for the numerical solution of elliptic partial differential equations', in *Sparse Matrix Computations* (J.R. Bunch and D.J. Rose, eds.), Academic Press (New York) 309–332.

E.J. Craig (1955), 'The N-step iteration procedures', *J. Math. Phys.* **34**, 64–73.

J. Cullum and R.A. Willoughby (1985), *Lanczos Algorithms for Large Symmetric Eigenvalue Computations, Volume 1, Theory*, Birkhäuser (Basel).

J. Cullum and R.A. Willoughby (1986), 'A practical procedure for computing eigenvalues of large sparse nonsymmetric matrices', in *Large Scale Eigenvalue Problems* (J. Cullum and R.A. Willoughby, eds.), North-Holland (Amsterdam) 193–240.

P. Deuflhard, R.W. Freund and A. Walter (1990), 'Fast secant methods for the iterative solution of large nonsymmetric linear systems', *IMPACT Comput. Sci. Eng.* **2**, 244–276.

A. Draux (1983), *Polynômes Orthogonaux Formels – Applications, (Lecture Notes in Mathematics 974)* Springer (Berlin).

I.S. Duff, A.M. Erisman and J.K. Reid (1986), *Direct Methods for Sparse Matrices*, Oxford University Press (Oxford).

M. Eiermann (1991), 'Fields of values and iterative methods', Preprint, IBM Scientific Center, Heidelberg.

M. Eiermann, W. Niethammer and R.S. Varga (1985), 'A study of semiiterative methods for nonsymmetric systems of linear equations', *Numer. Math.* **47**, 505–533.

T. Eirola and O. Nevanlinna (1989), 'Accelerating with rank-one updates', *Lin. Alg. Appl.* **121**, 511–520.

S.C. Eisenstat (1983a), 'A note on the generalized conjugate gradient method', *SIAM J. Numer. Anal.* **20**, 358–361.

S.C. Eisenstat (1983b), 'Some observations on the generalized conjugate gradient method', in *Numerical methods, Proceedings, Caracas 1982 (Lecture Notes in Mathematics 1005)* (V. Pereyra and A. Reinoza, eds), Springer (Berlin) 99–107.

S.C. Eisenstat, H.C. Elman and M.H. Schultz (1983), 'Variational iterative methods for nonsymmetric systems of linear equations', *SIAM J. Numer. Anal.* **20**, 345–357.

H.C. Elman (1982), 'Iterative methods for large sparse nonsymmetric systems of linear equations', Ph.D. Dissertation, Yale University, New Haven.

V. Faber and T. Manteuffel (1984), 'Necessary and sufficient conditions for the existence of a conjugate gradient method', *SIAM J. Numer. Anal.* **21**, 352–362.

V. Faber and T. Manteuffel (1987), 'Orthogonal error methods', *SIAM J. Numer. Anal.* **24**, 170–187.

B. Fischer and R.W. Freund (1990), 'On the constrained Chebyshev approximation problem on ellipses', *J. Approx. Theory* **62**, 297–315.

B. Fischer and R.W. Freund (1991), 'Chebyshev polynomials are not always optimal', *J. Approx. Theory* **65**, 261–272.

R. Fletcher (1976), 'Conjugate gradient methods for indefinite systems', in *Numerical Analysis Dundee 1975 (Lecture Notes in Mathematics 506)* (G.A. Watson, ed.), Springer (Berlin) 73–89.

R.W. Freund (1983), 'Über einige cg-ähnliche Verfahren zur Lösung linearer Gleichungssysteme', Doctoral Dissertation, Universität Würzburg.

R.W. Freund (1989a), 'Pseudo-Ritz values for indefinite Hermitian matrices', RIACS Technical Report 89.33, NASA Ames Research Center, Moffett Field.

R.W. Freund (1989b), 'Conjugate gradient type methods for linear systems with complex symmetric coefficient matrices', RIACS Technical Report 89.54, NASA Ames Research Center, Moffett Field.

R.W. Freund (1990), 'On conjugate gradient type methods and polynomial preconditioners for a class of complex nonHermitian matrices', *Numer. Math.* **57**, 285–312.

R.W. Freund (1991a), 'Krylov subspace methods for complex nonHermitian linear systems', Habilitation Thesis, Universität Würzburg.

R.W. Freund (1991b), 'A transpose-free quasi-minimal residual algorithm for non-Hermitian linear systems', RIACS Technical Report 91.18, NASA Ames Research Center, Moffett Field.

R.W. Freund (1991c), 'Quasi-kernel polynomials and their use in nonHermitian matrix iterations', RIACS Technical Report 91.20, NASA Ames Research Center, Moffett Field.

R.W. Freund (1992), 'Conjugate gradient-type methods for linear systems with complex symmetric coefficient matrices', *SIAM J. Sci. Stat. Comput.* **13**, to appear.

R.W. Freund and N.M. Nachtigal (1991), 'QMR: a quasi-minimal residual method for nonHermitian linear systems', *Numer. Math.*, to appear.

R.W. Freund and St. Ruscheweyh (1986), 'On a class of Chebyshev approximation problems which arise in connection with a conjugate gradient type method', *Numer. Math.* **48**, 525–542.

R.W. Freund and T. Szeto (1991), 'A quasi-minimal residual squared algorithm for nonHermitian linear systems', Technical Report, RIACS, NASA Ames Research Center, Moffett Field.

R.W. Freund and H. Zha (1991), 'Simplifications of the nonsymmetric Lanczos process and a new algorithms for solving indefinite symmetric linear systems', Technical Report, RIACS, NASA Ames Research Center, Moffett Field.

R.W. Freund, G.H. Golub and M. Hochbruck (1991a) 'Krylov subspace methods for nonHermitian *p*-cyclic matrices', Technical Report, RIACS, NASA Ames Research Center, Moffett Field.

R.W. Freund, M.H. Gutknecht and N.M. Nachtigal (1991b), 'An implementation of the look-ahead Lanczos algorithm for nonHermitian matrices', Technical Report 91.09, RIACS, NASA Ames Research Center, Moffett Field.

V.M. Fridman (1963), 'The method of minimum iterations with minimum errors for a system of linear algebraic equations with a symmetrical matrix', *USSR Comput. Math. and Math. Phys.* **2**, 362–363.

P.E. Gill, W. Murray, D.B. Ponceleón and M.A. Saunders (1990), 'Preconditioners for indefinite systems arising in optimization', Technical Report SOL 90-8, Stanford University.

G.H. Golub and D.P. O'Leary (1989), 'Some history of the conjugate gradient and Lanczos algorithms: 1948–1976', *SIAM Review* **31**, 50–102.

G.H. Golub and C.F. Van Loan (1989), *Matrix Computations*, second edition, The Johns Hopkins University Press (Baltimore).

G.H. Golub and R.S. Varga (1961), 'Chebyshev semi-iterative methods, successive overrelaxation iterative methods, and second order Richardson iterative methods', *Numer. Math.* **3**, 147–168.

W.B. Gragg (1974), 'Matrix interpretations and applications of the continued fraction algorithm', *Rocky Mountain J. Math.* **4**, 213–225.

W.B. Gragg and A. Lindquist (1983), 'On the partial realization problem', *Lin. Alg. Appl.* **50**, 277–319.

A. Greenbaum, L. Greengard and G.B. McFadden (1991), 'Laplace's equation and the Dirichlet-Neumann map in multiply connected domains', Technical Report, Courant Institute of Mathematical Sciences, New York University.

M.H. Gutknecht (1990a), 'The unsymmetric Lanczos algorithms and their relations to Padé approximation, continued fractions, and the QD algorithm', in *Proc. Copper Mountain Conference on Iterative Methods*.

M.H. Gutknecht (1990b), 'A completed theory of the unsymmetric Lanczos process and related algorithms, Part I', *SIAM J. Matrix Anal. Appl.*, to appear.

M.H. Gutknecht (1990c), 'A completed theory of the unsymmetric Lanczos process and related algorithms, Part II', IPS Research Report No. 90–16, ETH Zürich.

M.H. Gutknecht (1991), 'Variants of BiCGStab for matrices with complex spectrum', IPS Research Report No. 91–14, ETH Zürich.

P. Hanrahan, D. Salzman and L. Aupperle (1991), 'A rapid hierarchical radiosity algorithm', *Computer Graphics (Proc. SIGGRAPH '91)* **25**, 197–206.

M.T. Heath, E. Ng and B.W. Peyton (1991), 'Parallel algorithms for sparse linear systems', *SIAM Review* **33**, 420–460.

G. Heinig and K. Rost (1984), 'Algebraic methods for Toeplitz-like matrices and operators', Birkhäuser (Basel).

M.R. Hestenes and E. Stiefel (1952), 'Methods of conjugate gradients for solving linear systems', *J. Res. Natl Bur. Stand.* **49**, 409–436.

R.W. Hockney and J.W. Eastwood (1981), *Computer Simulation Using Particles*, McGraw-Hill (New York).

W.D. Joubert (1990), 'Generalized conjugate gradient and Lanczos methods for the solution of nonsymmetric systems of linear equations', Ph.D. Dissertation, The University of Texas at Austin.

W.D. Joubert and D.M. Young (1987), 'Necessary and sufficient conditions for the simplification of generalized conjugate-gradient algorithms', *Lin. Alg. Appl.* **88/89**, 449–485.

I.M. Khabaza (1963), 'An iterative least-square method suitable for solving large sparse matrices', *Comput. J.* **6**, 202–206.

S. Kung (1977), 'Multivariable and multidimensional systems: analysis and design', Ph.D. Dissertation, Stanford University.

C. Lanczos (1950), 'An iteration method for the solution of the eigenvalue problem of linear differential and integral operators', *J. Res. Natl Bur. Stand.* **45**, 255–282.

C. Lanczos (1952), 'Solution of systems of linear equations by minimized iterations', *J. Res. Natl Bur. Stand.* **49**, 33–53.

S.E. Laux (1985), 'Techniques for small-signal analysis of semiconductor devices', *IEEE Trans. Electron Dev.* **ED-32**, 2028–2037.

D.G. Luenberger (1969), 'Hyperbolic pairs in the method of conjugate gradients', *SIAM J. Appl. Math.* **17**, 1263–1267.

T.A. Manteuffel (1977), 'The Tchebychev iteration for nonsymmetric linear systems', *Numer. Math.* **28**, 307–327.

T.A. Manteuffel (1978), 'Adaptive procedure for estimating parameters for the nonsymmetric Tchebychev iteration', *Numer. Math.* **31**, 183–208.

T.A. Manteuffel (1991), 'Recent developments in iterative methods for large sparse linear systems', Seminar presentation, Numerical Analysis/Scientific Computing Seminar, Stanford University.

N.M. Nachtigal (1991), 'A look-ahead variant of the Lanczos algorithm and its application to the quasi-minimal residual method for nonHermitian linear systems', Ph.D. Dissertation, Massachusetts Institute of Technology, Cambridge.

N.M. Nachtigal, S.C. Reddy and L.N. Trefethen (1990a), 'How fast are nonsymmetric matrix iterations?', *SIAM J. Matrix Anal. Appl.*, to appear.

N.M. Nachtigal, L. Reichel and L.N. Trefethen (1990b), 'A hybrid GMRES algorithm for nonsymmetric linear systems', *SIAM J. Matrix Anal. Appl.*, to appear.

C.C. Paige and M.A. Saunders (1975), 'Solution of sparse indefinite systems of linear equations', *SIAM J. Numer. Anal.* **12**, 617–629.

C.C. Paige and M.A. Saunders (1982), 'LSQR: an algorithm for sparse linear equations and sparse least squares', *ACM Trans. Math. Softw.* **8**, 43–71.

P.N. Parlett (1990), 'Reduction to tridiagonal form and minimal realizations', Preprint, University of California at Berkeley.

B.N. Parlett, D.R. Taylor and Z.A. Liu (1985), 'A look-ahead Lanczos algorithm for unsymmetric matrices', *Math. Comput.* **44**, 105–124.

D. Rapoport (1978), 'A nonlinear Lanczos algorithm and the stationary Navier-Stokes equation', Ph.D. Dissertation, New York University.

J.K. Reid (1971), 'On the method of conjugate gradients for the solution of large sparse systems of linear equations', in *Large Sparse Sets of Linear Equations* (J.K. Reid, ed.), Academic Press, New York, 231–253.

V. Rokhlin (1985), 'Rapid solution of integral equations of classical potential theory', *J. Comput. Phys.* **60**, 187–207.

A. Ruhe (1987), 'Closest normal matrix finally found!', *BIT* **27**, 585–598.

Y. Saad (1980), 'Variations of Arnoldi's method for computing eigenelements of large unsymmetric matrices', *Lin. Alg. Appl.* **34**, 269–295.

Y. Saad (1981), 'Krylov subspace methods for solving large unsymmetric linear systems', *Math. Comput.* **37**, 105–126.

Y. Saad (1982), 'The Lanczos bi-orthogonalization algorithm and other oblique projection methods for solving large unsymmetric systems', *SIAM J. Numer. Anal.* **19**, 485–506.

Y. Saad (1984), 'Practical use of some Krylov subspace methods for solving indefinite and nonsymmetric linear systems', *SIAM J. Sci. Stat. Comput.* **5**, 203–227.

Y. Saad (1989), 'Krylov subspace methods on supercomputers', *SIAM J. Sci. Stat. Comput.* **10**, 1200–1232.

Y. Saad and M.H. Schultz (1985), 'Conjugate gradient-like algorithms for solving nonsymmetric linear systems', *Math. Comput.* **44**, 417–424.

Y. Saad and M.H. Schultz (1986), 'GMRES: a generalized minimal residual algorithm for solving nonsymmetric linear systems', *SIAM J. Sci. Stat. Comput.* **7**, 856–869.

P.E. Saylor and D.C. Smolarski (1991), 'Implementation of an adaptive algorithm for Richardson's method', *Lin. Alg. Appl.* **154–156**, 615–646.

P. Sonneveld (1989), 'CGS, a fast Lanczos-type solver for nonsymmetric linear systems', *SIAM J. Sci. Stat. Comput.* **10**, 36–52.

G. Starke and R.S. Varga (1991), 'A hybrid Arnoldi-Faber iterative method for nonsymmetric systems of linear equations', Technical Report ICM-9108-11, Kent State University, Kent.

E. Stiefel (1955), 'Relaxationsmethoden bester Strategie zur Lösung linearer Gleichungssysteme', *Comm. Math. Helv.* **29**, 157–179.

J. Stoer (1983), 'Solution of large linear systems of equations by conjugate gradient type methods', in *Mathematical Programming – The State of the Art* (A. Bachem, M. Grötschel and B. Korte, eds.), Springer (Berlin) 540–565.

J. Stoer and R.W. Freund (1982), 'On the solution of large indefinite systems of linear equations by conjugate gradient algorithms', in *Computing Methods in Applied Sciences and Engineering V* (R. Glowinski and J.L. Lions, eds.), North-Holland (Amsterdam) 35–53.

D.B. Szyld and O.B. Widlund (1989), 'Variational analysis of some conjugate gradient methods', Technical Report CS-1989-28, Duke University, Durham.

D.R. Taylor (1982), 'Analysis of the look ahead Lanczos algorithm', Ph.D. Dissertation, University of California at Berkeley.

L.N. Trefethen (1991), 'Pseudospectra of matrices', in *Proceedings of the 14th Dundee Biennial Conference on Numerical Analysis* (D.F. Griffiths and G.A. Watson, eds.), to appear.

H.A. Van der Vorst (1990), 'Bi-CGSTAB: A fast and smoothly converging variant of Bi-CG for the solution of nonsymmetric linear systems', Preprint, University of Utrecht.

P.K.W. Vinsome (1976), 'Orthomin, an iterative method for solving sparse sets of simultaneous linear equations', in *Proc. Fourth Symposium on Reservoir Simulation*, Society of Petroleum Engineers of AIME, Los Angeles.

V.V. Voevodin (1983), 'The problem of a nonselfadjoint generalization of the conjugate gradient method has been closed', *USSR Comput. Math. Math. Phys.* **23**, 143–144.

C. Vuik (1990), 'A comparison of some GMRES-like methods', Technical Report, Delft University of Technology, Delft.

R. Weiss (1990), 'Convergence behaviour of generalized conjugate gradient methods', Doctoral Dissertation, Universität Karlsruhe.

O. Widlund (1978), 'A Lanczos method for a class of nonsymmetric systems of linear equations', *SIAM J. Numer. Anal.* **15**, 801–812.

J.H. Wilkinson (1965), *The Algebraic Eigenvalue Problem*, Oxford University Press (Oxford).

D.M. Young and K.C. Jea (1980), 'Generalized conjugate gradient acceleration of nonsymmetrizable iterative methods', *Lin. Alg. Appl.* **34**, 159–194.

Acta Numerica (1991), *pp.* 101–139

Problems with different time scales

Heinz-Otto Kreiss

Deparment of Mathematics
University of California at Los Angeles
Los Angeles, CA 90024 USA
E-mail: kreiss@math.ucla.edu

CONTENTS

1. An example

In this section we discuss a very simple problem. Consider the scalar initial value problem

$$\varepsilon y' = ay + e^{it}, \quad t \geq 0,$$
$$y(0) = y_0. \tag{1.1}$$

Here $\varepsilon > 0$ is a small constant and $a = a_1 + ia_2$, a_1, a_2 real, is a complex number with $|a| = 1$. We can write down the solution of (1.1) explicitly. It is

$$y = y^{\mathrm{S}} + y^{\mathrm{F}},$$

where

$$y^{\mathrm{S}} = -\frac{1}{a - i\varepsilon} e^{it}$$

is the forced solution and

$$y^{\mathrm{F}} = \left(y_0 + \frac{1}{a - i\varepsilon} \right) e^{a/\varepsilon\, t}$$

is a solution of the homogeneous equation

$$\varepsilon v' = av.$$

y^S varies on the time scale '1' while y^F varies on the much faster scale $1/\varepsilon$. We say that y^S, y^F vary on the slow and fast scale, respectively. We use also the phrase: y^S and y^F are the slow and the fast part of the solution, respectively.

There are three different possibilities.

Case 1. $a_1 \gg \varepsilon$. In this case y^F grows rapidly and dominates the solution. We are not interested in this case.

Case 2. $a_1 \gg -\varepsilon$. Now y^F decays rapidly. Therefore, outside a boundary layer, the solution of (1.1) is essentially represented by y^S.

Case 3. $a_1 = 0$. For general initial data both scales are present for all times. However, if

$$y_0 + \frac{1}{a - i\varepsilon} = 0,$$

then $y^F = 0$ and the solution varies on the slow scale only.

In this survey article we shall mainly discuss the third case.

In applications like meteorology, oceanography and plasma physics one is often only interested in solutions, which vary on the slow scale. However, the data are such that the fast scale is present anyway. Therefore we shall develop a theory, which leads to a systematic way to 'initialize' the data such that the fast scale is not excited. This theory is based on a very simple principle. If $y(t)$ varies on the slow time scale, then

$$d^\nu y(t)/dt^\nu \approx \mathcal{O}(1), \quad \nu = 0, 1, 2, \ldots, p,$$

where $p > 1$ is a suitable number. Therefore our principle as follows.

Choose the initial data $y(0) = y_0$ such that at $t = 0$

$$d^\nu y(0)/dt^\nu \approx \mathcal{O}(1), \quad \nu = 0, 1, 2, \ldots, p. \tag{1.2}$$

We shall call this procedure *the bounded derivative principle*.

Let us apply the principle to our example. We think of ε as a small parameter, which approaches 0, and we want to choose the initial data such that the derivatives at $t = 0$ are bounded independently of ε.

$dy(0)/dt$ is bounded independently of ε, if and only if

$$ay(0) + 1 = \mathcal{O}(\varepsilon), \quad \text{i.e.} \quad y(0) = -1/a + \mathcal{O}(\varepsilon).$$

Thus the initial data are determined up to terms of order $\mathcal{O}(\varepsilon)$.

For the second derivative we have

$$\varepsilon^2 y' = \varepsilon a y' + \varepsilon e^{it} = a^2 y + a e^{it} + i\varepsilon e^{it}.$$

Therefore $d^2y(0)/dt^2$ is bounded independently of ε, if and only if

$$y(0) = -1/a - i\varepsilon/a^2 + \mathcal{O}(\varepsilon^2).$$

Thus the initial data are determined up to terms of order $\mathcal{O}(\varepsilon^2)$.

An easy calculation shows that the initial data are determined up to terms of order $\mathcal{O}(\varepsilon^p)$, if and only if the first p derivatives are bounded indepentently of ε.

Earlier we have shown that $y^F \equiv 0$, if

$$y(0) = -\frac{1}{a - i\varepsilon} = -\frac{1}{a} \sum_{\nu=0}^{\infty} \left(\frac{i\varepsilon}{a}\right)^{\nu}.$$

The bounded derivative principle gives us the first p terms in the power series expansion.

We want to prove that for general nonlinear systems the bounded derivative principle lets us determine the slow solution to any order.

The bounded derivative principle is very much connected with asymptotic expanstions. To discuss the connection we consider the slightly more general equation

$$\begin{aligned} \varepsilon y' &= ia(t)y + f(t), \\ y(0) &= y_0, \end{aligned} \qquad (1.3)$$

where $a(t), f(t) \in C^\infty(t)$ and $a(t) \geq a_0$, $a_0 = \text{constant} > 0$.

First we shall show that the bounded derivative principle is valid. The construction will be generalized to systems in the next section.

If $y(t)$ is a slow solution, then

$$y(t) \approx \varphi_0(t) =: i\, f(t)/a(t).$$

This suggests the substitution

$$y(t) = \varphi_0(t) + y_1(t). \qquad (1.4)$$

Introducing (1.4) into (1.3) gives us

$$\begin{aligned} \varepsilon y_1' &= ia(t)y_1 + \varepsilon f_1(t), \quad f_1(t) =: -\varphi_0'(t), \\ y_1(0) &= y_{10} =: y_0 - \varphi_0(0). \end{aligned} \qquad (1.5)$$

(1.5) is of the same form as (1.3). However, the forcing is reduced to order $\mathcal{O}(\varepsilon)$. We can repeat the process. After p steps we obtain

$$y(t) = \psi_{p-1}(t) + y_p(t), \quad \psi_{p-1}(t) = \sum_{j=0}^{p-1} \varepsilon^j \varphi_j(t), \qquad (1.6)$$

where $y_p(t)$ solves

$$\begin{aligned} \varepsilon y_p' &= ia(t)y_p + \varepsilon^p f_p(t), \\ y_p(0) &= y_0 - \psi_{p-1}(0). \end{aligned} \qquad (1.7)$$

The solution of (1.7) can be written as

$$y_p = \bar{y} + \bar{\bar{y}},$$

where $\bar{y}$ is the forced solution satisfying

$$
\begin{aligned}
\varepsilon \bar{y}' &= ia(t)\bar{y} + \varepsilon^p f_p(t), \\
\bar{y}(0) &= 0,
\end{aligned}
\tag{1.8}
$$

and $\bar{\bar{y}}$ solves

$$
\begin{aligned}
\varepsilon \bar{\bar{y}}' &= ia(t)\bar{\bar{y}}, \\
\bar{\bar{y}}(0) &= \bar{\bar{y}}_{p0}.
\end{aligned}
\tag{1.9}
$$

By Duhamel's principle

$$|d^j \bar{y}/dt^j| \leq \text{constant} \times \varepsilon^{p-j-1}$$

in any finite time interval $0 \leq t \leq T$. Thus $\bar{y}$ has $p-1$ derivatives bounded independently of ε.

Now apply the bounded derivative principle to (1.9). An easy calculation shows that $\bar{\bar{y}}(t)$ and therefore also $y(t)$ have $p-1$ derivatives bounded independently of ε at $t = 0$ if and only if

$$\bar{\bar{y}}(0) = \mathcal{O}(\varepsilon^{p-1}), \quad \text{i.e.} \quad y(0) = \psi_{p-1}(0) + \mathcal{O}(\varepsilon^{p-1}). \tag{1.10}$$

If (1.10) holds, then $\bar{\bar{y}}(t)$ and therefore also

$$y(t) = \psi_{p-1}(t) + \bar{y}(t) + \bar{\bar{y}}(t) = \psi_{p-1}(t) + \mathcal{O}(\varepsilon^{p-1}), \tag{1.11}$$

have $p-1$ derivatives bounded independently of ε in any finite time interval. This shows that the bounded derivative principle is valid.

(1.10) and (1.11) also show that equation (1.3) has essentially a unique slow solution and that $\psi_{p-1}(t)$ represents the first p terms of its asymptotic expansion. One can determine the initial data either by the bounded derivative principle or by calculating the asymptotic expansion in a neighbourhood of $t = 0$ and use $\psi_{p-1}(0)$ as initial data.

We have calculated the asymptotic expansion by substitution. Instead we can also determine it by the iteration

$$\varepsilon(y^{(n-1)})' = iay^{(n)} + f, \quad y^{(-1)} \equiv 0, \ n = 0, 1, 2, \ldots .$$

An easy calculation shows that

$$y^{(p)} = \psi_p.$$

Our construction depends heavily on the assumption that $a(t), f(t)$ have derivatives of order $\mathcal{O}(1)$ and that $a(t) \geq a_0 > 0$. If, for example,

$$f(t) = \begin{cases} 0 & \text{for } 0 \leq t < \frac{1}{2} \\ 1 & \text{for } t \geq \frac{1}{2}, \end{cases}$$

then the asymptotic expansion tells us that the slow solution is given by

$$y(t) = \begin{cases} 0 & \text{for } 0 \le t < \frac{1}{2} \\ \frac{2i}{a(t)} + \mathcal{O}(\varepsilon) & \text{for } t \ge \frac{1}{2}. \end{cases}$$

Thus the solution of (1.3) with initial data $y(0) = 0$ will become highly oscillatory for $t \ge \frac{1}{2}$. Correspondingly, if

$$a(t) = (t - t_0)a_1(t),$$

then a solution, which is slow for $t < t_0$, becomes in general highly oscillatory for $t > t_0$.

2. Systems of ordinary differential equations

2.1. Form of the systems and assumptions

In applications the systems are real and often have the form

$$w_t = \frac{1}{\varepsilon}A_1(t)w + f_1(w,t), \quad 0 < \varepsilon \le \varepsilon_0, \tag{2.1}$$

i.e. the large part of the right-hand side is a linear function of w. We assume that $A_1(t)$ has constant rank, i.e. there is a smooth transformation $S(t)$ such that

$$S^{-1}(t)A_1(t)S(t) = \begin{pmatrix} A(t) & 0 \\ 0 & 0 \end{pmatrix}, \quad \det A \ne 0.$$

Changing the dependent variables accordingly, we obtain a system of the form

$$\begin{aligned} \varepsilon y' &= (A(t) + \varepsilon C(v,y,t))y + f(v,t), \quad 0 < \varepsilon \le \varepsilon_0, \\ v' &= g(v,y,t), \end{aligned} \tag{2.2}$$

where

$$y(t), f(v,t) \in \mathbb{R}^m, \quad v(t), g(v,y,t) \in \mathbb{R}^n, \quad A(t), C(v,y,t) \in \mathbb{R}^{m \times m}.$$

We want to show that the results of the previous section can be generalized to systems (2.2). We follow here closely the presentation in Kreiss and Lorenz (1991). (See also Kreiss (1979) and Sacker (1965).)

To be precise, we shall use the following terminology

Definition 2.1 Let $w(t, \varepsilon)$ denote a function defined for $0 \le t \le T$, $0 < \varepsilon \le \varepsilon_0$. We say that it is slow to order p in $0 \le t \le T$ if $w \in C^p(0, T)$ and if

$$\sup_{0 < \varepsilon \le \varepsilon_0} \max_{0 \le t \le T} |\partial^j w/\partial t^j| < \infty, \quad j = 0, 1, \ldots, p. \tag{2.3}$$

We say that w is slow if (2.3) holds for any p.

Our main assumption is

Assumption 2.1

(i) For all $t \geq 0$

$$A(t) + A^*(t) \leq 0, \quad \det A(t) \neq 0.$$

(ii)

$$A(t), A^{-1}(t), C(v, y, t), f(v, t), g(v, y, t)$$

are C^∞-functions of their arguments with bounded derivatives. They may also depend on ε but we assume that the bounds are uniform in ε.

2.2. The bounded derivative principle and asymptotic expansions

We shall separate the fast and slow variables by a suitable substitution. If $y(t), v(t)$ is a slow solution, then $\varepsilon y'(t) = \mathcal{O}(\varepsilon)$ and to the first approximation

$$y(t) = -A^{-1}(t)f(v, t).$$

This suggests the substitution

$$y(t) = \Phi_0(v, t) + y_1(t), \quad \Phi_0(v, t) = -A^{-1}(t)f(v, t).$$

Introducing (2.3) into (2.2) gives us

$$\varepsilon y_1'(t) + \varepsilon(\partial\Phi_0(v, t)/\partial v)g(v, \Phi_0 + y_1, t) + \varepsilon\Phi_{0t}(v, t)$$
$$= A(t)y_1(t) + \varepsilon C(v, \Phi_0 + y_1, t)(\Phi_0(v, t) + y_1(t))$$
$$v'(t) = g(v, \Phi_0 + y_1, t),$$

i.e.

$$\varepsilon y_1' = (A(t) + \varepsilon C_1(v, y_1, t))y_1 + \varepsilon f_1(v, t)$$
$$v' = g_1(v, y_1, t), \qquad (2.4)$$

where

$$f_1(v, t) = -(\partial\Phi_0(v, t)/\partial v)g(v, \Phi_0, t) - \Phi_{0t}(v, t).$$

Thus $y_1(t)$ satisfies a differential equation of the same form as $y(t)$ but the forcing is reduced to order $\mathcal{O}(\varepsilon)$.

We can repeat the process and obtain

Theorem 2.1 One can construct slow functions $\Phi_0(v, t), \Phi_1(v, t), \ldots$ with the following properties. If one substitutes

$$y(t) = \psi_{p-1}(v, t) + y_p(t),$$
$$\psi(v, t) = \Phi_0(v, t) + \varepsilon\Phi_1(v, t) + \cdots + \varepsilon^{p-1}\Phi_{p-1}(v, t),$$

into (2.2), then $y_p(t), v(t)$ satisfy a system

$$\varepsilon y_p' = (A(t) + \varepsilon C_p(v, y_p, t))y_p + \varepsilon^p f_p(v, t)$$
$$v' = g_p(v, y_p, t), \qquad (2.5)$$

where C_p, f_p, g_p have uniformly bounded derivatives with respect to all variables. Differentiating (2.5) we obtain immediately

Theorem 2.2 $y_p(t), v(t)$ and therefore also $y(t), v(t)$ are slow to order p in any fixed time interval $0 \le t \le T$, if and only if

$$y_p(0) = \mathcal{O}(\varepsilon^p), \quad \text{i.e.} \quad d^j y_p(0)/dt^j = \mathcal{O}(\varepsilon^{p-j}), \quad j = 0, 1, \ldots, p. \qquad (2.6a)$$

(2.6a) holds if and only if for the original variables

$$d^j y(0)/dt^j = \mathcal{O}(1) \quad \text{for} \quad j = 0, 1, \ldots, p. \qquad (2.6b)$$

Thus the bounded derivative principle is valid.

As in the previous section we can use the last theorem to initialize the data, i.e. for a given $v_0 = v(0)$ find $y_0 = y(0)$ such that the solution is slow to order p. We find the relations by enforcing (2.6b). Essentially we have to calculate $\psi_{p-1}(v(0), 0)$. The process can become quite complicated. Therefore it is often easier to determine the relations by iteration. We have

Theorem 2.3 Let $y^{(0)}(t) \equiv 0$. p iterations of

$$\begin{aligned}
\varepsilon(y^{(n-1)})' &= (A(t) + \varepsilon C_p(v^{(n)}, y^{(n)}, t))y^{(n)} + f(v^{(n)}, t), \\
(v^{(n)})' &= g(v^{(n)}, y^{(n)}, t), \quad v^{(n)}(0) = v_0, \ n = 0, 1, \ldots, p-1,
\end{aligned}$$

determines a solution of (2.2), which is slow to order p and which, for a given v_0, is unique up to terms of order $\mathcal{O}(\varepsilon^p)$.

We can solve this iteration numerically in a neighbourhood of $t = 0$ and use the resulting $y^{(p-1)}(0)$ as initial data to solve the system (2.2) in large time intervals.

Under the following additional assumptions the estimates can be extended to all times.

Assumption 2.2 There is a constant $\beta > 0$ and an integer $q_0 \ge 1$ such that

$$C_{q_0}(v, y, t) + C^*_{q_0}(v, y, t) \le -\beta \varepsilon^{q_0-1} I$$

for all v, y, ε.

(For the proof see Kreiss and Lorenz (1991).)

2.3. Existence of a slow manifold

Theorem 2.2 tells us that we can choose $y(0)$ as a function of $v(0)$ such that the resulting solution is slow to order p. In general the relationship between y and v depends on p. If we want the solution to be slow to order $p+1$, then we have to change the relation by terms of order ε^p.

For practical purposes this result is completely satisfactory. However, an interesting mathematical question is: Can we determine $y(0)$ as a function of $v(0)$ such that the resulting solution is slow to any order?

There is one trivial case, where this is so. If $f(v,t)$ and all its partial derivatives with respect to v and t vanish at $t = 0$, then the initial data

$$y(0) = 0, \quad v(0) = v_0$$

guarantee solutions, which are slow to any order. Also, for given v_0 the slow solution is unique to any order in ε.

In Kreiss and Lorenz (1991) we have reduced the general case to the above by constructing a substitution

$$y = \Phi(v, t, \varepsilon) + \tilde{y}, \quad 0 \le t \le T, \tag{2.7}$$

such that $f(v,t) \equiv 0$. We have also given conditions such that the substitution exists for all times. (See also Sacker (1965), Kopell (1985), Fenichel (1985).)

(2.7) shows that the slow solution forms a manifold represented by

$$
\begin{aligned}
y &= \Phi(v, t, \varepsilon) \\
v' &= g(v, y, t).
\end{aligned}
\tag{2.8}
$$

2.4. Interaction between the fast and the slow scale

Consider the system (2.2) and choose the initial data by

$$v(0) = v_0, \quad y(0) = \Phi(v_0, 0, \varepsilon)$$

to obtain the slow solution $v^S(t), y^S(t)$. Now perturb $y(0)$ and consider

$$v(0) = v_0, \quad y(0) = \Phi(v_0, 0, \varepsilon) + \delta$$

and denote the resulting solution by $v_\delta(t), y_\delta(t)$. In general the fast scale is excited and $y^S(t)$ will be of order $\mathcal{O}(|\delta|)$. We want to show that the effect on the slow part of the solution is much smaller than $|\delta|$.

Theorem 2.4 Let $0 \le t \le T$ be a fixed time interval. For sufficiently small $\varepsilon, |\delta|$ there is a constant c_0 such that

$$|v^S(t) - v_\delta(t)| \le c_0(\varepsilon|\delta| + |\delta|^2).$$

Proof. We shall only indicate the proof. For more details see Kreiss and Lorenz (1991). Without restriction we can assume that the system (2.2) has the form

$$
\begin{aligned}
\varepsilon y' &= (A(t) + \varepsilon C(v, y, t))y \\
v' &= g(v, y, t).
\end{aligned}
\tag{2.9}
$$

Otherwise we perform the substitution (2.7). The initial data for $v^S(t), y^S(t)$ are

$$v^S(0) = v_0, \quad y^S(0) = 0, \quad \text{i.e.} \quad y^S(t) \equiv 0,$$

and for $v_\delta(t), y_\delta(t)$

$$v_\delta(0) = v_0, \quad y_\delta(t) = \delta.$$

To first approximation $y_\delta(t)$ satisfies

$$\varepsilon y'_\delta(t) = (A(t) + \varepsilon C(v^S, 0, t))y_\delta, \quad \text{i.e.} \quad |y_\delta| = \mathcal{O}(\delta),$$

and $w = v^S - v_\delta$ solves

$$\begin{aligned}
w' &= (\partial g(v^S, 0, t)/\partial v)w + \partial g(v^S, 0, t)/\partial y y_\delta(t), \\
w(0) &= 0.
\end{aligned} \tag{2.10}$$

Therefore

$$\begin{aligned}
w(t) &= \int_0^t S(t, \xi)(\partial g(v^S, 0, \xi)/\partial y)y_\delta(\xi)\, d\xi \\
&= \varepsilon \int_0^t S(t, \xi)(\partial g(v^S, 0, \xi)/\partial y)(A + \varepsilon C)^{-1}y'_\delta(\xi)\, d\xi.
\end{aligned}$$

Here $S(t, \xi)$ is the solution operator of the homogeneous equation

$$u' = (\partial g(v^S, 0, t)/\partial v)u.$$

Integration by parts shows that the last integral is of order $\varepsilon\delta$. Also, we have neglected only terms of order $\mathcal{O}(\varepsilon\delta) + \mathcal{O}(\delta^2)$ and therefore the theorem follows. $\square$

The last theorem is important in applications. Often one is only interested in slow solutions. In these cases one has to prepare the initial data in such a way that the fast scale is not excited. Practically one can never remove the fast scale completely. The theorem says that the effect of the fast scale on the slow scale can be neglected, provided a moderate amount of data preparation has been performed. Also, the fast scale can be removed by post-filtering.

3. Numerical methods for ordinary differential equations

3.1. An example

We consider equation (1.3)

$$\begin{aligned}
\varepsilon y' &= i\, a(t)y + f(t) \\
y(0) &= y_0.
\end{aligned} \tag{3.1}$$

To begin with we want only to calculate the slow solution. There are a number of possibilities.

Asymptotic expansion, i.e.

$$y^S(t) = \frac{i f(t)}{a(t)} + \frac{\varepsilon}{a(t)}\left(\frac{f(t)}{a(t)}\right)' + \cdots + \mathcal{O}(\varepsilon^2). \tag{3.2}$$

Difference approximation. If we are willing to use a time step $k \ll \varepsilon$, then any of the standard explicit techniques can be used. However, we want only to calculate the slow solution and therefore we only want to resolve the slow scale, i.e. we want to use a time step k with $\varepsilon \ll k \ll 1$. Therefore we have to use an implicit method. It has to be stable on the imaginary axis and therefore the order of accuracy of a stable multi-step method is restricted to one or two. We shall discuss the *implicit Euler scheme* and *the midpoint rule*.

Let $k > 0$ denote the time step, $t_j = jk$, $j = 0, 1, \ldots$, the grid and denote by $u_j = u(jk)$ the values of u on the grid. Then the *implicit Euler* scheme has the form

$$\varepsilon(u_{n+1} - u_n) = k(ia_{n+1}u_{n+1} + f_{n+1}). \tag{3.3}$$

As for the continuous case we can derive an asymptotic expansion. Let

$$u_n = \frac{if_n}{a_n} + \tilde{u}_n.$$

Then $\tilde{u}_n$ is the solution of

$$\varepsilon(\tilde{u}_{n+1} - \tilde{u}_n) = k\left(ia_{n+1}\tilde{u}_{n+1} - iD_-\left(\frac{f_{n+1}}{a_{n+1}}\right)\right), \qquad D_- g_{n+1} =: \frac{g_{n+1} - g_n}{k}.$$

Repeating the process we obtain an asymptotic expansion of the slow discrete solution

$$u_n^S = \frac{if_n}{a_n} + \frac{\varepsilon}{a_n} D_-\left(\frac{f_n}{a_n}\right) + \mathcal{O}(\varepsilon^2). \tag{3.4}$$

Thus

$$\begin{aligned} |y^S(t_n) - u_n^S| &= \frac{\varepsilon}{|a_n|}\left|\left(\frac{f(t_n)}{a(t_n)}\right)' - D_-\left(\frac{f_n}{a_n}\right)\right| + \mathcal{O}(\varepsilon k^2) \\ &= \mathcal{O}(\varepsilon k). \end{aligned}$$

If we have chosen the initial data by

$$y_0 = y^S(0), \tag{3.5}$$

then the fast part of the discrete solution satisfies

$$\begin{aligned} v_{n+1} &= \kappa v_n, \qquad \kappa = \frac{1}{1 - i(k/\varepsilon)}, \\ v_0 &= y_0 - u^S(0) = \mathcal{O}(\varepsilon k). \end{aligned}$$

Thus the fast part of the discrete solution is, in general, not zero. However, $|\kappa| \ll$ and therefore

$$v_n = \kappa^n v_0 \tag{3.6}$$

shows that v_n converges rapidly to zero regardless of how we choose the initial data u_0. Thus the implicit Euler method will always determine the slow solution, if $k \gg \varepsilon$.

The midpoint rule is given by

$$\varepsilon(u_{n+1} - u_n) = \frac{k}{2}(i\,a_{n+1}u_{n+1} + f_{n+1} + i\,a_n u_n + f_n). \qquad (3.7)$$

Now the discrete slow solution satisfies

$$u_n^S = y^S(t_n) + \mathcal{O}(\varepsilon k^2). \qquad (3.8)$$

The fast part is the solution of

$$v_{n+1} = \kappa v_n, \quad \kappa = \frac{1 + \frac{1}{2}ik/\varepsilon}{1 - \frac{1}{2}ik/\varepsilon} = -\exp\left(\frac{-4i\varepsilon}{k}\right) + \mathcal{O}\left(\frac{\varepsilon^3}{k^3}\right) \approx -1,$$

$$v_0 = u_0 - u_0^S. \qquad (3.9)$$

In this case the fast part will not be damped. Instead it will oscillate like a ± 1 wave. If we choose $u_0 = y^S(0)$, then $v_0 = \mathcal{O}(\varepsilon k^2)$ is small.

The following local smoothing procedure can be used to decrease the amplitude of the fast wave, even if v_0 is large. (See Lindberg (1971).)

1   Starting with u_0 calculate u_1, u_2.

2   Determine new initial data at $t = k$ by

$$\tilde{u}_1 = u_1 + \tfrac{1}{4}(u_2 - 2u_1 + u_0).$$

Repeat the process starting at $t = k$. We have

$$u_j = u_j^S + \kappa^j(u_0 - u_0^S).$$

Therefore

$$\begin{aligned} \tilde{u}_1 &= u_1^S + \tfrac{1}{4}(u_2^S - 2u_1^S + u_0^S) + (\kappa + \tfrac{1}{4}(\kappa - 1)^2)(u_0 - u_0^S) \\ &= u_1^S + \mathcal{O}(k^2) + \mathcal{O}(\varepsilon/k)(u_0 - u_0^S). \end{aligned}$$

Thus the amplitude of the fast solution has been reduced by a factor $\mathcal{O}(\varepsilon/k)$. Generalizations are treated in Majda (1983).

Richardson extrapolation. As we have said in the beginning: If ε is very small, then it is uneconomical to use an explicit method. However, in applications the systems can be very large and it can therefore be quite expensive to solve the linear systems connected with the implicit methods. We know that the slow solution can be expanded into an asymptotic series in ε.

$$u_n^S(k) = \Phi_0 + \varepsilon\Phi_1 + \mathcal{O}(\varepsilon^2).$$

Therefore we can use *Richardson extrapolation*. We change ε to a more moderate value $\varepsilon^* \gg \varepsilon$ and calculate

$$\begin{aligned} u_n^S(\varepsilon^*) &= \Phi_0 + \varepsilon^*\Phi_1 + \mathcal{O}(\varepsilon^{*2}) \\ u_n^S(2\varepsilon^*) &= \Phi_0 + 2\varepsilon^*\Phi_1 + \mathcal{O}(\varepsilon^{*2}) \end{aligned}$$

Then

$$\begin{aligned} \varepsilon^*\Phi_1 &= u_n^S(2\varepsilon^*) - u_n^S(\varepsilon^*) + \mathcal{O}(\varepsilon^{*2}) \\ \Phi_0 &= u_n^S(\varepsilon^*) - (u_n^S(2\varepsilon^*) - u_n^S(\varepsilon)) + \mathcal{O}(\varepsilon^{*2}), \end{aligned}$$

i.e.

$$u_n^S(\varepsilon) = 2u_n^S(\varepsilon^*) - u_n^S(2\varepsilon^*) + \frac{\varepsilon}{\varepsilon^*}(u_n^S(2\varepsilon^*) - u_n^S(\varepsilon^*)) + \mathcal{O}(\varepsilon^{*2}).$$

For moderate values of ε^* we may be able to use an explicit method. The main difficulty is that the initial data have to properly initialized, because Richardson extrapolation does not work for the fast part of the solution.

Until now we have concentrated on calculating the slow solution. If we also want to calculate the fast part of the solution, then we have two possibilities.

1   Use a difference method and resolve the fast scale, i.e. choose $k \ll \varepsilon$.
2   Solve the homogeneous equation

$$\begin{aligned} \varepsilon v' &= ia(t)v, \\ v(0) &= v_0, \end{aligned}$$

analytically:

$$v(t) = \exp\left[\left(\frac{i}{\varepsilon}\right)\int_0^t a(\xi)\,d\xi\right]v(0).$$

3.2. *Slow solutions of fast systems*

We consider systems

$$\begin{aligned} \varepsilon y' &= A(t)y + f(t) \\ y(0) &= y_0, \end{aligned} \tag{3.10}$$

where $A(t)$, $f(t)$ satisfy Assumption 1.1. We can calculate the slow solution in the same way as for Example 3.1.

Asymptotic expansion.

$$y^S(t) = -A^{-1}(t)f(t) + \varepsilon A^{-1}(t)(A^{-1}(t)f(t))' + \mathcal{O}(\varepsilon^2).$$

Implicit difference approximation. The implicit Euler scheme has the form

$$\varepsilon(u_{n+1} - u_n) = k(A_{n+1}u_{n+1} + f_{n+1}).$$

For the slow part of its solution we have the asymptotic expansion

$$u_n^S = -A_n^{-1}f_n + \varepsilon A_n^{-1}D_-(A_n^{-1}f_n) + \mathcal{O}(\varepsilon^2),$$

and the fast part satisfies

$$v_{n+1} = \left(I - \frac{k}{\varepsilon}A_{n+1}\right)^{-1}v_n = -\frac{\varepsilon}{k}A_{n+1}^{-1}\left(I - \frac{\varepsilon}{k}A_{n+1}^{-1}\right)v_n.$$

Thus, if $\varepsilon \ll k$, then

$$\left|\frac{\varepsilon}{k}A_{n+1}^{-1}\right| \ll 1$$

and v_n converges rapidly to zero, i.e. the implicit Euler scheme gives us the slow solution regardless of the initial data.

The midpoint rule. The same arguments as in the scalar case show that the slow part of the solution can be described by an asymptotic expansion and that the fast part becomes a ± 1-wave, which can be damped by a local smoothing.

Richardson extrapolation. The possibility of Richardson extrapolation depends only on the existence of asymptotic expansions. Therefore we can also use it here.

3.3. Slow solution of the full system

We consider the system (2.2). All the methods discussed in the previous section can be generalized.

Asymptotic expansions. By Theorem 2.3 the first term $(v^{(0)}, y^{(0)})$ is the solution of

$$\begin{aligned}
0 &= A(t)y^{(0)} + f(v^{(0)}, t) \\
(v^{(0)})' &= g(v^{(0)}, y^{0}, t) \\
v^{(0)}(0) &= v_0.
\end{aligned} \tag{3.11}$$

Higher order terms are obtained by the iteration

$$\begin{aligned}
\varepsilon(y^{(j)})' &= A(t)y^{(j+1)} + \varepsilon C(v, y^{(j)}, t)y^{(j)} + f(v, t) \\
(v^{(j+1)})' &= g(v^{(j+1)}, y^{(j+1)}, t) \\
v^{(j+1)}(0) &= v_0.
\end{aligned} \tag{3.12}$$

The differential equations can be solved by any standard method.

Implicit difference methods. We can use *the backward Euler scheme* or *the midpoint rule* for the complete system. However, for less work we obtain better accuracy, if we apply these schemes to the fast part (*y*-variables) only. For example, we can use the Euler scheme for the *y*-variables and a stable *Adam's method* for

$$\varepsilon(\tilde{y}_{n+1} - \tilde{y}_n) = k(A_{n+1} + \varepsilon C_{n+1})\tilde{y}_{n+1} + \tilde{f}_{n+1}), \quad \tilde{f}_{n+1} = f(\tilde{v}_{n+1}, t_{n+1}),$$

$$\tilde{v}_{n+1} = \tilde{v}_n + \sum_j \beta_j \tilde{g}_{n-j}, \quad \tilde{g}_{n-j} = g(\tilde{v}_{n-j}, \tilde{y}_{n-j}, t_{n-j}). \qquad (3.13)$$

We can also develop the solutions of (3.13) into an asymptotic expansion and compare it with (3.11) and (3.12). This results in a satisfactory error estimate. However, the stability of the method has not been investigated.

In applications the system has often the form (2.1) and one uses a combination of *leap-frog* and the midpoint rule to solve it, i.e.

$$\tilde{w}_{n+1} - \tilde{w}_{n-1} = \frac{k}{\varepsilon}(A_{1\,n+1}\tilde{w}_{n+1} + A_{1\,n}\tilde{w}_n) + 2k\tilde{f}_{1\,n}. \qquad (3.14)$$

We shall give a truncation error and stability analysis of the method.

Let $w(x, t)$ be a slow solution of (2.1) and introduce it into (3.14). We obtain

$$w_{n+1} - w_{n-1} - \frac{k}{\varepsilon}(A_{1\,n+1}w_{n+1} + A_{1\,n}w_n) - 2kf_{1\,n}$$
$$= w_{n+1} - w_{n-1} - k(w'_{n+1} + w'_{n-1}) + k(f_{1\,n+1} + f_{1\,n-1} - 2f_{1\,n})$$
$$= ck^3 w'''_n + \mathcal{O}(k^4).$$

Thus the method is second order accurate. To discuss the stability we linearize (3.14) and freeze coefficients, i.e. we consider

$$\tilde{v}_{n+1} - \tilde{v}_{n-1} = \frac{k}{\varepsilon}(A_1\tilde{v}_{n+1} + A_1\tilde{v}_{n-1}) + 2kB\tilde{v}_n, \qquad (3.15)$$

where A, B are constant matrices. The stability follows from

Theorem 3.3 Assume that

$$A_1 + A_1^* \leq 0, \quad B = -B^*, \quad |kB| \leq 1 - \delta,$$

then

$$\delta(|\tilde{v}_{n+1}|^2 + |\tilde{v}_n|^2) \leq 2(|\tilde{v}_1|^2 + |\tilde{v}_0|^2). \qquad (3.16)$$

Proof. Multiplying (3.15) by $\tilde{v}_{n+1} + \tilde{v}_{n-1}$ gives us

$$|\tilde{v}_{n+1}|^2 - |\tilde{v}_{n-1}|^2 = \frac{k}{\varepsilon}\langle \tilde{v}_{n+1} + \tilde{v}_{n-1}, A_1(\tilde{v}_{n+1} + \tilde{v}_{n-1})\rangle$$
$$+ 2k\langle \tilde{v}_{n+1}, B\tilde{v}_n\rangle + 2k\langle \tilde{v}_{n-1}, B\tilde{v}_n\rangle$$
$$\leq 2k\langle \tilde{v}_{n+1}, B\tilde{v}_n\rangle + 2k\langle \tilde{v}_{n-1}, B\tilde{v}_n\rangle,$$

i.e.

$$L_{n+1} \quad =: \quad |\tilde{v}_{n+1}|^2 + |\tilde{v}_n|^2 - 2k \operatorname{Re} \langle \tilde{v}_{n+1}, B\tilde{v}_n \rangle$$
$$\leq \quad |\tilde{v}_n|^2 + |\tilde{v}_{n-1}|^2 - 2k \operatorname{Re} \langle \tilde{v}_n, B\tilde{v}_{n-1} \rangle = L_n.$$

Therefore

$$L_n \leq L_1.$$

Observing that

$$\langle v, Bw \rangle \leq \tfrac{1}{2}|B|(|v|^2 + |w|^2)$$

(3.16) follows. $\square$

The last theorem tells us that the combination of leap-frog and the mid-point rule is stable, if the slow part is oscillatory $(B = -B^*)$. If this is not the case, then the weak instability of the leap-frog scheme can cause difficulties.

It is very desirable to prove stability of other combinations, for example, the combination of the implicit Euler scheme with a Runge–Kutta method . Also, it is important to investigate when such a combination automatically determines the slow solution.

Richardson extrapolation. As in the previous section the method depends only on the existence of an asymptotic expansion. Therefore it can be used here.

3.4. Highly oscillatory solutions of linear systems

Highly oscillatory problems have been studied for a long time, and a large number of perturbation techniques have been developed: multi-scaling, averaging and the near identity transformation (see, for example, Bogoliubov and Mitropolsky (1961), Nayfeh (1973), Hoppensteadt and Miranker (1976), Kevorkian and Cole (1981), Neu (1980)). For the most part these tools are difficult to implement numerically. We feel that effective numerical techniques are only available for special problems. We will discuss such methods in the next two sections.

In this section we consider linear systems

$$\varepsilon y' \quad = \quad A(t)y, \quad 0 < \varepsilon \leq \varepsilon_0,$$
$$y(0) \quad = \quad y_0. \tag{3.17}$$

We make

Assumption 3.1 $A(t), A^{-1}(t) \in C^\infty$ and their derivatives are bounded independently of ε. The eigenvalues of A are distinct and

$$\operatorname{Re} \lambda \leq 0.$$

If one solves (3.17) by difference approximation, then one has to use a time step $k \ll \varepsilon$. We want to show that one can solve the system by analytic means.

This assumption implies that there is a nonsingular transformation

$$S(t) = (s_1(t), \ldots, s_n(t)) \in C^\infty, \quad s_j(t) \text{ eigenvalues of } A,$$

varying on the slow scale such that

$$S^{-1}AS = \begin{pmatrix} \lambda_1 & & 0 \\ & \ddots & \\ 0 & & \lambda_n \end{pmatrix} =: \Lambda.$$

Introducing into (3.17) a new variable $y_1 = S(t)y$ gives us

$$\varepsilon \, dy_1/dt = (\Lambda + \varepsilon B)y_1, \quad B = -S^{-1} \, dS/dt.$$

Now we can find a slowly varying transformation S_1 such that

$$(I+\varepsilon S_1)^{-1}(\Lambda+\varepsilon B)(I+\varepsilon S_1) = \begin{pmatrix} \lambda_1 + \varepsilon\lambda_{11} & & 0 \\ & \ddots & \\ 0 & & \lambda_n + \varepsilon\lambda_{n1} \end{pmatrix} =: \Lambda+\varepsilon\Lambda_1.$$

The change of variables

$$y_1 = (I + \varepsilon S_1)y_2$$

gives us

$$dy_2/dt = (\Lambda + \varepsilon\Lambda_1 + \varepsilon^2 B_1)y_2.$$

This process can be continued. Thus we can diagonalize the system to any order of $\mathcal{O}(\varepsilon^p)$. Neglecting the $\mathcal{O}(\varepsilon^p)$-terms we obtain scalar equations, whose solutions can be written down explicitly. We have proved

Theorem 3.1 The solution of (3.17) can be calculated analytically to any order in ε.

If the eigenvalues change multiplicity, then difficulties arise. An initial discussion can be found in Scheid (1982).

We consider now systems

$$dy/dt = A(t)y + F(t, \varepsilon),$$

where A is slowly varying and F has the property that

$$\int_0^t F(\eta, \varepsilon) \, d\eta = \mathcal{O}(\varepsilon).$$

For example, this is the case if

$$F = e^{it/\varepsilon}g(t), \quad g(t) \text{ slowly varying.}$$

By Duhamel's principle we can write the solution as

$$y(t) = S(t,0)y(0) + \int_0^t S(t,\xi)F(\xi)\,d\xi,$$

where $S(t,\xi)$ denotes the solution operator of the homogeneous differential equation. It is a slowly vaying function of t,ξ. Therefore integration by parts gives us

$$\int_0^t S(t,\xi)F(\xi)\,d\xi = S(t,\xi)\int_0^\xi F(\eta)\,d\eta\Big|_0^t - \int_0^t \frac{\partial S(t,\xi)}{\partial \xi}\int_0^\xi F(\eta)\,d\eta\,d\xi = \mathcal{O}(\varepsilon).$$

Thus F changes the solution by an $\mathcal{O}(\varepsilon)$-term. One can derive an asymptotic expansion, if more about F is known.

Numerical methods based on these results are exploited in Amdursky and Ziv (1977), Fatunla (1980), Gautschi (1961), Miranker (1981) and Scheid (1982).

3.5. Highly oscillatory solutions of nonlinear equations

We start with a number of examples.

$$y' = \frac{i\lambda}{\varepsilon}y + y^2,$$
$$y(0) = y_0. \tag{3.18}$$

We can calculate the solution of (3.11) explicitly. Introducing a new variable by

$$y = \exp\left(\frac{i\lambda}{\varepsilon}t\right)\tilde{y}$$

gives us

$$\tilde{y}' = \exp\left(\frac{i\lambda}{\varepsilon}t\right)\tilde{y}^2,$$
$$\tilde{y}(0) = y_0. \tag{3.19}$$

Therefore

$$\int_0^t \frac{\tilde{y}'}{\tilde{y}^2}\,dt = -\frac{i\varepsilon}{\lambda}\left[\exp\left(\frac{i\lambda}{\varepsilon}t\right) - 1\right],$$

i.e.

$$\tilde{y} = \frac{1}{\frac{1}{y_0} + \frac{i\varepsilon}{\lambda}\left[\exp\left(\frac{i\lambda}{\varepsilon}t\right) - 1\right]} = \frac{y_0}{1 + \frac{i\varepsilon}{\lambda}y_0\left[\exp\left(\frac{i\lambda}{\varepsilon}t\right) - 1\right]}$$
$$= y_0\left\{1 - \frac{i\varepsilon}{\lambda}y_0\left[\exp\left(\frac{i\lambda}{\varepsilon}t\right) - 1\right]\right\} + \mathcal{O}(\varepsilon^2). \tag{3.20}$$

Thus the nonlinear term changes the solution only by $\mathcal{O}(\varepsilon)$ in arbitrarily long time intervals.

It is also useful to calculate the solution in another way. (3.17) gives us

$$\tilde{y} - y_0 = \int_0^t \exp\left(\frac{i\lambda}{\varepsilon}\xi\right) \tilde{y}^2 \, d\xi$$

$$= -\frac{i\varepsilon}{\lambda} \exp\left(\frac{i\lambda}{\varepsilon}\xi\right) \tilde{y}^2 \Big|_0^t + \frac{2i\varepsilon}{\lambda} \int_0^t \exp\left(\frac{i\lambda}{\varepsilon}\xi\right) \tilde{y}\tilde{y}' \, d\xi$$

$$= \frac{i\varepsilon}{\lambda} \left[\exp\left(\frac{i\lambda}{\varepsilon}t\right) \tilde{y}^2(t) - \tilde{y}^2(0)\right] + \frac{2i\varepsilon}{\lambda} \int_0^t \exp\left(\frac{2i\lambda}{\varepsilon}\xi\right) \tilde{y}^3 \, d\xi.$$

The last integral can again be treated by integration by parts. Therefore

$$\tilde{y}(t) + \frac{i\varepsilon}{\lambda} \exp\left(\frac{i\lambda}{\varepsilon}t\right) \tilde{y}^2(t) = y_0 + \frac{i\varepsilon}{\lambda} y^2(0) + \mathcal{O}(\varepsilon^2),$$

i.e.

$$\tilde{y} = y_0 \left\{1 - \frac{i\varepsilon}{\lambda} y_0 \left[\exp\left(\frac{i\lambda}{\varepsilon}t\right) - 1\right]\right\} + \mathcal{O}(\varepsilon^2),$$

and we again obtain (3.18).

Now consider

$$y' = \frac{i\lambda_1}{\varepsilon} y + v, \quad v' = \frac{i\lambda_2}{\varepsilon} v + v^2. \tag{3.21}$$

The change of variables

$$y = \exp\left(\frac{i\lambda_1}{\varepsilon}t\right) \tilde{y}, \quad v = \exp\left(\frac{i\lambda_2}{\varepsilon}t\right) \tilde{v}$$

gives us

$$\tilde{y}' = \exp\left(\frac{i(\lambda_2 - \lambda_1)t}{\varepsilon}\right) \tilde{v}, \quad \tilde{v}' = \exp\left(\frac{i\lambda_2}{\varepsilon}t\right) \tilde{v}^2.$$

By (3.19)

$$\tilde{v} = v_0 + \sum_{j=1}^{\infty} \varepsilon^j \beta^j \exp\left(\frac{ij\lambda_2}{\varepsilon}t\right).$$

Therefore

$$\tilde{y}' = \exp\left(\frac{i}{\varepsilon}(\lambda_2 - \lambda_1)t\right) v_0 + \sum_{j=1}^{\infty} \varepsilon^j \beta^j \exp\left(\frac{i}{\varepsilon}[(j+1)\lambda_2 - \lambda_1]t\right).$$

If $\nu\lambda_2 - \lambda_1 \neq 0$ for all $\nu = 1, 2, \ldots$, then

$$\tilde{y}(t) = y_0 + \mathcal{O}(\varepsilon).$$

However, if $\nu\lambda_2 = \lambda_1$, then resonance occurs and

$$\tilde{y}(t) = \begin{cases} y(0) + \varepsilon^{\nu-1}\beta^{\nu-1}t & \text{if } \nu > 1, \\ y(0) + tv_0 & \text{if } \nu = 1. \end{cases}$$

Thus the solution is not bounded.

Now we can discuss systems

$$y' = \frac{i}{\varepsilon}\Lambda y + P(y),$$
$$y(0) = y_0, \qquad\qquad (3.22)$$

where P is a polynomial in y. Introducing new variables by

$$y = \exp\left(\frac{i}{\varepsilon}\Lambda t\right)\tilde{y}$$

gives us

$$\tilde{y}' = \exp\left(-\frac{i}{\varepsilon}\Lambda t\right)P\left[\exp\left(\frac{i}{\varepsilon}\Lambda t\right)\tilde{y}\right],$$
$$\tilde{y}(0) = y_0. \qquad\qquad (3.23)$$

The right-hand side of (3.22) consists of expressions

$$\exp\left[\frac{i}{\varepsilon}\left(\sum \alpha_j \lambda_j\right)t\right]p(\tilde{y}), \qquad\qquad (3.24)$$

where the α_j are integers and p is a polynomial in y. There are two possibilities.

1 $\underline{\tau = \sum \alpha_j \lambda_j = 0 \text{ for some terms.}}$ In this case (3.22) has the form

$$\tilde{y}' = Q_0(\tilde{y}) + Q_1(\tilde{y}), \qquad\qquad (3.25)$$

where Q_0, Q_1 contain the terms with and without exponentials, respectively. Our result in the last section tells us that we commit an error of order $\mathcal{O}(\varepsilon)$, if we neglect Q_1. Thus y is to first approximation the solution of

$$\tilde{y}' = Q_1(\tilde{y}), \quad \tilde{y}(0) = y_0, \qquad\qquad (3.26)$$

i.e. in general $\tilde{y}(t)$ does not stay close to y_0.

2 $\underline{\tau = \sum \alpha_j \lambda_j \neq 0 \text{ for all terms.}}$ Integration by parts gives us

$$\begin{aligned}
\tilde{y}(t) &= y_0 + \sum_\tau \int_0^t \exp\left(\frac{i\tau}{\varepsilon}\xi\right)p_\tau(\tilde{y})\,d\xi \\
&= y_0 - i\varepsilon \sum_\tau \frac{1}{\tau}\exp\left(\frac{i\tau}{\varepsilon}\xi\right)p_\tau(\tilde{y})\big|_0^t + i\varepsilon \sum_\tau \frac{1}{\tau}\int_0^t \exp\left(\frac{i\tau}{\varepsilon}\xi\right)\frac{\partial p_\tau}{\partial \tilde{y}}\tilde{y}'\,d\xi \\
&= y_0 - i\varepsilon \sum_\tau \frac{1}{\tau}\exp\left(\frac{i\tau}{\varepsilon}\xi\right)p_\tau(\tilde{y})\big|_0^t + i\varepsilon \sum_\tau \frac{1}{\tau}\int_0^t \exp\left(\frac{i\tau}{\varepsilon}\xi\right)\tilde{p}_\tau(\tilde{y})\,d\xi.
\end{aligned}$$
$$(3.27)$$

The integrals in (3.26) are over terms of type (3.23) and therefore we can repeat the previous arguments. If some of the terms are not of exponential type, then they will in general be of order $\mathcal{O}(\varepsilon t)$.

If all the terms are of exponential type, then we can use integration by parts to reduce them to (at least) $\mathcal{O}(\varepsilon^2 t)$.

We obtain

Theorem 3.2 Assume that for all integers α_j the linear combinations $\sum \alpha_j \lambda_j$ do not vanish. Then

$$\tilde{y} = y_0 + \mathcal{O}(\varepsilon)$$

in time intervals $0 \leq t \leq T$. $T = \mathcal{O}(\varepsilon^p)$ for any p.

There are no difficulties in extending the results and techniques to more general equations

$$y' = \frac{1}{\varepsilon} \Lambda(t) y + P(y, t).$$

Here $\Lambda(t)$ is slowly varying and $P(y, t)$ is a polynomial in y with slowly varying coefficients in time.

The numerical methods based on these results are exploited in Kreth (1977), Miranker and Wabba (1976), Miranker and van Veldhuisen (1978) and Scheid (1982).

3.6. Calculations of solutions, which contain both a fast and a slow part

One can solve these problems by brute force, i.e. use a time step $k \ll \varepsilon$. If one instead wants to calculate only the slow scale, i.e. $\varepsilon \ll k \ll 1$, then one has to combine analytic techniques with numerical methods. Very little is known about how to do this, see however (Petzold, 1981) for a different approach.

If the system has the form (2.1), splitting techniques have been used: Assume that we know the solution at time t_n, then we calculate the solution of

$$
\begin{aligned}
(w^{(1)}(t))' &= f_1(w^{(1)}, t), \quad t_n \leq t \leq t_{n+1}, \\
w^{(1)}(t_n) &= w(t_n)
\end{aligned}
$$

at $t = t_{n+1}$ to obtain $w^{(1)}(t_{n+1})$.

The next step is to solve

$$
\begin{aligned}
(w(t))' &= \frac{1}{\varepsilon} A(t) w(t), \quad t_n \leq t \leq t_{n+1}, \\
w(t_n) &= w^{(1)}(t_{n+1})
\end{aligned}
$$

analytically, using the results of Section 3.3. This gives us $w(t_{n+1})$. It is not at all clear what the accuracy of this procedure is. We believe it has to be modified before it is generally useful, because in general the error is $\mathcal{O}(1)$.

Assume now that the system has the form (2.2). If the fast part of the solution is small, then by Theorem 2.2 the effect of the fast part on the slow scale is one order of magnitude smaller. Therefore we can calculate the slow part of the solution first, and then treat the fast part as a perturbation, i.e. we have to solve

$$\varepsilon(y^F)' = (A(t) + \varepsilon C_1(v^S, y^S, t))y^F$$

by analytic techniques as described in Section 3.3. The next step is to determine the effect of the fast scale on the slow scale. We believe that progress can be made along these lines, but no results are available yet.

4. Partial differential equations

4.1. General theory

Let $0 < \varepsilon \leq \varepsilon_0$ be a small constant, $x = (x_1, \ldots, x_s)$ be a point in the real s-dimensional Euclidian space R_s, e_j the unit vector in the x_j direction and $u = (u^{(1)}(x,t), \ldots, u^{(n)}(x,t))^T$ a vector function with n components. We consider systems

$$u_t = \varepsilon^{-1} P_0(\partial/\partial x)u + P_1(u, x, t, \varepsilon, \partial/\partial x)u + F(x, t) \tag{4.1}$$

with periodic boundary conditions

$$u(x + 2\pi e_j) = u(x), \quad j = 1, 2, \ldots, s$$

and smooth periodic initial data

$$u(x, 0) = f(x). \tag{4.2}$$

Here $F(x, t)$ is a smooth function of x, t with derivatives of order $\mathcal{O}(1)$, and the coefficients of

$$P_0 = \sum_{j=1}^{s} A_j \partial/\partial x_j, \quad A_j = A_j^* \text{ constant matrices,}$$

$$P_1 = \sum_{j=1}^{s} B_j(u, x, t, \varepsilon)\partial/\partial x_j, \quad B_j = B_j^* \text{ smooth functions of all variables,}$$

$$\tag{4.3}$$

are real symmetric matrices.

We want to prove that the bounded derivative principle is valid. We follow Browning and Kreiss (1982) closely. (See also Klainerman and Majda (1982) and Kreiss (1980).)

Theorem 4.1 Assume that p time derivatives at $t = 0$ are bounded independently of ε. Then the same is true in a time interval $0 \leq t \leq T$, where $T > 0$ does not depend on ε.

Proof. We consider first the system

$$w_t = P_1(w, x, t, \varepsilon, \partial/\partial x)w. \tag{4.4}$$

Let

$$(u, v) = \int \langle u, v \rangle \mathrm{d}x, \quad \|u\|^2 = (u, u)$$

denote the usual L_2-scalar product and norm. In the usual way we can derive *a priori* estimates by constructing differential inequalities for

$$\frac{\partial}{\partial t} \|D^{|j|}w\|^2 = 2 \operatorname{Re}(D^{|j|}w, D^{|j|}(P_1 w)), \tag{4.5}$$

where $j = (j_1, \ldots, j_s)$, $|j| = \sum j_i$, is a multi-index, and

$$D^{|j|}w = \partial^{j_1}/\partial x_1^{j_1}, \ldots, \partial^{j_s}/\partial x_s^{j_s}.$$

(See, for example, Kreiss and Lorenz (1989).)

If we consider all expressions $\|D^{|j|}w\|^2$ with $|j| \leq [s/2] + 2$, then we can obtain a closed system of differential inequalities. The solutions of this system are bounded in some time interval $0 \leq t \leq T$, where $T > 0$ depends on the initial data but not on ε. Thus we obtain bounds of the first $[s/2] + 2$ space derivatives. Higher order derivatives can then be estimated in the same time interval.

Now consider the system (4.1). Corresponding to (4.5) we have

$$\frac{\partial}{\partial t} \|D^{|j|}u\|^2 = 2\varepsilon^{-1} \operatorname{Re}(D^{|j|}u, P_0(\partial/\partial x)D^{|j|}u) + 2 \operatorname{Re}(D^{|j|}u, D^{|j|}(P_1 u)). \tag{4.6}$$

P_0 is a first-order operator with constant symmetric matrix coefficients. Thus

$$2 \operatorname{Re}(D^{|j|}u, P_0(\partial/\partial x)D^{|j|}u) \equiv 0,$$

and we also obtain for u the relationships (4.5). Therefore we can estimate all derivatives independently of ε in the same time interval, where we can estimate the derivatives $D^{|j|}w$. In particular, if we can estimate $D^{|j|}w$ for all times, then the same is true for u.

To obtain estimates of time derivatives we differentiate (4.1) with respect to t. $v = u_t$ satisfies

$$v_t = \varepsilon^{-1}P_0(\partial/\partial x)v + P_1(u, x, t, \varepsilon, \partial/\partial x)v + F_1.$$

Here F_1 depends on x, t and on u and its derivatives. Therefore $\operatorname{Re}(v, P_0 v) = 0$ implies

$$\frac{\partial}{\partial t}\|v\|^2 = 2\varepsilon^{-1} \operatorname{Re}(v, P_0(\partial/\partial x)v) + (v, P_1 v) + (v, F_1)$$

$$\leq \text{constant} \times (\|v\|^2 + \|F_1\|^2).$$

Thus, if $v(x,0)$ is bounded independently of ε, then v stays bounded as long as the space derivatives stay bounded. Higher time derivatives are estimated correspondingly. This proves the theorem. $\square$

We will now make the connection with the theory for ordinary differential equations. For simplicity we assume that the coefficients of P_1 are polynomials in u and do not depend on x, t explicitly.

We have seen that for smooth initial data the solution of the differential equation is smooth in space. Therefore we can develop it into a rapidly convergent Fourier series

$$u(x,t) = \sum_{\omega} \hat{u}(\omega, t) \exp(i\langle \omega, x\rangle), \quad \langle \omega, x\rangle = \sum_{j} \omega_j x_j. \tag{4.7}$$

Introducing (4.7) into (4.1) and neglecting all frequencies with $|\omega| > N$ gives a system of ordinary differential equations

$$\frac{d}{dt}\hat{u}(\omega, t) = \varepsilon^{-1} P_0(i\omega)\hat{u}(\omega, t) + \hat{G}_{\omega}(\hat{u}, t). \tag{4.8}$$

$P_0(i\omega)$ is a skew Hermitean matrix and therefore there is a unitary matrix $U(i\omega)$ such that

$$U^*(i\omega)P_0(i\omega)U(i\omega) = \begin{pmatrix} \hat{R}(i\omega) & 0 \\ 0 & 0 \end{pmatrix}, \quad \det|\hat{R}| \neq 0. \tag{4.9}$$

Thus we can transform (4.8) into the form (2.2).

We will now formalize the process. We make

Assumption 4.1 There is a constant $\delta > 0$ such that for all ω

$$|\hat{R}^{-1}(i\omega)| \leq \delta^{-1}.$$

L_2 consists of all functions

$$f = \sum_{\omega} \hat{f}(\omega) \exp(i\langle \omega, x\rangle), \quad \sum |\hat{f}(\omega)|^2 < \infty,$$

which can be expanded into a Fourier series. Let Q denote the projection

$$f^I = Qf = \sum_{\omega} \hat{U}(i\omega) \begin{pmatrix} I_{\omega} & 0 \\ 0 & 0 \end{pmatrix} \hat{U}^*(i\omega) \exp(i\langle \omega, x\rangle) \hat{f}(\omega).$$

Here I_{ω} is the unit matrix of the same dimension as $\hat{R}(i\omega)$, Q splits L_2 into two subspaces L_2^I, L_2^{II} defined by

$$f^I = Qf, \quad f^{II} = (I - Q)f, \quad f = f^I + f^{II}.$$

Note that Q commutes with P_0, i.e. $QP_0 = P_0 Q$ because

$$QP_0 u = \sum_{\omega} \hat{U}(i\omega) \begin{pmatrix} I_{\omega} & 0 \\ 0 & 0 \end{pmatrix} \hat{U}^*(i\omega)\hat{U}(i\omega) \begin{pmatrix} \hat{R}(i\omega) & 0 \\ 0 & 0 \end{pmatrix}$$

$$\times U^*(\mathrm{i}\omega)\hat{u}(\omega)\exp(\mathrm{i}\langle\omega,x\rangle)$$

$$= \sum_\omega \hat{U}(\mathrm{i}\omega)\begin{pmatrix}\hat{R}(\mathrm{i}\omega) & 0 \\ 0 & 0\end{pmatrix}\hat{U}^*(\mathrm{i}\omega)\hat{u}(\omega)\exp(\mathrm{i}\langle\omega,x\rangle)$$

$$= \sum_\omega \hat{U}(\mathrm{i}\omega)\begin{pmatrix}\hat{R}(\mathrm{i}\omega) & 0 \\ 0 & 0\end{pmatrix}\hat{U}^*(\mathrm{i}\omega)\hat{U}(\mathrm{i}\omega)\begin{pmatrix}I_\omega & 0 \\ 0 & 0\end{pmatrix}$$

$$\times \hat{U}^*(\mathrm{i}\omega)\hat{u}(\omega)\exp(\mathrm{i}\langle\omega,x\rangle)$$

$$= P_0 Q u.$$

Also,

$$P_0 u^{\mathrm{II}} = P_0(I - Q)u$$

$$= \sum_\omega \hat{U}(\mathrm{i}\omega)\begin{pmatrix}\hat{R}(\mathrm{i}\omega) & 0 \\ 0 & 0\end{pmatrix}\left[I - \begin{pmatrix}I_\omega & 0 \\ 0 & 0\end{pmatrix}\right]\hat{U}^*(\mathrm{i}\omega)\hat{u}(\omega)\exp(\mathrm{i}\langle\omega,x\rangle)$$

$$= 0.$$

We can define the inverse of P_0 on L_2^{I} uniquely by

$$P_0^{-1}u^{\mathrm{I}} = \sum_\omega \hat{U}(\mathrm{i}\omega)\begin{pmatrix}\hat{R}^{-1}(\mathrm{i}\omega) & 0 \\ 0 & 0\end{pmatrix}\hat{U}^*(\mathrm{i}\omega)\hat{u}(\omega)\exp(\mathrm{i}\langle\omega,x\rangle),$$

and (4.9) gives us

$$\|P_0^{-1}u^{\mathrm{I}}\| \le \delta^{-1}\|u^{\mathrm{I}}\|. \tag{4.10}$$

Using the projection Q, we can now write the system (4.1) in the form

$$u_t^{\mathrm{I}} = \varepsilon^{-1}P_0 u^{\mathrm{I}} + (P_1(u,\partial/\partial x)u)^{\mathrm{I}} + F^{\mathrm{I}}, \tag{4.11a}$$

$$u_t^{\mathrm{II}} = (P_1(u,\partial/\partial x)u)^{\mathrm{II}} + F^{\mathrm{II}}, \quad u = u^{\mathrm{I}} + u^{\mathrm{II}}, \tag{4.11b}$$

which is the generalization of (2.2) to a partial differential equation.

We can now show that if u is slow to order p, then u^{I} is determined by u^{II} up to terms of order $\mathcal{O}(\varepsilon^p)$. $\partial u/\partial t$ is bounded independently of ε if and only if

$$P_0 u^{\mathrm{I}} = \mathcal{O}(\varepsilon),$$

hence

$$u^{\mathrm{I}} = \varepsilon u_1^{\mathrm{I}}, \quad u_1^{\mathrm{I}} = \mathcal{O}(1), \tag{4.12}$$

i.e. u has to first approximation no component in L_2^{I}. Therefore, to first approximation, the solution to our problem is given by

$$u^{\mathrm{I}} \equiv 0, \quad u_t^{\mathrm{II}} = (P_1(u^{\mathrm{II}},\partial/\partial x)u^{\mathrm{II}})^{\mathrm{II}} + F^{\mathrm{II}}.$$

Differentiating (4.11) with respect to t and assuming that (4.12) holds gives us

$$u_{tt}^{\mathrm{I}} = \varepsilon^{-1}P_0 u_t^{\mathrm{I}} + \mathcal{O}(1)$$

$$= \varepsilon^{-1}P_0\left(P_0u_1^{\mathrm{I}} + (P_1(u^{\mathrm{II}}, \partial/\partial x)u^{\mathrm{II}})^{\mathrm{I}} + F^{\mathrm{I}}\right) + \mathcal{O}(1),$$

$$u_{tt}^{\mathrm{II}} = \mathcal{O}(1). \tag{4.13}$$

Therefore the second time derivative is bounded independently of ε if and only if

$$P_0u_1^{\mathrm{I}} + (P_1(u^{\mathrm{II}}, \partial/\partial x)u^{\mathrm{II}})^{\mathrm{I}} + F^{\mathrm{I}} = \mathcal{O}(\varepsilon).$$

Thus u^{I} is determined by u^{II} up to terms of order $\mathcal{O}(\varepsilon^2)$. This process can be continued, and we obtain the desired relation between u^{I} and u^{II}.

As in Section 2 we can also derive the asymptotic expansion by the iteration

$$(u^{(n-1)})_t^{\mathrm{I}} = \varepsilon^{-1}P_0(u^{(n)})^{\mathrm{I}} + (P_1(u^{(n)}, \partial/\partial x)u^{(n)})^{\mathrm{I}} + F^{\mathrm{I}}, \quad (u^{(-1)})^{\mathrm{I}} \equiv 0,$$

$$(u^{(n)})_t^{\mathrm{II}} = (P_1(u^{(n)}, \partial/\partial x)u^{(n)})^{\mathrm{II}} + F^{\mathrm{II}}, \quad n = 0, 1, 2, \dots. \tag{4.14}$$

It again gives us the relationship between u^{I} and u^{II}. For meteorological applications there are many papers, which describe how to obtain this relationship in practice. (See, for example, Kasahara (1982), Leith (1980), Machenhauer (1977).)

One can generalize the results considerably. However, the theory becomes much more complicated, if $P_0 = P_0(x, t, \partial/\partial x)$ depends on x, t or if one wants to treat the initial boundary value problem. Details can be found in Browning and Kreiss (1982), Kreiss (1980) and Tadmor (1982). Numerical methods are discussed in Guerra and Gustafsson (1982), Gustafsson (1980a), Gustafsson (1980b), Gustaffson and Kreiss (1983).

4.2. The wave equation

We consider in this section the wave equation written as a first-order system

$$\varepsilon u_t = \begin{pmatrix} 0 & 1 \\ 1 & 0 \end{pmatrix} u_x + F$$

$$u(x, 0) = f. \tag{4.15}$$

Here $u, F, f \in C^\infty$ are vector-valued 2π-periodic functions. (4.15) is of the form (4.3) with

$$P_0(\mathrm{i}\omega) = \mathrm{i}\begin{pmatrix} 0 & 1 \\ 1 & 0 \end{pmatrix}\omega, \quad P_1 \equiv 0.$$

Thus the theory applies.

(4.15) implies

$$\varepsilon\frac{\partial}{\partial t}\int_0^{2\pi} u\,\mathrm{d}x = \int_0^{2\pi} F\,\mathrm{d}x,$$

i.e.

$$\int_0^{2\pi} u(x,t)\,dx - \int_0^{2\pi} u(x,0)\,dx = \frac{1}{\varepsilon}\int_0^t\int_0^{2\pi} F(x,\xi)\,dx\,d\xi.$$

If

$$\int_0^{2\pi} F\,dx = \mathcal{O}(1),$$

then the mean value of the solution becomes unbounded for $\varepsilon \to 0$. For simplicity we assume that

$$\int_0^{2\pi} f(x)\,dx = 0, \quad \int_0^{2\pi} F(x,t)\,dx \equiv 0. \tag{4.16}$$

Then

$$\int_0^{2\pi} u(x,t)\,dx \equiv 0. \tag{4.17}$$

We shall now derive the asymptotic expansion. We proceed in the same manner as for the ordinary diferential equations. u has one derivative bounded independently of ε, if

$$\begin{pmatrix} 0 & 1 \\ 1 & 0 \end{pmatrix} u_x + F = \mathcal{O}(\varepsilon).$$

This suggests the substitution

$$u = u_1 + \varphi_0, \tag{4.18}$$

where φ_0 is the solution of

$$\begin{pmatrix} 0 & 1 \\ 1 & 0 \end{pmatrix} \varphi_{0x} = -F, \quad \int_0^{2\pi} \varphi_0\,dx = 0.$$

Introducing (4.18) into (4.15) gives us

$$\varepsilon u_{1t} = \begin{pmatrix} 0 & 1 \\ 1 & 0 \end{pmatrix} u_{1x} - \varepsilon\varphi_{0t}, \quad \int_0^{2\pi} u_1\,dx = 0,$$
$$u_1(x,0) = f(x) - \varphi_0(x,0). \tag{4.19}$$

(4.19) is of the same form as (4.15) with the forcing reduced to $\mathcal{O}(\varepsilon)$. Therefore we can repeat the process and we obtain the slow solution

$$u^S = \sum_{j=0}^{p-1} \varepsilon^j \varphi_j + \mathcal{O}(\varepsilon^p).$$

The fast part u^F is the solution of

$$\varepsilon v_t^F = \begin{pmatrix} 0 & 1 \\ 1 & 0 \end{pmatrix} v_x^F, \quad \int_0^{2\pi} v^F\,dx = 0,$$
$$v^F(x,0) = f(x) - u^S(x,0), \tag{4.20}$$

i.e.

$$v^{\mathrm{F}}(x,t) = \sum_j a_j \begin{pmatrix} 1 \\ 1 \end{pmatrix} \exp\left[\mathrm{i}j\left(x + \frac{t}{\varepsilon}\right)\right] + b_j \begin{pmatrix} 1 \\ -1 \end{pmatrix} \exp\left[\mathrm{i}j\left(x - \frac{t}{\varepsilon}\right)\right],$$

where the a_j, b_j are determined by the initial data.

In applications the equations often do not appear as symmetric hyperbolic problems. As an example we consider instead of (4.15)

$$\begin{pmatrix} u \\ p \end{pmatrix}_t = \begin{pmatrix} 0 & 1 \\ \varepsilon^{-2} & 0 \end{pmatrix}\begin{pmatrix} u \\ p \end{pmatrix}_x + \begin{pmatrix} F \\ \varepsilon^{-2}G \end{pmatrix},$$

$$u(x,0) = f, \quad p(x,0) = g. \tag{4.21}$$

We can symmetrize the equations by introducing a new variable

$$\tilde{p} = \varepsilon p$$

and obtain

$$\begin{pmatrix} u \\ \tilde{p} \end{pmatrix}_t = \frac{1}{\varepsilon}\begin{pmatrix} 0 & 1 \\ 1 & 0 \end{pmatrix}\begin{pmatrix} u \\ \tilde{p} \end{pmatrix}_x + \begin{pmatrix} F \\ \varepsilon^{-1}G \end{pmatrix},$$

$$u(x,0) = f, \quad \tilde{p}(x,0) = \varepsilon g, \tag{4.22}$$

which is of the same form as (4.15). Therefore we can again write down the asymptotic expansion of the slow part of the solution and obtain

$$u^{\mathrm{S}} = \varphi_0^{(1)} + \varepsilon\varphi_1^{(1)} + \cdots, \quad \tilde{p}^{\mathrm{S}} = \varepsilon\varphi_0^{(2)} + \varepsilon^2\varphi_0^{(2)} + \cdots,$$

i.e. we also have bounded asymptotic expansions in the original variables $u, p = \varepsilon^{-1}\tilde{p}$. The fast part of the solution is again determined by the homogeneous equation (4.20) with initial data

$$u^{\mathrm{F}}(x,0) = f(x) - u^{\mathrm{S}}(x,0), \quad \tilde{p}^{\mathrm{F}}(x,0) = \varepsilon g(x) - \tilde{p}^{\mathrm{S}}(x,0).$$

Now we assume that the data have not been initialized. Then

$$u^{\mathrm{F}}(x,0) = \mathcal{O}(1), \quad \tilde{p}^{\mathrm{F}}(x,0) = \mathcal{O}(\varepsilon).$$

However, at later times energy from u^{F} will be transferred to $\tilde{p}^{\mathrm{F}}$ and therefore

$$u^{\mathrm{F}}(x,t) = \mathcal{O}(1), \quad \tilde{p}^{\mathrm{F}}(x,t) = \mathcal{O}(1).$$

Then we obtain in the original variables

$$u^{\mathrm{F}}(x,t) = \mathcal{O}(1), \quad p(x,t) = \varepsilon^{-1}p^{\mathrm{F}}(x,t) = \mathcal{O}(\varepsilon^{-1}),$$

and the amplitude of p will be amplified by a factor ε^{-1}.

For moderate values of ε this is not a problem. However, if ε becomes very small, it can cause a lot of trouble in numerical calculations. For example:

1   If the data are initialized analytically but the problem is solved numerically, then the initialization of the difference approximation is, due to truncation errors, different from the analytic initialization.

2   Rapid time changes in F can trigger large waves on the fast scale.

3   If ε is very small, then rounding errors can also cause difficulties.

Numerical methods. If one is only interested in the slow solution, then using the asymptotic expansion or Richardson extrapolation are efficient methods. (Observe that one uses Richardson extrapolation with moderate values of ε and therefore one can also treat nonsymmetric systems.)

If one is interested in both the fast and the slow part of the solution, then one can use asymptotic expansion for the slow part of the solution and solve (4.20) analytically.

5. Applications

5.1. Low Mach number flow

A slightly simplified version of the Euler equations for low Mach number flow in two space dimensions is given by

$$\begin{aligned} \mathbf{u}_t + (\mathbf{u} \cdot \nabla)\mathbf{u} + \nabla p &= F \\ M^2(p_t + (\mathbf{u} \cdot \nabla)p) + \nabla \cdot u &= G \end{aligned} \tag{5.1}$$

with initial data

$$\mathbf{u}(x, y, 0) = \mathbf{f}(x, y), \quad p(x, y, 0) = g(x, y). \tag{5.2}$$

Here $0 < M^2 \ll 1$ is the Mach number and

$$\mathbf{u} = (u(x, y, t), v(x, y, t)), \quad p = p(x, y, t)$$

denote the velocity field and the pressure, respectively. We are interested in 2π-periodic solutions.

We can also write (5.1) in component form

$$\begin{pmatrix} u \\ v \\ \tilde{p} \end{pmatrix}_t + \begin{pmatrix} u & 0 & M^{-1} \\ 0 & u & 0 \\ M^{-1} & 0 & u \end{pmatrix} \begin{pmatrix} u \\ v \\ \tilde{p} \end{pmatrix}_x + \begin{pmatrix} v & 0 & 0 \\ 0 & v & M^{-1} \\ 0 & M^{-1} & v \end{pmatrix} \begin{pmatrix} u \\ v \\ \tilde{p} \end{pmatrix}_y = \begin{pmatrix} F_1 \\ F_2 \\ g \end{pmatrix}. \tag{5.3}$$

We have introduced

$$\tilde{p} = Mp \tag{5.4}$$

as a new variable such that the system is symmetric hyperbolic. Thus (5.1) has the same difficulties as (4.21).

The symbol of the large part

$$P_0(i\omega) = M^{-1} \begin{pmatrix} 0 & 0 & i\omega_1 \\ 0 & 0 & i\omega_2 \\ i\omega_1 & i\omega_2 & 0 \end{pmatrix} \tag{5.5}$$

has rank two. The general theory tells us that there is one slow variable. In the general theory the slow variable can only be defined via Fourier transform. However, in this case we can also identify it in physical space. It is the vorticity $\nabla \times u =: v_x - u_y$, because differentiating the second equation of (5.3) with respect to x and the first with respect to y and subtracting them gives us

$$\xi_t + u\xi_x + v\xi_y + (u_x + v_y)\xi = F_{2x} - F_{1y}. \tag{5.6}$$

The fast variables are the pressure and the dilutation $\nabla \cdot \mathbf{u} =: u_x + v_y$.

We will now describe the results in Kreiss *et al.* (1991). (See also Klainerman and Majda (1982).) We derive an asymptotic expansion of the slow part, starting from (5.1). If the derivatives are bounded independently of M, then the leading term must satisfy

$$\mathbf{U}_t + (\mathbf{U} \cdot \nabla)\mathbf{U} + \nabla P = F$$
$$\nabla \cdot \mathbf{U} = G \tag{5.7}$$

with initial data defined by

$$\nabla \cdot \mathbf{U}(x, y, 0) = G(x, y, 0), \quad \nabla \times \mathbf{U}(x, y, 0) = \nabla \times \mathbf{f}(x, y). \tag{5.8}$$

Defining new variables by

$$\mathbf{u} = \mathbf{U} + \mathbf{u}', \quad p = P + p', \tag{5.9}$$

we obtain from (5.1)

$$\mathbf{u}'_t + (\mathbf{U} \cdot \nabla)\mathbf{u}' + (\mathbf{u}' \cdot \nabla)\mathbf{U} + (\mathbf{u} \cdot \nabla)\mathbf{u}' + \nabla p' = 0$$
$$M^2\left(p'_t + (\mathbf{U} \cdot \nabla)p' + (\mathbf{u}' \cdot \nabla)P + (\mathbf{u}' \cdot \nabla)p'\right) + \nabla \cdot \mathbf{u}' = M^2 G_1 \tag{5.10}$$

with

$$G_1 = -(p_t + (\mathbf{U} \cdot \nabla)P).$$

First we determine the slow part of $\mathbf{u}', p'$ and write

$$\mathbf{u}' = M^2\mathbf{U}_1 + \mathbf{u}'', \quad p' = M^2 P_1 + p'', \tag{5.11}$$

where

$$\mathbf{U}_{1t} + (\mathbf{U} \cdot \nabla)\mathbf{U}_1 + (\mathbf{U}_1 \cdot \nabla)\mathbf{U} + \nabla P_1 = 0$$
$$\nabla \cdot \mathbf{U}_1 = G_1. \tag{5.12}$$

The initial data for (5.11) are given by

$$\nabla \cdot \mathbf{U}_1(x, y, 0) = G_1(x, y, 0), \quad \nabla \times \mathbf{U}_1(x, y, 0) = 0.$$

Now we introduce $\mathbf{u}'', p''$ as new variables into (5.10) and repeat the procedure. We obtain

Theorem 5.1 We can expand the slow part of the solution of (5.1) into a series

$$u = U + M^2U_1 + \cdots + M^{2l}U_l + u_R =: U^{(l)} + u_R$$
$$p = P + M^2P_1 + \cdots + M^{2l}P_l + p_R = P^{(l)} + p_R,$$

where U_j, P_j satisfy linearized incompressible equations and their derivatives are bounded independently of ε. The remainder u_R, p_R are the solution of

$$u_{Rt} + (U^{(l)} \cdot \nabla)u_R + (u_R \cdot \nabla)U^{(l)} + (u_R \cdot \nabla)u_R + \nabla p_R = M^{4l}F_{l+1}$$
$$M^2\left(p_{Rt} + (U^{(l)} \cdot \nabla)p_R + (u_R \cdot \nabla)P^{(l)} + (u_R \cdot \nabla)p_R\right) + \nabla \cdot u_R = M^{2l+2}G_l.$$

$$(5.13)$$

The initial data

$$u_R(x, y, 0) = f(x, y) - U^{(l)}(x, y, 0), \quad p_R(x, y, 0) = G(x, y, 0) - P^{(l)}(x, y, 0)$$

satisfy

$$\nabla \times u_R(x, y, 0) = 0.$$

One can prove (see Kreiss *et al.* (1991))

Theorem 5.2 In any finite time interval $0 \leq t \leq T$

$$\nabla \times u_R(x, y, t) = \mathcal{O}(M).$$

Thus u_R, p_R represent the fast part of the solution.

To discuss their behaviour and to simplify the notation we introduce new variables by

$$\tau = t/M, \quad q = Mp_R, \quad v = u_R, \quad U^{(l)} = U, \quad P^{(l)} = P.$$

We also neglect the forcing. Then (5.13) becomes

$$v_\tau + M((U \cdot \nabla)v + (v \cdot \nabla)U) + \nabla q = 0$$
$$q_\tau + M(U \cdot \nabla)q + M^2(v \cdot \nabla)P + \nabla \cdot v = 0,$$
$$v(x, y, 0) = f(x, y) - U(x, y, 0), \quad q(x, y, 0) = M(G(x, y, 0) - P(x, y, 0)).$$

$$(5.14)$$

If we neglect terms multiplied by M, (5.14) becomes

$$v_\tau + \nabla q = 0$$
$$q_\tau + \nabla \cdot v = 0. \tag{5.15}$$

(5.14) can be rewritten as the wave equations for q and the dilutation $\nabla \cdot v$. Here we see again that the amplitude of the fast waves are amplified by the factor $1/M$ because $p_R = (1/M)q$. If $|v(x, y, 0)| \approx 1$, then q grows on the fast time scale from $\mathcal{O}(M)$ to $\mathcal{O}(1)$, i.e. p_R becomes $\mathcal{O}(1/M)$. Large fast waves can only be avoided, if $\mathbf{v}(x, y, 0) = \mathcal{O}(M)$. Since the vorticity is $\mathcal{O}(M)$, this can only be achieved, if the dilutation $\nabla \cdot \mathbf{v}(x, y, 0) = \mathcal{O}(M)$.

To solve this problem numerically we can use the asymptotic expansion. If we only want to determine the slow solutions, then we can also use Richardson extrapolation (see Johansson (1991)). If we are also interested in the fast part of the solution, then we have to solve (5.14). Here one should use (5.15) locally, because we can determine its solution analytically.

We have only discussed the periodic problem. However, one can also treat the initial boundary value problem (see Kreiss *et al.* (1991)).

5.2. Atmospheric motions

In this section we consider three-dimensional atmospheric motions and discuss results presented in Browning and Kreiss (1986, 1987). The corresponding results for oceanographic flows can be found in Browning *et al.* (1990). In Cartesian coordinates x, y, z are directed eastward, northward and upward, respectively, and the Eulerian equations have the form (see Kasahara (1974))

$$
\begin{aligned}
&ds/dt = 0, \\
&d/dt = \partial/\partial t + u\,\partial/\partial x + v\,\partial/\partial y + w\,\partial/\partial z, \\
&dp/dt + \gamma p(u_x + v_y + w_z) = 0, \quad \gamma = 1.4, \\
&\rho\,du/dt + p_x - f\rho v = 0, \\
&\rho\,dv/dt + p_y + f\rho u = 0, \quad \rho = sp^{-1}\gamma, \\
&\rho\,dw/dt + p_z + \rho g = 0.
\end{aligned}
\tag{5.16}
$$

Here s is the entropy, p the pressure, ρ the density, u, v, w the velocity components in the x, y, z directions, respectively, and $g \approx 10$ m s^{-2} the gravity acceleration. We make the β-plan approximation, i.e. the Coriolis force f is given by

$$
f = 2\Omega(\sin\theta_0 + \gamma/r\,\cos\theta_0), \quad 2\Omega = 10^{-4}\text{ s}^{-1}, \quad r = 10^7\text{ m}, \tag{5.17}
$$

where r is the radius, Ω the angular speed for the earth, and θ_0 the latitude of the coordinate origin.

We have to introduce scaled variables before we can apply our theory. We change the variables in such a way that the variables and their first derivatives are of order $\mathcal{O}(1)$.

$$
\begin{aligned}
x &= L_1 x', \quad y = L_2 y', \quad z = D y', \quad t = T t', \\
u &= U u', \quad v = V v', \quad w = W w'.
\end{aligned}
\tag{5.18}
$$

Density and pressure can be written in the form

$$
p = P_0(p_0(z) + S_1 p'), \quad \rho = R_0(\rho_0(z) + S_1 \rho'), \quad 0 < S_1 \ll 1, \tag{5.19}
$$

where

$$P_0 \partial p_0/\partial z + gR_0\rho_0 = 0,$$
$$P_0 = 10^5 \text{ kg m}^{-1} \text{ s}^{-2}, \quad R_0 = 1 \text{ kg m}^{-3}.$$

Equations (5.19) express the fact that a number of digits of the pressure and density are independent of x and y and that p and ρ are to the first approximation in the hydrostatic balance. Equations (5.18) also imply that

$$
\begin{aligned}
s &= R_0 P_0^{-1/\gamma} \rho_0(z)(p_0(z))^{-1/\gamma}(1 + S_1\rho'/\rho)(1 + S_1 p'/p_0)^{-1/\gamma} \\
&= R_0 P_0^{-1/\gamma} s_0(z)(1 + S_1 s'), \\
s_0(z) &= \rho_0(z)(p_0(z))^{-1/\gamma}, \quad s' = \rho'/\rho_0 - (1/\gamma)p'/p_0 + \mathcal{O}(S_1). \quad (5.20)
\end{aligned}
$$

We assume that the scales in the x, y directions are the same, that $\partial u/\partial t$ and $\partial v/\partial t$ balance the horizontal convection terms, and that the Coriolis force has a strong influence. This leads to the following relations:

$$
\begin{aligned}
U &= V, \quad L_1 = L_2 = L, \\
UT/L &= 1, \quad 2\Omega T = S_1 P_0/(R_0 U^2). \quad (5.21)
\end{aligned}
$$

These relationships are not valid for special types of motions like jet streams, ultralong waves and small-scale problems. For the treatment of these cases we refer to Browning and Kreiss (1986). Introducing the scaled variables into (5.16) gives us

$$\frac{ds'}{dt'} + S_1^{-1} S_2 \tilde{s}(z)(1 + S_1 s')w' = 0,$$

$$\frac{dp'}{dt'} + S_1^{-1} p_0 \left[\gamma \left(1 + \frac{S_1 p'}{p_0} \right)(u_x + v_y + S_2 w_z) + S_2 \tilde{p}(z)w' \right] = 0,$$

$$\frac{du'}{dt'} + S_3 \left[\rho_0^{-1} \left(1 + \frac{S_1 \rho'}{\rho_0} \right)^{-1} p'_{x'} - f'v' \right] = 0,$$

$$\frac{dv'}{dt'} + S_3 \left[\rho_0^{-1} \left(1 + \frac{S_1 \rho'}{\rho_0} \right)^{-1} p'_{y'} + f'u' \right] = 0,$$

$$\frac{dw'}{dt'} + S_1^{-1} S_4 \rho_0^{-1} \left(1 + \frac{S_1 \rho'}{\rho_0} \right)^{-1} (p'_{z'} - \gamma^{-1}\tilde{p}(z)\rho' + S_5\rho_0 s' + \mathcal{O}(S_1)) = 0,$$

$$(5.22)$$

where

$$\tilde{p}(z) = (\ln p_0)_z, \quad \tilde{s}(z) = (\ln s_0)_z;$$

typically,

$$\tilde{p} \approx -1.3, \quad -3 \le \tilde{s} \le -1;$$

$$d/dt = \partial/\partial t' + u' \, \partial/\partial x' + v' \, \partial/\partial y' + S_2 w' \, \partial/\partial z';$$

and the parameters S_i are given by

$$S_2 = D^{-1}TW, \quad S_3 = 2\Omega T,$$
$$S_4 = TP_0(DR_0W)^{-1}, \quad S_5 = gDP_0^{-1}R_0. \tag{5.23}$$

We now choose the parameters according to the so-called large-scale dynamics:

$$L = 10^6 \text{ m}, \quad D = 10^4 \text{ m}, \quad U = V = 10 \text{ m s}^{-1}, \quad S_1 = 10^{-2}, \quad W = 10^{-2} \text{ m s}^{-1}. \tag{5.24}$$

Introducing these values into (5.23) and (5.24), we obtain, dropping the prime notation,

$$\frac{ds}{dt} + \tilde{s}(z)w = 0,$$

$$\frac{\varepsilon^2}{\gamma p_0}\frac{dp}{dt} + u_x + v_y + \varepsilon Lw = 0, \quad \varepsilon = 10^{-1},$$

$$\varepsilon\frac{du}{dt} + \rho_0^{-1}p_x - fv = 0,$$

$$\varepsilon\frac{dv}{dt} + \rho_0^{-1}p_y + fu = 0,$$

$$\rho_0\varepsilon^6\frac{dw}{dt} - L^*p + \rho_0 s = 0.$$

Here

$$d/dt = \partial/\partial t + u\,\partial/\partial x + v\,\partial/\partial y + \varepsilon w\,\partial/\partial z,$$
$$Lw = w_z + \gamma^{-1}\tilde{p}(z)w, \quad L^*p = -p_z + \gamma^{-1}\tilde{p}(z)p.$$

For simplicity only, we have neglected terms of order $\mathcal{O}(S_1)$. Also, by (5.17),

$$f = f_0 + \varepsilon\beta y.$$

We will only consider the mid-latitude case $f_0 \approx 1$.

For ease of discussion we simplify the equations slightly by replacing

$$\tilde{s}(z) \to -1, \quad (\gamma p_0)^{-1} \to 1, \quad f \to 1, \quad lw \to w_z, \quad \rho_0 \to 1, \quad L^*p \to p_z, \tag{5.25}$$

and obtain

$$\frac{ds}{dt} - w = 0,$$

$$\varepsilon^2\frac{dp}{dt} + u_x + v_y + \varepsilon w_z = 0,$$

$$\varepsilon\frac{du}{dt} + p_x - v = 0,$$

$$\varepsilon\frac{dv}{dt} + p_y + u = 0,$$

$$\eta \frac{dw}{dt} + p_z + s = 0. \tag{5.26}$$

Here we have replaced ε^6 by η, and we think of η as another small parameter.

We will now discuss the initialization. The first time derivatives are bounded independently of ε, if

$$p_z + s = \mathcal{O}(\eta), \tag{5.27a}$$
$$p_y + u = \mathcal{O}(\varepsilon), \quad p_x - v = \mathcal{O}(\varepsilon), \tag{5.27b}$$
$$u_x + v_y + \varepsilon w_z = \mathcal{O}(\varepsilon^2). \tag{5.27c}$$

If we replace $\mathcal{O}(\eta)$ and $\mathcal{O}(\varepsilon)$ by zero in (5.27a) and (5.27b), then the resulting equations are called the *hydrostatic assumption* and *geostrophic approxima-tion*, respectively.

The second time derivatives are bounded independently of ε, if

$$\frac{d}{dt}(p_z + s) = \mathcal{O}(\eta), \tag{5.28a}$$

$$\frac{d}{dt}(p_y + u) = \mathcal{O}(\varepsilon), \quad \frac{d}{dt}(p_x - v) = \mathcal{O}(\varepsilon), \tag{5.28b}$$

$$\frac{d}{dt}(u_x + v_y + \varepsilon w_z) = \mathcal{O}(\varepsilon^2) \tag{5.28c}$$

(5.28) gives us

$$\begin{aligned}
\varepsilon^2 \frac{d}{dt}(p_z + s) &= \varepsilon^2 \left(\frac{dp}{dt}\right)_z - \varepsilon^2 H_1 + \varepsilon^2 \frac{ds}{dt} \\
&= (u_x + v_y + \varepsilon w_z)_z + \varepsilon^2 w - \varepsilon^2 H_1 \\
&= \mathcal{O}(\varepsilon^2 \eta), \tag{5.29}
\end{aligned}$$

where

$$H_1 = u_z p_x + v_z p_y + \varepsilon w_z p_z = u_z v - v_z u + \mathcal{O}(\varepsilon).$$

Thus (5.28a) gives us an improvement of (5.27c). If we replace the $\mathcal{O}(\varepsilon\eta)$ term by zero, then the resulting relation is called *Richardson's equation*.

Correspondingly, (5.28b) and (5.28c) lead to improvements in (5.27b) and (5.27c), respectively.

The primitive equations, that are often used in weather prediction models, are given by

$$\begin{aligned}
\frac{ds}{dt} - w &= 0, \\
\varepsilon \frac{du}{dt} + p_x + u &= 0, \\
\varepsilon \frac{dv}{dt} + p_y - v &= 0, \\
p_z + s &= 0,
\end{aligned}$$

$$(u_x + v_y + \varepsilon w_z)_z + \varepsilon^2 w - \varepsilon^2 H_1 = 0.$$

Their mathematical properties are discussed in Browning and Kreiss (1985) and Oliger and Sundström (1978).

Instead of pursuing the initialization to make more and more time derivatives bounded independently of ε it is easier to achieve this by iteration. We shall first derive a set of equations, which will determine the slow solution to order $\mathcal{O}(\varepsilon)$.

Ler $\xi = v_x - u_y$ denote the horizontal vorticity and use the notation

$$\frac{\mathrm{d}_H}{\mathrm{d}t} = \partial/\partial t + u\,\partial/\partial x + v\,\partial/\partial y.$$

(5.27) gives us the balance equation

$$\Delta_2 p = \xi + \mathcal{O}(\varepsilon), \quad \Delta_2 = \partial^2/\partial x^2 + \partial^2/\partial y^2. \tag{5.30}$$

Differentiating the horizontal momentum equations results in

$$\varepsilon\Big(\frac{\mathrm{d}_H\xi}{\mathrm{d}t} + \xi(u_x + v_y)\Big) + u_x + v_y = \mathcal{O}(\varepsilon^2).$$

Therefore, by (5.27c),

$$\frac{\mathrm{d}_H\xi}{\mathrm{d}t} = w_z + \mathcal{O}(\varepsilon). \tag{5.31}$$

The first equation of (5.26) tells us that

$$w_z = \frac{\mathrm{d}s_z}{\mathrm{d}t} + u_z s_x + v_z s_y + \mathcal{O}(\varepsilon),$$

i.e. by (5.27a) and (5.27b),

$$\begin{aligned}
\frac{\mathrm{d}_H}{\mathrm{d}t}(\xi - s_x) &= u_z s_x + v_z s_y + \mathcal{O}(\varepsilon) \\
&= -u_z p_{xz} - v_z p_{yz} + \mathcal{O}(\varepsilon) = \mathcal{O}(\varepsilon). \tag{5.32}
\end{aligned}$$

By (5.30) and (5.27c) we can also write (5.32) as

$$\frac{\mathrm{d}_H}{\mathrm{d}t}\Delta p = \mathcal{O}(\varepsilon).$$

Therefore the slow solution satisfies to a first approximation

$$\begin{aligned}
\frac{\mathrm{d}_H}{\mathrm{d}t}\Delta p &= 0, \\
p_z + s &= 0, \quad p_y + u = 0, \quad p_x - v = 0, \\
w &= \frac{\mathrm{d}_H s}{\mathrm{d}t}. \tag{5.33}
\end{aligned}$$

Higher order approximations are obtained by iteration.

Remark. If we had not made the simplification (5.25), then (5.33) would be slightly more complicated. It would still be a well posed problem.

We have assumed that we can apply our theory. We will now discuss this question. We can symmetrize the equations by introducing new variables

$$\sqrt{\varepsilon}\, p = \tilde{p}, \quad \sqrt{\eta}\, w = \tilde{w},$$

and obtain

$$\frac{ds}{dt} - \frac{1}{\sqrt{\eta}}\tilde{w} = 0,$$

$$\frac{d\tilde{p}}{dt} + \frac{1}{\varepsilon^{3/2}}(u_x + v_y) + \frac{1}{\sqrt{\varepsilon\eta}}\tilde{w}_z = 0,$$

$$\frac{du}{dt} + \frac{1}{\varepsilon^{3/2}}\tilde{p}_x - \frac{1}{\varepsilon}v = 0,$$

$$\frac{dv}{dt} + \frac{1}{\varepsilon^{3/2}}\tilde{p}_y + \frac{1}{\varepsilon}u = 0,$$

$$\frac{d\tilde{w}}{dt} + \frac{1}{\sqrt{\varepsilon\eta}}\tilde{p}_z + \frac{1}{\sqrt{\eta}}s = 0, \qquad (5.34)$$

where now

$$\frac{d}{dt} = \frac{\partial}{\partial t} + u\frac{\partial}{\partial x} + v\frac{\partial}{\partial y} + \frac{\varepsilon}{\sqrt{\eta}}\tilde{w}\frac{\partial}{\partial z}. \qquad (5.35)$$

Therefore our theory applies, provided

$$\frac{\varepsilon}{\sqrt{\eta}} \le \text{constant}, \qquad (5.36)$$

otherwise the term $(\varepsilon/\sqrt{\eta})\tilde{w}\,\partial/\partial z$ becomes large. We shall assume that

$$\eta = \varepsilon^2. \qquad (5.37)$$

In Browning *et al.* (1990) we have proved

Theorem 5.3 Assume that (5.26) with $\eta \gg \varepsilon^2$ has a solution with derivatives bounded independently of ε. We commit an error of order $\mathcal{O}(\varepsilon^2)$, if we change η to ε^2 and solve the new system with the same initial data.

Thus we can use the system (5.26) with $\eta = \varepsilon^2$ to obtain the desired slow solution up to terms of order $\mathcal{O}(\varepsilon^2)$, provided it exists. Again we can use Richardson extrapolation to approximate it to higher order.

The question, whether for $\eta \ll \varepsilon^2$ the system (5.26) has slow solutions, is not clear. Numerical calculations seem to indicate that it is so. However, if one linearizes (5.26) around a slow solution $U, V, \ldots$, then the linearized equations are unstable, if the sheer U_z, V_z is large compared with $\varepsilon/\sqrt{\eta}$, i.e. if we locally freeze the coefficients, then there are waves, which grow like $\exp(\alpha t)$, $\alpha = (|U_z| + |V_z|)\varepsilon/\sqrt{\eta}$. Further investigations are necessary.

There are other applications. For example, in Browning and Kreiss (1982) some problems in plasma physics are discussed, in Browning *et al.* (1980) and

Browning and Kreiss (1982) the shallow water equations are treated and in Raviart (1991) approximative models of Maxwell's equation are investigated.

REFERENCES

V. Amdursky and A. Ziv (1977), 'On the numerical solution of stiff linear systems of the oscillatory type', *SIAM J. Appl. Math.* **33**, 593–606.

N. Bogoliubov and Y. A. Mitropolsky (1961), *Asymptotic Methods in the Theory of Nonlinear Oscillations,* Gordon and Breach (New York).

G. Browning, A. Kasahara, and H.-O. Kreiss (1980), 'Initialization of the primitive equations by the bounded derivative principle', *J. Atmos. Sci.* **37**, 1424–1436.

G. Browning and H.-O. Kreiss (1982), 'Initialization of the shallow water equations with open boundaries by the bounded derivative method', *Tellus* **34**, 334-351.

G. Browning and H.-O. Kreiss (1982), 'Problems with different time scales for nonlinear partial differential equations' *SIAM J. Appl. Math.* **42**, 704–718.

G. Browning and H.-O. Kreiss (1985), 'Numerical problems connected with weather prediction', in *Progress and Supercomputing in Computational Fluid Dynamics,* (E.M. Murman and S.S. Abarbanel, eds), Birkhauser (Boston) 377–394.

G. L. Browning and H.-O. Kreiss (1986), 'Scaling and computation of smooth atmospheric motions', *Tellus* **38A**, 295–313.

G. L. Browning and H.-O. Kreiss (1987), 'Reduced systems for the shallow water equations', *J. Atmos. Sci.* **44**, 2813–2822.

G. L. Browning, W. R. Holland, H.-O. Kreiss and S. J. Worley (1990), 'An accurate hyperbolic system for approximately hydrostatic and incompressible oceanographic flows', *Dyn. of Atmos. Oceans* **14**, 303–332.

S. O. Fatunla, (1980), 'Numerical Integrators for stiff and highly oscillatory differential equations', *Math. Comput.* **34**, 373–390.

N. Fenichel (1985), 'Persistence and smoothness of invariant manifolds for flows', *Indiana U. Math. J.* **21**, 193–226.

W. Gautschi, (1961), 'Numerical integration of ordinary differential equations based on trigonometric polynomials', *Numer. Math.* **3**, 381–397.

J. Guerra, and B. Gustafsson (1982), *A Semi-implicit Method for Hyperbolic Problems with Different Time Scales,* Report No. 90, Department of Computer Sciences, Uppsala University.

B. Gustafsson (1980a), 'Asymptotic expansions for hyperbolic systems with different time scales', *SIAM J. Numer. Anal.* **17**, No. 5, 623–634.

B. Gustafsson, (1980b), 'Numerical solution of hyperbolic systems with different time scales using asymptotic expansions', *J. Comput. Phys.* **36**, 209–235.

B. Gustafsson and H.-O. Kreiss (1983), 'Difference approximations of hyperbolic problems with different time scales. I. The reduced problem', *SIAM J. Numer. Anal.* **20**, 46–58.

F. C. Hoppensteadt and W. L. Miranker, (1976), 'Differential equations having rapidly changing solutions: analytic methods for weakly nonlinear systems', *J. Diff. Eqns* **22**, 237–249.

C. Johansson (1991), 'The numerical solution of low Mach number flow in confined regions by Richardson extrapolation', to appear.

A. Kasahara (1974), 'Various vertical coordinate systems used for numerical weather prediction', *Mon. Weather Rev.* **102**, 509–522.

A. Kasahara (1982), 'Nonlinear normal mode initialization and the bounded derivative method', *Rev. Geophys. Space Phys.* **20**, No.3, 385–397.

J. Kevorkian and J. D. Cole (1981), *Perturbation Methods in Applied Mathematics*,. Springer-Verlag (New York)

S. Klainerman and A. Majda (1982), 'Compressible and incompressible fluids', *Comm. Pure Appl. Math.* **35**, 629–651.

N. Kopell (1985), 'Invariant manifolds and the initialization problem for atmospheric equations', *Physica* **14D**, 203–215.

H.-O. Kreiss (1979), 'Problems with different time scales for ordinary differential equations', *SIAM J. Numer. Anal.* **16**, 980–998.

H.-O. Kreiss (1980), 'Problems with different time scales for partial differential equations', *Comm. Pure Appl. Math.* **33**, 399–439.

H.-O. Kreiss and J. Lorenz (1989), *Initial-Boundary Value Problems and the Navier–Stokes Equations*, Academic Press (New York).

H.-O. Kreiss J. Lorenz and M. Naughton (1991), 'Convergence of the solutions of the compressible to the solutions of the Incompressible Navier–Stokes equations', *Adv. Appl. Math.* **12**, 187–214.

H.-O. Kreiss and J. Lorenz (1991), 'Manifolds of slow solutions for highly oscillatory problems', to appear.

H. Kreth (1977), 'Time-discretizations for nonlinear evolution equations', *Lecture Notes in Mathematics*, Vol. 679, Springer (Berlin).

C. E. Leith (1980), 'Nonlinear normal mode initialization and quasi-geostrophic theory', *J. Atmos. Sci.* **37**, 954–964.

B. Lindberg (1971), 'On smoothing and extrapolation for the trapezoidal rule', *BIT* **11**, 29–52.

B. Machenhauer (1977), 'On the dynamics of gravity oscillations in a shollow water model, with applications to normal mode initialization', *Beitr. Phys. Atmos.* **50**, 253–271.

G. Majda (1984), 'Filtering techniques for oscillatory stiff ODE's', *SIAM J. Numer. Anal.* **21**, 535–566.

W. L. Miranker and G. Wabba (1976), 'An averaging method for the stiff highly oscillatory problem', *Math. Comput.* **30**, 383–399.

W. L. Miranker and M. van Veldhuisen (1978), 'The method of envelopes', *Math. Comput.* **32**, 453–498.

W. L. Miranker (1981), *Numerical Methods for Stiff Equations and Singular Perturbation Problems*, D. Reidel (Dordrecht, Holland).

J. C. Neu (1980), 'The method of near-identity transformations and its applications', *SIAM J. Appl. Math.* **38**, 189–200.

A. H. Nayfeh (1973), *Perturbation Methods*, John Wiley and Sons (New York).

J. Oliger and A. Sundström (1978), 'Theoretical and practical aspects of some initial-boundary value problems in fluid dynamics', *SIAM J. Appl. Math.* **35**, 419–446.

L. R. Petzold (1981), 'An efficient numerical method for highly oscillatory ordinary differential equations', *SIAM J. Numer. Anal.* **18**, 455–479.

P. A. Raviart (1991), *Approximative Models for Maxwell's Equations and Applications, Proceedings of the Third International Conference on Hyperbolic Problems, Uppsala*, (B. Engquist and B. Gustaffson, eds), Studentliteratur 792–804.

R. J. Sacker (1965), 'A new approach to the perturbation theory of invariant surfaces', *Comm. Pure Appl. Math.* **18**, 717–732.

R. E. Scheid (1982), 'The accurate numerical solution of highly oscillatory ordinary differential equations', Thesis, Caltech, Pasadena, CA 91125.

E. Tadmor (1982), 'Hyperbolic systems with different time scales', *Comm. Pure Appl. Math.* **35**, 839–866.

Acta Numerica (1991), *pp.* 141–198

Numerical methods for differential algebraic equations

Roswitha März

Humboldt-Universität
Fachbereich Mathematik
Postfach 1297, D-O-1086 Berlin, Germany
E-mail: ivs@mathematik.hu-berlin.dbp.de

CONTENTS

0. Introduction

Differential algebraic equations (DAE) are special implicit ordinary differential equations (ODE)

$$f(x'(t), x(t), t) = 0, \qquad (0.1)$$

where the partial Jacobian $f'_y(y, x, t)$ is singular for all values of its arguments.

These DAEs arise in various fields of applications. The most popular ones are simulation of electrical circuits, chemical reactions subject to invariants, vehicle system dynamics, optimal control of lumped-parameter systems, semi-discretization of partial differential equation systems and singular perturbation problems. For a fairly detailed survey of applications we refer to Brenan *et al.* (1989).

In the last few years, DAEs have developed into a highly topical subject in applied mathematics. There is a rapidly increasing number of contributions devoted to DAEs in the mathematical literature as well as in the fields of mechanical engineering, chemical engineering, system theory, etc. Frequently, other names such as semi-state equations, descriptor systems, singular systems are assigned to DAEs. In 1971 C.W. Gear proposed that DAEs should be handled numerically by backward differentiation formulae

(BDF). Since then, powerful codes which successfully simulate large circuits have been developed. For a long time DAEs were considered to be essentially similar to regular implicit ODEs in general. However, challenged by computation results that could not be brought into line with this supposition (e.g. Sincovec *et al.*, 1981), the mathematical community started investigating DAEs more thoroughly. With their famous paper, C.W. Gear *et al.* (1981) initiated a discussion on DAEs which will surely continue for a long time.

What kind of mathematical objects are DAEs? First of all, they are singular ODEs. Can they really be treated numerically like regular ODEs? Surely not in every case! How can one characterize single classes of problems for which methods that have proved their value for regular ODEs work well in other instances? What is the reason for their not working otherwise? How are appropriate numerical methods to be constructed then? All these questions can only be answered when more is known about the mathematical nature of DAEs.

In 1984 W.C. Rheinboldt began regarding DAEs as differential equations on manifolds. This approach provided useful insights into the geometrical and analytical nature of these equations (e.g. Reich (1990), Rabier and Rheinboldt (1991)).

Assuming sufficient smoothness of all functions involved, the DAE

$$\left.\begin{array}{r} u' + g(u, v) = 0 \\ h(u, v) = 0 \end{array}\right\} \tag{0.2}$$

can be regarded as a vector field on

$$S_1 := \left\{ \begin{bmatrix} u \\ v \end{bmatrix} : h(u, v) = 0 \right\},$$

$$\left.\begin{array}{l} u' = -g(u, v) \\ v' = h_v'(u, v)^{-1} h_u'(u, v) g(u, v) \end{array}\right\}, \quad \begin{bmatrix} u \\ v \end{bmatrix} \in S_1,$$

provided that $h_v'(u, v)$ is nonsingular everywhere. All solutions belong to S_1, and each point of S_1 is passed by a solution.

The DAE

$$\left.\begin{array}{r} u' + g(u, v) = 0 \\ h(u) = 0 \end{array}\right\} \tag{0.3}$$

is more complicated. By differentiating twice and eliminating derivatives it can be checked that this system generates the vector field

$$\left.\begin{array}{l} u' = -g(u, v) \\ v' = (h'(u) g_v'(u, v))^{-1} \{ h''(u) g(u, v) + h'(u) g_u'(u, v) \} g(u, v) \end{array}\right\},$$

$$(u^T, v^T)^T \in S_2,$$

where now

$$S_2 := \left\{ \begin{bmatrix} u \\ v \end{bmatrix} : h(u) = 0, h'(u)g(u,v) = 0 \right\}$$

represents the state manifold. The nonsingularity of $h'(u)g'_v(u,v)$ has been assumed here.

Analogously, for the DAE

$$\left. \begin{aligned} v' + f(u,v,w) &= 0 \\ u' + g(u,v) &= 0 \\ h(u) &= 0 \end{aligned} \right\} \tag{0.4}$$

one can define a vector field on the manifold

$$\begin{aligned} S_3 \quad := \quad & \{(u^T, v^T, w^T)^T : h(u) = 0, h'(u)g(u,v) = 0, \\ & h''(u)g(u,v)g(u,v) + h'(u)(g'_u(u,v)g(u,v) + g'_v(u,v)f(u,v,w)) = 0)\} \end{aligned}$$

provided that $h'(u)g'_v(u,v)f'_w(u,v,w)$ remains nonsingular.

In these three cases one speaks of semi-explicit DAEs with index 1, 2 and 3, respectively. The special structure of equations (0.3) and (0.4) is called the Hessenberg form.

If these vector fields were not considered on the specified manifolds $S_i \subset \mathbb{R}^m$, but formally on $\mathbb{R}^m$, then the resulting regular ODEs could be integrated with the usual methods. Even if we start with consistent initial values, we will very swiftly drift away from S_2 and S_3 in (0.3) and (0.4), respectively. Hence, many authors are concerned with the development of very special methods for (0.3) and (0.4), thereby exploiting the geometry of these equations. There are important applications that have this form, e.g. the Euler–Lagrange formulation of constrained mechanical systems leads to the form (0.4).

Under the corresponding assumptions, a state manifold and a vector field can also be assigned to the general DAE (0.1). However, both are only defined implicitly and, in general, not available in practice. This has already been indicated by the simple case of equation (0.4) and S_3. More general approaches for the constructive use of geometry for numerical mathematics are not known to the author.

If we have a closer look at equation (0.2) it becomes obvious that, theoretically, in the neighbourhood of a consistent initial value $(u_0^T, v_0^T)^T \in S_1$ we could investigate the locally decoupled system

$$u' + g(u, \mathcal{S}(u)) = 0, \quad v = \mathcal{S}(u) \tag{0.5}$$

with $h(u, \mathcal{S}(u)) = 0$ instead of (0.2). Now it would be advantageous to integrate this regular ODE for the component u numerically and, then, simply to determine $v_j = \mathcal{S}(u_j)$ in each case. With suitable integration methods, this idea can even be realized in practice for general index-1 equations (0.1).

We would like to point out another aspect of the characterization of DAEs, which is fundamental, in particular, to the numerical treatment. For this, we consider the special equation of the form (0.3), which is perturbed by an inhomogeneity,

$$\left. \begin{array}{r} u' - v = 0 \\ u = p(t) \end{array} \right\} . \tag{0.6}$$

Here, the function p has to be differentiated, i.e. $v(t) = p'(t)$ has to be computed. Differentiation is one of the classical examples of ill posed problems. A corresponding inhomogeneous problem of the form (0.4) will require a second differentiation. The greater the number of differentiations, the more strongly ill posed the problems become.

Both (0.5) and (0.6) make clear that a natural approach to the solution is directed to $u \in C^1$, $v \in C$. In many applications one aims at reducing smoothness, which has, unfortunately, not yet been successfully taken into account in the interpretation of DAEs used to represent ODEs on manifolds.

In the present paper we characterize general DAEs (0.1) under possibly minimal smoothness demands, where the characterization aims at the numerical tractability. Since (from the present point of view) all the essentially new numerical difficulties in comparison with regular ODEs have already become for linear equations with variable coefficients, we devote most of our investigations to the analytical characterization and investigation of integration methods for linear equations.

To apply the results to nonlinear equations we slightly modify the standard arguments of discretization theory. The BDFs are studied in detail here because, on the one hand, they can be especially recommended just for DAEs and, on the other hand, they serve, in a certain sense, as model methods.

We want to emphasize that this paper does not aim at providing a survey of all the available results and methods. In particular, we do not enter into the details of the many nice but very special results for (0.3) and (0.4) (for this, see e.g. Hairer *et al.* (1989), Lubich (1990), Potra and Rheinboldt (1991), Simeon *et al.* (1991)). We focus our interest on exposing problems and showing constructive approaches for their solution, where we try to maintain a uniform concept of representation.

Altogether, many problems with respect to DAEs still remain open. An appropriate numerical treatment requires – provided it is to be more than only favourable intention – profound knowledge about the analytical background of this type of equation.

The paper is organized as follows. In Section 1 the reader becomes acquainted with the fact that additional stability conditions and weak instabilities may occur in the integration of linear constant coefficient DAEs. Section 2 is devoted to the analytical and geometrical foundations of general DAEs, where those of linear equations with time-dependent coefficients play a special role. In Section 3 the BDFs are discussed in detail, as already mentioned, as a model for constructing methods. Section 4 presents brief outlines on index reduction as well as on boundary value problems.

1. Analysing linear constant coefficient equations

Linear equations

$$Ax'(t) + Bx(t) = q(t) \tag{1.1}$$

with matrix coefficients $A, B \in L(\mathbb{R}^m)$, A singular, are easy to understand when taking into account the close relationship with matrix pencils $\{A, B\}$ (e.g. Gantmacher (1966)). In this section we explain some basic facts on how and for what reasons well-known discretization methods behave when applied to DAEs.

Definition The ordered pair of matrices $\{A, B\}$ forms a *regular matrix pencil* if the polynomial $p(\lambda) := \det(\lambda A + B)$ does not vanish identically. Otherwise, the pencil is called singular.

Weierstrass (cf. Grantmacher (1966)) has shown that a regular pencil $\{A, B\}$ can be transformed into $\{\tilde{A}, \tilde{B}\}$,

$$\left. \begin{array}{l} \tilde{A} := EAF = \text{diag}(I, J), \\ \tilde{B} := EAF = \text{diag}(W, I) \end{array} \right\} \tag{1.2}$$

by the use of suitable regular matrices $E, F \in L(\mathbb{R}^m)$. Thereby, $W \in L(\mathbb{R}^k)$, and $J \in L(\mathbb{R}^{m-k})$ is a nilpotent Jordan block matrix with chains

$$\begin{bmatrix} 0 & 1 & & \\ & 0 & \ddots & \\ & & \ddots & 1 \\ & & & 0 \end{bmatrix}.$$

Definition $\{\tilde{A}, \tilde{B}\}$ given by (1.2) is called the *Kronecker canonical normal form* of the regular pencil $\{A, B\}$. The *index* of a regular pencil is defined to be $\text{ind}(A, B) := \text{ind}(J) :=$ maximal Jordan chain order of J.

An equation of type (1.1) with a singular matrix pencil $\{A, B\}$ is somewhat incomplete. For these equations, the homogeneous initial value problem

$$Ax'(t) + Bx(t) = 0, \quad x(0) = 0$$

has more than countably many different solutions (see Griepentrog and März

(1986)). A typical example is

$$A = \begin{bmatrix} 1 & 0 \\ 0 & 0 \end{bmatrix}, \quad B = \begin{bmatrix} 1 & 1 \\ 0 & 0 \end{bmatrix}.$$

Singular matrix pencils in (1.1) indicate some defect in the modelling.

Here, we are interested in equations (1.1) with regular matrix pencils $\{A, B\}$ only. Using the transformation matrices E, F leading to the Kronecker normal form (cf. (1.2)) we may transform (1.1) equivalently into

$$\tilde{A}\tilde{x}'(t) + \tilde{B}\tilde{x}(t) = \tilde{q}(t), \tag{1.3}$$

where $\tilde{A}$, $\tilde{B}$ are given by (1.2), $\tilde{q}(t) := Eq(t)$, $\tilde{x}(t) := F^{-1}x(t)$. In more detail, (1.3) reads

$$u'(t) + Wu(t) = p(t) \tag{1.4}$$
$$Jv'(t) + v(t) = r(t), \tag{1.5}$$

where u, v and p, r are the related components of $\tilde{x}$ and $\tilde{q}$, respectively. Now, the decoupled system (1.4), (1.5) is said to be the *Kronecker normal form of equation (1.1)*. Moreover the index of equation (1.1) can also be traced back to $\mathrm{ind}(A, B) =: \mu$.

In accordance with the Jordan structure of J, equation (1.5) decouples into parts such as

$$\begin{bmatrix} 0 & 1 & & \\ & \ddots & \ddots & \\ & & & 1 \\ & & & 0 \end{bmatrix} w'(t) + w(t) = s(t) \tag{1.6}$$

of dimension $\gamma \le \mu$.

If $\gamma = 1$, then (1.6) simply yields

$$w(t) = s(t).$$

If $\gamma = 2$, then (1.6) represents

$$\left. \begin{array}{l} w_2'(t) + w_1(t) = s_1(t) \\ w_2(t) = s_2(t) \end{array} \right\}, \tag{1.7}$$

which leads to

$$w(t) = \begin{pmatrix} s_1(t) - s_2'(t) \\ s_2(t) \end{pmatrix}.$$

For $\gamma = 3$ we have

$$\left. \begin{array}{l} w_2'(t) + w_1(t) = s_1(t) \\ w_3'(t) + w_2(t) = s_2(t) \\ w_3(t) = s_3(t) \end{array} \right\}, \tag{1.8}$$

hence

$$
w(t) = \left[\begin{array}{l} s_1(t) - (s_2(t) - s_3'(t))' \\ s_2(t) - s_3'(t) \\ s_3(t) \end{array} \right].
$$

In general, if μ denotes the index of our equation (1.1), then (1.5) contains at least one part (1.6) of dimension $\gamma = \mu$, and, in consequence, certain components of the right-hand side have to be differentiated $\mu - 1$ times.

Clearly, (1.4) is a regular linear ODE. For all continuous right-hand sides $p(\cdot) : \mathcal{I} \to \mathbb{R}^k$ there is a unique solution $u(\cdot) : \mathcal{I} \to \mathbb{R}^k$ passing through given $(u^0, t_0) \in \mathbb{R}^k \times \mathcal{I}$.

On the other hand, the solution of (1.5) may be expressed as

$$
v(t) = \sum_{j=0}^{\mu-1} (-1)^j (J^j r(t))^{(j)}.
$$

The initial value $v(t_0)$ is fixed completely, and for solvability we have to assume $r(\cdot) : \mathcal{I} \to \mathbb{R}^{m-k}$ to be as smooth as necessary. From this point of view, for $\mu > 1$, equation (1.5) represents a differentiation problem. It will be pointed out later that this causes numerical difficulties. (Recall the well known fact that differentiation represents an ill posed problem in the continuous function space!)

Clearly, initial value problems for (1.1) only become solvable for *consistent initial values*

$$
x(t_0) = F\tilde{x}(t_0) = F \left[\begin{array}{c} u^0 \\ v(t_0) \end{array} \right],
$$

where $u^0 \in \mathbb{R}^k$ is a free parameter, but $v(t_0)$ is determined as described earlier.

This is the second essential difference from regular ODEs and, when $\mu > 1$, this also entails considerable numerical problems, which have not yet been solved sufficiently.

At this point it should be emphasized again that the canonical normal form is used only to provide an immediate insight into the structure of (1.1). However, we do not think of transforming (1.1) into (1.5), (1.6) in practical computations!

Next we check what will happen when numerical integration methods approved for regular ODEs are applied to the singular ODE (1.1). First we consider the multi-step method

$$
\frac{1}{h} A \sum_{j=0}^{s} \alpha_j x_{\ell-j} + B \sum_{j=0}^{s} \beta_j x_{\ell-j} = q(\bar{t}_\ell), \tag{1.9}
$$

$$\bar{t}_\ell := \sum_{j=0}^{s} \beta_j t_{\ell-j}, \qquad \alpha_0 \neq 0,$$

where x_i is expected to approximate the true solution value $x(t_i)$. Again we decouple equations (1.9) according to the Kronecker canonical normal form by multiplying (1.9) by E and transforming

$$F^{-1} x_i = \tilde{x}_i = \begin{bmatrix} u_i \\ v_i \end{bmatrix}.$$

This yields

$$\frac{1}{h} \sum_{j=0}^{s} \alpha_j u_{\ell-j} + W \sum_{j=0}^{s} \beta_j u_{\ell-j} = p(\bar{t}_\ell) \qquad (1.10)$$

$$J \frac{1}{h} \sum_{j=0}^{s} \alpha_j v_{\ell-j} + \sum_{j=0}^{s} \beta_j v_{\ell-j} = r(\bar{t}_\ell). \qquad (1.11)$$

Formula (1.10) represents the given multi-step method applied to the inherent regular ODE within the singular system (1.1). On the other hand, (1.11) may be solved with respect to v_ℓ if the matrix

$$\alpha_0 J + h\beta_0 I$$

is nonsingular, that is for $\beta_0 \neq 0$.

In the index-1 case $J = 0$, and (1.11) simply becomes

$$\sum_{j=0}^{s} \beta_j v_{\ell-j} = r(\bar{t}_\ell). \qquad (1.12)$$

In März (1984, 1985) it was pointed out that, for the stability of the difference equation (1.12), it is necessary for the polynomial $\sum_{j=0}^{s} \beta_j \lambda^{s-j}$ to have all its roots within the interior of the complex unit circle. In particular, symmetric schemes (1.12) like, for example, the centred Euler scheme become unstable.

The best way to avoid error accumulations in (1.12) is to choose $\beta_0 = 1$, $\beta_1 = \cdots = \beta_s = 0$, e.g. to use the BDF.

For higher indexes $\mu > 1$ we only discuss the BDF. In index-2 parts such as (1.7) we have

$$\left. \begin{array}{r} \dfrac{1}{h} \displaystyle\sum_{j=0}^{s} \alpha_j w_{2,\ell-j} + w_{1,\ell} = s_1(t_\ell) \\[2mm] w_{2,\ell} = s_2(t_\ell) \end{array} \right\}, \qquad \ell \geq s,$$

thus

$$w_{1,\ell} = s_1(t_\ell) - \frac{1}{h} \sum_{j=0}^{s} \alpha_j s_2(t_{\ell-j}) \left. \begin{array}{l} \\ \\ \end{array} \right\}, \quad \ell \geq s,$$
$$w_{2,\ell} = s_2(t_\ell)$$

if we assume exact starting values $w_{2,i} = s_2(t_i)$, $i = 0, \ldots, s-1$ are available. Inexact starting values as well as round-off errors in the linear equation (1.9) to be solved for x_ℓ give rise to a weak instability, i.e. the errors are amplified by $1/h$. However, fortunately, only the component $w_{1,\ell}$ is affected by this.

Analogously, in index-3 parts, such as e.g. (1.8), we have

$$\frac{1}{h} \sum_{j=0}^{s} \alpha_j w_{2,\ell-j} + w_{1,\ell} = s_1(t_\ell) \left. \begin{array}{l} \\ \\ \\ \\ \end{array} \right\}, \quad \ell \geq s.$$
$$\frac{1}{h} \sum_{j=0}^{s} \alpha_j w_{3,\ell-j} + w_{2,\ell} = s_2(t_\ell)$$
$$w_{3,\ell} = s_3(t_\ell)$$

Using these formulae for $\ell \geq 2s$ together with exact values $w_{3,i} = s_3(t_i)$, $i = 0, \ldots, 2s - 1$, $w_{2,i} = s_2(t_i)$, $i = 0, \ldots, s - 1$, would lead to

$$w_{1,\ell} = s_1(t_\ell) - \frac{1}{h} \sum_{j=0}^{s} \alpha_j s_2(t_{\ell-j}) + \frac{1}{h^2} \sum_{j=0}^{s} \alpha_j \sum_{i=0}^{s} \alpha_i s_3(t_{\ell-j-i})$$

$$w_{2,\ell} = s_2(t_\ell) - \frac{1}{h} \sum_{j=0}^{s} \alpha_j s_3(t_{\ell-j})$$

$$w_{3,\ell} = s_3(t_\ell).$$

Of course, in practical computations $2s$ exact starting values are not available, thus the components $w_{1,\ell}$ and $w_{2,\ell}$ will be affected by instabilities of the type $1/h^2$ and $1/h$, respectively. It should be mentioned that these instabilities are due to the differentiations arising in (1.1), and in this sense they are very natural.

Now, let us turn shortly to implicit Runge–Kutta methods for (1.1). Given the Runge–Kutta tableau

$$\frac{c \mid A}{\mid \beta^T}$$

we have to solve the system

$$AX_i' + B(x_{\ell-1} + h \sum_{j=1}^{s} \alpha_{ij} X_j') = q(t_{\ell-1} + c_i h) \qquad i = 1, \ldots, s, \qquad (1.13)$$

and then to compute

$$x_\ell = x_{\ell-1} + h \sum_{j=1}^{s} \beta_j X_j'. \qquad (1.14)$$

Again we use the transformation to the Kronecker normal form. This gives

$$U_i' + W(u_{\ell-1} + h \sum_{j=1}^{s} \alpha_{ij} U_j') = p(t_{\ell-1} + c_i h), \qquad i = 1, \ldots, s, \qquad (1.15)$$

$$u_\ell = u_{\ell-1} + h \sum_{j=1}^{s} \beta_j U_j', \qquad (1.16)$$

$$JV_i' + v_{\ell-1} + h \sum_{j=1}^{s} \alpha_{ij} V_j' = r(t_{\ell-1} + c_i h), \qquad i = 1, \ldots, s, \qquad (1.17)$$

$$v_\ell = v_{\ell-1} + h \sum_{j=1}^{s} \beta_j V_j'. \qquad (1.18)$$

Clearly, (1.15) and (1.16) are nothing else but the given Runge–Kutta method applied to the regular inherent ODE (1.4). This part does not cause any new difficulties.

Equations (1.17) and (1.18) decouple further according to the Jordan chains in J (cf. (1.6)). For index-1 chains ($\gamma = 1$) we simply have

$$w_{\ell-1} + h \sum_{j=1}^{s} \alpha_{ij} W_j' = s(t_{\ell-1} + c_i h), \qquad i = 1, \ldots, s, \qquad (1.19)$$

$$w_\ell = w_{\ell-1} + h \sum_{j=1}^{s} \beta_j W_j'. \qquad (1.20)$$

Now it becomes clear that we have to use a nonsingular Runge–Kutta matrix $\mathbf{A}$ to make the system (1.19) solvable with respect to $W_1', \ldots, W_s'$. Denoting the elements of $\mathbf{A}^{-1}$ by $\hat{\alpha}_{ij}$, we obtain

$$\begin{aligned}
w_\ell &= w_{\ell-1} + \sum_{j=1}^{s} \beta_j \sum_{k=1}^{s} \hat{\alpha}_{jk}(r(t_{\ell-1} + c_k h) - w_{\ell-1}) \\
&= \varrho w_{\ell-1} + \sum_{j=1}^{s} \beta_j \sum_{k=1}^{s} \hat{\alpha}_{jk} r(t_{\ell-1} + c_k h)
\end{aligned}$$

with

$$\varrho = 1 - \beta^T \mathbf{A}^{-1}(1, \ldots, 1)^T. \qquad (1.21)$$

Recall that w_ℓ should approximate $w(t_\ell) = r(t_\ell)$. Obviously, $|\varrho| > 1$ would yield an unstable scheme. Choosing $\beta_j = \alpha_{sj}$, $j = 1, \ldots, s$, in the Runge–Kutta tableau we obtain $\varrho = 0$ and $w_\ell = r(t_{\ell-1} + c_s h)$.

Thus, the so-called IRK (DAE) (cf. Petzold (1986), Griepentrog and März

(1986)), i.e. s-stage Runge–Kutta methods, with

$$\beta_j = \alpha_{sj}, \qquad j = 1, \ldots, s, \qquad c_s = 1, \qquad \mathbf{A} \text{ nonsingular}, \qquad (1.22)$$

appears to be an appropriate tool for handling index-1 equations (1.1).

Next we investigate what happens with such a method if (1.5) contains an index-2 block (1.7). As earlier, we compute

$$w_{2,\ell} = s_2(t_{\ell-1} + h).$$

$$
\begin{aligned}
w_{1,\ell} &= s_1(t_{\ell-1} + h) - \frac{1}{h} \sum_{k=1}^{s} \hat{\alpha}_{s,k} s_2(t_{\ell-1} + c_k h) + \frac{1}{h} \sum_{k=1}^{s} \hat{\alpha}_{s,k} w_{2,\ell-1} \\
&= s_1(t_{\ell-1} + h) - \sum_{k=1}^{s} \hat{\alpha}_{s,k} \frac{1}{h} (s_2(t_{\ell-1} + c_k h) - s_2(t_{\ell-1})) \qquad (1.23)
\end{aligned}
$$

assuming the starting value $w_{2,\ell-1}$ to be consistent, i.e. $w_{2,\ell-1} = s_2(t_{\ell-1})$.

If the Runge–Kutta method has an inner order of consistency ≥ 1 we know the condition

$$\sum_{k=1}^{s} \hat{\alpha}_{sk} c_k = 1$$

is satisfied. Thus, (1.23) with a consistent starting value actually provides an approximation of $w_1(t_\ell) = s(t_\ell) - s_2'(t_\ell)$. However, we do not usually have consistent starting values, and the errors are unstably amplified by $1/h$.

Let us summarize what has been pointed out in this section:

1   Singular systems (1.1) of index μ are mixed regular differential equations (1.4) and equations (1.5) including $\mu - 1$ differentiations.
2   Consistent initial values are not easy to compute in practice.
3   Integration methods handle the inherent regular ODE (1.4) as expected.
4   To avoid singular coefficient matrices in the linear systems to be solved per integration step we should use implicit multi-step methods ($\beta_0 \neq 0$) and nonsingular Runge–Kutta matrices $\mathbf{A}$. Moreover, there have to be additional conditions to ensure stability in the related index-1 parts.
5   Errors in the starting values are amplified by $h^{1-\mu}$ in the best case, but only the components v_j are affected.

The decoupled system (1.4), (1.5) and also (1.10), (1.11) respectively (1.15)–(1.18) lead us to the idea that it would be nice to allow different approaches for the parts (1.4) and (1.5), respectively, say a possibly explicit higher order method for the regular ODE (1.4) and a BDF for (1.5).

Of course, this should be done without knowing the canonical normal form. Furthermore, we regard the linear constant coefficient equation (1.1) as the simplest model with which to give some hints as to how to proceed with more general equations.

2. Characterizing DAEs

2.1. *Linear equations with variable coefficients*

Consider the linear equation

$$A(t)x'(t) + B(t)x(t) = q(t), \tag{2.1}$$

where $A(\cdot), B(\cdot) : \mathcal{I} \to L(\mathbb{R}^m)$ are continuous matrix functions on the interval $\mathcal{I} \subseteq \mathbb{R}$, and $A(t)$, $t \in \mathcal{I}$, is singular.

The first classification of these singular ODEs was given by C. W. Gear and L. Petzold (1984).

Definition (2.1) is said to be a *global index* μ DAE if there exist regular matrix functions $E \in C(\mathcal{I}, L(\mathbb{R}^m))$, $F \in C^1(\mathcal{I}, L(\mathbb{R}^m))$ so that multiplying (2.1) by $E(t)$ and transforming $F(t)^{-1}x(t) = \tilde{x}(t)$ leads to the decoupled system

$$\begin{bmatrix} I & 0 \\ 0 & J \end{bmatrix} \tilde{x}'(t) + \begin{bmatrix} W(t) & 0 \\ 0 & I \end{bmatrix} \tilde{x}(t) = E(t)q(t), \tag{2.2}$$

where J is a constant nilpotent Jordan block matrix, $\mathrm{ind}(J) = \mu$.

Unfortunately, except for some interesting case studies, this *Kronecker canonical normal form* (2.2) as well as the transforms E, F are not available. Moreover, no way is known for relating this form to nonlinear equations. This is why we are looking for another way to characterize (2.1).

Denote by $N(t) := \ker A(t)$ the null space of $A(t)$, $t \in \mathcal{I}$, and assume this null space to be smooth, i.e. that there exists a matrix function $Q \in C^1(\mathcal{I}, L(\mathbb{R}^m))$ which projects $\mathbb{R}^m$ onto $N(t)$ for each $t \in \mathcal{I}$ (that is $Q(t)^2 = Q(t)$, $\mathrm{im}\,Q(t) = N(t)$).

If DAE (2.1) has a global index μ, then e.g. $Q(t) = F(t)\mathrm{diag}(0, Q_J)F(t)^{-1}$ represents such a projector function, where Q_J denotes a projector onto $\ker J$. In particular, for global index-1 equations (2.1), we simply have $J = 0$, hence $Q(t) = F(t)\mathrm{diag}(0, I)F(t)^{-1}$.

In the following we let Q denote any such a projector function, and we also use $P(t) := I - Q(t)$, $t \in \mathcal{I}$.

Since $A(t)Q(t) \equiv 0$, we may insert $A(t) \equiv A(t)P(t)$ into (2.1), and rewrite it as

$$A(t)\{(Px)'(t) - P'(t)x(t)\} + B(t)x(t) = q(t)$$

or

$$A(t)(Px)'(t) + (B(t) - A(t)P'(t))x(t) = q(t). \tag{2.3}$$

This makes clear that, in general, we should not ask for C^1 solutions of (2.3) and (2.1), respectively, but for solutions belonging to the function space

$$C_N^1 := \{x \in C(\mathcal{I}, \mathbb{R}^m) : Px \in C^1(\mathcal{I}, \mathbb{R}^m)\}.$$

Example $A(t) = \text{diag}(I,0)$ immediately leads to

$$\left.\begin{array}{r}x_1'(t) + B_{11}(t)x_1(t) + B_{12}(t)x_2(t) = q_1(t) \\ B_{21}(t)x_1(t) + B_{22}(t)x_2(t) = q_2(t)\end{array}\right\}, \qquad (2.4)$$

which is called a *semi-explicit DAE*.

Obviously, it is neither necessary nor useful that $x_2 \in C^1$!

Next we reformulate (2.3) to

$$A(Px)' + (B - AP')(Px + Qx) = q,$$

and then to

$$\{A + (B - AP')Q\}(P(Px)' + Qx) + (B - AP')Px = q. \qquad (2.5)$$

Denote $A_1 := A + B_0Q$, $B_0 := B - AP'$ and ask whether $A_1(t)$ is nonsingular for all $t \in \mathcal{I}$. If it is so, we multiply (2.5) by PA_1^{-1} and QA_1^{-1}, respectively. This yields the system

$$(Px)' - P'Px + PA_1^{-1}B_0Px = PA_1^{-1}q \qquad (2.6)$$
$$Qx + QA_1^{-1}B_0Px = QA_1^{-1}q, \qquad (2.7)$$

which decomposes into a regular explicit ODE for the nonnull space component Px and a simple derivative-free equation for determining the null space component Qx. The inherent ODE

$$u' - P'Pu + PA_1^{-1}B_0u = PA_1^{-1}q \qquad (2.8)$$

has the property that solutions starting in $\text{im}P(t_0)$ for some $t_0 \in \mathcal{I}$ remain in $\text{im}P(t)$ for all $t \in \mathcal{I}$, since multiplying (2.8) by Q yields

$$(Qu)' - Q'Qu = 0.$$

Consequently, if for any $q \in C(\mathcal{I}, \mathbb{R}^m)$, $u_0 \in \text{im}P(t_0)$, we denote the solution of (2.8) passing through (u_0, t_0) by $u \in C^1$, we obtain, with

$$\begin{aligned}x &:= u - QA_1^{-1}B_0u + QA_1^{-1}q \\ &= (I - QA_1^{-1}B_0)u + QA_1^{-1}q,\end{aligned} \qquad (2.9)$$

a C_N^1 solution of (2.1).

To be sure to address the initial condition to the respective component, we may state as follows

$$P(t_0)(x(t_0) - x^0) = 0. \qquad (2.10)$$

This means that $u(t_0) = P(t_0)x(t_0) = P(t_0)x^0$, i.e. $P(t_0)x^0$ plays the role of u_0. Now $x^0 \in \mathbb{R}^m$ can be chosen arbitrarily. In general, $x(t_0) = x^0$ cannot be expected to hold for the solution $x(\cdot)$ of the initial value problem (IVP) (2.1), (2.10), but

$$x(t_0) = (I - Q(t_0)A_1(t_0)^{-1}B_0(t_0))P(t_0)x^0 + Q(t_0)A_1(t_0)^{-1}q(t_0).$$

Lemma 2.1   Let $A, B, Q \in L(\mathbb{R}^m)$ be given, $N := \ker A \neq \{0\}$, $Q^2 = Q$, $\mathrm{im}\, Q = N$, $S := \{z \in \mathbb{R}^m : Bz \in \mathrm{im}\, A\}$.

Then the following three statements are equivalent:

(i) $\mathbb{R}^m = N \oplus S$
(ii) $\mathrm{ind}(A, B) = 1$
(iii) $A + BQ$ is nonsingular.

Moreover, if $G := A + BQ$ is nonsingular, then $G^{-1}BQ = Q$, $G^{-1}A = I - Q$, and $QG^{-1}B$ represents the projection onto N along S.

Proof.   The first part is given in Griepentrog and März (1986), Theorem A.13. Here we check the second part only.

Trivially, $G^{-1}BQ = G^{-1}(A + BQ)Q = Q$,

$$G^{-1}A = G^{-1}A(I - Q) = G^{-1}(A + BQ)(I - Q) = I - Q.$$

Then, we have for $Q_s := QG^{-1}B$

$$\begin{aligned} Q_s^2 &= QG^{-1}BQG^{-1}B = QG^{-1}B = Q_s, \\ Q_sQ &= QG^{-1}BQ = Q, \quad \text{i.e.} \quad \mathrm{im}\, Q_s = \mathrm{im}\, Q = N, \end{aligned}$$

and $Q_s z = 0$ implies

$$G^{-1}Bz = (I - Q)G^{-1}Bz,$$

thus $Bz = G(I - Q)G^{-1}Bz = AG^{-1}Bz \in \mathrm{im}\, A$. $\square$

Lemma 2.1 applies to our DAE in the following sense.

In addition to $N(t) =: N_0(t)$ introduce

$$\begin{aligned} S_0(t) &:= \{z \in \mathbb{R}^m : B_0(t)z \in \mathrm{im}\, A(t)\} \\ &= \{z \in \mathbb{R}^m : B(t)z \in \mathrm{im}\, A(t)\}. \end{aligned} \tag{2.11}$$

By Lemma 2.1, our matrix $A_1(t)$ is nonsingular if and only if

$$S_0(t) \oplus N_0(t) = \mathbb{R}^m. \tag{2.12}$$

If (2.12) holds, then

$$Q_s(t) := Q(t)A_1(t)^{-1}B_0(t) \tag{2.13}$$

projects $\mathbb{R}^m$ onto $N_0(t)$ along $S_0(t)$.

Definition   The DAE (2.1) is said to be *index-1 tractable* (or *transferable*) if A, B are continuous, $A(t)$ is singular but has a smooth null space, and $A_1(t)$ remains nonsingular for all $t \in \mathcal{I}$.

Theorem 2.2   Let (2.1) be transferable. Then

(i) For all $q \in C(\mathcal{I}, \mathbb{R}^m)$, $x^0 \in \mathbb{R}^m$, the IVP (2.1), (2.10) is uniquely solvable on $C_N^1(\mathcal{I}, \mathbb{R}^m)$.

(ii) $S_0(t_0)$ is the set of all consistent initial values at time $t_0 \in \mathcal{I}$ for the homogeneous equation, all IVPs $Ax' + Bx = 0$, $x(t_0) = x_0 \in S_0(t_0)$ are uniquely solvable.

Proof. It only remains to check the consistency of $x_0 \in S_0(t_0)$. In fact, solving the IVP $Ax' + Bx = 0$, $P(t_0)(x(t_0) - x_0) = 0$, $x_0 \in S_0(t_0)$, we derive

$$x(t_0) = (I - Q_s(t_0))P(t_0)x_0 = (I - Q_s(t_0))x_0 = x_0.$$

□

Remarks

1   The semi-explicit system (2.4) is transferable if $B_{22}(t)$ remains nonsingular. Here we simply have

$$Q = \mathrm{diag}(0, I), \quad A_1 = A + BQ = \begin{bmatrix} I & B_{12} \\ 0 & B_{22} \end{bmatrix}.$$

and, furthermore,

$$Q_s = \begin{bmatrix} 0 & 0 \\ B_{22}^{-1}B_{21} & I \end{bmatrix}.$$

2   Equation (2.2) in Kronecker canonical normal form is transferable if $J + Q_J$ is regular, that is if $\mu = 1$.

3   It may be easily checked whether each DAE (2.1) which has a global index $\mu = 1$ is also transferable, whereby even Q_s is continuously differentiable,

$$Q_s = F\mathrm{diag}(0, I)F^{-1}.$$

Theorem 2.3   Supposed (2.1) is transferable, the system

$$\left.\begin{array}{l} A(t_0)y_0 + B(t_0)x_0 = q(t_0) \\ Q(t_0)y_0 + P(t_0)(x_0 - x^0) = 0 \end{array}\right\} \tag{2.14}$$

is uniquely solvable with respect to x_0, y_0. x_0 is the fully consistent initial value related to (2.1), (2.10), $y_0 = (Px)'(t_0) - P'(t_0)x_0$.

Proof. Rewrite the first equation of (2.14) as

$$A(t_0)\{y_0 + P'(t_0)x_0\} + B_0(t_0)x_0 = q(t_0).$$

Rearrange this as

$$A_1(t_0)\{P(t_0)y_0 + P(t_0)P'(t_0)x_0 + Q(t_0)x_0\} + B_0(t_0)P(t_0)x^0 = q(t_0).$$

Now we decouple into

$$P(t_0)y_0 + P(t_0)P'(t_0)x_0 + P(t_0)A_1(t_0)^{-1}B_0(t_0)P(t_0)x^0$$
$$= P(t_0)A_1(t_0)^{-1}q(t_0)$$

$$Q(t_0)x_0 + Q_s(t_0)P(t_0)x^0 = Q(t_0)A_1(t)^{-1}q(t_0)$$

and compare those with (2.6), (2.7), in order to obtain

$$\begin{aligned} x_0 &= x(t_0), \\ y_0 &= (Px)'(t_0) - P'(t_0)x(t_0). \end{aligned}$$

Finally, the matrix

$$\begin{bmatrix} A(t_0) & B(t_0) \\ Q(t_0) & P(t_0) \end{bmatrix} \tag{2.15}$$

is nonsingular since $A(t_0) + B(t_0)Q(t_0)$ is so. $\square$

Now, let us turn to nontransferable DAEs (2.1), that is to those DAEs with a singular matrix $A_1(t)$.

Introduce new subspaces

$$\begin{aligned} N_1(t) &:= \ker A_1(t) \\ S_1(t) &:= \{z \in \mathbb{R}^m : B(t)P(t)z \in \operatorname{im} A_1(t)\} \\ &= \{z \in \mathbb{R}^m : B_0(t)P(t)z \in \operatorname{im} A_1(t)\}. \end{aligned} \tag{2.16}$$

and now assume that

$$N_1(t) \oplus S_1(t) = \mathbb{R}^m, \quad t \in \mathcal{I}$$

holds. Choose $Q_1(t)$ to be the projector onto $N_1(t)$ along $S_1(t)$, and let Q_1 be continuously differentiable. Note that for $B_1 := (B_0 - A_1(PP_1)')P$

$$S_1(t) = \{z \in \mathbb{R}^m : B_1(t)z \in \operatorname{im}(A_1(t))\}$$

holds. By Lemma 2.1, the matrix

$$A_2(t) := A_1(t) + B_1(t)Q_1(t), \quad t \in \mathcal{I}$$

becomes nonsingular and, finally,

$$Q_1(t) = Q_1(t)A_2(t)^{-1}B_1(t),$$

which implies that

$$Q_1(t)Q(t) = 0, \quad t \in \mathcal{I}, \tag{2.17}$$

is true. As a consequence, the products $P(t)P_1(t)$, $P(t)Q_1(t)$ are also projectors. Hence, it makes sense to look for a decomposition $x = PP_1x + PQ_1x + Qx$ of the solution. To this end, rewrite (2.1) again (cf. (2.5)) as

$$A_1(P(Px)' + Qx) + B_0Px = q,$$

then as

$$A_1\{(PP_1x)' + PQ_1(Px)' + Qx\} + (B_0 - A_1(PP_1)')Px = q$$

and finally as

$$A_2\{P_1(PP_1x)' + P_1PQ_1(Px)' + P_1Qx + Q_1x\} + B_1PP_1x = q. \qquad (2.18)$$

Multiplying (2.18) by $PP_1A_2^{-1}$, $QP_1A_2^{-1}$ and $Q_1A_2^{-1}$, respectively, and performing some technical calculations we obtain the system

$$(PP_1x)' - (PP_1)'PP_1x + PP_1A_2^{-1}B_1PP_1x = PP_1A_2^{-1}q \qquad (2.19)$$

$$\begin{aligned}
-(QQ_1x)' + Qx &= QP_1A_2^{-1}q - (QQ_1)'PQ_1x \\
&\quad -\{(QQ_1 - QP_1)' + QP_1A_2^{-1}B_1\}PP_1x \quad (2.20) \\
Q_1x &= Q_1A_2^{-1}q. \qquad (2.21)
\end{aligned}$$

Clearly, (2.19) represents a regular ODE for the component PP_1x, (2.21) simply determines Q_1x, but to obtain the null space component Qx we have to insert $Q_1x = Q_1A_2^{-1}q$ into the term $(QQ_1x)'$, i.e. we have to differentiate $QQ_1A_2^{-1}q$ once.

Multiplying the ODE

$$u' - (PP_1)'u + PP_1A_2^{-1}B_1u = PP_1A_2^{-1}q \qquad (2.22)$$

by $I - PP_1$ leads to $((I - PP_1)u)' + (PP_1)'(I - PP_1)u = 0$. Therefore, $u(t_0) \in \operatorname{im} P(t_0)P_1(t_0)$ implies $u(t) \in \operatorname{im} P(t)P_1(t)$ for all $t \in \mathcal{I}$.

Example Consider the semi-explicit DAE (2.4) with $B_{22}(t) \equiv 0$ and assume that $B_{21}(t)B_{12}(t)$ is nonsingular. We have

$$A_1(t) = \begin{bmatrix} I & B_{12}(t) \\ 0 & 0 \end{bmatrix},$$

$$S_1(t) = \left\{ \begin{pmatrix} u \\ v \end{pmatrix} \in \mathbb{R}^m : B_{21}(t)u = 0 \right\}.$$

Now $\begin{pmatrix} u \\ v \end{pmatrix} \in N_1(t) \cap S_1(t)$ implies $u = -B_{12}(t)v$, $B_{21}(t)u = 0$, that is $v = 0$, $u = 0$. Then, compute

$$\begin{aligned}
Q_1 &= \begin{bmatrix} B_{12}(B_{21}B_{12})^{-1}B_{21} & 0 \\ -(B_{21}B_{12})^{-1}B_{21} & 0 \end{bmatrix}, \\
PP_1 &= \begin{bmatrix} I - B_{12}(B_{21}B_{12})^{-1}B_{21} & 0 \\ 0 & 0 \end{bmatrix}.
\end{aligned}$$

It should be mentioned that this kind of equation is often discussed, and it is said to be an index-2 DAE in Hessenberg form. The simplest system of this type is (cf. (1.7))

$$\left. \begin{aligned} x_1' + x_2 &= q_1 \\ x_1 &= q_2 \end{aligned} \right\}.$$

Let us turn back to the general equation (2.1). If $A_2(t)$ is also singular, we proceed analogously using new subspaces and projectors. More precisely, for given $A, B \in L(\mathcal{I}, L(\mathbb{R}^m))$, we define the chain of matrix functions

$$
\begin{aligned}
A_0 &:= A, \quad B_0 := B - AP', \\
A_{i+1} &:= A_i + B_iQ_i, \\
B_{i+1} &:= (B_i - A_{i+1}(P_0P_1 \cdots P_{i+1})')P_i, \quad i \geq 0,
\end{aligned}
\tag{2.23}
$$

where $P_j = I - Q_j$, and $Q_j(t)$ projects onto $N_j(t) := \ker A_j(t)$, $t \in \mathcal{I}$, $j \geq 0$. Introduce further

$$
\begin{aligned}
S_j(t) &:= \{z \in \mathbb{R}^m : B_j(t)z \in \operatorname{im}A_j(t)\} \\
&= \{z \in \mathbb{R}^m : B_{j-1}(t)P_{j-1}(t)z \in \operatorname{im}A_j(t)\}, \quad j \geq 1.
\end{aligned}
$$

Definition The ordered pair $\{A, B\}$ of continuous matrix functions (and also the DAE (2.1)) is said to be *index-μ tractable* if all matrices $A_j(t)$, $t \in \mathcal{I}$, $j = 0, \ldots, \mu - 1$, within the chain (2.23) are singular with smooth null spaces, and $A_\mu(t)$ remains nonsingular on $\mathcal{I}$.

Theorem 2.4 If the DAE (2.1) has the global index μ, then this DAE is also index-μ tractable.

Proof. We refer to Hansen (1990), where this assertion is verified by means of a very complicated induction. $\square$

Theorem 2.5 Let $\{A, B\}$ be index-μ tractable. Then the IVP (2.1),

$$
P_0(t_0) \ldots P_{\mu-1}(t_0)(x(t_0) - x^0) = 0
\tag{2.24}
$$

is uniquely solvable on $C_N^1(\mathcal{I}, \mathbb{R}^m)$ for any given $x^0 \in \mathbb{R}^m$ and sufficiently smooth right-hand sides q, in particular for all $q \in C^{\mu-1}(\mathcal{I}, \mathbb{R}^m)$.

Proof. The assertion follows from the previous explanations for $\mu = 1$ and $\mu = 2$. It is proved in März (1989) for $\mu = 3$, and for $\mu > 3$ in Griepentrog and März (1989) and Hansen (1989). $\square$

Remark The solution of an index-μ-tractable DAE, $\mu > 1$ decomposes in the following way:

$$
x = P_0 \ldots P_{\mu-1}x + P_0 \ldots P_{\mu-2}Q_{\mu-1}x + \cdots + P_0Q_1x + Q_0x.
$$

Thereby $P_0 \ldots P_{\mu-1}x \in C^1$ solves the inherent regular ODE,

$$
P_0 \ldots P_{\mu-2}Q_{\mu-1}x \in C^1
$$

is given by the 'algebraic' part. The components $P_0 \ldots P_{\mu-j}Q_{\mu-j+1}x \in C^1$ include derivatives of order $j - 2$ for $j = 3, \ldots, \mu$ and, finally, $Q_0x \in C$ includes a $(\mu - 1)$ derivative.

When investigating discretizations we are often interested in compact intervals $\mathcal{I}$, and in the properties of the maps representing our IVPs and

BVPs. Let $\mathcal{I} := [t_0, T]$, and $\{A, B\}$ be index-μ tractable. Let $\Pi_\mu := P_0(t_0) \cdots P_{\mu-1}(t_0)$, $M_\mu := \mathrm{im}(\Pi_\mu) \subseteq \mathbb{R}^m$. Then introduce the linear map

$$\mathcal{L} : C_N^1(\mathcal{I}, \mathbb{R}^m) \to C(\mathcal{I}, \mathbb{R}^m) \times M_\mu =: C \times M_\mu$$

by defining

$$\mathcal{L}x := (A(Px)' + B_0 x, \Pi_\mu x(t_0)), \quad x \in C_N^1(\mathcal{I}, \mathbb{R}^m). \tag{2.25}$$

The function space $C_N^1(\mathcal{I}, \mathbb{R}^m)$ completed with its natural norm

$$\|x\| := \|x\|_\infty + \|(Px)'\|_\infty, \quad x \in C_N^1,$$

becomes a Banach space. Note that the topology of this space is independent of the choice of projector function.

The map $\mathcal{L}$ is bounded, but does a bounded inverse exist?

Theorem 2.6 Let $\{A, B\}$ be index-μ tractable, $\mathcal{I} = [t_0, T]$. Then

(i) it holds that

$$\|x\| \leq K \left\{ \sum_{j=0}^{\mu-1} \|q^{(j)}\|_\infty + |\Pi_\mu x(t_0)| \right\} \tag{2.26}$$

for all solutions x corresponding to sources $q \in C^{\mu-1}(\mathcal{I}, \mathbb{R}^m)$;

(ii) the map $\mathcal{L}$ is injective;

(iii) $\mathcal{L}$ is surjective for $\mu = 1$, but for $\mu > 1$ $\mathrm{im}(\mathcal{L})$ becomes a nonclosed proper subset within $C \times M_\mu$.

Proof. The first assertion is obvious for $\mu = 1$ and $\mu = 2$ (cf. (2.6), (2.7) respectively (2.19)–(2.21)). In general, it can be verified by decoupling the DAE (Griepentrog and März 1989, Hansen 1989).

The injectivity of $\mathcal{L}$ is given by Theorem 2.5. Moreover, for $\mu = 1$, Theorem 2.2 provides solvability for all continuous right-hand sides q, i.e. $\mathrm{im}(\mathcal{L}) = C \times M_\mu$.

In the higher index cases, that is for $\mu \geq 2$, we have to assume that certain components of q are continuously differentiable for solvability. However, the set of these functions is not closed in the continuous function space, but it is a nonclosed proper subset. $\square$

Remarks

1   Inequality (2.26) is somewhat liberal. It could be stated more strictly but would take immense technical effort. To do this by means of the decoupling technique, those parts of q which have to be differentiated have to be described precisely. In particular for $\mu = 2$, the system (2.19)–(2.21) makes this transparent. There we have

$$\mathrm{im}(\mathcal{L}) = \{q \in C : Q_1 A_2^{-1} q \in C^1\} \times M_2,$$

and

$$\|x\| \leq \tilde{K}\{\|q\|_\infty + \|(Q_1 A_2^{-1} q)'\|_\infty + |\Pi_2 x(t_0)|\}. \qquad (2.27)$$

2   The inequalities

$$\|x\|_\infty \leq \bar{K}\left\{\sum_{j=1}^{\mu-1} \|q^{(j)}\|_\infty + |x(t_0)|\right\} \qquad (2.28)$$

are used in Hairer *et al.* (1989) to define the so-called perturbation index. In our framework, (2.28) as well as its sharper form (2.26) appear as secondary effects.

Corollary 2.7   If $\mu = 1$, then the inverse of $\mathcal{L}$ is bounded, and $\mathcal{L}$ is a homeomorphism. If $\mu > 1$, then the inverse of $\mathcal{L}$ becomes unbounded.

Proof.   Since $\mathcal{L}$ is acting in Banach spaces, this assertion follows immediately from Theorem 2.6. $\square$

In other words, higher index DAEs ($\mu > 1$) become *ill posed* in Tichonov's sense in the given setting, i.e. the solutions do not depend continuously on the inputs. This has bad consequences for the numerical treatment. The unboundedness of $\mathcal{L}^{-1}$ makes the related discretized maps unstable.

At this point it should be recalled that the explanations in Section 1 concerning integration methods confirm the expected instability. On the other hand, in certain cases they cause us to be optimistic as they are only weak instabilities and we are to be able to handle them.

We conclude this section by emphasizing once more that the described decoupling of (2.1) should be understood as an appropriate technique for analysing large classes of DAEs and the precise behaviour of numerical methods.

It should also be possible to compute the projectors $Q_j(t)$ and matrices $A_i(t)$ at certain points t in order to formulate the initial conditions and organize a numerical index-testing. However, in general the decoupling technique is not aimed at representing a numerical method.

2.2. *DAEs as vector fields on manifolds*

The most frequently used notion of an index of a general nonlinear DAE

$$f(x', x, t) = 0 \qquad (2.29)$$

is the *differentiation index*, which goes back to the work of S.L. Campbell on linear DAEs with smooth coefficients (e.g. Campbell (1987), Brenan *et al.* (1989)).

Assuming f and the respective solutions to be smooth enough we form the system

$$
\left.
\begin{aligned}
f(x', x, t) &= 0 \\
\frac{\mathrm{d}}{\mathrm{d}t} f(x', x, t) &= \frac{\partial}{\partial x'} f(x', x, t) x'' + \cdots = 0 \\
&\;\vdots \\
\frac{\mathrm{d}^\mu}{\mathrm{d}t^\mu} f(x', x, t) &= \frac{\partial}{\partial x'} f(x', x, t) x^{(\mu+1)} + \cdots = 0
\end{aligned}
\right\}
\tag{2.30}
$$

by differentiating μ times. Consider (2.30) as a system in separate dependent variables $x', x'', \ldots, x^{(\mu+1)}$ with x, t as independent variables.

Definition The DAE (2.29) has the *differentiation index* μ if there exists an integer μ such that system (2.30) can be solved for $x' = H(x, t)$, H continuously differentiable, and μ is the smallest integer having this property.

We do not recommend carrying out this procedure in order to obtain the underlying regular ODE $x' = H(x, t)$ in practice. This ODE does not give a good reflection of the qualitative behaviour of the original equation.

Example (Führer and Leimkuhler, 1989) The inherent regular ODE of the DAE

$$
\left.
\begin{aligned}
x'_1 - x_2 + ax_1^2 &= 0 \\
x_2 - ax_1^2 &= 0
\end{aligned}
\right\}
\tag{2.31}
$$

is $x'_1 = 0$, and the origin represents a stable equilibrium (all solutions are stationary here). By differentiating once we formally obtain the system

$$
\left.
\begin{aligned}
x'_1 - x_2 + ax_1^2 &= 0 \\
x_2 - ax_1^2 &= 0 \\
x''_1 - x'_2 + 2ax_1 x'_1 &= 0 \\
x'_2 - 2ax_1 x'_1 &= 0
\end{aligned}
\right\},
$$

which leads to

$$
\left.
\begin{aligned}
x'_1 &= x_2 - ax_1^2 \\
x'_2 &= 2ax_1(x_2 - ax_1^2)
\end{aligned}
\right\},
\tag{2.32}
$$

but now the origin is no longer stable.

System (2.30) suggests the idea of defining a compound function or a derivative array

$$
F_\mu(\bar{y}_\mu, x, t) :=
\begin{bmatrix}
f(y_1, x, t) \\
\frac{\partial}{\partial x'} f(y_1, x, t) y_2 + \cdots \\
\vdots \\
\frac{\partial}{\partial x'} f(y_1, x, t) y_{\mu+1} + \cdots
\end{bmatrix},
\tag{2.33}
$$

where $\bar{y}_\mu := (y_1^T, \ldots, y_{\mu+1}^T)^T \in \mathbb{R}^{(\mu+1)m}$.

If we assume the Jacobian $H_\mu(\bar{y}_\mu, x, t) := \partial F_\mu(\bar{y}_\mu, x, t)/\partial \bar{y}_\mu$ has constant rank, we can form the constraint manifold of order μ

$$S_\mu := \{(x, t) \in \mathbb{R}^m \times \mathbb{R} : F_\mu(\bar{y}_\mu, x, t) = 0 \text{ for a } \bar{y}_\mu \in \mathbb{R}^{(\mu+1)m}\}$$

as well as

$$\begin{aligned}
M_\mu(x, t) &:= \{\bar{y}_\mu \in \mathbb{R}^{(\mu+1)m} : F_\mu(\bar{y}_\mu, x, t) = 0\}, \\
M_\mu^1(x, t) &:= \{y_1 \in \mathbb{R}^m : \bar{y}_\mu \in M_\mu(x, t)\} \text{ for } (x, t) \in S_\mu.
\end{aligned}$$

Definition (Griepentrog, 1991) $f \in C^{\mu+1}(\mathbb{R}^m \times \mathbb{R}^m \times \mathbb{R}, \mathbb{R}^m)$ is called an *index-μ mapping* if S_μ is nonempty and $M_\mu(t, x)$ is a singleton for all $(t, x) \in S_\mu$, and if μ is the smallest integer with these properties.

Clearly, DAE (2.29) has the differentiation index μ if f is an index-μ mapping. However, now it becomes transparent that this DAE represents a vector field defined on S_μ, namely

$$v(x, t) := y_1 \in M_\mu^1(x, t) \quad \text{for } (x, t) \in S_\mu.$$

By definition, $f(v(x, t), x, t) = 0$ holds for $(x, t) \in S_\mu$. The solution of each IVP

$$x'(t) = v(x(t), t), \quad (x(t_0), t_0) \in S_\mu \tag{2.34}$$

evolves in the manifold and solves the DAE. More precisely, the following assertion is proved in Griepentrog (1991).

Theorem 2.8   If (2.29) is an index-μ DAE, then all solutions proceed in the differentiable constraint manifold S_μ. A vector field $v(x, t)$ is defined on S_μ and has the following properties:

(i) v is continuously differentiable; and
(ii) the solutions of (2.29) are identical with the solutions of the IVPs (2.34).

Remark Griepentrog (1991) describes both the manifold S_μ and the vector field v in detail by means of the rank theorem. These investigations are closely related to the differential-geometric concepts of regular DAEs in Reich (1990) and Rabier and Rheinboldt (1991). However, these studies are still in an early phase, but they are very promising and are aimed at making the results of differential geometry applicable to numerical treatment.

Return shortly to the trivial example (2.31). Now, it appears to be an index-1 equation, whereby

$$S_1 := \{(x, t) \in \mathbb{R}^2 \times \mathbb{R} : x_2 - ax_1^2 = 0\},$$

and $M_1^1(x, t) = \{0\}$ for all $(x, t) \in S_1$.

We would like to direct attention to an essential detail of this index definition as well as of Theorem 2.8, namely the condition that the Jacobian of the compound function (2.33) has constant rank. If this property is lost, different singularities may arise, as is illustrated by the next two easy examples, which model certain RC circuits (Chua and Deng, 1989).

Examples Consider the DAEs

$$x_1' - x_2 = 0, \quad x_1 - x_2^3 = 0 \tag{2.35}$$

and

$$x_1' + x_2 = 0, \quad x_1 - x_2^3 = 0. \tag{2.36}$$

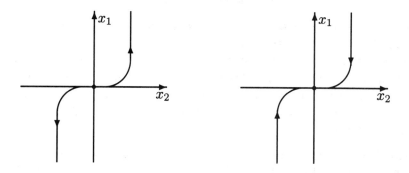

In both cases the Jacobian $H_1(\bar{y}_1, x, t) = \partial F_1(\bar{y}_1, x, t)/\partial \bar{y}_1$ has constant rank 3 for $x_2 \neq 0$, but it suffers from a rank deficiency at $x_2 = 0$. In any case, the origin becomes a stationary solution. Besides the trivial solution, (2.35) has the solution $x_1(t) = (\frac{2}{3}t)^{3/2}$, $x_2(t) = (\frac{2}{3}t)^{1/2}$, which starts at the origin. On the other hand, (2.36) has no solutions coming out of the origin, but $x_1(t) = (1 - \frac{2}{3}t)^{3/2}$, $x_2(t) = (1 - \frac{2}{3}t)^{1/2}$ starts at $(1,1)$ for $t = 0$, and ends, for $t = \frac{3}{2}$, at the origin.

2.3. Many open questions are left

What do the differentiation index and the tractability index have to do with each other. At first glance seemingly nothing. Let us consider the matter in the case of homogeneous linear DAEs in Kronecker canonical normal form (1.4), (1.5). In this case we obtain

$$F_1(y_1, y_2, x, t) = \begin{bmatrix} I & 0 & 0 & 0 \\ 0 & J & 0 & 0 \\ W & 0 & I & 0 \\ 0 & I & 0 & J \end{bmatrix} \begin{bmatrix} u' \\ v' \\ u'' \\ v'' \end{bmatrix} + \begin{bmatrix} W & 0 \\ 0 & I \\ 0 & 0 \\ 0 & 0 \end{bmatrix} \begin{bmatrix} u \\ v \end{bmatrix},$$

where

$$y_1 =: \begin{bmatrix} u' \\ v' \end{bmatrix} \quad y_2 =: \begin{bmatrix} u'' \\ v'' \end{bmatrix} \quad x =: \begin{bmatrix} u \\ v \end{bmatrix},$$

$$S_1 := \{(x, t) \in \mathbb{R}^m \times \mathbb{R} : v \in \mathrm{im}J\},$$

and $M_1^1(x,t)$ is a singleton if and only if

$$Jv' = 0, \quad v' + Jv'' = 0$$

implies $v' = 0$, that is $J = 0$, $\operatorname{ind} J = 1$.

In general, for linear DAEs (2.1), the different index notations are related to different smoothness requirements with respect to the coefficients A, B, q; however, they are identical in essence.

Theorem 2.9 Each linear DAE (2.1) with a differentiation index μ_D also has a global Kronecker normal form index $\mu_K = \mu_D$. Each DAE (2.1) having a global index μ_K is tractable with index $\mu_T = \mu_K$.

For the technically expensive proof we refer to Hansen (1990), Griepentrog and März (1989) and Griepentrog (1991).

For linear equations, the concept of index-μ tractability seems to be the most general one. But, how is the index-μ tractability to be defined for non-linear DAEs? First, this concept is based upon another notion of solution, which can also be reasonably applied to nonlinear equations (2.29) under certain assumptions.

Assumption 2.10 Let the function $f \in C(\mathcal{G}, \mathbb{R}^m)$, where $\mathcal{G} = \mathbb{R}^m \times \mathcal{D} \times \mathcal{I} \subseteq \mathbb{R}^m \times \mathbb{R}^m \times \mathbb{R}$ is an open set and $f_y'(y,x,t)$, $f_x'(y,x,t) \in L(\mathbb{R}^m)$ exist for all $(y,x,t) \in \mathcal{G}$, and $f_y', f_x' \in C(\mathcal{G}, L(\mathbb{R}^m))$.

Suppose that the null space of $f_y'(y,x,t)$ is independent of (y,x), i.e.

$$N(t) := \ker f_y'(y,x,t), \quad (y,x,t) \in \mathcal{G}. \tag{2.37}$$

Let $N(t)$ be smooth in t. Let $Q \in C^1(\mathcal{I}, L(\mathbb{R}^m))$ denote the corresponding projector function onto N and set $P := I - Q$.

In most applications we know, the null space $N(t)$ is kept constant. Due to Assumption 2.10, the identity

$$f(y,x,t) - f(P(t)y,x,t)$$
$$= \int_0^1 f_y'(sy + (1-s)P(t)y, x, t)Q(t)ds = 0, \quad (y,x,t) \in \mathcal{G}$$

becomes true. Consequently, (2.29) may be rewritten as

$$f((Px)'(t) - P'(t)x(t), x(t), t) = 0, \tag{2.38}$$

hence the function space to which the solutions of (2.29) should belong again appears to be

$$C_N^1(\mathcal{I}_0, \mathbb{R}^m) := \{x \in C(\mathcal{I}_0, \mathbb{R}^m) : Px \in C^1(\mathcal{I}_0, \mathbb{R}^m)\},$$

where $\mathcal{I}_0 \subseteq \mathcal{I}$ is a certain interval.

This space seems to be very natural. If $x_* \in C_N^1(\mathcal{I}_0, \mathbb{R}^m)$ is any given function whose trajectory remains in $\mathcal{G}$, then the equation linearized along $x_*(t)$ has continuous coefficients

$$A_*(t) \;=\; f_y'(\zeta(t)), \quad B_*(t) = f_x'(\zeta(t)),$$
$$\zeta(t) \;:=\; ((Px_*)'(t) - P'(t)x_*(t), x_*(t), t) \in \mathcal{G},$$

and the null space of $A_*(t)$ is again $N(t)$.

It should also be mentioned that the given nontrivial solutions of the examples (2.35) and (2.36) do not belong to

$$C^1([0, \infty), \mathbb{R}^2),$$

but to

$$C_N^1([0, \infty), \mathbb{R}^2).$$

Does it make sense to define the notion of index-μ tractability via linearization?

Definition Suppose (2.29) satisfies Assumption 2.10, and $x_* \in C_N^1(\mathcal{I}_0, \mathbb{R}^m)$ is given, $\mathcal{T}_* := \{\zeta(t) : t \in \mathcal{I}_0\} \subset \mathcal{G}$. The DAE (2.29) is said to be *transferable or index-1 tractable* around x_* if the pair $\{A_*, B_*\}$ is index-1 tractable.

Lemma 2.11 $\{A_*, B_*\}$ becomes index-1 tractable if and only if the matrix

$$G(y, x, t) := f_y'(y, x, t) + f_x'(y, x, t)Q(t) \qquad (2.39)$$

remains nonsingular for all (y, x, t) from a neighbourhood $\mathcal{N} \subseteq \mathcal{G}$ of $\mathcal{T}_*$.

Proof. Let $\{A_*, B_*\}$ be index-1 tractable, that means that $A_1 := A_* + (B_* - A_*P')Q$ is nonsingular, then $G_* := A_* + B_*Q = A_1 + A_*P'Q = A_1(I + PP'Q)$ is also nonsingular. Next, $G(y, x, t)$ becomes nonsingular for all $(y, x, t) \in \mathcal{T}_*$, because of

$$G_*(t) = G(\zeta(t)), \quad t \in \mathcal{I}.$$

Since G depends continuously on its arguments, there is a neighbourhood $\mathcal{N}$ of $\mathcal{T}_*$ where $G(y, x, t)$ remains nonsingular. Now the assertion is evident. $\square$

Theorem 2.12 Let $x_* \in C_N^1([t_0, T], \mathbb{R}^m)$ solve the DAE (2.29). Let (2.29) be transferable around x_*. Then, for any given $q \in C([t_0, T], \mathbb{R}^m)$, $x^0 \in \mathbb{R}^m$ the IVP

$$\begin{aligned} f(x'(t), x(t), t) &= \,'q(t) \\ P(t_0)(x(t_0) - x^0) &= 0 \end{aligned} \right\} \qquad (2.40)$$

is uniquely solvable on $C_N^1([t_0, T], \mathbb{R}^m)$, provided that $\|q\|_\infty$ and $|P(t_0)(x^0 - x_*(t_0))|$ are sufficiently small.

Moreover,

$$\|x - x_*\| \le K\{\|q\|_\infty + |P(t_0)(x^0 - x_*(t_0))|\}$$

is valid with constant K.

Proof. Denote shortly $C_N^1 := C_N^1([t_0, T], \mathbb{R}^m)$, $C := C([t_0, T], \mathbb{R}^m)$. For $x \in C_N^1$, $q \in C$, $\beta \in \mathrm{im}(P(t_0))$ with

$$\|x - x_*\| \leq \varrho, \quad \|q\|_\infty \leq \varrho, \quad |\beta - \beta_*| \leq \varrho,$$
$$\beta_* := P(t_0)x_*(t_0), \quad \varrho > 0 \quad \text{sufficiently small},$$

we define the map $\mathcal{F}$ by

$$\mathcal{F}(x, q, \beta) := (f((Px)'(\cdot) - P'(\cdot)x(\cdot), x(\cdot), \cdot) - q(\cdot), P(t_0)x(t_0) - \beta).$$

$\mathcal{F}$ maps a ball within $C_N^1 \times C \times \mathrm{im}P(t_0)$ into $C \times \mathrm{im}P(t_0)$, it is continuously differentiable and, in particular for $z \in C_N^1$,

$$\mathcal{F}'_x(x_*, 0, \beta_*)z = (A_*((Pz)' - P'z) + B_*z, P(t_0)z(t_0)).$$

holds. Trivially, $\mathcal{F}(x_*, 0, \beta_*) = 0$. Due to Corollary 2.7 (cf. also Theorem 2.2), $\mathcal{F}'_x(x_*, 0, \beta_*)$ is a homeomorphism, hence it remains to apply the Implicit Function Theorem. $\square$

Remarks

1   To obtain the fully consistent initial value $x_0 := x(t_0)$ related to the IVP (2.40) the system

$$\left.\begin{array}{c} f(y_0, x_0, t_0) - q(t_0) = 0 \\ P(t_0)(x_0 - x^0) + Q(t_0)y_0 = 0 \end{array}\right\}$$

will be helpful (cf. Theorem 2.3). The Jacobian of this system is nonsingular because of the index-1 requirement.

2   Let $(y_0, x_0, t_0) \in \mathcal{G}$ be given, and let $G(y_0, x_0, t_0)$ be nonsingular. Rewrite $f(y, x, t) = f(w, u + Q(t)w, t) =: \tilde{f}(w, u, t)$ where new variables $w = P(t)y + Q(t)x$, $u = P(t)x$ are introduced.

Clearly, since $\tilde{f}(w_0, u_0, t_0) = 0$, and $\tilde{f}'_w(w_0, u_0, t_0) = G(y_0, x_0, t_0)$ is nonsingular, due to the Implicit Function Theorem there exists a continuous function $w(u, t)$ with continuous partial Jacobian $w'_u(u, t)$, satisfying $\tilde{f}(w, u, t) = 0$. Then, it is easy to check that (cf. (2.9))

$$x(t) = u(t) + Q(t)w(u(t), t) \tag{2.41}$$

represents a solution of $f(x', x, t) = 0$ passing through (x_0, t_0), whereby u denotes the solution of the inherent regular IVP

$$\begin{aligned} u'(t) - P'(t)u(t) &= P(t)(I + P'(t))w(u(t), t), \tag{2.42} \\ u(t_0) &= P(t_0)x_0. \end{aligned}$$

3   Supposing that $G(y, x, t)$ remains nonsingular for all $(y, x, t) \in \mathcal{G}$, we immediately know that

$$S_1 := \{(x, t) \in \mathcal{D} \times \mathcal{I} : f(y, x, t) = 0 \text{ for a } y \in \mathbb{R}^m\}$$

is the manifold of consistent initial values. Comparing this with the matters related to understanding DAEs as vector fields on manifolds we feel the smoothness demanded there to be difficult to realise.

4   In our examples (2.35), (2.36) we compute $G(y, x, t)$ to be equal to

$$\begin{bmatrix} 1 & -1 \\ 0 & -3x_2^2 \end{bmatrix} \text{ respectively } \begin{bmatrix} 1 & 1 \\ 0 & -3x_2^2 \end{bmatrix}.$$

This shows that the transferability matrix may be used as a tool for detecting singularities numerically.

Unfortunately, the situation becomes much more complicated for higher indexes. According to Theorem 2.6 and Corollary 2.7 linear DAEs with an index-μ-tractable coefficient pair $\{A_*, B_*\}$, $\mu > 1$, result in ill posed IVPs, i.e. they have discontinuous inverse mappings in the given topologies. Now, the standard arguments used in Theorem 2.12 no longer apply because the derivative $\mathcal{F}'_x(x_*, 0, \beta_*)$ does not have a continuous inverse.

By means of the following example we want to elucidate that Theorem 2.12 cannot be saved for the index-2 case even if $P(t_0)P_1(t_0)$ is appropriately used instead of $P(t_0)$ (cf. Theorem 2.5), and if only q from C^1 is admissible.

Example (Chua and Deng, 1989; März, 1991) Consider the system

$$x'_1 = x_3^2, \quad x'_2 = -x_3, \quad x_3^3 + x_2 x_3 + x_1 = 0. \tag{2.43}$$

$x_*(t) = (2(\frac{t}{6} + 1)^3, -3(\frac{t}{6} + 1)^2, \frac{t}{6} + 1)^T$ solves this DAE, and

$$A_*(t) = \begin{bmatrix} 1 & 0 & 0 \\ 0 & 1 & 0 \\ 0 & 0 & 0 \end{bmatrix}, \quad B_*(t) = \begin{bmatrix} 0 & 0 & -\frac{1}{3}t - 2 \\ 0 & 0 & 1 \\ 1 & \frac{1}{6}t + 1 & 0 \end{bmatrix}$$

form an index-2-tractable pair $\{A_*, B_*\}$. However, e.g.,

$$x(t) = (2 + t, -3 - t, 1)^T$$

represents another solution, and $x_*(0) = x(0)$ holds, i.e. certain bifurcation phenomena arise. The respective matrix (2.39)

$$G(y, x, t) = \begin{bmatrix} 1 & 0 & -2x_3 \\ 0 & 1 & 1 \\ 0 & 0 & x_2 + 3x_3^2 \end{bmatrix}$$

is nonsingular for $x_2 + 3x_3^2 \neq 0$. Thus, equation (2.43) represents an index-1 DAE everywhere, where $x_2 + 3x_3^2 \neq 0$ holds. The whole thing should be understood as an index-1 DAE with a singularity at $x_2 + 3x_3^2 = 0$.

We might possibly overcome these problems by making the following definition: The nonlinear DAE (2.29) is called index-μ tractable around x_* if for *all* $x \in \{\tilde{x} \in C^1_N : \|\tilde{x} - x_*\| < \varrho\}$, ϱ sufficiently small, the respective pairs $\{A, B\}$ are index-μ tractable. However, how can this be checked?

Finally, we are also interested in conditions that can be treated numerically, such as those provided by Lemma 2.11 for instance. So far, statements have only been successfully made for index-2 equations (cf. Lemma 3.5) and for special index-3 equations (e.g. März, 1989). In these cases it has also been successfully proved that the differentiation index and the tractability index coincide identically except for smoothness.

On the other hand, the approach of considering higher index DAEs as differential equations on manifolds seems to be easier to grasp and handle. In particular, this is true for DAEs with a special structure, e.g. those of Hessenberg form. In this respect, interesting results are to be expected. However, a uniform analysis of DAEs with natural smoothness is still out of sight.

3. Numerical integration methods

3.1. General remarks on the BDF

The integration method used most frequently for regular as well as for singular implicit equations

$$f(x'(t), x(t), x(t)) = 0 \tag{3.1}$$

is the BDF. It is well known that there are powerful codes like DASSL (cf. Brenan *et al.* (1989)) which treat large classes of DAEs well.

On the other hand, the following example shows that BDFs may fail even in very simple cases. Thus, in this section we try to clarify the related problems together with possible ways out.

Example The DAE

$$\begin{bmatrix} 0 & 1 & 0 \\ 0 & \eta t & 1 \\ 0 & 0 & 0 \end{bmatrix} x'(t) + \begin{bmatrix} 1 & 0 & 0 \\ 0 & \eta + 1 & 0 \\ 0 & \eta t & 1 \end{bmatrix} x(t) = q(t) \tag{3.2}$$

has global index-3 for all parameter values $\eta \in \mathbb{R}$. The leading coefficient matrix also has constant null space and constant image space. Table 1 shows results generated by BDFs with different constant step-sizes h and consistent starting values for parameter values $\eta = 0, -0.5$ and 2.0, respectively.

The exact solution is

$$x_1(t) = \mathrm{e}^{-t} \sin t, \quad x_2(t) = \mathrm{e}^{-2t} \sin t, \quad x_3(t) = \mathrm{e}^{-t} \cos t,$$

and $[0, 0.1]$ is the integration interval. The absolute errors at $t = 0.1$ arising in the components of the solution belonging to the null space $\ker A(t)$ and the other one are given separately.

Note that for $\eta = 0$ we have a linear constant coefficient equation as discussed in Section 1. Obviously, the null space component is particularly

Table 1

$h \approx$	$\eta = -0.5$		$\eta = 0$		$\eta = 2.0$	
	Px	Qx	Px	Qx	Px	Qx
BDF_2						
2.5-2	3.-2	3.+0	3.-4	4.-3	1.-4	5.-2
3.1-3	3.+7	2.+10	5.-6	6.-5	1.-5	2.-5
7.8-4	1.+43	3.+46	3.-7	4.-6	1.-6	2.-6
3.9-4	—	—	8.-8	1.-6	2.-7	5.-7
1.9-4	—	—	2.-8	2.-7	6.-8	1.-7
9.7-5	—	—	5.-9	9.-9	1.-8	3.-8
BDF_3						
2.5-2	—	—	1.-5	2.-2	1.-3	8.-2
3.1-3	—	—	3.-8	1.-7	3.-4	1.-2
7.8-4	—	—	4.-10	2.-9	1.-1	1.+2
3.9-4	—	—	5.-11	1.-8	3.+3	4.+6
1.9-4	—	—	5.-12	3.-8	6.+12	2.+16
9.7-5	—	—	9.-13	1.-7	1.+32	1.+36
BDF_6						
3.1-3	—	—	2.-13	3.-9	—	—
7.8-4	—	—	5.-13	2.-8	—	—
3.9-4	—	—	4.-12	3.-8	—	—
1.9-4	—	—	2.-12	2.-7	—	—
9.7-5	—	—	4.-12	4.-6	—	—

$(\sigma = 1.-16)$

affected by round-off errors. Furthermore, order expectations do not become true in practice.

Before we investigate the BDF applied to DAEs (3.1) we describe this class of DAEs in more detail. Assume DAE (3.1) satisfies Assumption 2.10 and, in particular,

$$N(t) := \ker f_y'(y, x, t), \quad (y, x, t) \in \mathcal{G}, \tag{3.3}$$

Let $Q \in C^1(\mathcal{I}, L(\mathbb{R}^m))$ denote the corresponding projector function onto N, $P := I - Q$. Recall that Assumption 2.10 allows equation (3.1) to be rewritten as

$$f((Px)'(t) - P'(t)x(t), x(t), t) = 0, \tag{3.4}$$

thus the function space to which the solutions of (3.4) should belong again appears to be

$$C_N^1(\mathcal{I}_0, \mathbb{R}^m) := \{x \in C(\mathcal{I}_0, \mathbb{R}^m) : Px \in C^1(\mathcal{I}_0, \mathbb{R}^m)\},$$

where $\mathcal{I}_0 \subseteq \mathcal{I}$.

We now ask how integration methods approximate solutions of (3.4) and (3.1). For this purpose, we assume here that solutions exist, say on the interval $\mathcal{I}_0 = [t_0, T]$. However, it should be mentioned once more that a comprehensive analysis of nonlinear DAEs is only in its infancy. Very interesting problems remain to be solved. In particular, solvability is closely related to the description of the set of consistent initial values.

Assume $x_* \in C^1_N := C^1_N([t_0, T], \mathbb{R}^m)$ solves DAE (3.1). Let $\mathcal{B}(x_*, \varrho_0) \subseteq C^1_N$ denote a small ball around x_* within C^1_N such that $x(t) \in \mathcal{D}$ for $t \in [t_0, T]$, and for all $x \in \mathcal{B}(x_*, \varrho_0)$. Introduce the map

$$\mathcal{F} : \mathcal{B}(x_*, \varrho_0) \subseteq C^1_N \to C := C([t_0, T])$$

by means of

$$(\mathcal{F}x)(t) := f((Px)'(t) - P'(t)x(t), x(t), t),$$
$$t \in [t_0, T], \ x \in \mathcal{B}(x_*, \varrho_0). \tag{3.5}$$

The map is continuously differentiable; and its Frechet derivative at x_* is given by

$$(\mathcal{F}'(x_*)z)(t) := A_*(t)((Pz)'(t) - P'(t)z(t)) + B_*(t)z(t),$$
$$t \in [t_0, T], \ z \in C^1_N, \tag{3.6}$$

where

$$A_*(t) := f'_y(\zeta(t)), \ B_*(t) := f'_x(\zeta(t)),$$
$$\zeta(t) := ((Px_*)'(t) - P'(t)x_*(t), x_*(t), t).$$

Let the interval $[t_0, T]$ be partitioned by

$$\pi : t_0 < t_1 < \cdots < t_N = T.$$

Denote by h, $\underline{h}$ the maximal and minimal step-sizes of π, respectively, and $h_j := t_j - t_{j-1}$. Given starting values $x_0, \ldots, x_{s-1}$, we apply the variable step-size BDF to (3.1), i.e.

$$f\left(\frac{1}{h_j} \sum_{i=0}^{s} \alpha_{ji} x_{j-i}, x_j, t_j\right) = 0, \quad j = s, \ldots, N, \tag{3.7}$$

expecting x_j to become an approximation of the true solution value $x_*(t_j)$.

Introduce the map

$$
\mathcal{F}_\pi z \;:=\;
\begin{bmatrix}
z_0 - x_0 \\[4pt]
z_{s-1} - x_{s-1} \\[4pt]
f\left(\dfrac{1}{h_s}\displaystyle\sum_{i=0}^{s}\alpha_{si} z_{s-i},\, z_s,\, t_s\right) \\[6pt]
\vdots \\[6pt]
f\left(\dfrac{1}{h_N}\displaystyle\sum_{i=0}^{s}\alpha_{Ni} z_{N-i},\, z_N,\, t_N\right)
\end{bmatrix},
$$

$$
z \in \mathbb{R}^{m(N+1)}, \quad |z_j - x_*(t_j)| < \varrho_0, \quad j = 0, 1, \ldots, N,
$$

which represents the discretized map corresponding to the BDF. $\mathcal{F}_\pi$ acts within $\mathbb{R}^{m(N+1)}$. Then denote

$$
x_\pi^* =
\begin{bmatrix}
x_*(t_0) \\
\vdots \\
x_*(t_N)
\end{bmatrix}
\in \mathbb{R}^{m(N+1)}
$$

and compute the Jacobian

$$
\mathcal{F}_\pi'(x_\pi^*) =
\begin{bmatrix}
I & & & & & \\
& \ddots & & & & \\
& & I & & & \\
\dfrac{\alpha_{ss}}{h_s}A_s^* & \cdots & \dfrac{\alpha_{s1}}{h_s}A_s^* & F_s^* & & \\
& \ddots & & & \ddots & \\
& & \dfrac{a_{Ns}}{h_N}A_N^* & \cdots & \dfrac{a_{N1}}{h_N}A_N^* & F_N^*
\end{bmatrix}
\in L(\mathbb{R}^{m(N+1)}),
$$

whereby

$$
A_j^* \;:=\; f_y'(\eta_j), \qquad F_j^* \;:=\; \frac{\alpha_{jo}}{h_j}A_j^* + f_x'(\eta_j),
$$

and

$$
\eta_j \;:=\; \left(P(t_j)\frac{1}{h_j}\sum_{i=0}^{s}\alpha_{ji}x_*(t_{j-i}),\, x_*(t_j),\, t_j\right).
$$

Complete $\mathbb{R}^{m(N+1)}$ with respect to the norms

$$
\|z\|_\infty \;:=\; \max\{|z_i| : i = 0, 1, \ldots, N\},
$$

$$
\|z\|_\pi \;:=\; \|z\|_\infty + \max\left\{\left|\frac{1}{h_j}\sum_{i=0}^{s}\alpha_{ji}P(t_{j-i})z_{j-i}\right| : j = s, \ldots, N\right\},
$$

which are consistent with the norms of C and C_N^1, respectively. We then use the matrix norms

$$\|G\|_\pi := \max\{\|Gz\|_\pi : \|z\|_\infty = 1\},$$
$$\|G\|_\pi := \max\{\|Gz\|_\infty : \|z\|_\pi = 1\}, \qquad G \in L(\mathbb{R}^{m(N+1)}).$$

Let $\mathcal{B}_\pi(x_\pi^*, \varrho) := \{z \in \mathbb{R}^{m(N+1)} : \|z - x_\pi^*\|_\pi < \varrho\}$.

In the following we use grids π belonging to a given grid class Π, e.g. the class of locally uniform grids with given constants c_1, c_2, $h_{\max}$, such that $c_1 h_{j-1} \leq h_j \leq c_2 h_{j-1}$, $h \leq h_{\max}$, for all j and all $\pi \in \Pi$. The smallest grid class in which we are interested is the set Π_{equ} of all sufficiently fine equidistant grids; however we always assume $\Pi_{\mathrm{equ}} \subseteq \Pi$.

Definition The BDF (3.7) is stable for (3.1) on grid class Π if there exist constants $S > 0$, $\varrho > 0$ such that for arbitrary $\pi \in \Pi$ the inequality

$$\|z - \bar{z}\|_\pi \leq S \|\mathcal{F}_\pi z - \mathcal{F}_\pi \bar{z}\|_\infty \tag{3.8}$$

is satisfied for all $z, \bar{z} \in \mathcal{B}_\pi(x_\pi^*, \varrho)$.

Definition The BDF (3.7) is weakly unstable for (3.1) on Π if the inequality

$$\|z - \bar{z}\|_\pi \leq S \underline{h}^{-\gamma} \|\mathcal{F}_\pi z - \mathcal{F}_\pi \bar{z}\|_\infty \tag{3.9}$$

is valid for all $z, \bar{z} \in \mathcal{B}_\pi(x_\pi^*, \varrho_\pi)$, $\pi \in \Pi$, where $S > 0$, $\gamma > 0$ are constants but $\varrho_\pi > 0$ may depend on the chosen grid π; γ is said to be the order of instability.

Then, introduce the *local discretization error* $\tau_\pi := \mathcal{F}_\pi x_\pi^*$. Clearly, its first components τ_j, $j = 0, \ldots, s - 1$, represent the errors in the starting values, but

$$\tau_j = f\left(\frac{1}{h_j} \sum_{i=0}^{s} \alpha_{ji} x_*(t_{j-i}), x_*(t_j), t_j\right) \qquad \text{for } j = s, \ldots, N.$$

Surely, the point of interest is the so-called *global error*

$$\varepsilon_\pi := x_\pi^* - x_\pi$$

where $x_\pi \in \mathbb{R}^{m(N+1)}$ consists of the components $x_0, x_1, \ldots, x_N \in \mathbb{R}^m$. Note that $\varepsilon_j = \tau_j$ for $j = 0, 1, \ldots, s - 1$.

Recall some standard arguments from discretization theory (e.g. Keller (1975)), which we apply and modify, appropriately.

1. First of all, if the x_j, $j \geq s$ in (3.7) exist and $x_\pi \in \mathcal{B}_\pi(x_\pi^*, \varrho)$, then stability implies the error estimate

$$\|\varepsilon_\pi\|_\pi \leq S \|\tau_\pi\|_\infty,$$

hence

$$\max_{j \geq 0} |x_*(t_j) - x_j| \leq S\{\max_{j \leq s-1} |x_*(t_j) - x_j| + \max_{j \geq s} |\tau_j|\}. \tag{3.10}$$

2   Then, it is sufficient for stability that there exists a uniform bound S_1,

$$\|(\mathcal{F}'_\pi(x^*_\pi))^{-1}\|_\pi \leq S_1, \quad \pi \in \Pi. \tag{3.11}$$

3   Assuming stability, for sufficiently small $h_{\max}$ and ϱ, the equation

$$\mathcal{F}_\pi z = 0$$

has exactly one solution x_π on $\mathcal{B}_\pi(x^*_\pi, \varrho)$, which can be computed by the Newton method.

4   (3.11) can be proved by permuting linearization and discretization, and using the Banach lemma.

Under which conditions do these standard arguments remain valid when the BDF is applied to DAEs?

In Section 1 we have learnt that, in higher index cases, some weak instabilities should be expected. Is it possible to carry over these standard arguments then? How can weak instabilities be distinguished?

3.2. On the BDF applied to linear DAEs

In any case, the behaviour of the BDF applied to linear DAEs plays a crucial role. This is why we investigate this question in more detail. It should not be surprising that stability and instability, respectively, depend on the index of the DAE. In the following we will point out that certain time-dependent subspaces are also responsible for exponential instabilities.

Let us turn to the special case when the BDF is applied to linear DAEs, that is

$$A(t_j)\frac{1}{h_j}\sum_{i=0}^{s}\alpha_{ji}x_{j-i} + B(t_j)x_j = q(t_j), \qquad j \geq s, \tag{3.12}$$

where starting values $x_0, \dots, x_{s-1}$ are given. For the local error τ_ℓ we now derive

$$
\begin{aligned}
\tau_j &= A(t_j)\frac{1}{h_j}\sum_{i=0}^{s}\alpha_{ji}x_*(t_{j-i}) + B(t_j)x_*(t_j) - q(t_j)\\
&= A(t_j)\left\{\frac{1}{h_j}\sum_{i=0}^{s}\alpha_{ji}x_*(t_{j-i}) - (Px_*)'(t_j) + P'(t_j)x_*(t)\right\};
\end{aligned}
\tag{3.13}
$$

consequently, the local error belongs to a subspace,

$$\tau_j \in \mathrm{im}A(t_j), \quad j \geq s,$$

which is characteristic of DAEs, and we will make use of this later.

To obtain x_j for $j \geq s$, we have to solve the linear system

$$F_j x_j = A(t_j)\frac{1}{h_j}\sum_{i=1}^{s}\alpha_{ji}x_{j-i} + q(t_j), \tag{3.14}$$

with the coefficient matrix

$$F_j := \frac{1}{h_j}\alpha_{j0}A(t_j) + B(t_j). \tag{3.15}$$

Is this matrix F_j nonsingular? Clearly, nonsingularity is necessary for the BDF to work well. Note that for $\eta = -1$ in (3.2), all related coefficient matrices F_j are singular. However, as example (3.2) also shows, even if the BDF becomes formally feasible, i.e. if all matrices F_j are nonsingular, the BDF may fail.

Equation $\mathcal{F}_\pi x = 0$, which represents the BDF, now has the special form $\mathcal{F}_\pi x = \mathcal{L}_\pi x - q_\pi$, where

$$\mathcal{L}_\pi := \begin{bmatrix} I & & & & & & \\ & \ddots & & & & & \\ & & I & & & & \\ \frac{\alpha_{ss}}{h_s}A_s & \cdots & \frac{\alpha_{s1}}{h_s}A_s & F_s & & & \\ & \ddots & & & \ddots & & \\ & & \frac{\alpha_{Ns}}{h_N}A_N & \cdots & \frac{\alpha_{N1}}{h_N}A_N & F_N \end{bmatrix} \in L(\mathbb{R}^{m(N+1)}), \tag{3.16}$$

$A_j := A(t_j)$, $j \geq 0$, $(q_\pi)_j := x_j$ for $j = 0, \dots, s-1$, $(q_\pi)_j = q(t_j)$ for $j \geq s$.

Clearly, $\mathcal{L}_\pi$ is nonsingular if $F_s, \dots, F_N$ are. Moreover, in the case of linear DAEs, the stability inequality (3.8) simplifies to

$$\|\mathcal{L}_\pi^{-1}\|_\pi \leq S. \tag{3.17}$$

It should be mentioned that $\mathcal{L}_\pi$ is to be understood as a discretized map $\mathcal{L}$ given in (2.25). By Corollary 2.7, we may expect $\mathcal{L}_\pi$ to be stable for index-1-tractable DAEs only.

Next we are going to prove the characteristic inequalities of $\mathcal{L}_\pi^{-1}$.

Case 1: Assume (2.1) to be transferable (index-1 tractable).
Recall from Section 2 that $A_1 := A + (B - AP')Q$ becomes a nonsingular matrix function. In addition, by Lemma 2.1, $G := A + BQ$ also remains nonsingular; and furthermore $G = A_1 + AP'Q = A_1(I + PP'Q)$. First of all, it is easy to check that F_j is nonsingular if

$$I + \alpha_{j0}\frac{1}{h_j}P(t_j)G(t_j)^{-1}B(t_j) =: H_j$$

is, i.e. at least for small h_j. Then we compute

$$F_j^{-1} = \left\{Q(t_j) + \frac{1}{\alpha_{j0}}h_jP_s(t_j)H_j^{-1}P(t_j)\right\}G(t_j)^{-1}, \tag{3.18}$$

$\mathrm{cond}(F_j) \sim h_j^{-1}$.

Since $F_s, \ldots, F_N$ are nonsingular, so is $\mathcal{L}_\pi$. Next, by decoupling $\mathcal{L}_\pi z = w$ in a similar way as (2.1) in Section 2, we generate a uniform bound for $\mathcal{L}_\pi$ on an appropriate grid class Π. $\mathcal{L}_\pi z = w$ means in detail that $z_j = w_j$, $j = 0, \ldots, s-1$, and

$$A(t_j)\frac{1}{h_j}\sum_{i=0}^{s}\alpha_{ji}z_{j-i} + B(t_j)z_j = w_j, \qquad j = s, \ldots, N. \qquad (3.19)$$

Denote $u_j := P(t_j)z_j$, $v_j := Q(t_j)z_j$. Multiply (3.19) by $P(t_j)A_1(t_j)^{-1}$ and $Q(t_j)A_1(t_j)^{-1}$, respectively. This yields

$$\left.\begin{array}{c}\dfrac{1}{h_j}\displaystyle\sum_{i=o}^{s}\alpha_{ji}P(t_j)(u_{j-i}+v_{j-i}) + (PA_1^{-1}B)(t_j)u_j = (PA_1^{-1})(t_j)w_j \\[2mm] v_j + Q_s(t_j)u_j = (QA_1^{-1})(t_j)w_j.\end{array}\right\} \quad (3.20)$$

Clearly, if the projector function P is constant, then this formula fits the system (2.6), (2.7) very well. In particular, the first equation simplifies to

$$\frac{1}{h_j}\sum_{i=0}^{s}\alpha_{ji}Pu_{j-i} + (PA_1^{-1}B)(t_j)u_j = (PA_1^{-1})(t_j)w_j.$$

For $w = q_\pi$, this is in fact the same expression we would obtain by applying the BDF to the regular ODE inherent in (2.1)

$$u' + PA_1^{-1}Bu = PA_1^{-1}q.$$

If P' does not vanish, there arises some additional feedback between the components in (3.20). Because

$$P(t_j)(u_{j-i}+v_{j-i}) = u_{j-i} + (P(t_j) - P(t_{j-i}))(u_{j-i}+v_{j-i})$$

$$= u_{j-i} + \int_0^1 P'(t_{j-i}+s(t_j-t_{j-i}))ds(t_j-t_{j-i})(u_{j-i}+v_{j-i})$$

we are able to rearrange the first equation in (3.20) to

$$\frac{1}{h_j}\sum_{i=0}^{s}\alpha_{ji}u_{j-i} + \sum_{i=1}^{s}D_{ji}(u_{j-i}+v_{j-i}) + (PA_1^{-1}B)(t_j)u_j = (PA_1^{-1})(t_j)w_j,$$

where the coefficient matrices are uniformly bounded.

Theorem 3.1 Let the given variable step-size BDF applied to a regular explicit ODE be stable on the grid class Π. Let the DAE (2.1) be index-1 tractable. Then, there is a bound S such that $\mathcal{L}_\pi$ is bijective, and

$$\|\mathcal{L}_\pi^{-1}\|_\infty \le \|\mathcal{L}_\pi^{-1}\|_\pi \le S \quad \text{for all } \pi \in \Pi. \qquad (3.21)$$

Proof. By standard arguments we easily obtain

$$\max_{j\ge s}|u_j| \le S_1 \max_{j\ge 0}|w_j|,$$

therefore

$$\max_{j \geq s} |v_j| \leq S_2 \max_{j \geq 0} |w_j|$$

and

$$\max_{j \geq s} \left| \frac{1}{h_j} \sum_{i=0}^{s} \alpha_{ji} u_{j-i} \right| \leq S_3 \max_{j \geq 0} |w_j|.$$

$\square$

Note that Theorem 3.1 implies the error estimation

$$\max_{j \geq s} |x_*(t_j) - x_j| \leq S \left\{ \max_{j \leq s-1} |x_*(t_j) - x_j| + \max_{j \geq s} |\tau_j| \right\}, \qquad (3.22)$$

which is well known in the case of regular ODEs.

Case 2: Assume (2.1) to be index-2-tractable.
We begin this part by quoting the nice linear index-2 DAE from Gear and Petzold (1984), which was constructed to illustrate the instability of Euler's backward rule.

Example The DAE

$$\begin{bmatrix} 0 & 0 \\ 1 & \eta t \end{bmatrix} x'(t) + \begin{bmatrix} 1 & \eta t \\ 0 & 1+\eta \end{bmatrix} x(t) = q(t) \qquad (3.23)$$

has the global index-2 for all parameter values $\eta \in \mathbb{R}$. Compute here (cf. (2.19)–(2.21))

$$Q(t) = \begin{bmatrix} 0 & -\eta t \\ 0 & 1 \end{bmatrix}, \quad A_1(t) = \begin{bmatrix} 0 & 0 \\ 1 & 1+\eta t \end{bmatrix}$$

$$Q_1(t) = \begin{bmatrix} 1+\eta t & \eta t(1+\eta t) \\ -1 & -\eta t \end{bmatrix}, \quad P(t)P_1(t) = 0.$$

The backward Euler rule applied to (3.23) gives for $\eta \neq -1$

$$x_{1,j} = q_1(t_j) - \eta t_j x_{2,j},$$

$$x_{2,j} = \frac{\eta}{1+\eta} x_{2,j-1} + \frac{1}{1+\eta} \left\{ q_2(t_j) - \frac{1}{h_j}(q_1(t_j) - q_1(t_{j-1})) \right\},$$

but the exact solution is

$$x_1(t) = q_1(t) - \eta t x_2(t),$$
$$x_2(t) = q_2(t) - q_1'(t).$$

Careful further investigation will reveal that the backward Euler rule for this problem is weakly unstable but convergent if $\eta > -0.5$, and exponentially unstable for all $\eta < -0.5$, $\eta \neq -1$. For $\eta = -1$ the backward Euler rule does not work at all.

In the following we exclude such situations where the behaviour of a numerical method depends essentially on parameter values, all of which belong to the same category, by restricting the class of DAEs (2.1) to those with constant $\ker A(t)$ and $P' = 0$, respectively.

Let us turn back to the BDF applied to (2.1), that is to formula (3.12). The first problem to be solved is the nonsingularity of F_j given by (3.15). The question may be answered by the use of the decoupling technique described in Section 2. Supposing that $P' = 0$, and

$$\tilde{H}(t_j) := I + h_j \frac{1}{\alpha_{jo}} (PP_1 G_2^{-1} B)(t_j)$$

becomes nonsingular (which happens at least for small h_j), the matrix F_j will also be nonsingular, and

$$F_j^{-1} = \left(\left\{ QP_1 + P_1 Q_1 + \alpha_{jo} \frac{1}{h_j} QQ_1 + h_j \frac{1}{\alpha_{jo}} (I - QP_1 G_2^{-1} B) \tilde{H}^{-1} PP_1 \right\} G_2^{-1} \right)(t_j) \tag{3.24}$$

$\operatorname{cond}(F_j) \sim h_j^{-2}$.

Expression (3.24) is evaluated in März (1990, Lemma 3.1). Thereby, $G_2 := A_1 + B_0 PQ_1$ is used instead of A_2 in Section 2. Due to Lemma 2.1 (cf. (2.16)), both A_2 and G_2 are nonsingular simultaneously. More precisely, $A_2 = G_2(I - P_1(PP_1)'Q_1)$, $(I - P_1(PP_1)'Q_1)^{-1} = (I + P_1(PP_1)'Q_1)$ are valid.

It should be mentioned that the term QQ_1 within (3.24) does not vanish principally as a matter of index-2 tractability. This is true independently of possible special structural forms of the DAE itself. However, if the DAE has a special form, e.g. Hessenberg form, then, employing the special structure, we can look for an appropriate scaling of F_j.

Next we decompose the system $\mathcal{L}_\pi z = w$ (cf. (3.16), (3.19)) to gain information about $\mathcal{L}_\pi^{-1}$, once again using the projector technique. Multiplying (3.19) by $(PP_1 G_2^{-1})(t_j)$, $(QP_1 G_2^{-1})(t_j)$ and $(Q_1 G_2^{-1})(t_j)$, respectively, we derive

$$\frac{1}{h_j} PP_1(t_j) \sum_{i=0}^{s} \alpha_{ji} z_{j-i} \; + \; PP_1(t_j) G_2(t_j)^{-1} B(t_j) PP_1(t_j) z_j =$$

$$= PP_1(t_j) G_2(t_j)^{-1} w_j, \tag{3.25}$$

$$-\frac{1}{h_j} QQ_1(t_j) \sum_{i=0}^{s} \alpha_{ji} z_{j-i} + Q z_j \; + \; QP_1(t_j) G_2(t_j)^{-1} B(t_j) PP_1(t_j) z_j =$$

$$= QP_1(t_j) G_2(t_j)^{-1} w_j, \tag{3.26}$$

$$Q_1(t_j) z_j = Q_1(t_j) G_2(t_j)^{-1} w_j, \tag{3.27}$$

for $j \geq s$. Recall that $z_j = w_j$ for $j = 0, \ldots, s-1$. Note that we also use P,

Q here as constant projector matrices. Inserting

$$\frac{1}{h_j}PP_1(t_j)\sum_{i=0}^{s}\alpha_{ji}z_{j-i} = \frac{1}{h_j}\sum_{i=0}^{s}\alpha_{ji}PP_1(t_{j-i})z_{j-i}+$$

$$+ \sum_{i=0}^{s}\alpha_{ji}P\frac{1}{h_j}(P_1(t_j)-P_1(t_{j-i}))\{PP_1(t_{j-i})z_{j-i}+PQ_1(t_{j-i})z_{j-i}\}$$

in equation (3.25), and taking into account (3.27), we are able to prove the inequality

$$\max_{j\geq 0}|PP_1(t_j)z_j| \leq S_1\max_{j\geq 0}|w_j|$$

by standard arguments. Trivially,

$$\max_{j\geq 0}|Q_1(t_j)z_j| \leq S_2\max_{j\geq 0}|w_j|$$

also becomes true due to (3.27). Moreover, (3.26), (3.27) yield

$$
\begin{aligned}
Qz_j &= \frac{1}{h_j}QQ_1(t_j)\sum_{i=0}^{s}\alpha_{ji}\{PP_1(t_{j-i})+PQ_1(t_{j-i})\}z_{j-i} \\
&\quad -QP_1(t_j)G_2(t_j)^{-1}B(t_j)PP_1(t_j)z_j + QP_1(t_j)G_2(t_j)^{-1}w_j \\
&= \frac{1}{h_j}QQ_1(t_j)\sum_{i=0}^{s}\alpha_{ji}Q_1(t_{j-i})G_2(t_{j-i})^{-1}\tilde{w}_{j-i} \quad\quad (3.28)\\
&\quad +\sum_{i=0}^{s}\alpha_{ji}Q\frac{1}{h_j}(Q_1(t_j)-Q_1(t_{j-i}))PP_1(t_{j-i})z_{j-i} \\
&\quad -QP_1(t_j)G_2(t_j)^{-1}B(t_j)PP_1(t_j)z_j + QP_1(t_j)G_2(t_j)^{-1}w_j,
\end{aligned}
$$

for $j \geq s$, where we introduce, for more convenience,

$$
\begin{aligned}
\tilde{w}_j &:= w_j && \text{for } j \geq s, \\
\tilde{w}_j &:= G_2(t_j)w_j && \text{for } j \leq s-1.
\end{aligned}
\quad\quad (3.29)
$$

Now, we can estimate

$$|Qz_j| \leq \frac{1}{h_j}|QQ_1(t_j)\sum_{i=0}^{s}\alpha_{ji}Q_1(t_{j-i})G_2(t_{j-i})^{-1}\tilde{w}_{j-i}| + S_3\max_{j\geq 0}|w_j|. \quad (3.30)$$

Since (3.19) immediately implies

$$\frac{1}{h_j}\sum_{i=0}^{s}\alpha_{ji}Pz_{j-i} = PA(t_j)^+\{-B(t_j)z_j + w_j\}, \quad\quad j \geq s, \quad\quad (3.31)$$

it follows that

$$\|z\|_\pi \leq S_4\left\{\max_{j\geq 0}|w_j| + \max_{j\geq s}\frac{1}{h_j}\left|QQ_1(t_j)\sum_{i=0}^{s}\alpha_{ji}Q_1(t_{j-i})G_2(t_{j-i})^{-1}\tilde{w}_{j-i}\right|\right\}$$

$$(3.32)$$

becomes valid.

Consider now expression (3.14) again. Solving this equation in practice, instead of the values x_j, only certain $\tilde{x}_j$ satisfying

$$F_j\tilde{x}_j = -A(t_j)\frac{1}{h_j}\sum_{i=1}^{s}\alpha_{ji}\tilde{x}_{j-i} + q(t_j) + \delta_j, \qquad j \geq s, \qquad (3.33)$$

where $\tilde{x}_j := x_j$, $j = 0, \ldots, s-1$, are generated. The δ_j represent round-off errors (but also errors that arise later when solving nonlinear equations).

Then, if we put $z_j = x_*(t_j) - \tilde{x}_j$, $j \geq 0$, $w_j = \tau_j - \delta_j$ results for $j \geq s$, and $w_j = x_*(t_j) - x_j$ for the starting phase $j = 0, \ldots, s-1$. Because $\tau_j \in \operatorname{im} A(t_j)$ (cf. (3.13)) we obtain for all $k \geq s$

$$\begin{aligned}
Q_1(t_k)G_2(t_k)^{-1}\tilde{w}_k &= Q_1(t_k)G_2(t_k)^{-1}w_k \\
&= Q_1(t_k)G_2(t_k)^{-1}(\tau_k - \delta_k) \\
&= Q_1(t_k)G_2(t_k)^{-1}(A(t_k)A(t_k)^+\tau_k - \delta_k).
\end{aligned}$$

However, on the other hand,

$$\begin{aligned}
G_2^{-1}A &= G_2^{-1}(A + (B - AP')Q)P = G_2^{-1}A_1P \\
&= G_2^{-1}(A_1 + BPQ_1)P_1P = P_1P
\end{aligned}$$

holds, thus $Q_1G_2^{-1}A = Q_1P_1P = 0$.

Consequently,

$$Q_1(t_k)G_2(t_k)^{-1}\tilde{w}_k = -Q_1(t_k)G_2(t_k)^{-1}\delta_k \qquad (3.34)$$

for $k \geq s$, which appears to be characteristic of these DAEs.

Finally, collect this result in

Theorem 3.2 Let the given BDF applied to regular explicit ODEs become stable on the grid class II. Let the DAE (2.1) be index-2 tractable, and, additionally, let $P' = 0$.

(i) Then $\mathcal{L}_\pi$ is bijective, and

$$\|\mathcal{L}_\pi^{-1}\|_\infty \leq \|\mathcal{L}_\pi^{-1}\|_\pi \leq S\underline{h}^{-1}, \qquad \pi \in \Pi, \qquad (3.35)$$

is true with a certain constant $S > 0$.

(ii) The following precise error estimates hold:

$$\max_{j\geq s}|P(x_*(t_j) - \tilde{x}_j| \leq S_P\{\max_{j\leq s-1}|P(x_*(t_j) - x_j)| + \max_{j\geq s}|\tau_j - \delta_j|\}$$

$$(3.36)$$

and

$$|Q(x_*(t_j) - \tilde{x}_j)| \leq$$

$$\leq S_Q\left\{\max_{j\leq s-1}|P(x_*(t_j) - x_j)| + \max_{j\geq s}|\tau_j - \delta_j|\right\} +$$

$$+\frac{1}{h_j}\left|QQ_1(t_j)\sum_{i=0}^{s}\alpha_{ji}Q_1(t_{j-i})G_2(t_{j-i})^{-1}\tilde{\delta}_{j-i}\right|,\qquad(3.37)$$

where $\tilde{\delta}_j := G_2(t_j)(x_*(t_j) - x_j)$, $j = 0,\ldots,s-1$, reflect the errors in the starting values, and

$$\tilde{\delta}_j := \delta_j \quad\text{for } j \geq s.$$

Remarks

1   If exact values $x_j = x_*(t_j)$, $j = 0,\ldots,s-1$, are used in the starting phase, (3.36), (3.37) immediately imply, for $\delta_j = 0$, $j \geq s$,

$$\max_{j\geq 0}|x_*(t_j) - x_j| \leq \bar{S}\max_{j\geq s}|\tau_j|,\qquad(3.38)$$

hence the BDF converges formally with the expected order. However, practical computations cannot be managed in such a way that all $\tilde{\delta}_j$ vanish in reality.

2   Expression (3.28) shows that, for small h_j, Qz_j behaves in fact mainly as

$$\frac{1}{h_j}QQ_1(t_j)\sum_{i=0}^{s}\alpha_{ji}Q_1(t_{j-i})G_2(t_{j-i})^{-1}\tilde{w}_{j-i}.$$

In this sense, (3.37) and (3.35) cannot be improved. We really have to deal with a weak instability. Fortunately, this instability does not affect the nonnull space components at all (cf. (3.36)).

Case 3: Assume (2.1) to be index-3 tractable.
First recall our example (3.2) to illustrate that a restriction to the class of DAEs with constant null space $\ker A(t)$ will not do. It may be checked that, in (3.2), $(PQ_1Q_2)'(t)$ does not vanish identically. This seems to be the crucial point in the index-3 case. As in the previous parts we consider $\mathcal{L}_\pi$ given by (3.16).
    Recall that (cf. (3.16), (3.33))

$$\mathcal{L}_\pi x_\pi = q_\pi := \begin{bmatrix} x_0 \\ \vdots \\ x_{s-1} \\ q(t_s) \\ \vdots \\ q(t_N) \end{bmatrix}, \quad \mathcal{L}_\pi \tilde{x}_\pi - q_\pi = \delta_\pi := \begin{bmatrix} 0 \\ \vdots \\ 0 \\ \delta_s \\ \vdots \\ \delta_N \end{bmatrix}$$

$$\mathcal{L}_\pi x^*_\pi - q_\pi = \tau_\pi := \begin{bmatrix} x_*(t_0) - x_0 \\ \vdots \\ \dfrac{x_*(t_{s-1}) - x_{s-1}}{T_s} \\ \vdots \\ T_N \end{bmatrix},$$

thus $\mathcal{L}_\pi(x^*_\pi - \tilde{x}_\pi) = \tau_\pi - \delta_\pi$.

By the use of this projector technique we decompose (3.19) respectively $\mathcal{L}_\pi z = w$. Now we multiply (3.19) by $PP_1P_2A_3^{-1}$, $QP_1P_2A_3^{-1}$, $Q_1P_2A_3^{-1}$ and $Q_2A_3^{-1}$, respectively. We omit these straigthforward but very extensive evaluations here and mention only that we now have to insert expressions given by the BDF into each other twice to approximate first and second derivatives. This is why the BDF on the whole becomes active just for $j \geq 2s$. The first s steps have to be analysed separately. Let us formulate the results:

Theorem 3.3 Let the given BDF applied to regular explicit ODE become stable on the grid class II. Let the DAE (2.1) be index-3 tractable and, in addition, let $P' = 0$, $(PQ_1Q_2)' = 0$.

(i) Then $\mathcal{L}_\pi$ is bijective,

$$\|\mathcal{L}_\pi^{-1}\|_\infty \leq \|\mathcal{L}_\pi^{-1}\|_\pi \leq S\underline{h}^{-2}, \qquad \pi \in \Pi, \qquad (3.39)$$

holds with a certain constant S.

(ii) The following detailed error estimates become true with

$$\omega_\pi := \max_{j \geq s} |\tau_j - \tilde{\delta}_j| + \max_{j \leq s-1} |x_*(t_j) - x_j|, \quad \text{for } j \geq 2s :$$

$$|PP_1(t_j)(x_*(t_j) - \tilde{x}_j)| \leq S_1\omega_\pi, \qquad (3.40)$$

$$|PQ_1(t_j)(x_*(t_j) - \tilde{x}_j)| \leq S_2\omega_\pi \; + \; \frac{1}{h_j}\left|\sum_{i=0}^{s}\alpha_{ji}PQ_1Q_2A_3(t_{j-i})^{-1}\tilde{\delta}_{j-i}\right|$$

$$(3.41)$$

and

$$|Q(x_*(t_j) - \tilde{x}_j| \leq S_3\omega_\pi +$$

$$+ \frac{1}{h_j}\left|\sum_{i=0}^{s}\alpha_{ji}(QQ_1P_2A_3^{-1})(t_{j-i})(\tau_{j-i} - \tilde{\delta}_{j-i})\right|$$

$$+ \gamma\frac{1}{h_j}\left|\sum_{i=0}^{s}\alpha_{ji}\frac{1}{h_{j-i}}\sum_{k=0}^{s}\alpha_{j-i-k}PQ_1Q_2A_3(t_{j-i-k})^{-1}\tilde{\delta}_{j-i-k}\right| \; (3.42)$$

Thereby, $\gamma := \max |QQ_1(t)|$, and

$$\tilde{\delta}_j \ := \ \delta_j \quad \text{for } j \geq s,$$
$$\tilde{\delta}_j \ := \ A_3(t_j)(x_*(t_j) - x_j) \quad \text{for } j \leq s - 1.$$

Remarks

1   Now, the worst error sensitivity has order h_j^{-2}. It is again somewhat local and belongs to the null space component only.

2   Putting $\tilde{\delta}_j = 0$ in (3.40)–(3.42), i.e. using exact starting values and computing without any round-off error, (3.40), (3.41) provide for $j \geq 2s$

$$|P(x_*(t_j) - x_j)| \leq S_4 \max_{j \geq s} |\tau_j|, \tag{3.43}$$

but

$$|Q(x_*(t_j) - x_j)| \ \leq \ S_3 \max_{i \geq s} |\tau_j| +$$
$$+ \ \frac{1}{h_j} |\sum_{i=0}^{s} \alpha_{ji}(QQ_1 P_2 A_3^{-1})(t_{j-i})\tau_{j-i}|. \tag{3.44}$$

The last term in (3.44) is also troublesome. It reflects the new quality of the problem of index-3-tractable DAEs.

In constant step-size computations, the local error is smooth if the solution $x_*(t)$ itself is smooth enough. In this case, we again have $|x_*(t_j) - x_j| = \prime(h^s)$, $j \geq 2s$. However, step-size changes, and similarly the first s steps raise difficulties. In particular, the variable step backward Euler method does not converge since

$$\frac{1}{h_j}((QQ_1 P_2 A_3^{-1})(t_j)\tau_j - (QQ_1 P_2 A_3^{-1})(t_{j-1})\tau_{j-1} \approx$$
$$\approx -\frac{1}{2h_j}(QQ_1 Q_2)(t_j)(Px_*)''(t_j)(h_j - h_{j-1}).$$

Unfortunately, the backward Euler method also fails to provide accurate starting values.

3.3. On the BDF applied to nonlinear DAEs

Now, having provided information on $\mathcal{L}_\pi$ we continue the investigation of nonlinear DAEs started previously. In the following, we understand $\mathcal{L}_\pi$ to be related to the equation linearized in the solution x_*, i.e. (cf. (3.6))

$$\mathcal{L}_\pi = \mathcal{F}'(x_*)_\pi.$$

In other words, $\mathcal{L}_\pi$ represents the discretization of the linearization. On the other hand, $\mathcal{F}'_\pi(x_\pi^*)$ is the derivative of the discretized map $\mathcal{F}_\pi$ at x_π^*.

Compute

$$\mathcal{L}_\pi - \mathcal{F}'_\pi(x^*_\pi) = \begin{bmatrix} 0 & & & & & & \\ & \ddots & & & & & \\ & & 0 & & & & \\ \frac{\alpha_{ss}}{h_s}D_s & \cdots & \frac{\alpha_{s1}}{h_s}D_s & K_s & & & \\ & \ddots & & & \ddots & & \ddots \\ & & \frac{\alpha_{Ns}}{h_N}D_N & & & \frac{\alpha_{N1}}{h_N}D_N & K_N \end{bmatrix}, \qquad (3.45)$$

where

$$\begin{aligned}
D_j &:= A_*(t_j) - A^*_j = f'_y(\zeta(t_j)) - f'_y(\eta_j), \\
K_j &:= \frac{\alpha_{jo}}{h_j}A_*(t_j) + B_*(t_j) - \frac{\alpha_{jo}}{h_j}A^*_j - f'_x(\eta_j) \\
&= \frac{\alpha_{jo}}{h_j}(f'_y(\zeta(t_j)) - f'_y(\eta_j)) + f'_x(\zeta(t_j)) - f'_x(\eta_j)), \\
\zeta(t) &:= ((Px_*)'(t) - P'(t)x_*(t), x_*(t), t), \\
\eta_j &:= (P(t_j)\frac{1}{h_j}\sum_{i=0}^{s}\alpha_{ji}x^*_{j-i}, x^*_j, t_j), \quad x^*_\ell := x_*(t_\ell).
\end{aligned}$$

Further, we have for $j \geq s$

$$((\mathcal{L}_\pi - \mathcal{F}'_\pi(x^*))z)_j = D_j\frac{1}{h_j}\sum_{i=0}^{s}\alpha_{ji}z_{j-i} + (f'_x(\zeta(t_j)) - f'_x(\eta_j))z_j,$$

and $D_j = D_jP(t_j)$,

$$\begin{aligned}
D_j\frac{1}{h_j}\sum_{i=0}^{s}\alpha_{ji}z_{j-i} &= D_j\frac{1}{h_j}\sum_{i=0}^{s}\alpha_{ji}P(t_{j-i})z_{j-i} + \\
&\quad + D_j\sum_{i=0}^{s}\alpha_{ji}\frac{1}{h_j}(P(t_j) - P(t_{j-i}))z_{j-i}.
\end{aligned}$$

Consequently

$$\|(\mathcal{L}_\pi - \mathcal{F}'_\pi(x^*_\pi))z\|_\infty \leq \gamma_\pi\|z\|_\pi \qquad (3.46)$$

with a grid-dependent value

$$\gamma_\pi := c_0 \max_{j\geq s}\|D_j\| + \max_{j\geq s}\|f'_x(\zeta(t_j)) - f'_x(\eta_j)\|,$$

where c_0 denotes a certain constant. Clearly, γ_π becomes small for refined grids.

Note that for all quasi-linear DAEs of the form

$$A(t)x'(t) + g(x(t), t) = 0 \qquad (3.47)$$

we just obtain

$$\mathcal{L}_\pi = \mathcal{F}'_\pi(x^*_\pi). \tag{3.48}$$

In other words, the BDF discretization and linearization commute asymptotically in general, but for special DAEs (3.47) they commute exactly.

Next, supposing $\mathcal{F}'_\pi(x^*_\pi)$ to be bijective, we turn to the question as to whether the nonlinear equation $\mathcal{F}_\pi z = 0$ is solvable. To this end, we introduce the equivalent fixed point problem $E_\pi z = z$, where the map E_π acts in $\mathbb{R}^{m(N+1)}$,

$$E_\pi z := z - \mathcal{F}'_\pi(x^*_\pi)^{-1}\mathcal{F}_\pi z, \quad z \in B_\pi(x^*_\pi, \varrho_0).$$

As usual when we mean to apply Banach's Fixed Point Theorem, we state

$$E_\pi z - E_\pi \bar{z} = \mathcal{F}'_\pi(x^*_\pi)^{-1} \int_0^1 \{\mathcal{F}'_\pi(x^*_\pi) - \mathcal{F}'_\pi(sz + (1-s)\bar{z})\}\mathrm{d}s(z - \bar{z}) \tag{3.49}$$

and

$$\begin{aligned}
E_\pi z - x^*_\pi &= z - x^*_\pi - \mathcal{F}'_\pi(x^*_\pi)^{-1}(\mathcal{F}_\pi z - \mathcal{F}_\pi x^*_\pi + \mathcal{F}_\pi x^*_\pi) \\
&= \mathcal{F}'_\pi(x^*_\pi)^{-1} \int_0^1 \{\mathcal{F}'_\pi(x^*_\pi) - \mathcal{F}'_\pi(sz + (1-s)x^*_\pi)\}\mathrm{d}s(z - x^*_\pi) - \\
&\quad - \mathcal{F}'_\pi(x^*_\pi)^{-1}\mathcal{F}_\pi x^*_\pi, \tag{3.50}
\end{aligned}$$

for $z, \bar{z} \in B_\pi(x^*_\pi, \varrho_0)$.

Given a constant $\alpha < 1$, we choose $\varepsilon = \varepsilon(\pi)$ such that

$$\varepsilon \|\mathcal{F}'_\pi(x^*_\pi)^{-1}\|_\pi \leq \alpha < 1. \tag{3.51}$$

Moreover, since $\mathcal{F}'_\pi$ is continuous, there exists a $\varrho = \varrho(\varepsilon(\pi)) > 0$ so that

$$\|\mathcal{F}'_\pi(x^*_\pi) - \mathcal{F}'_\pi(y)\|_\pi \leq \varepsilon \quad \text{for all } y \in \overline{B_\pi(x^*_\pi, \varrho)}.$$

Hence, for $z, \bar{z} \in \overline{B_\pi(x^*_\pi, \varrho)}$ (3.49), (3.51) provide

$$\|E_\pi z - E_\pi \bar{z}\|_\pi \leq \alpha \|z - \bar{z}\|_\pi \tag{3.52}$$
$$\|E_\pi z - x^*_\pi\|_\pi \leq \alpha \|z - x^*_\pi\|_\pi + \|\mathcal{F}'_\pi(x^*_\pi)^{-1}\tau_\pi\|_\pi. \tag{3.53}$$

If we are sure to manage the inequality

$$\|\mathcal{F}'_\pi(x^*_\pi)^{-1}\tau_\pi\|_\pi \leq (1-\alpha)\varrho, \tag{3.54}$$

we know the map E_π to have a unique fixed point on $\overline{B(x^*_\pi, \varrho)}$. However, keep in mind that ε and ϱ, may both depend on the grid π.

The same arguments apply to the perturbed equation

$$\mathcal{F}_\pi z = \delta_\pi. \tag{3.55}$$

If we suppose the inequality

$$\|\mathcal{F}'_\pi(x^*_\pi)^{-1}(\tau_\pi - \delta_\pi)\|_\pi \leq (1-\alpha)\varrho$$

can be satisfied, equation (3.55) is uniquely solvable on $\overline{\mathcal{B}_\pi(x^*_\pi, \varrho)}$, and, for its solution $\tilde{x}_\pi$, the error estimate

$$\|x^*_\pi - \tilde{x}_\pi\|_\pi \leq \frac{1}{1-\alpha}\|\mathcal{F}'_\pi(x^*_\pi)^{-1}(\tau_\pi - \delta_\pi)\|_\pi \qquad (3.56)$$

is valid.

The BDF is said to be *feasible* in this case, i.e. if the nonlinear equations to be solved per step are locally uniquely solvable. Then, the Newton method may be applied, where (3.56) suggests how accurate the defects δ_j should be.

In the following $\Pi_0 \subseteq \Pi$ always denotes a grid class where the maximal step-sizes of all grids are sufficiently small.

Theorem 3.4 Let the given BDF applied to regular explicit ODEs be stable on the grid class Π. Let the DAE (3.1) satisfy Assumption 2.10, and let $x_* \in C^1_N$ solve this DAE. In addition, let $\{A_*, B_*\}$ be index-1 tractable. Then the BDF is feasible and stable on $\Pi_0 \subseteq \Pi$. The convergence order is the same as in case of regular ODEs.

Proof. By Theorem 3.1, there is a uniform bound S such that $\|\mathcal{L}^{-1}_\pi\|_\pi \leq S$ for all $\pi \in \Pi$. Choose sufficiently fine grids (cf. (3.46)) so that

$$\gamma_\pi S < 1, \qquad \pi \in \Pi_0.$$

Hence $\|\mathcal{L}_\pi - \mathcal{F}'_\pi(x^*)\|_\pi \leq \gamma_\pi$, $\|\mathcal{L}^{-1}_\pi\|_\pi \leq S$ imply the bijectivity of $\mathcal{F}'_\pi(x^*)$ as well as

$$\|\mathcal{F}'_\pi(x^*_\pi)^{-1}\|_\pi \leq \frac{S}{1-\gamma_\pi S} =: S_1.$$

Consequently, in (3.51) we need uniform $\varepsilon = \alpha/S_1$ and δ, respectively, for all $\pi \in \Pi_0$.

Moreover, (3.54) is easy to satisfy by choosing refined grids and sufficiently accurate starting values such that

$$\|\tau_\pi\|_\infty \leq \frac{1}{S_1}(1-\alpha)\varrho.$$

Moreover, for $z, \bar{z} \in \overline{\mathcal{B}_\pi(x^*_\pi, \varrho)}$ the matrix $\int_0^1 \mathcal{F}'_\pi(sz + (1-s)z)ds =: \mathcal{F}'_\pi[z, \bar{z}]$ is also nonsingular since

$$\|\mathcal{F}'_\pi(x^*) - \mathcal{F}'_\pi[z, \bar{z}]\|_\pi \leq \varepsilon, \qquad \|\mathcal{F}'_\pi[z, \bar{z}]^{-1}\|_\pi \leq \frac{S_1}{(1-\alpha)} =: S_2.$$

Hence, $z - \bar{z} = \mathcal{F}'_\pi[z, \bar{z}]^{-1}(\mathcal{F}_\pi z - \mathcal{F}_\pi \bar{z})$ implies stability immediately. Then, (3.10) (or (3.56)) provides convergence. $\square$

Remark Clearly, the nonlinear equations to be solved per step are locally uniquely solvable, and the Newton method can be applied. The same is true for the perturbed equations (cf. (3.56)).

In the case of an index-2-tractable matrix coefficient pair $\{A_*, B_*\}$ the situation becomes worse. Theorem 3.2 only provides for $\|\mathcal{L}_\pi^{-1}\|_\pi \leq S\underline{h}^{-1}$ for $\pi \in \Pi$. How does this affect nonlinear DAEs?

Lemma 3.5 Let Assumption 2.10 be fulfilled, and let $x_* \in C_N^1$ be given. In addition, let $P' = 0$. Then, let $A_1(y, x, t) := f_y'(y, x, t) + f_x'(y, x, t)Q$ be singular for all (y, x, t) belonging to a neighbourhood $\mathcal{N}$ of the trajectory $\mathcal{T}_*$ of x_* within $\mathbb{R}^m \times \mathbb{R}^m \times \mathbb{R}$, but with constant rank there. Moreover, let $A_1(\zeta(t))$ have a smooth null space. Furthermore let

$$\ker A_1(y, x, t) \cap S_1(y, x, t) = \{0\}, \qquad (3.57)$$

$$S_1(y, x, t) := \{z \in \mathbb{R}^m : f_x'(y, x, t)Pz \in \mathrm{im}A_1(y, x, t)\},$$

for all $(y, x, t) \in \mathcal{T}$. Then the DAE linearized in x_* is index-2 tractable.

Proof. We have $A_*(t) := f_y'(\zeta(t))$, $B_*(t) := f_x'(\zeta(t))$, furthermore $A_{*,1}(t) = A_*(t) + B_*(t)Q = A_1(\zeta(t))$, $S_{*,1}(t) := S_1(\zeta(t))$, $\ker(A_{*,1}(t)) \cap S_{*,1}(t) = \{0\}$. Since the null space of $A_1(\zeta(t))$ is assumed to depend continuously differentiably on t we are done. $\square$

Lemma 3.6 Let Assumption 2.10 be valid and let $x_* \in C_N^1$ solve the DAE (3.1). In addition, let $\mathrm{im}f_y'(y, x, t)$ be independent of y, i.e.

$$\mathrm{im}(f_y'(y, x, t,)) =: R(x, t).$$

Then, for the local errors τ_j generated by the BDF (3.7), the implication

$$\tau_j \in R(x_*(t_j), t_j), \quad j \geq s, \qquad (3.58)$$

becomes true.

Proof. Denote shortly

$$\mu_j := P(t_j)\frac{1}{h_j}\sum_{i=0}^s \alpha_{ji}x_*(t_{j-i}), \qquad \lambda_j := (Px_*)'(t_j) - P'(t_j)x_*(t_j).$$

Derive

$$\begin{aligned}
\tau_j &:= f(\eta_j) = f(\mu_j, x_*(t_j), t_j) \\
&= f(\mu_j, x_*(t_j), t_j) - f(\lambda_j, x_*(t_j), t_j) \\
&= \int_0^1 f_y'(s\mu_j + (1-s)\lambda_j, x_*(t_j), t_j)\mathrm{d}s(\mu_j - \lambda_j),
\end{aligned}$$

thus (3.58) is valid. $\square$

Theorem 3.7 Let Π be such a grid class where the quotient of the maximal and minimal step-sizes of any $\pi \in \Pi$ is bounded by a global constant K, i.e.

$$\overline{h} \cdot \underline{h}^{-1} \leq K, \qquad \pi \in \Pi.$$

Let the given BDF applied to regular explicit ODEs become stable on Π. Let DAE (3.1) satisfy all assumptions of Lemma 3.5 as well as Lemma 3.6; furthermore, let the partial Jacobians f_y', f_x' be Lipschitz with respect to (y, x). In addition, let

$$\left| (Px_*)'(t_j) - \frac{1}{h_j} \sum_{i=0}^{s} \alpha_{ji} Px_*(t_{j-i}) \right| \leq c_0 h^p, \qquad (3.59)$$

be valid for all $j \geq s$, $\pi \in \Pi$ and certain constants $c_0 > 0$, $p > 1$. Then the BDF is feasible and weakly instable on Π.

If we suppose the Q_1-components of the starting values to have the order of accuracy $p+1$ and the other ones order p, then the convergence order is p.

Proof. Because of the Lipschitz continuity of f_y' and f_x', $\mathcal{F}_\pi'$ also becomes Lipschitz continuous. In particular

$$\| \mathcal{F}_\pi'(x_\pi^*) - \mathcal{F}_\pi'(y) \|_\pi \leq L \| x_\pi^* - y \|_\pi \qquad (3.60)$$

is valid. Moreover, in (3.46) we may estimate (using the notation from the proof of Lemma 3.6)

$$\gamma_\pi \leq c_1 \max_{j \geq s} |\mu_j - \lambda_j| \leq c_2 h^p.$$

Since (cf. (3.46)) $\| \mathcal{L}_\pi - \mathcal{F}_\pi'(x_\pi^*) \|_\pi \leq c_2 h^p$ and, due to Theorem 3.2, $\| \mathcal{L}_\pi^{-1} \|_\pi \leq S\underline{h}^{-1}$, we may refine the grids in such a way that

$$c_2 S \underline{h}^{-1} h^p \leq c_2 S K h^{p-1} < 1.$$

Consequently, $\mathcal{F}_\pi'(x_\pi^*)$ becomes nonsingular, and

$$\| \mathcal{F}_\pi'(x_\pi^*)^{-1} \|_\pi \leq S\underline{h}^{-1}(1 - c_2 S K h^{p-1})^{-1}.$$

Next, by Lemma 3.6,

$$\tau_j \in R(x_*(t_j), t_j) = \operatorname{im} A_*(t_j), \qquad j \geq s.$$

Taking into account (3.32), this implies

$$\| \mathcal{L}_\pi^{-1} \tau_\pi \|_\pi \leq \tilde{S} \left\{ \max_{j \geq o} |\tau_j| + \right. \qquad (3.61)$$

$$\left. + \max_{s \leq j \leq 2s-1} \left| \frac{1}{h_j} QQ_1(t_j) \sum_{i=j+1-s}^{s} \alpha_{ji} Q_1(t_{j-i})(x_*(t_{j-i}) - x_{j-i}) \right| \right\}.$$

Choosing the starting values to be as accurate as necessary for

$$\frac{1}{h_j}|Q_1(t_j)(x_*(t_j) - x_j)| \leq c_3 h^p, \tag{3.62}$$

$$|x_*(t_j) - x_j| \leq c_3 h^p, \quad j = 0, \ldots, s-1,$$

to be satisfied, we obtain

$$\|\mathcal{L}_\pi^{-1}\tau_\pi\|_\pi \leq c_4 h^p.$$

Then, because

$$\mathcal{F}'_\pi(x^*_\pi)^{-1} = (I - \mathcal{L}_\pi^{-1}(\mathcal{L}_\pi - \mathcal{F}'_\pi(x^*_\pi)))^{-1}\mathcal{L}_\pi^{-1} \tag{3.63}$$

the inequality

$$\|\mathcal{F}'_\pi(x^*_\pi)^{-1}\tau_\pi\|_\pi \leq \|(I - \mathcal{L}_\pi^{-1}(\mathcal{L}_\pi - \mathcal{F}'_\pi(x^*_\pi)))^{-1}\| \, \|\mathcal{L}_\pi^{-1}\tau_\pi\|_\pi$$
$$\leq (1 - c_2 SK h^{p-1})^{-1} c_4 h^p$$

becomes true.

Next we show that both (3.51) and (3.54) may be satisfied. Given $0 < \alpha < 1$, choose $\varepsilon = \alpha(1/S)\underline{h}(1 - c_2 SK h^{p-1})$ to make E_π contractive. Since $\mathcal{F}'_\pi$ fulfils the Lipschitz condition (3.60), we may choose the related $\varrho(\varepsilon(\pi)) =: \varrho$ as $\varrho = 1/L\varepsilon$.

Finally, condition (3.54) becomes valid if

$$(1 - c_2 SK h^{p-1})^{-1} c_4 h^p \leq (1 - \alpha)\varrho = (1 - \alpha)\frac{1}{L}\alpha\frac{1}{S}\underline{h}(1 - c_2 SK h^{p-1})$$

is satisfied, or equivalently

$$c_4 h^{p-1} \leq \alpha(1 - \alpha)\frac{1}{LS}(1 - c_2 SK h^{p-1})^2,$$

but this can be managed by refining the grids.

By the same arguments as in Theorem 3.4, we derive

$$\|\mathcal{F}'_\pi[z, \bar{z}]^{-1}\|_\pi \leq \frac{1}{1 - \alpha}S\underline{h}^{-1}(1 - c_2 SK h^{p-1})^{-1},$$

and hence the BDF becomes weakly unstable. $\square$

Remarks

1   From (3.56), (3.63) we conclude the error estimate

$$\|x^*_\pi - \tilde{x}_\pi\|_\pi \leq S_1\|\mathcal{L}_\pi^{-1}(\tau_\pi - \delta_\pi)\|_\pi.$$

Taking (3.32) into consideration we are recommended to compute the Q_1 components of the starting values with a higher order accuracy than the remaining components (cf. also (3.61)). Moreover, the defects δ_j in the nonlinear equations should also be kept smaller in those components which do not belong to $\mathrm{im}(f'_y(\ldots, x_*(t_j), t_j) = R(x_*(t_j), t_j)$. This can be realized more easily if this subspace is kept constant.

2   Clearly, $Px_* \in C^s$ implies $p = s$ in (3.59).

3   Theorem 3.7 does not apply to the backward Euler method. It is not
     yet clear whether the condition $p > 1$ is a technical one for that large
     class of index-2 DAEs considered. However, Theorem 3.2 is also valid
     for the backward Euler method. The detailed error estimates (3.36),
     (3.37) show that the weak instabilities only affect certain components,
     and, moreover, only act locally. Inequality (3.35) does not comprise
     this situation precisely, but it represents a crude upper bound.

     Using the detailed information given by (3.37), (3.36) and (3.32) for
     investigating nonlinear equations requires much technical effort. For
     special DAEs in Hessenberg form (0.3) this is done in Gear *et al.* (1985),
     Lötstedt and Petzold (1986) and Brenan and Engquist (1988). The
     statements of Theorem 3.2 remain valid for these nonlinear DAEs. In
     particular, the backward Euler is proved to converge.

Analogously to Theorem 3.7, an assertion concerning index-3-tractable
DAEs could be proved using Theorem 3.3. In Lötstedt and Petzold (1986)
and Brenan and Engquist (1988), a careful detailed decoupling of nonlinear
index-3 DAEs in Hessenberg form (0.4) is carried out to obtain results similar
to those we have proved for the linear case (cf. (3.40)–(3.42)).

While we are optimistic about overcoming the practical problems arising
in large classes of index-2 equations, like error estimation and step-size control, the difficulties concerning the index-3 case seem to be more intractable.
As is shown by (3.42), the Q_2 components of the starting values should now
have order h^{p+2}, if the local error τ_j has order p. Moreover, the defects δ_j
of the nonlinear equations to be solved per integration step should be kept
small enough in the respective subspaces. Furthermore, remember that providing sufficiently accurate initial and starting values now becomes difficult.
The nonlinear systems to be solved are ill conditioned, namely $\text{cond}(F_i)$
behaves like h_j^{-3}.

For special nonlinear index-3 DAEs describing constrained mechanical
motion, BDF codes are reported to work (e.g. Petzold and Lötstedt (1986),
Führer (1988)) if the critical components are omitted from the error control,
and only the PP_1 component (cf. (3.40)) is controlled. This may be applied
if computing these PP_1 components only will do for practical reasons.

For a fairly detailed discussion of software for DAEs we refer to Brenan
et al. (1989) and Hairer and Wanner (1991).

3.4. Further integration methods

First of all, it should be mentioned that these results, which have been
proved for variable step-size BDFs, also apply to variable order variable
step-size BDFs in the same way.

For one-step methods, there is a natural extension to fully implicit DAEs

(3.1), namely

$$f\left(\frac{1}{h_j}\sum_{i=0}^{s}\alpha_{ji}x_{j-i}, \sum_{i=0}^{s}\beta_{ji}x_{j-i}, \bar{t}_j\right) = 0, \qquad t_j := \sum_{i=0}^{s}\beta_{ji}t_{j-i}. \qquad (3.64)$$

The consistency conditions are the same as those for the regular ODE case, but extra stability requirements are needed to ensure stability even in index-1 DAEs (cf. (1.12)). We do not recommend this method since it did not work well in experiments.

If the leading coefficient matrix has a constant range $\operatorname{im} f'_y(y, x, t) =: R$, and $S \in L(\mathbb{R}^m)$ denotes a projector onto R, $T := I - S$, we may formulate (März, 1985) a projected version of (3.64) as follows:

$$Sf\left(\frac{1}{h_j}\sum_{i=0}^{s}\alpha_{ji}x_{j-i}, \sum_{i=0}^{s}\beta_{ji}x_{j-i}, \bar{t}_j\right) + Tf(0, x_j, t_j) = 0. \qquad (3.65)$$

Applied to semi-linear DAEs

$$u' + g(u, v, t) = 0, \quad h(u, v, t) = 0 \qquad (3.66)$$

this simply means

$$\left.\begin{array}{c}\dfrac{1}{h_j}\displaystyle\sum_{i=0}^{s}\alpha_{ji}u_{j-i} + g\left(\displaystyle\sum_{i=0}^{s}\beta_{ji}u_{j-i}, \displaystyle\sum_{i=0}^{s}\beta_{ji}v_{j-i}, \bar{t}_j\right) = 0\\[2mm] h(u_j, v_j, t_j) = 0\end{array}\right\}. \qquad (3.67)$$

Moreover, linear multi-step methods may be formulated as

$$\left.\begin{array}{c}P(t_j)\left\{\dfrac{1}{h_j}\displaystyle\sum_{i=0}^{s}\alpha_{ji}x_{j-i} - \displaystyle\sum_{i=0}^{s}\beta_{ji}y_{j-i}\right\} - y_j = 0\\[2mm] f(y_j, x_j, t_j) = 0\end{array}\right\}. \qquad (3.68)$$

Method (3.68) is motivated by the equivalent formulation of (3.1)

$$\left.\begin{array}{c}P(t)\{x'(t) - y(t)\} - y(t) = 0\\ f(y(t), x(t), t) = 0\end{array}\right\}. \qquad (3.69)$$

When applied to the semi-explicit system (3.66), this linear multi-step method leads to

$$\left.\begin{array}{c}\dfrac{1}{h_j}\displaystyle\sum_{i=0}^{s}\alpha_{ji}u_{j-i} - \displaystyle\sum_{i=0}^{s}\beta_{ji}g((u_{j-i}, v_{j-i}, t_{j-i}) = 0\\[2mm] h(u_j, v_j, t_j) = 0\end{array}\right\}. \qquad (3.70)$$

All these methods are considered on general nonequidistant partitions π : $t_0 < t_1 < \cdots < t_N = T$. Our notation does not only allow for variable step-size, but also for formulae of different order and type, which is the common situation in many ODE codes.

Theorem 3.8 Let the methods considered be stable on the grid class Π for regular ODEs. Let DAE (3.1) satisfy Assumption 2.10, and let $x_* \in C^1_N$ solve this DAE. Furthermore, let $\{A_*, B_*\}$ be index-1 tractable. In addition, let the partial Jacobian $f'_y(y, x, t)$ have a constant null space N when applying the linear multi-step method (3.68), but a constant range R when applying the one-step method (3.65).

Then both methods are feasible and stable on $\Pi_0 \subseteq \Pi$. The order of consistency is the same as for regular ODEs.

Proof. Taking into account that (3.69) is again an index-1-tractable DAE, we apply the same arguments as those used for Theorem 3.4 in both cases. $\square$

Remarks

1   In both methods, the choice $\beta_{jo} = 0$ is allowed. In particular, in (3.70) one can take advantage of such 'explicit' methods.

2   The linear multi-step method is also proved to be stable (by the same arguments) for a time varying null space $N(t)$. However, then a certain order reduction may occur. This is caused by a somewhat inexact realization of the subspace structure of the DAE. More precisely, if x_* is smooth enough, we have

$$
\tau_j = P(t_j) \sum_{i=0}^{s} \left\{ \frac{1}{h_j} \alpha_{ji}(Px_*)(t_{j-i}) - \beta_{ji}(Px_*)'(t_{j-i}) \right\}
$$
$$
+ P(t_j) \sum_{i=0}^{s} \left\{ \frac{1}{h_j} \alpha_{ji} Q(t_{j-i}) x_*(t_{j-i}) + \beta_{ji} P'(t_{j-i}) x_*(t_{j-i}) \right\}
$$
$$
= P(t_j) \sum_{i=0}^{s} \left\{ \frac{1}{h_j} \alpha_{ji}(Px_*)(t_{j-i}) - \beta_{ji}(Px_*)'(t_{j-i}) + \right.
$$
$$
\left. + \left(\frac{1}{h_j} \alpha_{ji} Q(t_{j-i}) - \beta_{ji} Q'(t_{j-i}) \right) x_*(t_{j-i}) \right\}.
$$

Again, the conditions

$$
\sum_{i=0}^{s} \alpha_{ji} = 0, \quad \sum_{i=0}^{s} \alpha_{ji}(t_{j-i} - t_j) = h_j \sum_{i=0}^{s} \beta_{ji}
$$

turn out to be necessary and sufficient for the consistency at all. However, for order 2 we need two more conditions, the expected one

$$
\sum_{i=0}^{s} \{ \alpha_{ji}(t_{j-i} - t_j)^2 - 2h_j \beta_{ji}(t_{j-i} - t_j) \} = 0
$$

as well as

$$\sum_{i=0}^{s}\{\alpha_{ji}(t_{j-i}-t_j)^2 - h_j\beta_{ji}(t_{j-i}-t_j)\}Q'(t_j) = 0.$$

Hence, e.g. the trapezoidal rule has order 1 only in this case.

3   The methods (3.65) and (3.68) naturally generate values x_j belonging to the state manifold of the DAE, which turns out to be a favourable property.

Among the Runge–Kutta methods

$$\left.\begin{aligned}x_j &= x_{j-1} + h_j\sum_{i=1}^{s} b_i X_i',\\ f(X_i', x_{j-1} + h_j\sum_{k=1}^{s} a_{ik}X_k', t_{j-1}+c_i h_j) &= 0,\\ i &= 1,\ldots,s,\end{aligned}\right\} \qquad (3.71)$$

those with the coefficients $b_i = a_{si}$, $i = 1,\ldots,s$, $c_s = 1$ and a nonsingular matrix (a_{ik}) automatically provide values x_j belonging to the state manifold, if the method is applied to an index-1 DAE (3.1) with constant null space (e.g. Griepentrog and März (1986)). Then, the method maintains the order which it has for regular ODEs.

As we have learnt in Section 1, explicit Runge–Kutta methods (those having $a_{ij} = 0$ for $i \le j$) are not suited for DAEs.

A fairly detailed discussion of general implicit Runge–Kutta methods as well as of extrapolation methods for index-1 DAEs is given in Brenan et al. (1989). As already indicated in Section 1, additional stability conditions have to be fulfilled, and one has to put up with order reduction.

A comprehensive exposition of Runge–Kutta methods for index-2 and index-3 DAEs in Hessenberg form (cf. (0.3), (0.4)) one can find in Hairer et al. (1989). A good work in which the well-known extrapolation methods, for example, are extended can be found in Deuflhard et al. (1987), Lubich (1990) and Hairer et al. (1989). All these methods extensively use the special structure of the given Hessenberg form DAEs. In particular, the problems caused by weak instability have been overcome e.g. by special error control in the nonlinear equations and by projections onto the given manifolds, respectively.

The projected implicit Runge–Kutta methods (Ascher and Petzold, 1990) also use these ideas.

4. Brief remarks on related problems

4.1. Index reduction

From the point of view of computational tractability, it is desirable for the DAE to have an index which is as small as possible. The procedure for determining the differentiation index described in Section 2.2 is an index reduction method (cf. Griepentrog (1991)), in fact. However, in Section 2.2 it was mentioned that the attained system does not reflect the stability behaviour of the original DAE well.

A different method for reducing the index of a DAE is presented in Mrziglod (1987) and Čistjakov (1982). Instead of replacing constraints by differential equations, in their method suitable differential equations are deleted. This method works for linear DAEs, but it is not clear to what kind of nonlinear DAEs it may be applicable.

A very useful idea for reducing the index is proposed in Gear *et al.* (1985) for the special index-3 Hessenberg system

$$u'(t) - v(t) = 0, \tag{4.1}$$

$$v'(t) + g(u(t), v(t), t) + h'_u(u(t), t)^T w(t) = 0, \tag{4.2}$$

$$h(u(t), t) = 0, \tag{4.3}$$

which results from the Euler–Lagrange formulation of a constrained mechanical system. As mentioned earlier, the system with the differentiated constraint

$$h'_u(u(t), t)v(t) + h'_t(u(t), t) = 0 \tag{4.4}$$

instead of (4.3) would cause the numerical solution to drift away from the constraint manifold. Note that the system (4.1), (4.2), (4.4) has index 2. To stabilize the obtained index-2 system, an additional Lagrange multiplyer z is introduced, and (4.3) is summed again. The resulting system

$$\left. \begin{array}{r} u'(t) - v(t) + h'_u(u(t), t)^T z(t) = 0 \\ v'(t) + g(u(t), v(t), t) + h'_u(u(t), t)^T w(t) = 0 \\ h'_u(u(t), t)v(t) + h'_t(u(t), t) = 0 \\ h(u(t), t) = 0 \end{array} \right\} \tag{4.5}$$

is index-2 tractable. Is it easy to check that any solution of (4.5) has a trivial component z. Note that Führer and Leimkuhler (1990) took advantage of this fact to create a skilful special BDF modification to the Euler–Lagrange equations.

In Section 2 we pointed out that higher index DAEs lead to ill posed IVPs in the naturally given topologies (cf. Corollary 2.7). Hence, we may treat them as such, i.e. use some regularization procedure. At first glance this approach might appear a heavy gun, which is true insofar as standard regularization techniques (Tikhonov regularization, least-squares collocation) are concerned. However, different special parametrizations may be created, which are closely connected with the structure of the DAEs and the source of their ill posedness. As usual, the regularized equations represent singularly perturbed index-1-tractable DAEs and ODEs, respectively.

For instance, the index-2 DAE

$$x_1' - x_2 = 0, \quad x_1 = q$$

may be approximated by the index-1 system

$$x_1' - x_2 = 0, \quad \varepsilon x_1' + x_1 = q, \quad x_1(0) = q(0).$$

For general DAEs $f(x'(t), x(t), t) = 0$ the same regularization method provides

$$f(x'(t), x(t) + \varepsilon P(t)(Px)'(t), t) = 0.$$

We refer to Hanke (1990, 1991) for a comprehensive survey on methods, convergence results, asymptotic expansions etc.

4.2. Boundary value problems

Let us consider the linear equation

$$A(t)x'(t) + Bx(t) = q(t), \quad t \in [t_0, T], \tag{4.6}$$

once again. Now we are interested in a solution of (4.6) that satisfies the boundary condition

$$D_1 x(t_0) + D_2 x(T) = d \tag{4.7}$$

with given matrices $D_1, D_2 \in L(\mathbb{R}^m)$, $d \in M := \text{im}(D_1, D_2)$. According to the discussion of linear IVPs in section 2.1 we determine a fundamental solution matrix $X(\cdot)$ by

$$AX' + BX = 0 \tag{4.8}$$
$$\Pi_\mu(X(t_0) - I) = 0, \tag{4.9}$$

where $\Pi_\mu := P_0(t_0) \dots P_{\mu-1}(t_0)$, and the coefficient matrix pair $\{A, B\}$ is supposed to be index-μ tractable. From Theorem 2.5, the fundamental solution matrix is uniquely determined, the columns of X belong to C_N^1.

Now, (4.8), (4.9) immediately imply that

$$X(t) = X(t)\Pi_\mu \tag{4.10}$$

holds, i.e. $X(t)$ is singular for all t. Moreover, even $\ker X(t) = \ker \Pi_\mu$ is true. However, what about the so-called shooting matrix

$$K := D_1 X(t_0) + D_2 X(T). \tag{4.11}$$

Trivially, K becomes singular, too.

Theorem 4.1 Let $\{A, B\}$ be index-μ tractable and q sufficiently smooth. Then the BVP (4.6), (4.7) is uniquely solvable for each $d \in M$ if and only if

$$\text{im}K = M, \quad \ker K = \ker \Pi_\mu \tag{4.12}$$

are valid.

Proof. By standard arguments, we have to consider the linear system

$$Kz = d - D_2\tilde{x}(T),$$

where $\tilde{x} \in C_N^1$ denotes that solution of (4.6) which satisfies $\Pi_\mu \tilde{x}(t_0) = 0$.

Clearly, $\ker \Pi_\mu \subseteq \ker K$ holds. Furthermore, $Kz = 0$, $\Pi_\mu z \neq 0$ would imply that $X(t)\Pi_\mu z$ has to become a nontrivial solution of the homogeneous BVP. $\square$

Remarks

1   Theorem 4.1 generalizes facts that are well known for regular ODEs ($M = \mathbb{R}^m$) and index-1-tractable DAEs (Griepentrog and März, 1986), respectively.

2   The relations (4.12) mean that the *boundary conditions are stated well*; in particular, the number of linearly independent boundary conditions is rank Π_μ.

3   The *whole* BVP is well posed in the naturally given topologies if and only if $\mu = 1$, and if (4.12) is satisfied (cf. Corollary 2.7).

Linear and nonlinear BVPs in transferable (index-1-tractable) DAEs are well understood. Classical arguments apply for discretizations by finite differences (Griepentrog and März, 1986) and spline-collocation (Degenhardt, 1991; Ascher, 1989), respectively. In particular, it is possible to trace the stability question of the BVP back to that of the IVPs. However, for the latter we refer to the typical explanations in Section 3, which are carried out in the same manner for certain one-step methods, for example in Griepentrog and März (1986). Furthermore, dichotomy is considered in its relationship to the conditioning of the BVP in Lentini and März (1990a,b).

Of course, the singular shooting equation causes numerical difficulties. This is why modified shooting techniques yielding isolatedly solvable nonlinear shooting equations have been proposed (Lamour, 1991a,b). The basic idea is to combine the shooting equation and the equation for calculating consistent initial values (cf. Theorem 2.3). For instance, simple shooting for a linear BVP leads to the system

$$\begin{aligned} A(t_0)y_0 \;+\; B(t_0)x_0 &= q(t_0), \\ Q(t_0)y_0 &= 0, \qquad Kx_0 = d - D_2\tilde{x}(T). \end{aligned}$$

Surely, Theorem 4.1 suggests to apply shooting methods also to higher index DAEs. This will work, supposed we are able to integrate the IVPs.

We are looking forward to related results for general index-2-tractable DAEs.

REFERENCES

U. Ascher (1989), 'On numerical differential algebraic problems with application to semi-conductor device simulations', *SIAM J. Numer. Anal.* **26**, 517–538.

U. Ascher and L.R. Petzold (1990), *Projected implicit Runge–Kutta methods for differential-algebraic equations*, Preprint, Lawrence Livermore National Laboratory

K.E. Brenan and B.E. Engquist (1988), 'Backward differentiation approximations of nonlinear differential/algebraic systems', and Supplement, *Math. Comput.* **51**, 659–676, S7–S16.

K.E. Brenan, S.L. Campbell and L.R. Petzold (1989), *Numerical Solution of Initial-Value Problems in Differential-Algebraic Equations*, North-Holland (Amsterdam).

S.L. Campbell (1987), 'A general form for solvable linear time varying singular systems of differential equations', *SIAM J. Math. Anal.* **18**, 1101–1115.

L.O. Chua and A. Deng (1989), 'Impasse points. Part I: Numerical aspects', *Int. J. Circuit Theory Applics* **17**, 213–235.

V.F. Čistjakov (1982), 'K metodam rešenija singuljarnych linejnych sistem obyknovennych differencial'nych uravnenij', in *Vyroždennyje Sistemy Obyknovennych Differencial'nych Uravnenij* (Ju.E. Bojarincev, ed.) Nauka Novosibirsk, 37–65.

A. Degenhardt (1991), 'A collocation method for boundary value problems of transferable DAEs', *Numer. Math.*, to appear.

P. Deuflhard, E. Hairer, J. Zugck (1987), 'One-step and extrapolation methods for differential-algebraic systems' *Numer. Math.* **51**, 501–516.

C. Führer (1988), Differential-algebraische Gleichungssysteme in mechanischen Mehrkörpersystemen. Theorie, numerische Ansätze und Anwendungen, Dissertation, Techn. Univ. München, Fak. für Mathematik und Informatik.

C. Führer and B. Leimkuhler (1989), Formulation and numerical solution of the equations of constrained mechanical motion, DFVLR-Forschungsbericht 89-08, Deutsche Forschungs- und Versuchsanstalt für Luft- und Raumfahrt, Oberpfaffenhofen.

C. Führer and B.J. Leimkuhler (1990), A new class of generalized inverses for the solution of discretized Euler–Lagrange equations, in *NATO Advanced Research Workshop on Real-Time Integration Methods for Mechanical System Simulation* (Snowbird, Utah 1989) (E. Haug and R. Deyo, eds.) Springer (Berlin).

F.R. Gantmacher (1966), *Teorija matric*, Nauka (Moskva).

C.W. Gear (1971), 'The simultaneous numerical solution of differential-algebraic equations', *IEEE Trans. Circuit Theory*, **CT-18**, 89–95.

C.W. Gear and L.R. Petzold (1984), 'ODE methods for the solution of differential/algebraic systems', *SIAM J. Numer. Anal.* **21**, 716–728.

C.W. Gear, H.H. Hsu and L. Petzold (1981), Differential-algebraic equations revisited, *Proc. ODE Meeting, Oberwolfach, Germany*, Institut für Geom. und Praktische Mathematik, Technische Hochschule Aachen, Bericht 9, Germany.

C.W. Gear, B. Leimkuhler and G.K. Gupta (1985), 'Automatic integration of Euler–Lagrange equations with constraints', *J. Comput. Appl. Math.* **12 & 13**, 77–90.

E. Griepentrog (1991), Index reduction methods for differential-algebraic equations, Preprint 91-12, Humboldt-Univ. Berlin, Fachbereich Mathematik.

E. Griepentrog and R. März (1986), *Differential-Algebraic Equations and their Numerical Treatment (Teubner Texte zur Mathematik 88)* Teubner (Leipzig).

E. Griepentrog and R. März (1989), 'Basic properties of some differential-algebraic equations', *Z. Anal. Anwend.* **8** (1), 25–40.

E. Hairer and G. Wanner (1991), *Solving Ordinary Differential Equations II: Stiff and Differential-Algebraic Problems (Springer Series in Computational Mathematics 14)* Springer (Berlin).

E. Hairer, Ch. Lubich and M. Roche (1989), *The Numerical Solution of Differential-Algebraic Systems by Runge–Kutta Methods (Lecture Notes in Mathematics 1409)* Springer (Berlin).

M. Hanke (1990) Regularization methods for higher index differential-algebraic equations, Preprint 268, Humboldt-Univ. Berlin, Fachbereich Mathematik.

M. Hanke (1991), 'On the asymptotic representation of a regularization approach to nonlinear semiexplicit higher-index differential-algebraic equations', *IMA J. Appl. Math.* **46**, 225–245.

B. Hansen (1989), Comparing different concepts to treat differential-algebraic equations, Preprint 220, Humboldt-Univ. Berlin, Sektion Mathematik.

B. Hansen (1990), Linear time-varying differential-algebraic equations being tractable with the index k, Preprint 246, Humboldt-Univ. Berlin, Sektion Mathematik.

H.B. Keller (1975), 'Approximation methods for nonlinear problems with application to two-point boundary value problems', *Math. Comput.* **29** (130), 464–474.

H.B. Keller and A.B. White (1975), 'Difference methods for boundary value problems in ordinary differential equations', *SIAM J. Numer. Anal.* **12**, 791–802.

R. Lamour (1991a), 'A well-posed shooting method for transferable DAE's', *Numer. Math.* **59**.

R. Lamour (1991b), Oscillations in differential-algebraic equations, Preprint 272, Humboldt-Universität Berlin, Fachbereich Mathematik.

M. Lentini and R. März (1990a), 'The conditioning of boundary value problems in transferable differential-algebraic equations', *SIAM J. Numer. Anal.* **27**, 1001–1015.

M. Lentini and R. März (1990b), 'Conditioning and dichotomy in transferable differential-algebraic equations', *SIAM J. Numer. Anal.* **27**, 1519–1526.

P. Lötstedt and L. Petzold (1986), 'Numerical solution of nonlinear differential equations with algebraic constraints I: Convergence results for backward differentiation formulas', *Math. Comput.* **46**, 491–516.

Ch. Lubich (1990), Extrapolation integrators for constrained multibody systems, Report, Univ. Innsbruck.

R. März (1984), 'On difference and shooting methods for boundary value problems in differential-algebraic equations', *ZAMM* **64** (11), 463–473.

R. März (1985), 'On initial value problems in differential-algebraic equations and their numerical treatment', *Computing* **35**, 13–37.

R. März (1989), 'Some new results concerning index-3 differential-algebraic equations', *J. Math. Anal. Applics* **140** (1), 177–199.

R. März (1990), 'Higher-index differential-algebraic equations: Analysis and numerical treatment', *Banach Center Publ.* **24**, 199–222.

R. März (1991), On quasilinear index 2 differential algebraic equations, Preprint 269, Humboldt-Universität Berlin, Fachbereich Mathematik.

T. Mrziglod (1987), Zur Theorie und Numerischen Realisierung von Lösungs-
methoden bei Differentialgleichungen mit angekoppelten algebraischen Gleichungen, Diplomarbeit, Universität zu Köln.

L.R. Petzold (1986), 'Order results for implicit Runge–Kutta methods applied to
differential/algebraic systems', *SIAM J. Numer. Anal.* **23**, 837–852.

L. Petzold and P. Lötstedt (1986), 'Numerical solution of nonlinear differential
equations with algebraic constraints II: Practical implications', *SIAM J. Sci.
Stat. Comput.* **7**, 720–733.

F.A. Potra and W.C. Rheinboldt (1991), 'On the numerical solution of Euler–
Lagrange equations', *Mech. Struct. Machines* **19** (1).

P.J. Rabier and W.C. Rheinboldt (1991), 'A general existence and uniqueness theory for implicit differential-algebraic equations', *Diff. Int. Eqns* **4** (3), 563–582.

S. Reich (1990), Beitrag zur Theorie der Algebrodifferentialgleichungen, Dissertation (A), Technische Universität Dresden.

W.C. Rheinboldt (1984), 'Differential-algebraic systems as differential equations on
manifolds', *Math. Comput.* **43**, 473–482.

B. Simeon, C. Führer and P. Rentrop (1991), 'Differential-algebraic equations in
vehicle system dynamics', *Surv. Math. Ind.* **1** (1), 1–37.

R.F. Sincovec, A.M Erisman, E.L. Yip and M.A. Epton (1981), 'Analysis of descriptor systems using numerical algorithms', *IEEE Trans. Aut. Control*, **AC-26**,
139–147.

Acta Numerica (1991), *pp.* 199–242

Theory of algorithms for unconstrained optimization

Jorge Nocedal
Department of Electrical Engineering and Computer Science
Northwestern University
Evanston, IL 60208 USA
E-mail nocedal@jorge.eecs.nwu.edu

CONTENTS

1. Introduction

A few months ago, while preparing a lecture to an audience that included engineers and numerical analysts, I asked myself the question: from the point of view of a user of nonlinear optimization routines, how interesting and practical is the body of theoretical analysis developed in this field? To make the question a bit more precise, I decided to select the best optimization methods known to date – those methods that deserve to be in a subroutine library – and for each method ask: what do we know about the behaviour of this method, *as implemented in practice*? To make my task more tractable, I decided to consider only algorithms for unconstrained optimization.

I was surprised to find that remarkable progress has been made in the last 15 years in the theory of unconstrained optimization, to the point that it is reasonable to say that we have a good understanding of most of the techniques used in practice. It is reassuring to see a movement towards practicality: it is now routine to undertake the analysis under realistic assumptions, and to consider optimization algorithms as they are implemented in practice. The depth and variety of the theoretical results available to us today have made unconstrained optimization a mature field of numerical analysis.

Nevertheless there are still many unanswered questions, some of which are fundamental. Most of the analysis has focused on global convergence and rate of convergence results, and little is known about average behaviour, worst case behaviour and the effect of rounding errors. In addition, we do not have theoretical tools that will predict the efficiency of methods for large scale problems.

In this article I will attempt to review the most recent advances in the theory of unconstrained optimization, and will also describe some important open questions. Before doing so, I should point out that the value of the theory of optimization is not limited to its capacity for explaining the behaviour of the most widely used techniques. The question posed in the first paragraph: 'what do we know about the behaviour of the most popular algorithms?' is not the only important question. We should also ask how useful is the theory when designing new algorithms, i.e. how well can it differentiate between efficient and inefficient methods. Some interesting analysis will be discussed in this regard. We will see that the weaknesses of several classical algorithms that have fallen out of grace, such as the Fletcher–Reeves conjugate gradient method and the Davidon–Fletcher–Powell variable metric method, are fairly well understood. I will also describe several theoretical studies on optimization methods that have not yet enjoyed widespread popularity, but that may prove to be highly successful in the future.

I have used the terms 'theoretical studies' and 'convergence analysis', without stating precisely what I mean by them. In my view, convergence results fall into one of the four following categories.

1   Global convergence results. The questions in this case are: will the iterates converge from a remote starting point? Are all cluster points of the set of iterates solution points?

2   Local convergence results. Here the objective is to show that there is a neighourhood of a solution and a choice of the parameters of the method for which convergence to the solution can be guaranteed.

3   Asymptotic rate of convergence. This is the speed of the algorithm, as it converges to the solution (which is not necessarily related to its speed away from the solution).

4   Global efficiency or global rate of convergence. There are several measures; one of them estimates the function reduction at every iteration. Another approach is to study the worst case global behaviour of the methods.

Most of the literature covers results in categories 1–3. Global efficiency results, category 4, can be very useful but are difficult to obtain. Therefore it is common to restrict these studies to convex problems (Nemirovsky and Yudin, 1983), or even to strictly convex quadratic objective functions (Powell, 1986). Global efficiency is an area that requires more attention and where important new results can be expected.

To be truly complete, the four categories of theoretical studies mentioned above should also take into account the effect of rounding errors, or noise in the function (Hamming, 1971). However, we will not consider these aspects here, for this would require a much more extensive survey. The term global optimization is also used to refer to the problem of finding the global minimum of a function. We will not discuss that problem here, and reserve the term 'global convergence' to denote the properties described in 1.

2. The most useful algorithms for unconstrained optimization

Since my goal is to describe recent theoretical advances for practical methods of optimization, I will begin by listing my selection of the most useful optimization algorithms. I include references to particular codes in subroutine libraries instead of simply referring to mathematical algorithms. However the routines mentioned below are not necessarily the most efficient implementations available, and are given mainly as a reference. Most of the algorithms listed here are described in the books by Dennis and Schnabel (1983), Fletcher (1987) and Gill et al. (1981).

1   *The conjugate gradient method, or extensions of it.* Conjugate gradient methods are useful for solving very large problems and can be particularly effective on some types of multiprocessor machines. An efficient code implementing the Polak–Ribière version of the conjugate gradient method, with restarts, is the routine VA14 of the Harwell subroutine library (Powell, 1977). A robust extension of the conjugate gradient method, requiring a few more vectors of storage, is implemented in the routine CONMIN (Shanno and Phua, 1980).

2   *The BFGS variable metric method.* Good line search implementations of this popular variable metric method are given in the IMSL and NAG libraries. The BFGS method is fast and robust, and is currently being used to solve a myriad of optimization problems.

3   *The partitioned quasi-Newton method for large scale optimization.* This method, developed by Griewank and Toint (1982c), is designed for partially separable functions. These types of functions arise in numerous applications, and the partitioned quasi-Newton method takes good advantage of their structure. This method is implemented in the Harwell routine VE08, and will soon be superseded by a more general routine of the Lancelot package which is currently being developed by Conn, Gould and Toint.

4   *The limited memory BFGS method for large scale optimization.* This
     method resembles the BFGS method but avoids the storage of matrices.
     It is particularly useful for large and unstructured problems. It is
     implemented in the Harwell routine VA15 (Liu and Nocedal, 1989).
5   *Newton's method.* A good line search implementation is given in the
     NAG library, whereas the IMSL library provides a trust region im-
     plementation (Dennis and Schnabel, 1983; Gay, 1983). A truncated
     Newton method for large problems, which requires only function and
     gradients, is given by Nash (1985).
6   *The Nelder–Meade simplex method for problems with noisy functions.*
     An implementation of this method is given in the IMSL library.

In the following sections I will discuss recent theoretical studies on many of
these methods. I will assume that the reader is familiar with the fundamental
techniques of unconstrained optimization, which are described, for example,
in the books by Dennis and Schnabel (1983), Fletcher (1987) and Gill *et
al.* (1981). We will concentrate on line search methods because most of
our knowledge on trust region methods for unconstrained optimization was
obtained before 1982, and is described in the excellent survey papers by
Moré and Sorensen (1984) and Moré (1983). However in Section 8 we will
briefly compare the convergence properties of line search and trust region
methods.

3. The basic convergence principles

One of the main attractions of the theory of unconstrained optimization is
that a few general principles can be used to study most of the algorithms.
In this section, which serves as a technical introduction to the paper, we
describe some of these basic principles. The analysis that follows gives us a
flavour of what theoretical studies on line search methods are, and will be
frequently quoted in subsequent sections.

Our problem is to minimize a function of n variables,

$$\min f(x), \tag{3.1}$$

where f is smooth and its gradient g is available. We consider iterations of
the form

$$x_{k+1} = x_k + \alpha_k d_k, \tag{3.2}$$

where d_k is a search direction and α_k is a step-length obtained by means of a
one-dimensional search. In conjugate gradient methods the search direction
is of the form

$$d_k = -g_k + \beta_k d_{k-1}, \tag{3.3}$$

where the scalar β_k is chosen so that the method reduces to the linear conjugate gradient method when the function is quadratic and the line search is exact. Another broad class of methods defines the search direction by

$$d_k = -B_k^{-1} g_k \qquad (3.4)$$

where B_k is a nonsingular symmetric matrix. Important special cases are given by

$$
\begin{aligned}
B_k &= I \qquad \text{(the steepest descent method)} \\
B_k &= \nabla^2 f(x_k) \qquad \text{(Newton's method)}.
\end{aligned}
$$

Variable metric methods are also of the form (3.4), but in this case B_k is not only a function of x_k, but depends also on B_{k-1} and x_{k-1}.

All these methods are implemented so that d_k is a descent direction, i.e. so that $d_k^T g_k < 0$, which guarantees that the function can be decreased by taking a small step along d_k. For the Newton-type methods (3.4) we can ensure that d_k is a descent direction by defining B_k to be positive definite. For conjugate gradient methods obtaining descent directions is not easy and requires a careful choice of the line search strategy. Throughout this section we will assume that the optimization method is of the form (3.2) where d_k is a descent direction.

The convergence properties of line search methods can be studied by measuring the goodness of the search direction and by considering the length of the step. The quality of the search direction can be studied by monitoring the angle between the steepest descent direction $-g_k$ and the search direction. Therefore we define

$$\cos \theta_k := -g_k^T d_k / \|g_k\| \, \|d_k\|. \qquad (3.5)$$

The length of the step is determined by a line search iteration. A strategy that will play a central role in this paper consists in accepting a positive step-length α_k if it satisfies the two conditions:

$$
\begin{aligned}
f(x_k + \alpha_k d_k) &\leq f(x_k) + \sigma_1 \alpha_k g_k^T d_k \qquad &(3.6) \\
g(x_k + \alpha_k d_k)^T d_k &\geq \sigma_2 g_k^T d_k, \qquad &(3.7)
\end{aligned}
$$

where $0 < \sigma_1 < \sigma_2 < 1$. The first inequality ensures that the function is reduced sufficiently, and the second prevents the steps from being too small. We will call these two relations the *Wolfe conditions*. It is easy to show that if d_k is a descent direction, if f is continuously differentiable and if f is bounded below along the ray $\{x_k + \alpha d_k | \alpha > 0\}$, then there always exist step-lengths satisfying (3.6)–(3.7) (Wolfe, 1969, 1971). Algorithms that are guaranteed to find, in a finite number of iterations, a point satisfying the Wolfe conditions have been developed by Lemaréchal (1981), Fletcher (1987) and Moré and Thuente (1990).

This line search strategy allows us to establish the following useful result due to Zoutendijk. At first, the result appears to be obscure, but its power and simplicity will soon become evident. We will give a proof so that the reader can have a clear idea of how it depends on the properties of the function and line search. This result was essentially proved by Zoutendijk (1970) and Wolfe (1969, 1971). The starting point of the algorithm is denoted by x_1.

Theorem 3.1 Suppose that f is bounded below in $\mathbb{R}^n$ and that f is continuously differentiable in a neighourhood $\mathcal{N}$ of the level set $\mathcal{L} := \{x : f(x) \leq f(x_1)\}$. Assume also that the gradient is Lipschitz continuous, i.e. there exists a constant $L > 0$ such that

$$\|g(x) - g(\tilde{x})\| \leq L\|x - \tilde{x}\|, \tag{3.8}$$

for all $x, \tilde{x} \in \mathbb{N}$. Consider any iteration of the form (3.2), where d_k is a descent direction and α_k satisfies the Wolfe conditions (3.6)–(3.7). Then

$$\sum_{k \geq 1} \cos^2 \theta_k \|g_k\|^2 < \infty. \tag{3.9}$$

Proof. From (3.7) we have that

$$(g_{k+1} - g_k)^T d_k \geq (\sigma_2 - 1) g_k^T d_k.$$

On the other hand, the Lipschitz condition (3.8) gives

$$(g_{k+1} - g_k)^T d_k \leq \alpha_k L \|d_k\|^2.$$

Combining these two relations we obtain

$$\alpha_k \geq \left(\frac{\sigma_2 - 1}{L}\right) g_k^T d_k / \|d_k\|^2. \tag{3.10}$$

Using the first Wolfe condition (3.6) and (3.10), we have

$$f_{k+1} \leq f_k + \sigma_1 \left(\frac{\sigma_2 - 1}{L}\right) (g_k^T d_k)^2 / \|d_k\|^2.$$

We now use definition (3.5) to write this relation as

$$f_{k+1} \leq f_k + c \cos^2 \theta_k \|g_k\|^2,$$

where $c = \sigma_1(\sigma_2 - 1)/L$. Summing this expression and recalling that f is bounded below we obtain

$$\sum_{k=1}^{\infty} \cos^2 \theta_k \|g_k\|^2 < \infty,$$

which concludes the proof. □

We shall call inequality (3.9) the *Zoutendijk condition*. Let us see how Zoutendijk's condition can be used to obtain global convergence results.

Suppose that an iteration of the form (3.2) is such that

$$\cos \theta_k \geq \delta > 0, \tag{3.11}$$

for all k. Then we conclude directly from (3.9) that

$$\lim_{k \to \infty} \|g_k\| = 0. \tag{3.12}$$

In other words, if the search direction does not tend to be orthogonal to the gradient, then the sequence of gradients converges to zero. This implies, for example, that the method of steepest descent, with a line search satisfying the Wolfe conditions, gives (3.12), since in this case we have $\cos \theta_k = 1$ for all k. Thus to make the steepest descent method 'globally convergent' it is only necessary to perform an adequate line search.

For line search methods of the form (3.2), the limit (3.12) is the best type of global convergence result that can be obtained – we cannot guarantee that the method converges to minimizers, but only that it is attracted by stationary points.

Consider now the Newton-type method (3.2),(3.4), and assume that the condition number of the matrices B_k is uniformly bounded, i.e. that for all k

$$\|B_k\| \|B_k^{-1}\| \leq \Delta,$$

for some constant $\Delta > 0$. Then from (3.5) we have that

$$\cos \theta_k \geq 1/\Delta.$$

As before, we use Zoutendijk's condition (3.9) to obtain the global convergence result (3.12). We have therefore shown that Newton's method or the variable metric methods are globally convergent if the matrices B_k are positive definite (which is needed for the descent condition), if their condition number is uniformly bounded, and if the line search satisfies the Wolfe conditions. For a more thorough discussion see Ortega and Rheinboldt (1970).

For some algorithms, such as conjugate gradient methods, it is not possible to show the limit (3.12), but only a weaker result, namely

$$\liminf_{k \to \infty} \|g_k\| = 0. \tag{3.13}$$

We can also obtain this type of result from Zoutendijk's condition (3.9), but this time the method of proof is contradiction. Suppose that (3.13) does not hold, which means that the gradients remain bounded away from zero, i.e. there exists $\gamma > 0$ such that for all k

$$\|g_k\| \geq \gamma. \tag{3.14}$$

Then from (3.9) we conclude that

$$\cos \theta_k \to 0. \tag{3.15}$$

In other words, the algorithm can only fail, in the sense of (3.14), if the whole sequence $\{\cos\theta_k\}$ converges to 0. Therefore to establish (3.13) it suffices to show that a subsequence $\{\cos\theta_{k_j}\}$ is bounded away from zero.

For example, any line search method can be made globally convergent, in the sense of (3.13), by interleaving steepest descent steps. To be more precise, consider any method of the form (3.2) where d_k is a descent direction for all k, and where α_k is chosen to satisfy the Wolfe conditions. Suppose, in addition, that at every m steps, where m is some pre-selected integer, we define $d_k = -g_k$. Since for these steepest descent steps $\cos\theta_k = 1$, the previous discussion shows that the limit (3.13) is obtained.

It would seem that designing optimization algorithms with good convergence properties is easy, since all we need to ensure is that the search direction does not tend to become orthogonal to the gradient, or that steepest descent steps are interleaved regularly. Indeed, since the gradient g_k is always available, we can compute $\cos\theta_k$ at every iteration and apply the following angle test: if $\cos\theta_k$ is less than some pre-selected constant, then we can modify the search direction by turning it towards the steepest descent direction. Such angle tests have been proposed many times in the literature, and ensure global convergence, but are undesirable for the following reasons.

In addition to global convergence we would like the methods to converge rapidly. After all, if all we want to achieve is global convergence we should be satisfied with the steepest descent method. It is well known, however, that steepest descent is very slow and that much faster algorithms can be designed. A classical result of Dennis and Moré states that the iteration (3.2) is superlinearly convergent if and only if

$$\alpha_k d_k = d_k^{\text{N}} + \mathcal{O}(\|d_k^{\text{N}}\|), \qquad (3.16)$$

where d_k^{N} is the Newton step (Dennis and Moré, 1974). Therefore to attain a fast rate of convergence it is necessary that we approximate the Newton direction asymptotically. An angle test may prevent us from doing so. For example, the BFGS variable metric method described in Section 5 can generate ill conditioned approximations B_k of the Hessian. It is difficult, however, to determine if this is undesirable or if the matrices B_k are approximating well an ill conditioned Hessian matrix. To decide this requires knowledge of the problem that we do not possess. We have learned that it is preferable not to interfere with the BFGS method and to let the matrices B_k evolve freely, because convergence is usually obtained and the rate is superlinear.

By far the most substantial argument against angle tests is this: the best implementations of the methods listed in Section 2 do not need them; it has been found that other types of safeguards are more effective. We will return to this.

Dennis and Moré (1977) prove a result that is of great practical value because it suggests how to estimate the initial trial value in the line search of a variable metric method. They show that for an iteration in which the the search directions approach the Newton direction, the step-length $\alpha_k = 1$ satisfies the Wolfe conditions for all large k, provided $\sigma_1 < \frac{1}{2}$. Thus the unit trial step-length should always be used in variable metric methods.

Let us summarize what we have discussed so far. Zoutendijk's condition plays a central role when studying the global convergence properties of line search methods. Most of the global convergence analyses use it explicitly or follow similar approaches. The Dennis–Moré (3.16) condition is fundamental to the study of rates of convergence. It states that a method is superlinearly convergent if and only if the direction and the length of the step approximate those of Newton's method, asymptotically. Many variable metric methods are superlinearly convergent, and this is proved by simply verifying that (3.16) holds.

So far, we have only talked about one type of line search, namely the one satisfying the Wolfe conditions, and it would be misleading to suggest that this is the only useful strategy. Indeed many convergence results can also be proved for other line searches, as we will discuss in later sections. A popular strategy, called backtracking, consists of successively decreasing the step-length, starting from an initial guess, until a sufficient function reduction is obtained; see, for example, Ortega and Rheinboldt (1970). A backtracking line search is easy to implement and is well suited for constrained problems.

Let us now discuss global efficiency analyses. One of the earliest results concerns the steepest descent method, with exact line searches, when applied to quadratic problems. This result is characteristic of global efficiency studies, which are established under very restrictive assumptions, and yet provide useful insight into the methods.

Suppose that f is the quadratic function

$$f(x) = \tfrac{1}{2}x^T A x, \qquad (3.17)$$

where A is symmetric and positive definite. Consider the steepest descent method with exact line searches

$$x_{k+1} = x_k - \alpha_k g_k, \qquad (3.18)$$

where

$$\alpha_k = g_k^T g_k / g_k^T A g_k. \qquad (3.19)$$

A simple computation (Luenberger, 1984) shows that

$$f_{k+1} = \left[1 - \frac{(g_k^T g_k)^2}{(g_k^T A g_k)(g_k^T A^{-1} g_k)}\right] f_k. \qquad (3.20)$$

This gives the function reduction at each iteration, and it is interesting that we have an equality. It is clear that the quotient in (3.20) can be bounded in terms of quantities involving only the matrix A. To do this, we use the

Kantorovich inequality to obtain (Luenberger, 1984)

$$\frac{g_k{}^T g_k}{(g_k{}^T A g_k)(g_k{}^T A^{-1} g_k)} \geq \frac{4\lambda_1 \lambda_n}{(\lambda_1 + \lambda_n)^2},$$

where $\lambda_1 \leq \cdots \leq \lambda_n$ are the eigenvalues of A. By substituting this in (3.20) we obtain the simple relation

$$f_{k+1} \leq \left[\frac{\lambda_n - \lambda_1}{\lambda_n + \lambda_1}\right]^2 f_k. \tag{3.21}$$

This is the worst-case global behaviour of the steepest descent method (3.18) - (3.19) on the quadratic problem (3.17), but it can be argued that this is also the average behaviour (Akaike, 1959). Note that this global efficiency result also shows that asymptotic rate of convergence of the sequence $\{f(x)\}$ is linear, with a constant that depends on the condition number of A. Clearly, if λ_n/λ_1 is large, the term inside the square brackets in (3.21) is close to 1 and convergence will be slow.

Does this analysis help our understanding of the steepest descent method with inexact line searches on general nonlinear functions? The answer is definitely 'yes'. If at the solution point x_* the Hessian matrix is positive definite then, near x_*, f can be approximated well by a strictly convex quadratic, and the previous analysis is relevant – except that an inexact line search can make matters worse. However, if the line search always performs one quadratic interpolation, then the step-length will be exact asymptotically, and one can show that the rate of convergence is linear with constant $[(\lambda_n - \lambda_1)/(\lambda_n + \lambda_1)]^2$, where $\lambda_1 \leq \cdots \leq \lambda_n$ are now the eigenvalues of the Hessian $\nabla^2 f(x_*)$.

This global efficiency result has been presented in some detail because it is illustrative of such studies in optimization methods: a simple model problem is chosen, and by direct computation, recurrence relations are established to determine the function reduction. Such relations are difficult to obtain for general nonlinear functions, but Nemirovsky and Yudin are able to derive several interesting results for convex functions. Their work is described in the book (Nemirovsky and Yudin, 1983) and in subsequent papers. We will now give a very brief description of their approach, to show its flavour.

Suppose that f is a strongly convex and continuously differentiable function. Suppose also that the gradient satisfies the Lipschitz condition (3.8) for all $x \in \mathbb{R}^n$. Let us denote a lower bound on the smallest eigenvalue of the Hessian $\nabla^2 f(x)$ by m. Nemirovsky and Yudin define the global estimate of the rate of convergence on an iterative method as a function $h(x_1 - x_*, m, L, k) :\to \mathbb{R}$ such that for any objective function f and for any $k \geq 1$ we have

$$f_k - f_* \leq c_1 h(x_1 - x_*, m, L, k),$$

where c_1 is a constant, k is the iteration number, L is the Lipschitz constant, and x_* is the solution point.

The faster the rate at which h converges to 0 as $k \to \infty$, the more efficient the method. Nemirovsky and Yudin (see also Nesterov (1988)) show that there is a lower bound on the rate of convergence of h.

Theorem 3.2 Consider an optimization method which, at every iteration k, evaluates the function f and gradient g at N_k auxiliary points whose convex hull has dimension less than or equal to l. Then for all k

$$h(x_1-x_*, m, L, k) \geq c_2\|x_1-x_*\|^2 \min\left[([l+1]k)^{-2}, \exp\left(-\sqrt{\frac{m}{L}}c_3 k(l-1)\right)\right],$$
$$(3.22)$$

where c_2 depends on m and L, and c_3 is a constant.

In this framework, a method is optimal if its efficiency mapping h is bounded above by the right-hand side of (3.22), where c_2 and c_3 are allowed to be any constants. Nemirovsky and Yudin show that the well known conjugate gradient and variable metric methods are not optimal, and Nesterov (1983) proposed a conjugate gradient method that achieves the optimal bound. In this theoretical framework optimization algorithms are ranked according to their worst case behaviour. We will discuss this in more detail in later sections.

This concludes our outline of some basic principles used in the theoretical analysis of optimization methods. Two classical books giving an exhaustive treatment of this subject are Ostrowski (1966) and Ortega and Rheinboldt (1970). Much of what is known about the theory of quasi-Newton methods is described in the survey paper by Dennis and Moré (1977) and in Dennis and Walker (1981). More recent survey papers include Dennis and Schnabel (1987), Schnabel (1989), Toint (1986a) and Powell (1985). In the following sections we focus on recent theoretical developments which are, to a great extent, not covered in these articles.

4. Conjugate gradient methods

The introduction of the conjugate gradient method by Fletcher and Reeves, in the 1960s, marks the beginning of the field of large scale nonlinear optimization. Here was a technique that could solve very large problems, since it requires storage of only a few vectors, and could do so much more rapidly than the steepest descent method. The definition of a large problem has changed drastically since then, but the conjugate gradient method has remained one of the most useful techniques for solving problems large enough to make matrix storage impractical. Numerous variants of the method of Fletcher and Reeves have been proposed over the last 20 years, and many theoretical studies have been devoted to them. Nevertheless, nonlinear con-

jugate gradient methods are perhaps the least understood methods of optimization.

The recent development of limited memory and discrete Newton methods
have narrowed the class of problems for which conjugate gradient methods
are recommended. Nevertheless, in my view, conjugate gradient methods
are still the best choice for solving very large problems with relatively inex-
pensive objective functions (Liu and Nocedal, 1989). They can also be more
suitable than limited memory methods on several types of multiprocessor
computers (Nocedal, 1990).

The theory of conjugate gradient methods for nonlinear optimization is
fascinating. Unlike the linear conjugate gradient method for the solution of
systems of equations, which is known to be optimal (in some sense), some
nonlinear conjugate gradient methods possess surprising, and sometimes,
bizarre properties. The theory developed so far offers fascinating glimpses
into their behaviour, but our knowledge remains fragmentary. I view the
development of a comprehensive theory of conjugate gradient methods as
one of the outstanding challenges in theoretical optimization, and I believe
that it will come to fruition in the near future. This theory would not
only be a significant mathematical accomplishment, but could result in the
discovery of a superior conjugate gradient method.

The original conjugate gradient method proposed by Fletcher and Reeves
(1964) is given by

$$d_k = -g_k + \beta_k^{\mathrm{FR}} d_{k-1}, \tag{4.1}$$

$$x_{k+1} = x_k + \alpha_k d_k, \tag{4.2}$$

where α_k is a step-length parameter, and where

$$\beta_k^{\mathrm{FR}} = \begin{cases} 0 & \text{for } k = 1, \\ \|g_k\|^2 / \|g_{k-1}\|^2 & \text{for } k \geq 2. \end{cases} \tag{4.3}$$

When applied to strictly quadratic objective functions this method reduces
to the linear conjugate gradient method provided α_k is the exact minimizer
(Fletcher, 1987). Other choices of the parameter β_k in (4.1) also possess this
property, and give rise to distinct algorithms for nonlinear problems. Many
of these variants have been studied extensively, and the best choice of β_k is
generally believed to be

$$\beta_k^{\mathrm{PR}} = g_k^T (g_k - g_{k-1}) / \|g_{k-1}\|^2, \tag{4.4}$$

and is due to Polak and Ribière (1969).

The numerical performance of the Fletcher–Reeves method (4.3) is somewhat erratic: it is sometimes as efficient as the Polak–Ribière method, but
it is often much slower. It is safe to say that the Polak–Ribière method is,
in general, substantially more efficient than the Fletcher–Reeves method.

In many implementations of conjugate gradient methods, the iteration

(4.1) is restarted every n steps by setting β_k equal to zero, i.e. taking a steepest descent step. This ensures global convergence, as was discussed in Section 3. However many theoretical studies consider the iteration without restarts (Powell, 1977, 1984a; Nemirovsky and Yudin, 1983), and there are good reasons for doing so. Since conjugate gradient methods are useful for large problems, it is relevant to consider their behaviour as $n \to \infty$. When n is large (say 10,000) we expect to solve the problem in less than n iterations, so that a restart would not be performed. We can also argue that we would like to study the behaviour of large sequences of unrestarted conjugate gradient iterations to discover patterns in their behaviour. We will see that this approach has been very successful in explaining phenomena observed in practice. Therefore in this section we will only consider conjugate gradient methods without restarts.

The first practical global convergence result is due to Al-Baali (1985) and applies to the Fletcher–Reeves method. To establish this result it is necessary that the line search satisfy the *strong Wolfe conditions*

$$f(x_k + \alpha_k d_k) \;\; \leq \;\; f(x_k) + \sigma_1 \alpha_k g_k^T d_k, \tag{4.5}$$
$$|g(x_k + \alpha_k d_k)^T d_k| \;\; \leq \;\; -\sigma_2 g_k^T d_k, \tag{4.6}$$

where $0 < \sigma_1 < \sigma_2 < \frac{1}{2}$. Note that if a step-length α_k satisfies the strong Wolfe conditions, then it satisfies the usual Wolfe conditions (3.6)–(3.7). Therefore Zoutendijk's result (3.9) will hold, provided we can show that the search directions of the Fletcher–Reeves method are descent directions. Al-Baali does this, obtaining the following global convergence result. Throughout this section we assume that the starting point is such that the level set $\mathcal{L} := \{x : f(x) \leq f(x_1)\}$ is bounded, that in some neighbourhood $\mathcal{N}$ of $\mathcal{L}$, the objective function f is continuously differentiable, and that its gradient is Lipschitz continuous.

Theorem 4.1   Consider the Fletcher–Reeves method (4.1)–(4.2), where the step-length satisfies the strong Wolfe conditions (4.5)–(4.6). Then there is a constant $c > 0$ such that

$$g_k^T d_k \leq -c \, \|g_k\|^2, \tag{4.7}$$

for all $k \geq 1$, and

$$\liminf_{k \to \infty} \|g_k\| = 0.$$

This result is interesting in many respects. The relation (4.7) is established by induction in a novel and elegant fashion. It shows that the strong Wolfe conditions are sufficient to ensure the descent property of the Fletcher–Reeves method. Prior to this result it was thought that an *ad hoc* and complicated line search would be required to guarantee descent. Relation

(4.7) appears to play an important role in conjugate gradient methods, and we will encounter it again later. This theorem is also attractive because it applies to the algorithm as implemented in practice, and because the assumptions on the objective function are not restrictive.

Theorem 4.1 can be generalized to other iterations related to the Fletcher–Reeves method. Touati-Ahmed and Storey (1990) show that Theorem 4.1 holds for all methods of the form (4.1)–(4.2), which satisfy the strong Wolfe conditions, and with any β_k such that $0 \leq \beta_k \leq \beta_k^{\text{FR}}$. Gilbert and Nocedal (1990) extend this to any method with $|\beta_k| \leq \beta_k^{\text{FR}}$, and show that this result is tight in the following sense: there exists a smooth function f, a starting point x_1 and values of β_k satisfying

$$|\beta_k| \leq c\beta_k^{\text{FR}},$$

for some $c > 1$, such that the sequence of gradient norms $\{\|g_k\|\}$ generated by (4.1)–(4.2) is bounded away from zero.

This is our first encounter with a negative convergence result for conjugate gradient methods. It shows that the choice of the parameter β_k is crucial. An analysis of conjugate gradient methods with inexact line searches, shows that unless β_k is carefully chosen, the length of the search direction d_k can grow without bound causing the algorithm to fail. In the results mentioned so far, only the size of β_k with respect to β_k^{FR} plays an important role in ensuring global convergence. We will see later that a more subtle property of β_k determines the efficiency of the iteration.

Powell (1977) has given some arguments that explain, at least partially, the poor performance of the Fletcher–Reeves method in some problems: if a very small step is generated away from the solution, then due to the definition (4.3), it is likely, that subsequent steps will also be very short. We will not give the supporting facts for this argument, but only mention that the analysis is simple, and also shows that the Polak–Ribière method would not slow down in these circumstances. This propensity for short steps, causes the Fletcher–Reeves algorithm to sometimes stall away from the solution, and this behaviour can be observed in practice. For example, I have observed that when solving the minimal surface problem (Toint, 1983) with 961 variables, the Fletcher–Reeves method generates tiny steps for hundreds of iterations, and is only able to terminate this pattern after a restart is performed.

Powell (1977) and Nemirovsky and Yudin (1983) give global efficiency results that provide further evidence of the inefficiency of the Fletcher–Reeves method. The simplest analysis is that of Powell, who shows that if the Fletcher–Reeves method, with exact line searches, enters a region in which the function is the two-dimensional quadratic

$$f(x) = \tfrac{1}{2}x^T x,$$

then the angle between the gradient g_k and the search direction d_k stays constant. Therefore, if this angle is close to $90°$ the method will converge very slowly. Indeed since this angle can be arbitrarily close to $90°$, the Fletcher–Reeves method can be slower than the steepest descent method. Powell also shows that the Polak–Ribière method behaves quite differently in these circumstances, for if a very small step is generated, the next search direction tends to the steepest descent direction, preventing a sequence of tiny steps from happening.

With all the arguments given in favour of the Polak–Ribière method, we would expect to be able to prove, for it, a global convergence result similar to Theorem 4.1. That this is not possible follows from a remarkable result of Powell (1984a). He shows that the Polak–Ribière method with exact line searches can cycle infinitely, without approaching a solution point. Since the step-length of Powell's example would probably be accepted by any practical line search algorithm, it appears unlikely that a satisfactory global convergence result will ever be found for the Polak–Ribière method.

Powell establishes his negative result by an algebraic *tour de force*. He assumes that the line search always finds the first stationary point, and shows that there is a twice continuously differentiable function of three variables and a starting point such that the sequence of gradients generated by the Polak–Ribière method stays bounded away from zero. Since Powell's example requires that some consecutive search directions become almost contrary, and since this can only be achieved (in the case of exact line searches) when $\beta_k < 0$, Powell (1986) suggests modifying the Polak–Ribière method by setting

$$\beta_k = \max\{\beta_k^{\mathrm{PR}}, 0\}. \tag{4.8}$$

Thus if a negative value of β_k^{PR} occurs, this strategy will restart the iteration along the steepest descent direction.

Gilbert and Nocedal (1990) show that this modification of the Polak–Ribière method is globally convergent both for exact and inexact line searches. If negative values of β_k^{PR} occurred infinitely often, global convergence would follow, as discussed in Section 3, because an infinite number of steepest descent steps would be taken. Thus Gilbert and Nocedal consider the case where $\beta_k^{\mathrm{PR}} > 0$ for all sufficiently large k, and show that in this case $\liminf \|g_k\| = 0$, provided the line search has the following two properties: (i) it satisfies the strong Wolfe conditions; and (ii) it satisfies (4.7) for some constant c. Gilbert and Nocedal discuss how to implement such a line search strategy for any conjugate gradient method with $\beta_k \geq 0$. We will now describe their analysis, which is quite different from that used by Al-Baali for the study of the Fletcher–Reeves method.

The use of inexact line searches in conjugate gradient methods requires careful consideration. In contrast with the Fletcher–Reeves method, the strong Wolfe conditions (4.5)–(4.6) no longer guarantee the descent property for the Polak–Ribière or other conjugate gradient methods. It turns out, however, that if β_k is always nonnegative it is possible to find a line search strategy that will provide the descent property. To see this note that from (4.1) we have

$$g_k^T d_k = -\|g_k\|^2 + \beta_k g_k^T d_{k-1}. \tag{4.9}$$

Therefore, to obtain descent for an inexact line search algorithm, one needs to ensure that the last term is not too large. Suppose that we perform a line search along the descent direction d_{k-1}, enforcing the Wolfe (or strong Wolfe) conditions, to obtain x_k. If $g_k^T d_{k-1} \le 0$, the nonnegativity of β_k implies that the sufficient descent condition (4.7) holds. On the other hand, if (4.7) is not satisfied then it must be the case that $g_k^T d_{k-1} > 0$, which means that a one-dimensional minimizer has been bracketed. It is then easy to apply a line search algorithm, such as that given by Lemaréchal (1981), Fletcher (1987) or Moré and Thuente (1990), to reduce $|g_k^T d_{k-1}|$ sufficiently and obtain (4.7). Note that the only condition imposed so far on β_k is that it be nonnegative.

To obtain global convergence for other conjugate gradient methods we need to impose another condition on β_k, and interestingly enough, it is the property that makes the Polak–Ribière method avoid the inefficiencies of the Fletcher–Reeves method. We say that a method has Property (∗) if a small step, $\alpha_{k-1} d_{k-1}$ in a region away from the solution implies that β_k will be small. A precise definition is given in Gilbert and Nocedal (1990). It isolates an important property of the Polak–Ribière method: the tendency to turn towards the steepest descent direction if a small step is generated away from the solution. The global convergence result of Gilbert and Nocedal is as follows.

Theorem 4.2 Consider any method of the form (4.1)–(4.2) with the following three properties: (i) $\beta_k \ge 0$ for all k; (ii) the line search satisfies the Wolfe conditions (3.6)–(3.7) and the sufficient descent condition (4.7); and (iii) Property (∗) holds. Then $\liminf \|g_k\| = 0$.

This is one of the most general convergence results known to date. However it is not clear if the restriction $\beta_k \ge 0$ is essential, in some way, and should always be imposed in conjugate gradient methods, or if it only simplifies the analysis. It is also not known if the cycling of the Polak–Ribière method predicted by Powell can occur in practice; to my knowledge it has never been observed. Lukšan (1991a) performed numerical tests with several conjugate gradient methods that restrict β_k^{PR} to be nonnegative, as well as methods that are constrained by β_k^{FR}. The results are interesting, but inconclusive, and more research is needed.

How fast is the convergence of conjugate gradient methods? Let us first answer this question under the assumption that exact line searches are made. Crowder and Wolfe (1972) show that the rate of convergence is linear, and give an example that shows that the rate cannot be Q-superlinear. Powell (1976b) studies the case in which the conjugate gradient method enters a region where the objective function is quadratic, and shows that either finite termination occurs, or the rate of convergence is linear. Cohen (1972) and Burmeister (1973) show that, for general objective functions, the rate of convergence is n-step quadratic, i.e.

$$\|x_{k+n} - x_*\| = \mathcal{O}(\|x_k - x_*\|^2),$$

and Ritter (1980) strengthens the result to

$$\|x_{k+n} - x_*\| = \mathcal{O}(\|x_k - x_*\|^2).$$

Powell (1983) gives a slightly better result and performs numerical tests on small problems to measure the rate observed in practice. Faster rates of convergence can be established (Schuller, 1974; Ritter, 1980), under the assumption that the search directions are uniformly linearly independent, but this does not often occur in practice. Several interesting results assuming asymptotically exact line searches are given by Baptist and Stoer (1977) and Stoer (1977). We will not discuss any of these rate of convergence results further because they are not recent and are described, for example, in Powell (1983).

Nemirovsky and Yudin (1983) devote some attention to the global efficiency of the Fletcher–Reeves and Polak–Ribière methods with exact line searches. For this purpose they define a measure of 'laboriousness' and an 'optimal bound' for it among a certain class of iterations. They show that on strongly convex problems, not only do the Fletcher–Reeves and Polak–Ribière methods fail to attain the optimal bound, but they also construct examples in which both methods are slower than the steepest descent method. Subsequently Nesterov (1983) presented an algorithm that attains this optimal bound. It is related to PARTAN – the method of parallel tangents (Luenberger, 1984), and is unlikely to be effective in practice, but this has not been investigated, to the best of my knowledge. Some extensions of Nesterov's algorithm have been proposed by Güler (1989).

Let us now consider extensions of the conjugate gradient method. Motivated by the inefficiencies of the Fletcher–Reeves method, and guided by the desire to have a method that cannot converge to point where the gradient is nonzero, Powell (1977) proposed a conjugate gradient method which restarts automatically using a three-term recurrence iteration introduced by Beale (1972). This method has been implemented in the Harwell routine VE04 and outperforms the Fletcher–Reeves and Polak–Ribière methods, but requires more storage. Shanno and Phua (1980) proposed a different extension of the conjugate gradient method that uses even more storage, and which resembles a variable metric iteration. It has been implemented in the highly successful and popular code CONMIN. This method, which is not simple to describe, also uses automatic restarts. The iteration is of the form

$$d_k = -H_k g_k,$$

where H_k is a positive definite and symmetric matrix. Since this ensures that the search directions are descent directions, the line search needs only to satisfy the usual Wolfe conditions (3.6)–(3.7). Shanno (1978a,b) shows that this algorithm is globally convergent, with inexact line searches, on strongly convex problems. The convergence properties on nonconvex problems are not known; in fact, CONMIN is related to the BFGS variable metric method, whose global convergence properties on nonconvex problems are not yet understood, as we will discuss in the next section.

It is interesting to note that for all the conjugate gradient methods described in this section, and for their extensions, increased storage results in fewer function evaluations. The Fletcher–Reeves method requires four n-vectors of storage, Polak–Ribière five, VE04 six and CONMIN seven. In terms of function evaluations, their ranking corresponds to the order in which they were just listed – with CONMIN at the top.

Are automatic restarts useful? This remains controversial. Gill and Murray (1979) speculate that the efficiency of VE04 and CONMIN is due to the fact that they make good use of the additional information they store, rather than to the effects of restarting. I agree with this assessment, and as we will see when we discuss limited memory methods, it is possible to design methods that are more effective than CONMIN and use no restarts. In my view, an undesirable feature of all the restarting criteria proposed so far is that they do not rule out the possibility of triggering a restart at every step, hence degrading the speed of convergence of the methods. Indeed, I have observed examples in which CONMIN restarts at every iteration and requires an excessive number of function evaluations.

I will end this section with a question that has intrigued me for some time: have we failed to discover the 'right' implementation of the conjugate gradient method? Is there a simple iteration of the form (4.1)–(4.2) which performs significantly better than all the methods proposed so far, and which has all the desirable convergence properties? Given the huge number of articles proposing new variations of the conjugate gradient method, without much success, the answer would seem to be 'no'. However I have always felt that the answer is 'yes' – but I could say no more.

5. Variable metric methods

We have seen that in order to obtain a superlinearly convergent method it is necessary to approximate the Newton step asymptotically – this is the principle (3.16) of Dennis and Moré. How can we do this without actually evaluating the Hessian matrix at every iteration? The answer was discovered by Davidon (1959), and was subsequently developed and popularized by Fletcher and Powell (1963). It consists of starting with *any* approximation to the Hessian matrix, and at each iteration, update this matrix by incorporating the curvature of the problem measured along the step. If this update is done appropriately, one obtains some remarkably robust and efficient methods, called variable metric methods. They revolutionized nonlinear optimization by providing an alternative to Newton's method, which is too costly for many applications. There are many variable metric methods, but since 1970, the BFGS method has been generally considered to be the most effective. It is implemented in all major subroutine libraries and is currently being used to solve optimization problems arising in a wide spectrum of applications.

The theory of variable metric methods is beautiful. The more we study them, the more remarkable they seem. We now have a fairly good understanding of their properties. Much of this knowledge has been obtained recently, and we will discuss it in this section. We will see that the BFGS method has interesting self-correcting properties, which account for its robustness. We will also discuss some open questions that have resisted an answer for many years. Variable metric methods, aside from being highly effective in practice, are intricate mathematical objects, and one could spend a lifetime discovering new properties of theirs. Ironically, our many theoretical studies of variable metric methods have not resulted in the discovery of new methods, but have mainly served to explain phenomena observed in practice. However it is hard to predict the future of this area, which has given rise to many surprising developments.

The BFGS method is a line search method. At the kth iteration, a symmetric and positive definite matrix B_k is given, and a search direction is computed by

$$d_k = -B_k^{-1} g_k. \tag{5.1}$$

The next iterate is given by

$$x_{k+1} = x_k + \alpha_k d_k, \tag{5.2}$$

where the step-size α_k satisfies the Wolfe conditions (3.6)–(3.7). It has been found that it is best to implement BFGS with a very loose line search: typical values for parameters in (3.6)–(3.7) are $\sigma_1 = 10^{-4}$ and $\sigma_2 = 0.9$. The Hessian approximation is updated by

$$B_{k+1} = B_k - \frac{B_k s_k s_k^T B_k}{s_k^T B_k s_k} + \frac{y_k y_k^T}{y_k^T s_k}, \tag{5.3}$$

where, as before,

$$y_k = g_{k+1} - g_k, \qquad s_k = x_{k+1} - x_k. \tag{5.4}$$

Note that the two correction matrices on the right-hand side of (5.3) have rank one. Therefore by the interlocking eigenvalue theorem (Wilkinson, 1965), the first rank-one correction matrix, which is subtracted, decreases the eigenvalues – we will say that it 'shifts the eigenvalues to the left'. On the other hand, the second rank-one matrix, which is added, shifts the eigenvalues to the right. There must be a balance between these eigenvalue shifts, for otherwise the Hessian approximation could either approach singularity or become arbitrarily large, causing a failure of the method.

A global convergence result for the BFGS method can be obtained by careful consideration of these eigenvalue shifts. This is done by Powell (1976a), who uses the trace and the determinant to measure the effect of the two rank-one corrections on B_k. He is able to show that if f is convex, then for any positive definite starting matrix B_1 and any starting point x_1, the BFGS method gives $\liminf \|g_k\| = 0$. If in addition the sequence $\{x_k\}$ converges to a solution point at which the Hessian matrix is positive definite, then the rate of convergence is superlinear.

This analysis has been extended by Byrd *et al.* (1987) to the restricted Broyden class of quasi-Newton methods in which (5.3) is replaced by

$$B_{k+1} = B_k - \frac{B_k s_k s_k^T B_k}{s_k^T B_k s_k} + \frac{y_k y_k^T}{y_k^T s_k} + \phi(s_k^T B_k s_k) v_k v_k^T, \tag{5.5}$$

where $\phi \in [0, 1]$, and

$$v_k = \left[\frac{y_k}{y_k^T s_k} - \frac{B_k s_k}{s_k^T B_k s_k} \right].$$

The choice $\phi = 0$ gives rise to the BFGS update, whereas $\phi = 1$ defines the DFP method – the first variable metric method proposed by Davidon, Fletcher and Powell (see e.g. Fletcher (1987)). Byrd *et al.* prove global and superlinear convergence on convex problems, for all methods in the restricted Broyden class, *except for DFP*. Their approach breaks down when $\phi = 1$, and leaves that case unresolved. Indeed the following question has remained unanswered since 1976, when Powell published his study on the BFGS method.

Open question I. Consider the DFP method with a line search satisfying the Wolfe conditions (3.6)–(3.7). Assume that f is strongly convex, which implies that there is a unique minimizer x_*. Do the iterates generated by the DFP method converge to x_*, for any starting point x_1 and any positive definite starting matrix B_1?

It is rather surprising that, even though the DFP method has been known for almost 30 years, we have little idea of what the answer to this basic question will turn out to be. DFP can be made to perform extremely poorly on convex problems, making a negative result plausible. On the other hand, the method has never been observed to fail; in fact even in the worst examples we can see the DFP method creeping towards a solution point. The most we can say is that the DFP method is globally convergent on convex functions if the line searches are exact (Powell, 1971, 1972), or that if it converges to a point, and line searches are exact, then the gradient at this point must be zero (Pu and Yu, 1988). It may also seem puzzling to the reader that global convergence has been established for $\phi = 0.999$, say, but not for $\phi = 1$. Wouldn't a continuity argument show that if the result holds for all $\phi < 1$ then it must also hold for $\phi = 1$? To answer this question, and to describe the self-correcting properties of the BFGS method, mentioned earlier, we will now discuss in some detail the convergence analyses of Powell, and Byrd $et\ al.$

Let us begin by considering only the BFGS method, and let us assume that the function f is strongly convex, i.e. that there exist positive constants m and M such that

$$m\|z\|^2 \le z^T G(x)z \le M\|z\|^2 \tag{5.6}$$

for all $z, x \in \mathbb{R}^n$, where G denotes the Hessian matrix of f. Computing the trace of (5.3) we obtain

$$\text{Tr}\,(B_{k+1}) = \text{Tr}\,(B_k) - \frac{\|B_k s_k\|^2}{s_k^T B_k s_k} + \frac{\|y_k\|^2}{y_k^T s_k}. \tag{5.7}$$

It turns out that the middle term on the right-hand side of this equation depends on $\cos\theta_k$, the angle between the steepest descent direction and the search direction, which was used extensively in Section 3. To see this, we first note that

$$f_{k+1} - f_k = g_k^T s_k + \tfrac{1}{2}s_k^T G(\xi_k)s_k,$$

for some ξ_k between x_{k+1} and x_k. Thus, using the first Wolfe condition (3.6) we have

$$\sigma_1 g_k^T s_k \ge g_k^T s_k + \tfrac{1}{2}s_k^T G(\xi_k)s_k. \tag{5.8}$$

Next we use (5.6) and the definition (3.5) of $\cos\theta_k$ to obtain

$$(1 - \sigma_1)\|g_k\|\|s_k\|\cos\theta_k \ge \tfrac{1}{2}m\|s_k\|^2,$$

which implies that

$$\|s_k\| \le c_2\|g_k\|\cos\theta_k, \tag{5.9}$$

where $c_2 = 2(1 - \sigma_1)/m$. Since $B_k s_k = -\alpha_k g_k$, using (5.9) we obtain

$$
\begin{aligned}
\frac{\|B_k s_k\|^2}{s_k^T B_k s_k} &= \frac{\alpha_k^2 \|g_k\|^2}{\alpha_k \|s_k\| \|g_k\| \cos \theta_k} \\
&= \frac{\alpha_k \|g_k\|}{\|s_k\| \cos \theta_k} \\
&\geq \frac{\alpha_k}{c_2 \cos^2 \theta_k}.
\end{aligned} \tag{5.10}
$$

We have thus shown that the term that tends to decrease the trace can be proportional to $\alpha_k / \cos^2 \theta_k$. Let us now consider the last term in the trace equation (5.7). From the definition of y_k we have that

$$
y_k = \bar{G} s_k, \tag{5.11}
$$

where

$$
\bar{G} = \int_0^1 G(x_k + \tau s_k) d\tau. \tag{5.12}
$$

Let us define $z_k = \bar{G}^{1/2} s_k$, where $\bar{G}^{1/2} \bar{G}^{1/2} = \bar{G}$. Then from (5.11) and (5.6)

$$
\begin{aligned}
\frac{y_k^T y_k}{y_k^T s_k} &= \frac{s_k^T \bar{G}^2 s_k}{s_k^T \bar{G} s_k} \\
&= \frac{z_k^T \bar{G} z_k}{z_k^T z_k} \\
&\leq M.
\end{aligned} \tag{5.13}
$$

Therefore the term that tends to increase the trace is bounded above for all k, on convex problems. We obtain from (5.7) and (5.10)

$$
\text{Tr}(B_{k+1}) \leq \text{Tr}(B_k) - \frac{\alpha_k}{c_2 \cos^2 \theta_k} + M. \tag{5.14}
$$

This relation allows insight into the behaviour of the BFGS method. The discussion that follows is not rigorous, but all the statements made here can be established rigorously.

Suppose for the moment that the step-lengths α_k are bounded below. If the algorithm produces iterations for which $\cos \theta_k$ is not very small, it will advance towards the solution, but some of the eigenvalues of $\{B_k\}$ could become large because the middle term on the right-hand side of (5.14) could be significantly smaller than M. If, as a result of having an excessively large Hessian approximation B_k, steps with very small $\cos \theta_k$ are produced, little progress may be achieved, but a self-correcting mechanism takes place: the middle term in (5.14) will be larger than M, thus decreasing the trace. This self-correction property is in fact very powerful. The smaller $\cos \theta_k$ is, the faster the reduction in the trace relation.

Suppose now that the step-lengths α_k tend to zero. It is easy to see

(Byrd *et al.*, 1987, p. 1179) that this is due to the existence of very small eigenvalues in B_k, which cannot be monitored by the means of the trace. Fortunately, it turns out that the BFGS update formula has a strong self-correcting property with respect to the determinant, which can be used to show that, in fact, α_k is bounded away from zero in mean. Indeed, the determinant of (5.3) is given by Pearson (1969)

$$\det(B_{k+1}) = \det(B_k) \frac{y_k^T s_k}{s_k^T B_k s_k}. \tag{5.15}$$

Note that when $s_k^T B_k s_k$ is small relative to $y_k^T s_k = s_k^T \bar{G} s_k$, the determinant increases, reflecting the fact that the small curvature of our model is corrected, thus increasing some eigenvalues.

In conclusion, the trace relation shows that, for strongly convex problems, the eigenvalues of the matrices B_k cannot become too large, and the determinant relation shows that they cannot become too small. This can be used to show that the method is convergent, and by verifying the Dennis–Moré condition (3.16), one deduces that the rate of convergence is superlinear.

Let us now consider the restricted Broyden class (5.5) with $\phi \in [0, 1)$. The analysis proceeds along similar lines. The trace relation is now (Byrd *et al.*, 1987)

$$\mathrm{Tr}(B_{k+1}) \leq \mathrm{Tr}(B_k) + M + \frac{\phi \alpha_k}{c_1} - \frac{(1 - \phi) \alpha_k}{c_2 \cos^2 \theta_k} + \frac{2 \phi M \alpha_k}{m c_1 \cos \theta_k}, \tag{5.16}$$

where $c_1 = (1 - \sigma_2)/M$. Note that the second and the third terms on the right-hand side of (5.16) produce a shift to the right in the eigenvalues, in the sense that they increase the trace. The fourth term on the right-hand side of (5.16) produces a shift to the left, which can be very strong when $\cos \theta_k$ is small. The last term can produce a shift in *either* direction. A crucial fact is that this last term, of uncertain sign, is inversely proportional to $\cos \theta_k$, whereas the negative fourth term is inversely proportional to $\cos^2 \theta_k$. Therefore, when $\cos \theta_k$ is tiny, we still have a guaranteed decrease in the trace relation. This can be used to show that the Hessian approximation B_k cannot grow without bound.

The determinant relation, for any $\phi \in [0, 1]$, can be shown to satisfy,

$$\det(B_{k+1}) \geq \det(B_k) \frac{y_k^T s_k}{s_k^T B_k s_k}, \tag{5.17}$$

which is essentially the same as for the BFGS update, and so we can reason as before to deduce that small eigenvalues are efficiently corrected. These arguments can be made rigorous, and can be used to establish global and superlinear convergence for any method in the restricted Broyden class using $\phi \in [0, 1)$.

Why does this analysis not apply to the DFP method? It turns out that small eigenvalues do not cause problems, because (5.17) holds when $\phi = 1$, showing that the method possesses the self-correcting property with respect to the determinant mentioned earlier. Therefore if very small eigenvalues occur, the DFP method will be able to increase them quickly. Difficulties, however, can arise due to large eigenvalues. Note that the fourth term on the right-hand side of (5.16), which plays a crucial role in preventing the trace from growing, is no longer present. The only term capable of decreasing the trace is the last term in (5.16). In addition to being of uncertain sign, this term is smaller in magnitude than the fourth term in (5.16), when $\cos\theta_k$ is small. Thus it is not certain that a shift to the left will occur and even if it does we cannot expect it to be as strong as for other methods in the Broyden class. Therefore we can expect the DFP method to either develop excessively large Hessian approximations B_k or, at the very least, to have difficulties in reducing a large initial Hessian approximation. Numerical tests confirm these observations, which also seem to agree with a global efficiency study of Powell (1986), which we discuss later on.

We have assumed all along that the Wolfe conditions are always satisfied. Are the good properties of the BFGS method strongly dependent on them? This question is of practical importance, because for problems with inequality constraints it is often not possible to satisfy the second Wolfe condition (3.7). Fortunately it is proved by Byrd and Nocedal (1989) that the BFGS updating formula has excellent properties as long as it perceives positive curvature – regardless of how large the function reduction or the change in the gradient are. We now formally state one of these properties.

Theorem 5.1 Let $\{B_k\}$ be generated by the BFGS formula (5.3) where B_1 is symmetric and positive definite and where for all $k \geq 1$ y_k and s_k are any vectors that satisfy

$$\frac{y_k^T s_k}{s_k^T s_k} \geq m > 0 \tag{5.18}$$

$$\frac{\|y_k\|^2}{y_k^T s_k} \leq M. \tag{5.19}$$

Then for any $p \in (0,1)$ there exists a constant β_1, such that, for any $k > 1$, the relation

$$\cos\theta_j \geq \beta_1 \tag{5.20}$$

holds for at least $\lceil pk \rceil$ values of $j \in [1, k]$.

This result states that, even though we cannot be sure that all the $\cos\theta_k$ will be bounded below, we can be sure that this is the case for most of them. This is enough to obtain certain global convergence results. For example, Theorem 5.1 can be used to show that the BFGS method using a backtracking line search is globally convergent on convex problems. Various results of this type have also been obtained by Werner (1978, 1989); see also Warth and Werner (1977).

The recent analysis on variable metric methods has not only produced new results but, as can be expected, has also provided simpler tools for performing the analysis. Byrd and Nocedal (1989) show that it is easier to work simultaneously with the trace and determinant relations. For this purpose they define, for any positive definite matrix B, the function

$$\psi(B) = \operatorname{tr} B - \ln \det B, \qquad (5.21)$$

where ln denotes the natural logarithm. It is easy to see that $\psi(B) > \ln[\operatorname{cond}(B)]$, so that global convergence can be established by analysing the behaviour of $\psi(B_k)$. Moreover the function ψ can also be used to establish superlinear convergence without having to explicitly verify the Dennis–Moré condition (3.16); this is explained in Byrd and Nocedal (1989).

5.1. Nonconvex objective functions

All the results for the BFGS method discussed so far depend on the assumption that the objective function f is convex. At present, few results are available for the case in which f is a more general nonlinear function. Even though the numerical experience of many years suggests that the BFGS method always converges to a solution point, this has not been proved.

Open question II. Consider the BFGS method with a line search satisfying the Wolfe conditions (3.6)–(3.7). Assume that f is twice continuously differentiable and bounded below. Do the iterates satisfy $\liminf \|g_k\| = 0$, for any starting point x_1 and any positive definite starting matrix B_1?

This is one of the most fundamental questions in the theory of unconstrained optimization, for BFGS is perhaps the most commonly used method for solving nonlinear optimization problems. It is remarkable that the answer to this question has not yet been found. Nobody has been able to construct an example in which the BFGS method fails, and the most general result available to us, due to Powell (1976a), is as follows.

Theorem 5.2 Suppose that f is differentiable and bounded below. Consider the BFGS method with a line search satisfying the Wolfe conditions (3.6)–(3.7). Then the limit $\liminf \|g_k\| = 0$ is obtained for any starting point x_1 and any positive definite starting matrix B_1 if

$$\left\{ \frac{y_k^T y_k}{y_k^T s_k} \right\} \qquad (5.22)$$

is bounded above for all k.

We showed earlier (see (5.13)) that in the convex case (5.22) is always bounded, regardless of how the step s_k is chosen. However in the nonconvex case, in which the Hessian matrix can be indefinite or singular, the quotient (5.22) can be arbitrarily large, and only the line search could control its size. It is not known if the Wolfe conditions ensure that (5.22) is bounded, and if not, it would be interesting to find a practical line search that guarantees this.

Now that the global behaviour of variable metric methods on convex problems is reasonably well understood, it is time that we made some progress in the case when f is a general nonlinear function. Unfortunately establishing any kind of practical results in this context appears to be extremely difficult.

The 1970s witnessed the development of a very complete *local* convergence theory for variable metric methods. The main results, due to Broyden *et al.* (1973) and Dennis and Moré (1974) have been used extensively for the analysis of both constrained and unconstrained methods, and are very well summarized in Dennis and Moré (1977) and Dennis and Schnabel (1983). A typical result is as follows. Suppose that x_* is a minimizer where the Hessian is positive definite. If x_1 is sufficiently close to x_* and B_1 is sufficiently close to $\nabla^2 f(x_*)$, then the iterates generated by the BFGS or DFP methods, with unit step-lengths, converge to x_* superlinearly.

Another interesting result of Dennis and Moré makes no assumptions on the Hessian approximations, and states that if the iterates generated by BFGS or DFP satisfy

$$\sum_{k=1}^{\infty} \|x_k - x_*\| < \infty,$$

then the rate of convergence is superlinear. Griewank and Toint (1982b) extended this result to the restricted Broyden class. A stronger result for BFGS is implicit in the analysis of Griewank (1991) and Byrd *et al.* (1990): if the iterates converge (in any way) then the convergence rate must be superlinear.

A more general local convergence theory for least change secant methods has been developed by Dennis and Walker (1981). This work is important because it unifies several local convergence analyses, and because it can be used to design methods for special applications. Recently, Martínez (1990) presented a theoretical framework that applies to some methods not covered by the theory of Dennis and Walker; see also Martínez (1991).

5.2. Global efficiency of the BFGS and DFP methods

Nemirovsky and Yudin (1983) note that to obtain efficiency measures of optimization methods on general objective functions appears to be an unproductive task, because only very pessimistic results can be established. Therefore they restrict their attention to convex problems, and make some

interesting remarks on the properties of the DFP method. They do not resolve the question of whether DFP is optimal, in their sense, but note that
DFP is not invariant under the scaling of f. They use this fact to show that,
by badly scaling f, the DFP method can develop very large Hessian approximations and advance slowly. Their construction exploits the weakness of
DFP with respect to a large Hessian approximation mentioned earlier.

Powell (1986) is able to obtain much insight into the global behaviour of
BFGS and DFP by focusing on a narrower class of problems. He considers
a strictly convex quadratic objective function of two variables, and studies
the DFP and BFGS methods with step-lengths of one. Since both methods
are invariant under a linear change of variables, he assumes without loss
of generality that $G(x_*) = I$, as this results when making the change of
variables from x to $x_* + G(x_*)^{1/2}(x - x_*)$. Therefore Powell considers the
objective function

$$f(u, v) = \tfrac{1}{2}(u^2 + v^2), \tag{5.23}$$

and analyses the behaviour of DFP and BFGS for different choices of the
starting point x_1 and the starting matrix B_1. Due to the special form of
the objective function, the secant equation $B_{k+1}s_k = y_k$, which is satisfied
at each iteration by both DFP and BFGS, takes the form

$$B_{k+1}(x_{k+1} - x_k) = (x_{k+1} - x_k).$$

This shows that B_k always has one unit eigenvalue, and can assume that for
all k,

$$B_k = \begin{pmatrix} 1 & 0 \\ 0 & \lambda_k \end{pmatrix}.$$

The DFP and BFGS iterations can be studied by measuring how fast λ_k
converges to 1. Powell derives recurrence relations expressing λ_{k+2} in terms
of λ_{k+1} and λ_k, and from them, estimates the total number of iterations
required to obtain the solution to a given accuracy. These recurrence relations can also be used to estimate the function reduction at each step, and to
predict how many iterations will be required before superlinear convergence
takes place.

The results show vast differences of performance between the DFP and
BFGS methods when the initial eigenvalue λ_1 is large. Powell shows that,
in this case, the number of iterations required by the DFP method to obtain
the solution with good accuracy can be of order λ_1. In contrast, the BFGS
method requires only $\log_{10} \lambda_1$ iterations, in the worst case. The analysis
shows that if λ_1 is large, and if the starting point is unfavourable, then the
DFP method may decrease λ_k by at most one at every iteration.

When λ_1 is small, both methods are very efficient. The BFGS method
requires only $\log_{10}(\log_{10})\lambda_1^{-1}$ iterations before superlinear convergence steps

take place, whereas for the DFP method this occurs after only one or two iterations.

This analysis depends heavily on the assumption that unit step-lengths are always taken. It is therefore relevant to ask if this is a reasonable assumption for problem (5.23). Powell shows that an algorithm using a backtracking line search, would accept the unit step-length in these circumstances. This would also be the case for other line search strategies that only demand a sufficient decrease in the function. However, a line search that requires the two Wolfe conditions may not accept the unit step-length in some iterations, if the initial eigenvalue λ_1 is large. Therefore Powell's analysis has some limitations, but the predictions of this analysis can be observed in some nonquadratic problems, as we now discuss.

Byrd *et al.* (1987) test methods in Broyden's class with a line search satisfying the Wolfe conditions. The objective function is strongly convex; it is the sum of a quadratic and a small quartic term. The problem has two variables and the starting matrix is chosen as a diagonal matrix with eigenvalues 1 and 10^4. The BFGS method obtained the solution to high accuracy in 15 iterations. It was able to decrease the trace of B_k from 10^4 to 3 in only 10 iterations. In contrast, the DFP method required 4041 iterations to obtain the solution (which is amazingly close to the estimate given by Powell). It took, for example, 3000 iterations for DFP to decrease the trace from 10^4 to 1100. These results agree closely with the theoretical predictions given earlier because the objective function is nearly quadratic – the quartic term is small.

What should we expect if we use $\phi = 0.999$ in this problem? Not surprisingly, we find that very many iterations are needed. However it is interesting that the number of iterations was 2223 – much less than for DFP. Thus a tiny change in ϕ, away from one, has a marked effect in performance.

5.3. Is BFGS the best variable metric method?

The search for a variable metric method that is more efficient than the BFGS method began in the 1970s and has not ceased. In fact a new burst of research has taken place in the last few years, and some of the new ideas may provide practical improvements in performance.

Davidon (1975) proposed a method in which B_{k+1} is chosen to be the member of the Broyden class that minimizes the condition number of

$$B_k^{-1} B_{k+1},$$

subject to preserving positive definiteness. The resulting value of ϕ_k sometimes lies outside $[0,1]$, and often coincides with the value of ϕ_k that defines the symmetric rank-one method. We will discuss the symmetric rank-one method in Section 6, and it suffices to say here that it possesses some important computational and theoretical properties. Unlike the symmetric rank-one method, however, Davidon's method is guaranteed to generate positive definite Hessian approximations B_k, and can be implemented without any safeguards. Nevertheless interest in the method died after numerical tests failed to show an improvement over the BFGS method, and since the theoretical study by Schnabel (1978) suggested that the advantages of using Davidon's approach were likely to be modest.

Recently several authors have taken a new look at the idea of deriving optimally conditioned updates, using different measures than the one proposed by Davidon. Dennis and Wolkowicz (1991) use the function

$$\omega(B) = \frac{\text{tr}B}{n\text{det}B},$$

to obtain a new class of updates. Fletcher (1991) notes that the optimal updates given by the ψ-function (5.21) are BFGS or DFP, depending on how the variational problem is posed. Other work in this area includes Al-Baali (1990), Lukšan (1991b), Nazareth and Mifflin (1991), Yuan (1991) and Hu and Storey (1991). A different approach, in which the secant equation is not imposed, has been investigated by Yuan and Byrd (1991). Even though these studies are interesting, it is too soon to know if any of these new methods can perform significantly better than the BFGS method.

The analysis of Section 5.2, on the two-dimensional quadratic, suggests that the BFGS method is better at correcting small eigenvalues than large ones. Could we modify the method so as to strengthen its ability to correct large eigenvalues? Some authors feel that this can be done by using negative values for the parameter ϕ_k in Broyden's class. It is easy to explain the reason for this conjecture. Note that if $\phi_k < 0$, the fourth term in the right-hand side of (5.16) remains negative and increases in magnitude, and the third term becomes negative. This suggests that, when $\phi_k < 0$, the algorithm is better able to correct large eigenvalues. Care should be taken because there is a negative value ϕ_k^c for which the update becomes singular (for values less than ϕ_k^c, the updated matrix becomes indefinite; see for example Fletcher (1987)). Zhang and Tewarson (1988) performed numerical tests with fixed negative values of ϕ_k, and their results show a moderate but consistent improvement over the BFGS method. They also prove that, for convex problems, global and linear convergence can be established for negative values of ϕ_k, provided that for all k,

$$(1 - \nu)\phi_k^c \leq \phi_k \leq 0, \tag{5.24}$$

where ν is an arbitrary constant in (0,1). However Byrd et al. (1990) show that this algorithm is not superlinearly convergent, in general. They show that designing a superlinearly convergent method which uses negative values of ϕ_k is possible, but is difficult to implement in practice.

One can also attempt to improve variable metric methods by introducing automatic scaling strategies that adjust the size of the matrix B_k. If properly done, this could alleviate, for example, the difficulties that DFP has with large eigenvalues. An idea proposed by Oren and Luenberger (1974) consists of multiplying B_k by a scaling factor ϑ_k before the update takes place. For example, for the BFGS method, the update would be of the form

$$B_{k+1} = \vartheta_k \left[B_k - \frac{B_k s_k s_k^T B_k}{s_k^T B_k s_k} \right] + \frac{y_k y_k^T}{y_k^T s_k}. \tag{5.25}$$

Several choices for ϑ_k have been proposed by Oren and Luenberger (1974), Oren (1982), and in the references cited in these papers. The choice

$$\vartheta_k = \frac{y_k^T s_k}{s_k^T B_k s_k} \tag{5.26}$$

is often recommended and has been tested in practice. The original motivation for self-scaling methods arises from the analysis of quadratic objective functions, and the main results also assume that exact line searches are performed. Disappointing numerical results were reported by several researches (see for example Shanno and Phua (1980)), and these results are explained by the analysis of Nocedal and Yuan (1991). They show that the method (5.25)–(5.26), using a line search that satisfies the Wolfe conditions, produces good search directions which allow superlinear convergence to take place if, in addition, the size of the step is correctly chosen. It turns out, however, that to estimate this step-size, it is normally necessary to use an extra function evaluation, which makes the approach inefficient. Nocedal and Yuan give an example in which the step-sizes needed for superlinear convergence alternate between $\frac{1}{2}$ and 2, and note that this type of behaviour can be observed in practice and is responsible for the relative inefficiency of the self-scaling method compared to the unscaled BFGS method.

For these reasons, the Oren–Luenberger scaling is now commonly applied only after the first iteration of a variable metric method. A quite different, and perhaps more promising strategy has been proposed by Powell (1987), and further developed by Lalee and Nocedal (1991) and Siegel (1991). Powell's idea is to work with the factorization

$$H_k = Z_k Z_k^T \tag{5.27}$$

of the inverse Hessian approximation H_k. This factorization has been used by Goldfarb and Idnani (1983) for quadratic programming and has the advantage that it can be used easily when inequality constraints are present. Powell shows that by introducing an orthogonal rotation that makes the first column of Z_k a multiple of s_k, the BFGS update of H_k can be obtained via a simple update to Z_k:

$$z_i^* = \begin{cases} s_k/\sqrt{s_k^T y_k} & i = 1, \\ z_i - (\frac{y_k^T z_i}{s_k^T y_k})s_k & i = 2, \ldots, n, \end{cases}$$

where z_i and z_i^* are the ith columns of Z_k and Z_{k+1} respectively. $Z_{k+1}Z_{k+1}^T$ gives H_{k+1}.

Note that the curvature information gathered during the most recent information is contained in the first column of Z_{k+1}, and that all other columns are obtained by a simple operation. Since in the BFGS update we wish to reduce the possibility of having an over-estimate of the Hessian, or equivalently an under-estimate of the inverse Hessian, Powell proposes to increase all columns of Z_{k+1} so that their norms are at least equal to a parameter which depends on the norm of the first column.

Lalee and Nocedal (1991) extend Powell's idea to allow scaling down columns that are too large, as well as scaling up those that are too small. They give conditions on the scaling parameters in order for the algorithm to be globally and superlinearly convergent. Siegel (1991) proposes a slightly different scaling strategy. At every iteration, he only scales up the last l columns of the matrix Z_k, where l is a nonincreasing integer. The parameter l does not change if the search direction d_k is in the span of the first $n - l$ columns of Z_k, or close to it. Otherwise, l is decreased by 1. These column scaling methods appear to work very well in practice, but there is not enough data yet to draw any firm conclusions.

6. The symmetric rank-one method

One of the most interesting recent developments in unconstrained optimization has been the resurgence of the symmetric rank-one method (SR1). Several new theoretical and experimental studies have reversed the general perception of this method. Instead of being considered 'fatally flawed', the SR1 method is now regarded by many researchers as a serious contender with the BFGS method for unconstrained problems, and as the most suitable quasi-Newton method for applications in which positive definite updates cannot be generated, such as constrained problems. The SR1 method remains controversial, and it is difficult to predict if the enthusiasm for this method is temporary, or if it will find a permanent place in optimization subroutine libraries.

The symmetric rank-one update is given by

$$B_{k+1} = B_k + \frac{(y_k - B_k s_k)(y_k - B_k s_k)^T}{s_k^T(y_k - B_k s_k)}. \tag{6.1}$$

It was first discovered by Davidon (1959) in his seminal paper on quasi-

Newton methods, and re-discovered by several authors. The SR1 method can be derived by posing the following simple problem. Given a symmetric matrix B_k and the vectors s_k and y_k, find a new symmetric matrix B_{k+1} such that $B_{k+1} - B_k$ has rank one, and such that

$$B_{k+1}s_k = y_k.$$

It is easy to see that if $(y_k - B_k s_k)^T s_k \neq 0$, then the unique solution is (6.1), whereas if $y_k = B_k s_k$ then the solution is $B_{k+1} = B_k$. However if $(y_k - B_k s_k)^T s_k = 0$ and $y_k \neq B_k s_k$, there is no solution to the problem, and this case clouds what is otherwise a clean and simple argument. To prevent the method from failing, one can simply set $B_{k+1} = B_k$ when the denominator in (6.1) is close to zero, but this could prevent the method from converging rapidly.

It was noted early on that the SR1 method has some very interesting properties, provided it does not break down. For example Fiacco and Mc-Cormick (1968) show that the SR1 method without line searches finds the solution of a strongly convex quadratic function in at most $n+1$ steps, if the search directions are linearly independent and if the denominator in (6.1) is always nonzero. In this case B_{n+1} equals the Hessian of the quadratic function. It is significant that this result does not require exact line searches, as is the case for the BFGS and DFP methods.

However, the fact that the denominator in (6.1) can vanish, introduces numerical instabilities and a possible breakdown of the method. Since this can happen even for quadratic functions, and since (6.1) does not always generate positive definite matrices, which complicates a line search implementation, the SR1 method fell out of favour. It was rarely used in practice, even though very good computational results had been obtained with safeguarded implementations (Dixon, 1972). The feeling in the early 1970s was that the method has some intrinsic weaknesses, and that the BFGS method was clearly preferable.

The revival of the SR1 method began, interestingly enough, during the development of the partitioned quasi-Newton method of Griewank and Toint (1982c). As we will discuss in the next section, the curvature condition $s^T y > 0$ cannot always be expected to hold for all element functions, and therefore the BFGS method cannot always be applied. Therefore the implementation of the partitioned quasi-Newton method by Toint (Harwell routine VE08) uses the SR1 update when BFGS cannot be applied. This happens often; in particular after an SR1 update has been applied all subsequent updates are performed by means of SR1. The partitioned quasi-Newton method performs very well in practice, giving a first indication of the success of the SR1 method – but this work drew less attention than it deserved.

The SR1 method came to the limelight with a sequence of papers by Conn *et al.* (1988a,b, 1991). The first two papers deal with trust region methods for bound constrained problems, and report better results for SR1 than for BFGS. The authors speculate that the success of SR1 may be due to its superior ability to approximate the Hessian matrix at the solution. This is investigated in the third paper, in which the following result is established.

Theorem 6.1 Suppose that f is twice continuously differentiable, and that its Hessian is bounded and Lipschitz continuous. Let $\{x_k\}$ be the iterates generated by the SR1 method and suppose that $x_k \to x_*$ for some $x_* \in \mathbb{R}^n$. Suppose in addition that, for all k,

$$|s_k^T(y_k - B_k s_k)| \geq r\|s_k\| \|y_k - B_k s_k\|, \tag{6.2}$$

for some $r \in (0, 1)$, and that the the steps s_k are uniformly linearly independent. Then

$$\lim_{k \to \infty} \|B_k - \nabla^2 f(x_*)\| = 0.$$

Condition (6.2) is often used in practice to ensure that the SR1 update is well behaved: if it is violated then the update is skipped. Conn *et al.* (1991) report that the assumption of uniform linear independence of the search directions holds in most of their runs, and that the Hessian approximations generated by the SR1 method are often more accurate than those generated by BFGS or DFP.

Osborne and Sun (1988) propose a modification in which the Hessian approximation is scaled before the SR1 update is applied. They analyse this method and report good numerical results. In an interesting recent paper, Khalfan *et al.* (1991) make further contributions to the theory of the SR1 method, and present numerical results that, to some extent, conflict with those of Conn *et al.* (1991). They consider both a line search and a trust region implementation and observe that, for the problems they tested, the Hessian approximations generated by the SR1 method are on the average only slightly more accurate than those produced by the BFGS method. They report that in about one third of their problems neither method produces close approximations to the Hessians at the solution.

These results suggest that the assumptions of Theorem 6.1 may not always be satisfied in practice . Therefore Khalfan *et al.* study whether the steps generated by the SR1 method are uniformly linearly independent and find that this is often not the case. They conclude that the efficiency of the SR1 method is unlikely to be due to the properties given in Theorem 6.1, and pursue an analysis that is not based on the linear independence assumption. They prove several results which we now describe.

The first result is related to the Dennis–Moré condition for superlinear convergence, and assumes that unit step-lengths are taken. It states that if

x_* is a minimizer such that $\nabla^2 f(x_*)$ is positive definite, and if

$$e_k \equiv \|x_k - x^*\|$$

and

$$\frac{\|(B_k - \nabla^2 f(x_*))s_k\|}{\|s_k\|},$$

are sufficiently small, then

$$\|x_k + s_k - x_*\| \leq c_1 \left[\frac{\|(B_k - \nabla^2 f(x_*))s_k\|}{\|s_k\|} e_k + c_2 e_k^2 \right]$$

where c_1 and c_2 are constants.

This bound suggests that some kind of quadratic rate is possible. To establish this, however, Khalfan et al. must assume that the matrices $\{B_k\}$ are positive definite and bounded. This appears, at first, to be a very unrealistic assumption, but the authors note that this is very often the case in their numerical tests. We now formally state this second result on the SR1 method.

Theorem 6.2 Suppose that the iterates generated by the SR1 method converge to x_* – a minimizer such that $\nabla^2 f(x_*)$ is positive definite. Assume that for all $k \geq 0$ the condition (6.2) is satisfied and that the matrices B_k are positive definite and uniformly bounded above in norm. Then the rate of convergence is $2n$-step q-quadratic, i.e.

$$\limsup_{k \to \infty} \frac{e_{k+2n}}{e_k^2} \leq \infty.$$

These new results are, of course, not as strong as the global convergence results described for the BFGS method, but one should keep in mind that the renewed interest in the SR1 method is very recent. Therefore substantial advances in this area can be expected.

7. Methods for large problems

Every function f with a sparse Hessian is partially separable, i.e. it can be written in the form

$$f(x) = \sum_{i=1}^{ne} f_i(x), \tag{7.1}$$

where each of the ne element functions f_i depends only on a few variables. This statement is proved by Griewank and Toint (1982a), and provides the foundation for their partitioned quasi-Newton method for large-scale optimization. The idea behind this method is to exploit the partially separable structure (7.1) and update an approximation B_k^i to the Hessian of each element function f_i. These matrices, which are often very accurate, can be assembled to define an approximation B_k to the Hessian of f. There is one complication: even if $\nabla^2 f(x_*)$ is positive definite, some of the element functions may be concave, so that the BFGS method cannot always be used. In this case Griewank and Toint use the SR1 update formula, and implement safeguards that skip the update if it is suspect.

The search direction of the partitioned quasi-Newton method, as implemented by Toint (1983), is determined by solving the system

$$\left(\sum_{i=1}^{ne} B_k^i \right) d_k = -g_k \tag{7.2}$$

inside a trust region, using a truncated conjugate gradient iteration. If a direction of negative curvature is detected, the conjugate gradient iteration is terminated, and d_k is set to this direction of negative curvature. After this, a line search is performed along d_k. This method is described and analysed by Griewank and Toint (1982b,c, 1984); the implementation just outlined corresponds to the Harwell routine VE08.

The partitioned quasi-Newton method performs very well in practice, and represents one of the major algorithmic advances in nonlinear optimization. We should note that many practical problems are directly formulated in the form (7.1), and that many other problems can be recast in that form. Thus the partitioned quasi-Newton method is of wide applicability.

To establish global convergence results, similar to those for the BFGS method on convex problems, it is necessary to assume that all the element functions f_i are convex. Under this assumption Griewank (1991) shows that the partitioned quasi-Newton method is globally convergent, even if the system (7.2) is solved inexactly. Griewank also relaxes the smoothness conditions on the gradients of the element functions f_i, and establishes rate of convergence results under the assumption that these gradients are only Lipschitzian, rather than differentiable. Griewank's analysis completely describes the behaviour of the partitioned quasi-Newton method in the convex case, and strengthens earlier work by Toint (1986b).

A very different approach for solving large problems ignores the structure of the problem, and uses the information of the last few iterations to define a variable metric approximation of the Hessian. This, so-called limited memory BFGS method, has proved to be very useful for solving certain large unstructured problems, and is in fact competitive with the partitioned quasi-Newton method on partially separable problems in which the number of variables entering into the element functions f_i exceeds 5 or 6 (Liu and Nocedal, 1989).

The limited memory BFGS method is very similar to the standard BFGS method – the only difference is in the matrix update. Instead of storing the matrices H_k that approximate the inverse Hessian, one stores a certain number, say m, of pairs $\{s_i, y_i\}$ that define them implicitly. The product $H_k g_k$, which defines the search direction, is obtained by performing a sequence of inner products involving g_k and the m most recent vector pairs $\{s_i, y_i\}$. This is done efficiently by means of a recursive formula (Nocedal, 1980). After computing the new iterate, we delete the oldest pair from the set $\{s_i, y_i\}$ and replace it by the newest one. Thus the algorithm always keeps the m most recent pairs $\{s_i, y_i\}$ to define the iteration matrix. It has been observed that scaling can be highly beneficial for large problems and several strategies for doing this have been studied by Gilbert and Lemaréchal (1989).

The limited memory BFGS method is suitable for large scale problems because it has been observed in practice that small values of m (say $m \in [3, 7]$) give satisfactory results. It is not understood why this method is as fast as the standard BFGS method on many problems. Another interesting open question is how to design a strategy for selecting the most useful corrections pairs – not simply the most recent ones –to improve the performance of the method.

Since the Dennis–Moré condition (3.16) cannot possibly hold for the limited memory BFGS method, its rate of convergence must be linear. Liu and Nocedal (1989) prove that the limited memory BFGS method is globally and linearly convergent on convex problems for any starting point, and for several useful scaling strategies. It is interesting to note that, as implemented by Liu and Nocedal, the method does not possess quadratic termination. A different limited memory method, that combines cycles of BFGS and conjugate gradient directions has been developed by Buckley and LeNir (1983).

Newton's method is, of course, the best method for solving many types of problems. Both line search and trust region implementations have been developed for the large-scale case; see Steihaug (1983), Nash (1985), O'Leary (1982) and Toint (1986a). The convergence properties of implementations of Newton's method in which the linear system

$$\nabla^2 f(x_k) d_k = -g_k \tag{7.3}$$

is solved inaccurately were first considered by Dembo *et al.* (1982) and by Bank and Rose (1981). Several interesting recent papers generalizing this work, and focusing on specific methods for solving the linear system (7.3), include Brown and Saad (1989, 1990), El Hallabi and Tapia (1989), Martínez (1990) and Eisenstat and Walker (1991). Nonmonotone Newton methods, i.e. methods in which function values are allowed to increase at some iterations, have been analysed by Grippo *et al.* (1990a,b); the numerical results appear to be very satisfactory. Nonmonotone methods may prove to be very useful for solving highly nonlinear problems.

8. Remarks on other methods

I have concentrated on recent theoretical studies on methods for solving general unconstrained minimization problems. Due to space limitations I have not discussed the solution of systems of nonlinear equations or nonlinear least squares. The Nelder–Meade method is known to fail, so that establishing a global convergence result for it is not possible. Recently there has been research on modifications of the Nelder–Meade method to improve its performance, and it is possible to establish global convergence for some of them. For a description of this work see Torczon (1991).

As mentioned earlier, I have not reviewed trust region methods because most of their theoretical studies (for unconstrained problems) are not recent and are reviewed by Moré and Sorensen (1984). Nevertheless, I would like to briefly contrast their properties with those of line search methods.

Trust region methods do not require the Hessian approximations B_k to be positive definite. In fact, very little is required to establish global convergence: it is only necessary to assume that the norm of the matrices $\|B_k\|$ does not increase at a rate that is faster than linear (Powell, 1984b). In contrast, for line search methods one needs to ensure that the condition number of the Hessian approximations $\|B_k\|$ does not grow too rapidly. This requires control on both the largest and smallest eigenvalues of B_k, making the analysis more complex than for trust region methods. It is also possible to show that for trust region methods the sequence of iterates always has an accumulation point at which the gradient is zero and the Hessian is positive semi-definite. This is better than the result $\liminf \|g_k\| = 0$ which is the most that can be proved for line search methods.

Thus the theory of trust region methods has several advantages over that of line search methods, but both approaches seem to perform equally well in practice. Line search methods are more commonly used because they have been known for many years and because they can be simpler to implement. At present, line search and trust region methods coexist, and it is difficult to predict if one of these two approaches will become dominant. This will depend on the theoretical and algorithmic advances that the future has in store.

Acknowledgements

I am very grateful to Marucha Lalee, Christy Hartung and Peihuang Lu for their help in the preparation of this article. I would like to acknowledge support from the Department of Energy, grant DE-FG02-87ER25047-A001, and the National Science Foundation, grant CCR-9101359.

REFERENCES

H. Akaike (1959), 'On a successive transformation of probability distribution and its application to the analysis of the optimum gradient method', *Ann. Inst. Statist. Math.* **11**, 1–17.

M. Al-Baali (1985), 'Descent property and global convergence of the Fletcher–Reeves method with inexact line search', *IMA J. Numer. Anal.* **5**, 121–124.

M. Al-Baali (1990), 'Variational Quasi-Newton Methods for Unconstrained Optimization', Technical Report, Department of Mathematics, University of Damascus (Syria).

R.E. Bank and D.J. Rose (1981) 'Global approximate Newton methods', *Numer. Math.* **37**, 279–295.

P. Baptist and J. Stoer (1977), 'On the relation between quadratic termination and convergence properties of minimization algorithms, Part II, Applications', *Numer. Math.* **28**, 367–392.

E.M.L. Beale (1972), 'A derivation of conjugate gradients', in *Numerical Methods for Nonlinear Optimization* (F.A. Lootsma, ed.), Academic Press (New York).

P.N. Brown and Y. Saad (1989), 'Globally convergent techniques in nonlinear Newton–Krylov algorithms', Technical Report UCRL-102434, Lawrence Livermore National Laboratory.

P.N. Brown and Y. Saad (1990), 'Hybrid Krylov methods for nonlinear systems of equations', *SIAM J. Sci. Stat. Comput.* **11**, 450–481.

G.G. Broyden, J.E. Dennis and J.J. Moré (1973), 'On the local and superlinear convergence of quasi-Newton methods', *J. Inst. Math. Applics* **12**, 223–246.

A. Buckley and A. LeNir (1983), 'QN-like variable storage conjugate gradients', *Math. Program.* **27**, 155–175.

W. Burmeister (1973), 'Die Konvergenzordnung des Fletcher–Powell Algorithmus', *Z. Angew. Math. Mech.* **53**, 693–699.

R.H. Byrd and J. Nocedal (1989), 'A tool for the analysis of quasi-Newton methods with application to unconstrained minimization', *SIAM J. Numer. Anal.* **26**, 727–739.

R. Byrd, D. Liu and J. Nocedal (1990), 'On the Behavior of Broyden's Class of Quasi-Newton Methods', Technical Report NAM 01, Northwestern University, EECS Dept.

R.H. Byrd, J. Nocedal and Y. Yuan (1987), 'Global convergence of a class of quasi-Newton methods on convex problems', *SIAM J. Numer. Anal.* **24**, 1171–1190.

R.H. Byrd, R.A. Tapia and Y. Zhang (1990), 'An SQP Augmented Lagrangian Bfgs Algorithm for Constrained Optimization', Technical Report, University of Colorado (Boulder).

A. Cohen (1972), 'Rate of convergence of several conjugate gradient algorithms', *SIAM J. Numer. Anal.* **9**, 248–259.

A.R. Conn, N.I.M. Gould, and Ph.L. Toint (1988a), 'Global convergence of a class of trust region algorithms for optmization with simple bounds', *SIAM J. Numer. Anal.* **25**, 433–460.

A.R. Conn, N.I.M. Gould, and Ph.L. Toint (1988b), 'Testing a class of methods for solving minimization problems with simple bounds on the variables', *Math. Comput.* **50**, 399–430.

A.R. Conn, N.I.M. Gould, and Ph.L. Toint (1991), 'Convergence of quasi-Newton matrices generated by the symmetric rank one update', *Math. Prog.* **2**, 177–195.

H.P. Crowder and P. Wolfe (1972), 'Linear convergence of the conjugate gradient method', *IBM J. Res. Dev.* **16**, 431–433.

W.C. Davidon (1959), 'Variable metric methods for minimization', Argonne National Lab Report (Argonne, IL).

W.C. Davidon (1975), 'Optimally conditioned optimization algorithms without line searches', *Math. Prog.* **9**, 1-30.

R.S. Dembo, S.C. Eisenstat, and T. Steihaug (1982), 'Inexact Newton methods', *SIAM J. Numer. Anal.* **19**, 400–408.

J.E. Dennis, Jr and J.J. Moré (1974), 'A characterization of superlinear convergence and its application to quasi-Newton methods', *Math. Comp.* **28**, 549–560.

J.E. Dennis, Jr and J.J. Moré (1977), 'Quasi-Newton methods, motivation and theory', *SIAM Review* **19**, 46–89.

J.E. Dennis, Jr and R.B. Schnabel (1983), *Numerical Methods for Unconstrained Optimization and Nonlinear Equations*, Prentice-Hall (Englewood Cliffs, NJ).

J.E. Dennis, Jr and R.B. Schnabel (1987), 'A view of unconstrained optimization', Technical Report CU-CS-376-87, University of Colorado (Boulder), to appear in: *Handbooks in Operations Research and Management Science vol. 1: Optimization* (G.L. Nemhauser, A.H.G. Rinnooy Kan, and M.J. Todd, eds), North-Holland (Amsterdam).

J.E. Dennis, Jr and H.F. Walker (1981), 'Convergence theorems for least-change secant update methods', *SIAM J. Numer. Anal.* **18**, 949–987; **19**, 443.

J.E. Dennis, Jr and H. Wolkowicz (1991), 'Sizing and least change secant methods', Technical Report CORR 90-02, Department of Combinatorics and Optimization, University of Waterloo, Waterloo, Canada.

L.C.W. Dixon (1972), 'The choice of step length, a crucial factor in the performance of variable metric algorithms', in *Numerical Methods for Nonlinear Optimization* (F.A. Lootsma, ed.), Academic Press (London).

S. Eisenstat and H. Walker (1990), 'Globally convergent inexact Newton methods', Yale Technical Report.

M. El Hallabi and R.A. Tapia (1989), 'A global convergence theory for arbitrary norm trust-region methods for nonlinear equations', Technical Report TR87-25, Rice University, Department of Mathematical Sciences.

A.V. Fiacco and G.P. McCormick (1968), *Nonlinear Programmming* , John Wiley & Sons (New York).

R. Fletcher (1987), *Practical Methods of Optimization vol. 1: Unconstrained Optimization* John Wiley & Sons (New York).

R. Fletcher (1991), 'A new variational result for quasi-Newton formulae', *SIAM J. Optimization* **1**(1), 18–21.

R. Fletcher and M.J.D. Powell (1963), 'A rapidly convergent descent method for minimization', *Comput. J.* **6**, 163–168.

R. Fletcher and C. Reeves (1964), 'Function minimization by conjugate gradients', *Comput. J.* **7**, 149–154.

D. Gay (1983), 'Subroutines for unconstrained minimization using a model/ trust-region approach', *ACM Trans. Math. Soft.* **9**(4), 503–524.

J.C. Gilbert and C. Lemaréchal (1989), 'Some numerical experiments with variable storage quasi-Newton algorithms', *Math. Program.* **45**, 407–436.

J.C. Gilbert and J. Nocedal (1990), 'Global convergence properties of conjugate gradient methods for optimization', Rapport de Recherche, INRIA (Paris).

P.E. Gill and W. Murray (1979), 'Conjugate-gradient methods for large-scale nonlinear optimization', Technical report SOL 79-15, Dept. of Operations Research, Stanford University.

P. E. Gill, W. Murray and M. H. Wright (1981), *Practical Optimization*, Academic Press (London).

D. Goldfarb, and A. Idnani (1983), 'A numerically stable dual method for solving strictly convex quadric programs', *Math. Prog.* **27**, 1–33.

A. Griewank (1991), 'The global convergence of partitioned BFGS on problems with convex decompositions and Lipschitzian gradients', *Math. Prog.* **50**, 141–175.

A. Griewank and Ph.L. Toint (1982a), 'On the unconstrained optimization of partially separable objective functions', in *Nonlinear Optimization 1981* (M.J.D. Powell, ed.), Academic Press (London), 301–312.

A. Griewank and Ph.L. Toint (1982b), 'Local convergence analysis of partitioned quasi-Newton updates', *Numer. Math.* **39**, 429–448.

A. Griewank and Ph.L. Toint (1982c), 'Partitioned variable metric updates for large structured optimization problems', *Numer. Math.* **39**, 119–137.

A. Griewank and Ph.L. Toint (1984), 'Numerical experiments with partially separable optimization problems', in *Numerical Analysis: Proceedings Dundee 1983 (Lecture Notes in Mathematics 1066)* (D.F. Griffiths, ed.), Springer Verlag, (Berlin), 203–220.

L. Grippo, F. Lampariello and S. Lucidi (1990a), ' A quasi-discrete Newton algorithm with a nonmonote stabilization technique', *J. Optim. Theory Appl.* **64**, 485–500.

L. Grippo, F. Lampariello and S. Lucidi (1990b), ' A class of nonmonotone stabilization methods in unconstrained optimization', Technical Report R-290, Consiglio Nazionale delle Ricerche.

O. Güler (1989), 'Optimal algorithms for smooth convex programming', Working Paper Series No. 89-17, Department of Management Sciences, The University of Iowa.

R.W. Hamming (1971), *Introduction to Applied Numerical Analysis* , McGraw-Hill (New York).

Y.F. Hu and C. Storey (1991), 'On optimally and near-optimally conditioned quasi-Newton updates', Technical Report A141, Department of Mathematical Sciences, Loughborough University of Technology (Leicestershire).

H. Khalfan, R.H. Byrd, and R.B. Schnabel (1990), 'A theoretical and experimental study of the symmetric rank one update', Technical Report CU-CS-489-90, University of Colorado (Boulder).

M. Lalee and J. Nocedal (1991), 'Automatic column scaling strategies for quasi-Newton methods', Report NAM 04, EECS Department, Northwestern University (Evanston, IL).

C. Lemaréchal (1981), 'A view of line searches', in *Optimization and Optimal Control (Lecture Notes in Control and Information Science 30)* (A. Auslander, W. Oettli and J. Stoer, eds) Springer Verlag (Berlin) 59–78.

D.C. Liu and J. Nocedal (1989), 'On the limited memory BFGS method for large scale optimization', *Math. Program.* **45**, 503–528.

D.G. Luenberger (1984), *Linear and Nonlinear Programming*, 2nd edition, Addison-Wesley (Reading, MA).

L. Lukšan (1991a), 'Computational experience with improved conjugate gradient methods for unconstrained minimization', Technical Report 488, Institute of Computer and Information Sciences, Czechoslovak Academy of Sciences (Prague).

L. Lukšan (1991b), 'On variationally derived scaling and preconvex variable metric updates', Technical Report 496, Institute of Computer and Information Sciences, Czechoslovak Academy of Sciences (Prague).

J.M. Martínez (1990), 'Local convergence theory of inexact Newton methods based on structured least change updates', *Math. Comp.* **55**, 143–168.

J.M. Martínez (1991), 'On the relation between two local convergence theories of least change update methods', Technical Report IMECC-UNICAMP.

J. Moré (1983), 'Recent developments in algorithms and software for trust region methods', in *Mathematical Programming, The State of the Art* (A. Bachem, M. Grotschel and G. Korte, eds), Springer-Verlag (Berlin), 256–287.

J. Moré and D.C. Sorensen (1984), 'Newton's method', in *Studies in Numerical Analysis* (G.H. Golub, ed.), The Mathematical Association of America (Providence RI), 29–82.

J. Moré, and D.J. Thuente (1990) 'On line search algorithms with guaranteed sufficient decrease', Mathematics and Computer Science Division Preprint MCS-P153-0590, Argonne National Laboratory (Argonne, IL).

S.G. Nash (1985), 'Preconditioning of truncated-Newton methods', *SIAM J. Sci. Stat. Comput.* **6**, 599–616.

J.L. Nazareth and R.B. Mifflin (1991), 'The least prior deviation quasi-Newton update', Technical Report, Department of Pure and Applied Mathematics, Washington State University.

A.S. Nemirovsky and D. B. Yudin (1983), *Problem Complexity and Method Efficiency*, Wiley (New York).

Y.E. Nesterov (1983), 'A method of solving a convex programming problem with convergence rate $O(1/k^2)$', *Sov. Math. Dokl.* **27**, 372–376.

Y.E. Nesterov (1988), 'On an approach to the construction of optimal methods of minimization of smooth convex functions', *Ekonom. i Matem.* **24**, 509–517.

J. Nocedal (1980), 'Updating quasi-Newton matrices with limited storage', *Math. Comput.* **35**, 773–782.

J. Nocedal (1990), 'The performance of several algorithms for large scal unconstrained optimization', in *Large-Scale Numerical Optimization* (T.F. Coleman and Y. Li, eds), SIAM (Philadelphia), 138–151.

J. Nocedal and Y. Yuan (1991), 'Analysis of a self-scaling quasi-Newton method', Technical Report NAM-02, Department of Electrical Engineering and Computer Science, Northwestern University (Evanston, IL).

D.P. O'Leary (1982), 'A discrete Newton algorithm for minimizing a function of many variables', *Math. Prog.* **23**, 20–33.

S.S. Oren (1982), 'Perspectives on self-scaling variable metric algorithms', *J. Opt. Theory Appl.* **37**, 137–147.

S.S. Oren and D.G. Luenberger (1974), 'Self-scaling variable metric(SSVM) Algoriths I: Criteria and sufficient conditions for scaling a class of algorithms', *Management Sci.* **20**, 845–862.

J.M. Ortega and W.C.Rheinboldt (1970), *Iterative Solution of Nonlinear Equations in Several Variables*, Academic Press (New York).

M.R. Osborne and L.P. Sun (1988), 'A new approach to the symmetric rank-one updating algorithm', Report NMO/01, Department of Statistics, IAS, Australian National University.

A. Ostrowski (1966), *Solution of Equations and Systems of Equations* , second edition, Academic Press (New York).

J.D. Pearson (1969), 'Variable metric methods of minimization', *Comput. J.* **12**, 171–178.

E. Polak and G. Ribière (1969), 'Note sur la convergence de methodes de directions conjugées', *Rev. Française Informat Recherche Operationelle, 3e Année* **16**, 35–43.

M.J.D. Powell (1971), 'On the convergence of the variable metric algorithm', *J. Inst. Math. Applics* **7**, 21–36.

M.J.D. Powell (1972), 'Some properties of the variable metric method', in *Numerical Methods for Nonlinear Optimization* (F.A. Lootsma, ed.), Academic Press (London).

M.J.D. Powell (1976a), 'Some global convergence properties of a variable metric algorithm for minimization without exact line searches', in *Nonlinear Programming, SIAM-AMS Proceedings, Vol. IX* , (R.W. Cottle and C.E. Lemke, eds), SIAM Publications (Philadelphia).

M.J.D. Powell (1976b), 'Some convergence properties of the conjugate gradient method', *Math. Program.* **11**, 42–49.

M.J.D. Powell (1977), 'Restart procedures of the conjugate gradient method', *Math. Program.* **2**, 241–254J.

M.J.D. Powell (1983), 'On the rate of convergence of variable metric algorithms for unconstrained optimization', Report DAMTP 1983/NA7, Department of Applied Mathematics and Theoretical Physics, University of Cambridge (Cambridge).

M.J.D. Powell (1984a), 'Nonconvex minimization calculations and the conjugate gradient method', in *Lecture Notes in Mathematics* vol. 1066, Springer-Verlag (Berlin), 122–141.

M.J.D. Powell (1984b), ' On the global convergence of trust region algorithms for unconstrained minimization', *Math. Program.* **29**, 297–303.

M.J.D. Powell (1985), 'Convergence properties of algorithms for nonlinear optimization', Report DAMTP 1985/NA1, Department of Applied Mathematics and Theoretical Physics, University of Cambridge (Cambridge).

M.J.D. Powell (1986), 'How bad are the BFGS and DFP methods when the objective function is quadratic?', *Math. Prog.* **34**, 34–47.

M.J.D. Powell (1987), 'Update conjugate directions by the BFGS formula', *Math. Prog.* **38**, 29–46.

D.G. Pu and W.C. Yu (1988), 'On the convergence property of the DFP algorithm', *J. Qüfu Normol University* **14**(3), 63–69.

K. Ritter (1980), 'On the rate of superlinear convergence of a class of variable metric methods', *Numer. Math.* **35**, 293–313.

R.B. Schnabel (1978), 'Optimal conditioning in the convex class of rank two updates', *Math. Prog.* **15**, 247–260.

R.B. Schnabel (1989), 'Sequential and parallel methods for unconstrained optimization', in *Mathematical Programming, Recent Developments and Applications* (M. Iri and K. Tanabe, eds), Kluwer (Deventer), 227–261.

G. Schuller (1974), 'On the order of convergence of certain quasi-Newton methods', *Numer. Math.* **23**, 181-192.

D.F. Shanno (1978a), 'On the convergence of a new conjugate gradient algorithm', *SIAM J. Numer. Anal.* **15**, 1247–1257.

D.F. Shanno (1978b), 'Conjugate gradient methods with inexact searches', *Math. Operations Res.* **3**, 244–256.

D.F. Shanno and K.H. Phua (1980), 'Remark on algorithm 500: minimization of unconstrained multivariate functions', *ACM Trans. on Math. Software* **6**, 618–622.

D. Siegel (1991), 'Modifying the BFGS update by a new column scaling technique', Technical Report DAMTP 1991/NA5, Department of Applied Mathematics and Theoretical Physics, University of Cambridge.

T. Steihaug (1983), 'The conjugate gradient method and trust regions in large scale optimization', *SIAM J. Numer. Anal.* **20**, 626–637.

J. Stoer (1977), 'On the relation between quadratic termination and convergence properties of minimization algorithms', *Numer. Math.* **28**, 343–366.

Ph.L. Toint (1983), 'VE08AD, a routine for partially separable optimization with bounded variables', Harwell Subroutine Library, AERE (UK).

Ph.L. Toint (1986a), 'A view of nonlinear optimization in a large number of variables', Technical Report 86/16, Facultés Universitaires de Namur.

Ph.L. Toint (1986b), 'Global convergence of the partioned BFGS algorithm for convex partially separable optimization', *Math. Prog.* **36**, 290–306.

V. Torczon (1991), 'On the convergence of the multidimensional search algorithm', *SIAM J. Optimization* **1**(1), 123–145.

D. Touati-Ahmed and C. Storey (1990), 'Efficient hybrid conjugate gradient techniques', *J. Optimization Theory Appl.* **64**, 379–397.

W. Warth and J. Werner (1977), 'Effiziente Schrittweitenfunktionen bei unrestringierten Optimierungsaufgaben', *Computing* **19**, 1, 59–72.

J. Werner (1978), 'Uber die globale konvergenz von Variable-Metric Verfahren mit nichtexakter Schrittweitenbestimmung', *Numer. Math.* **31**, 321–334.

J. Werner (1989), 'Global convergence of quasi-Newton methods with practical line searches', NAM-Bericht 67, Institut für Numerische und Angewandte Mathematik der Universität Göttingen.

J.H. Wilkinson (1965), *The Algebraic Eigenvalue Problem*, Oxford University Press (London).

P. Wolfe (1969), 'Convergence conditions for ascent methods', *SIAM Review* **11**, 226-235.

P. Wolfe (1971), 'Convergence conditions for ascent methods. II: Some corrections', *SIAM Review* **13**, 185–188.

Y. Yuan (1991), 'A modified BFGS algorithm for unconstrained optimization', *IMA J. Numer. Anal.* **11**(3), 325–332.

Y. Yuan and R. Byrd (1991), 'Nonquasi-Newton updates for unconstrained optimization', Technical Report, Department of Computer Science, University of Colorado (Boulder).

Y. Zhang and R.P. Tewarson (1988), 'Quasi-Newton algorithms with updates from the pre-convex part of Broyden's family', *IMA J. Numer. Anal.* **8**, 487–509.

G. Zoutendijk (1970), 'Nonlinear Programming, Computational Methods', in *Integer and Nonlinear Programming* (J. Abadie, ed.), North-Holland (Amsterdam), 37–86.

Acta Numerica (1991), *pp.* 243–286

Symplectic integrators for Hamiltonian problems: an overview

J. M. Sanz-Serna

Departamento de Matemática Aplicada y Computación
Facultad de Ciencias
Universidad de Valladolid
Valladolid, Spain
E-mail: sanzserna@cpd.uva.es

CONTENTS

1. Introduction

In the sciences, situations where dissipation is not significant may invariably be modelled by Hamiltonian systems of ordinary, or partial, differential equations. Symplectic integrators are numerical methods specifically aimed at advancing in time the solution of Hamiltonian systems. Roughly speaking, 'symplecticness' is a characteristic property possessed by the solutions of Hamiltonian problems. A numerical method is called symplectic if, when applied to Hamiltonian problems, it generates numerical solutions which inherit the property of symplecticness.

If the reader is expecting to find the definition of symplecticness in this introduction, I am sorry he is going to be disappointed. I have devoted Sections 2–4 to the task of explaining symplecticness in what I believe to be the simplest possible way. The fact that six pages are needed to define symplecticness should not be taken as implying that this notion is particularly difficult: for readers with a differential geometry background, symplecticness can be defined in one line. However, here and elsewhere in the article, I have tried to be understandable rather than brief. In particular I have tried hard to relate the concepts in a language accessible to numerical analysts. This has not always been easy, as the area of symplectic integration directly relates to both numerical analysis and to other branches of science, such as symplectic geometry, dynamical systems, classical mechanics and theoretical physics.

After the study of the notion of symplecticness in Sections 2–4, I define in Section 5 the concept of symplectic integrator. Symplectic integrators fall into two categories. Some of them are standard methods, such as Runge–Kutta or Runge–Kutta–Nyström methods, that just happen to achieve symplecticness through some balance in their coefficients. For a method of this kind to be symplectic it is necessary and sufficient that its coefficients satisfy some algebraic equations. This first category of symplectic methods is studied in Sections 6–8. A remarkable feature of the methods of this category is that, for them, an alternative formulation of the order conditions exists, whereby the order conditions are expressed in terms of unrooted rather than rooted trees.

The second category of symplectic methods consists of methods derived via a so-called generating function. Generating functions were introduced in the nineteenth century as a means for solving some problems in classical mechanics. They are at the root of the Hamilton–Jacobi method for integrating differential systems via the Hamilton–Jacobi partial differential equation. In Section 9 I present the necessary background on generating functions and in Section 10 I survey symplectic integrators based on generating functions. In Section 11, I return to the first category of symplectic methods (i.e. to

Runge–Kutta and related symplectic methods) with the goal of seeing them in the light of the Hamilton–Jacobi theory.

In Sections 12–14, I summarize the general properties of symplectic integrators. Section 15 is devoted to the practical performance of symplectic integration and the final Section 16 contains a few indications in connection with material, such as Hamiltonian partial differential equations, not covered in the main part of the paper.

The current interest in symplectic integration started with the work of, for example, Ruth (1983), Channell (1983), Menyuk (1984), Feng (1985, 1986a,b). Since then, several dozens of papers on the subject have been written. Some of these have been published in the physics literature, while others have appeared in numerical analysis journals and others are only available as manuscripts. Under these circumstances, I cannot claim to have supplied a list of references that covers all the relevant items. However I have done my best to present a fair view of the field from a numerical analyst's point of view.

Symplectic integration is a new field. As such, much of the material reported here is likely to be superseded soon by new developments. From a theoretical point of view the field has already witnessed some interesting contributions bringing together seemingly unrelated parts of mathematics such as Hamilton–Jacobi equations and graph theory. On the other hand, little has been undertaken in the construction of practical high-order methods and the design of serious symplectic software is still waiting consideration. The area of symplectic integration is one where much scope is left for newcomers. I would be glad if this paper helped in attracting some of them to the field.

2. Hamiltonian systems

2.1. Preliminaries

We start by describing the class of problems with which we shall be concerned and by introducing some notation. Let Ω be a domain (i.e. a nonempty, open, connected set) in the oriented Euclidean space $\mathbb{R}^{2d}$ of the points $(\mathbf{p}, \mathbf{q}) = (p_1, \ldots, p_d; q_1, \ldots, q_d)$. If H is a sufficiently smooth real function defined in Ω, then the Hamiltonian system of differential equations with Hamiltonian H is, by definition, given by

$$\frac{\mathrm{d}p_i}{\mathrm{d}t} = -\frac{\partial H}{\partial q_i}, \quad \frac{\mathrm{d}q_i}{\mathrm{d}t} = +\frac{\partial H}{\partial p_i}, \qquad 1 \le i \le d. \tag{2.1}$$

The integer d is called the *number of degrees of freedom* and Ω is the *phase space*. The exact amount of smoothness required of H will vary from place to place and will not be explicitly stated, but we throughout assume at least C^2 continuity, so that the right-hand side of the system (2.1) is C^1 and the standard existence and uniqueness theorems apply to the corresponding initial value problem. Sometimes, the symbol $\mathcal{S}_H$ will be used to refer to system (2.1). A good starting point for the theory of Hamiltonian problems is the textbook by Arnold (1989). MacKay and Meiss (1987) have compiled an excellent collection of important papers in Hamiltonian dynamics. For applications to celestial mechanics see Arnold (1988). More advanced results on symplectic geometry can be found in Arnold and Novikov (1990). For the early history of the work of Hamilton and Jacobi on Hamiltonian systems, see Klein (1926).

In applications to mechanics (Arnold, 1989), the $\mathbf{q}$ variables are *generalized coordinates*, the $\mathbf{p}$ variables the conjugated *generalized momenta* and H usually corresponds to the total *mechanical energy*.

Often the Hamiltonian has the special structure

$$H(\mathbf{p}, \mathbf{q}) = T(\mathbf{p}) + V(\mathbf{q}). \tag{2.2}$$

In mechanics T and V would represent the *kinetic* and *potential* energy, respectively. Hamiltonians of this form are called *separable*. A commonly occurring case has $T = \frac{1}{2}\mathbf{p}^T\mathbf{p}$, so that the Hamiltonian reads

$$H(\mathbf{p}, \mathbf{q}) = \frac{1}{2}\mathbf{p}^T\mathbf{p} + V(\mathbf{q}). \tag{2.3}$$

Of course one may also consider *nonautonomous* (time-dependent) Hamiltonians $H = H(\mathbf{p}, \mathbf{q}; t)$. By using such an H in (2.1), we obtain a nonautonomous Hamiltonian system. Most of the material that follows may easily be extended to cater for the nonautonomous case. However, for simplicity, we shall assume that, unless otherwise explicitly stated, *all Hamiltonians considered are autonomous*, i.e. time-independent.

2.2. The flow of a Hamiltonian system

If t is a real number, we denote by $\phi_{t,H}$ the flow of the system $\mathcal{S}_H$ introduced in (2.1). Recall that, by definition, $\phi_{t,H}$ is a transformation mapping Ω into itself, in such a way that for $(\mathbf{p}^0, \mathbf{q}^0)$ in Ω, $(\mathbf{p}, \mathbf{q}) = \phi_{t,H}(\mathbf{p}^0, \mathbf{q}^0)$ is the value at time t of the solution of (2.1) that at time $t = 0$ has the initial condition $(\mathbf{p}^0, \mathbf{q}^0)$ (see e.g. Section 1.4 of the contribution by Arnold and Ili'yashenko to the book by Anosov and Arnold (1988) or Chapter 1 in Guckenheimer and Holmes (1983)). Therefore, if in

$$(\mathbf{p}, \mathbf{q}) = \phi_{t,H}(\mathbf{p}^0, \mathbf{q}^0) \tag{2.4}$$

t varies and $(\mathbf{p}^0, \mathbf{q}^0)$ is seen as fixed, then we recover the solution of (2.1) with initial condition $(\mathbf{p}^0, \mathbf{q}^0)$. The key point is that we will mainly be interested in seeing t in (2.4) as a fixed parameter and $(\mathbf{p}^0, \mathbf{q}^0)$ as a variable, so that we are defining a map of Ω into itself. In fact this is not quite true. The point $\phi_{t,H}(\mathbf{p}^0, \mathbf{q}^0)$ is defined only if the solution of (2.1) with initial condition $(\mathbf{p}^0, \mathbf{q}^0)$ exists at time t, which, for given $(\mathbf{p}^0, \mathbf{q}^0)$, is not necessarily the case if $|t|$ is large: solutions may reach the boundary of Ω in a finite time and exist only for bounded intervals of time. Thus, for given $t \neq 0$, the domain of definition of $\phi_{t,H}$ may be strictly smaller than Ω.

A simple example is provided by the the *harmonic oscillator*, the Hamiltonian system with $d = 1$, $\Omega = \mathbb{R}^2$ and $H = \frac{1}{2}p_1^2 + \frac{1}{2}q_1^2$. If we use the notation p and q for the dependent variables, and identify the point (p, q) with the column vector $[p, q]^T$, the system $\mathcal{S}_H$ reads

$$\frac{d}{dt}\begin{bmatrix} p \\ q \end{bmatrix} = A \begin{bmatrix} p \\ q \end{bmatrix}, \qquad A = \begin{bmatrix} 0 & -1 \\ 1 & 0 \end{bmatrix}, \qquad (2.5)$$

and the t-flow is simply the mapping that rotates points in $\mathbb{R}^2$ by an angle of t radians around the origin:

$$\begin{bmatrix} p^0 \\ q^0 \end{bmatrix} \mapsto \exp(At)\begin{bmatrix} p^0 \\ q^0 \end{bmatrix} = \begin{bmatrix} \cos t & -\sin t \\ \sin t & \cos t \end{bmatrix}\begin{bmatrix} p^0 \\ q^0 \end{bmatrix}. \qquad (2.6)$$

For nonlinear Hamiltonians, in general, an explicit representation of the flow cannot be found in terms of elementary functions.

3. Area-preserving transformations

3.1. Preservation of area by one degree of freedom Hamiltonian flows

The idea of symplectic integration revolves around the use of symplectic transformations. In our experience, some numerical analysts find difficulties when first coming across the notion of symplecticness and tend to confuse symplectic integrators with energy-preserving integrators or with integrators whose stability function has unit modulus on the imaginary axis. It is therefore important that we devote some time to understanding symplecticness. It is best to start with the one degree of freedom case, where symplecticness is nothing but preservation of area. We then assume in this section that $d = 1$ and use the notation p and q to refer to the dependent variables p_1 and q_1 respectively.

For each real t, the flow $\phi_{t,H}$ is an *area-preserving* transformation in Ω, in the sense that, for each bounded subdomain $\Sigma \subset \Omega$ for which $\phi_{H,t}(\Sigma)$ is defined, it holds true that Σ and $\phi_{H,t}(\Sigma)$ have the same (oriented) area. To see this, it is enough, after recalling Liouville's theorem (see e.g. Section 3.5, Chapter 1 in the article by Arnold and Il'yashenko in Anosov and Arnold (1988)), to observe that the vector field $[-\partial H/\partial q, \partial H/\partial p]^T$ that features in (2.1) is divergence free because

$$\frac{\partial}{\partial p}\left(-\frac{\partial H}{\partial q}\right) + \frac{\partial}{\partial q}\left(\frac{\partial H}{\partial p}\right) = 0.$$

In the harmonic oscillator example (2.5) the area-preserving property of the flow, i.e. of the rotation (2.6), is evident.

The area-preserving property of the flow has a marked impact on the long-time behaviour of the solutions of Hamiltonian problems. Clearly asymptotically stable equilibria or limit cycles (Guckenheimer and Holmes, 1983) cannot occur: in their neighbourhoods the area would have to shrink. The Poincaré recurrence holds (Arnold, 1989): under suitable assumptions and as t increases, each point in Ω being moved by $\phi_{t,H}$ returns repeatedly to the vicinity of its initial position.

In fact, all properties specific to the Hamiltonian dynamics can be derived from the preservation of area property. This is no surprise because the area-preserving character of the flow, which was shown earlier to hold for Hamiltonian systems, actually holds *only* for Hamiltonian systems. More precisely, assume that Ω is simply connected, i.e. it has no holes, and suppose that

$$\frac{dp}{dt} = f(p,q), \qquad \frac{dq}{dt} = g(p,q), \tag{3.1}$$

is a smooth differential system whose flow is area-preserving. Then (3.1) is actually a Hamiltonian system $\mathcal{S}_H$ for a suitable H. There is nothing deep about this. By Liouville's theorem the vector field $[f,g]^T$ is divergence free, so that

$$\frac{\partial}{\partial p}(-f) = \frac{\partial}{\partial q}g.$$

But this is just the necessary and sufficient condition for the field $[g,-f]^T$ to be the gradient of a scalar function H, i.e. for (3.1) to coincide with $\mathcal{S}_H$.

If Ω is not simply connected, then systems with area-preserving flows are, in general, only locally Hamiltonian: in each ball $B \subset \Omega$ they coincide with a Hamiltonian system $\mathcal{S}_{H_B}$ but, globally, the system may not be Hamiltonian because the various H_B cannot be patched together. A typical example is given by the area-preserving system

$$\frac{dp}{dt} = \frac{p}{p^2+q^2}, \qquad \frac{dq}{dt} = \frac{q}{p^2+q^2}$$

defined in $\Omega = \mathbb{R}^2\backslash(0,0)$. In each ball in Ω the system is Hamiltonian with H given by a branch of the argument of the point (p,q). The system is not Hamiltonian because of course the argument cannot be defined as a smooth single-valued function in $\mathbb{R}^2\backslash(0,0)$.

3.2. Checking preservation of area: Jacobians

Let $(p^*,q^*) = \psi(p,q)$ be a $\mathcal{C}^1$ transformation defined in a domain Ω. According to the standard rule for changing variables in an integral, ψ is area-preserving if and only if the Jacobian determinant is identically 1:

$$\forall (p,q) \in \Omega, \quad \frac{\partial p^*}{\partial p}\frac{\partial q^*}{\partial q} - \frac{\partial p^*}{\partial q}\frac{\partial q^*}{\partial p} = 1. \tag{3.2}$$

It is a trivial exercise in matrix multiplication to check that this relationship can be rewritten as

$$\forall (p,q) \in \Omega, \quad \psi'^T J \psi' = \frac{\partial(p^*, q^*)}{\partial(p, q)}^T J \frac{\partial(p^*, q^*)}{\partial(p, q)} = J, \qquad (3.3)$$

where

$$J = \begin{bmatrix} 0 & 1 \\ -1 & 0 \end{bmatrix}$$

and $\psi' = \partial(p^*, q^*)/\partial(p, q)$ is the Jacobian matrix of the transformation. Going from (3.2) to (3.3) may appear to be just a matter of complicating things. This is not so: the matrix J is a very important character in this play. If $\mathbf{v}$ and $\mathbf{w}$ are vectors in the plane, then $\mathbf{v}^T J \mathbf{w}$ is the oriented area of the parallelogram they determine. Now, let us fix a point (p, q) in Ω and construct a parallelogram $\mathcal{P}$ having a vertex at (p, q) and having as sides two small vectors $\mathbf{v}$ and $\mathbf{w}$ (i.e. the vertices are the points (p, q), $(p, q) + \mathbf{v}$, $(p, q) + \mathbf{w}$, $(p, q) + \mathbf{v} + \mathbf{w}$). Then $\psi(\mathcal{P})$ is a parallelogram with curved sides, which can be approximated by the parallelogram $\mathcal{P}^*$ based at $\psi(p, q)$ with sides $\psi'\mathbf{v}$, $\psi'\mathbf{w}$. In fact, by the very definition of ψ', $\psi(\mathcal{P})$ and $\mathcal{P}^*$ differ in terms higher than linear in $\mathbf{v}$ and $\mathbf{w}$. Now $\mathcal{P}^*$ and $\mathcal{P}$ have the same area if and only if

$$\mathbf{v}^T \psi'^T J \psi' \mathbf{w} = \mathbf{v}^T J \mathbf{w}.$$

Clearly, the last relationship holds for all parallelograms $\mathcal{P}$ in Ω if and only if (3.3) holds. The conclusion is that (3.3) means that, at each point $(p, q) \in \Omega$, the linear transformation ψ' maps parallelograms based at (p, q) into parallelograms based at $\psi(p, q)$ without altering the oriented area.

3.3. Checking preservation of area: differential forms

Differential forms in Ω provide an alternative language with which to express the considerations made in the preceding subsection. A detailed study of the meaning and properties of differential forms is definitely outside the scope of this paper (the interested reader is referred to Arnold (1989, Chapter 7)). However the algebraic manipulations required to prove conservation of area via differential forms are as a rule easier than those required to prove conservation of area via (3.3). It is therefore advisable to comment, albeit briefly, on differential forms. Our treatment will be merely formal and we shall not explain why differential 2-forms are ways of measuring two-dimensional areas. We see a differential 1-form in Ω as a formal combination $P(p, q)\mathrm{d}p + Q(p, q)\mathrm{d}q$ where P and Q are smooth real-valued functions defined in Ω. For instance, the differentials $\mathrm{d}p^*$ and $\mathrm{d}q^*$ of the components of the transformation ψ considered earlier are differential 1-forms

$$\mathrm{d}p^* = \frac{\partial p^*}{\partial p}\mathrm{d}p + \frac{\partial p^*}{\partial q}\mathrm{d}q, \qquad \mathrm{d}q^* = \frac{\partial q^*}{\partial p}\mathrm{d}p + \frac{\partial q^*}{\partial q}\mathrm{d}q.$$

Two differential 1-forms ω and ω' give rise, via the exterior product $\wedge$, to a new entity $\omega \wedge \omega'$ called a differential 2-form. The exterior product is bilinear, so that, for instance,

$$\mathrm{d}p^*\wedge\mathrm{d}q^* = \frac{\partial p^*}{\partial p}\frac{\partial q^*}{\partial p}\mathrm{d}p\wedge\mathrm{d}p + \frac{\partial p^*}{\partial p}\frac{\partial q^*}{\partial q}\mathrm{d}p\wedge\mathrm{d}q + \frac{\partial p^*}{\partial q}\frac{\partial q^*}{\partial p}\mathrm{d}q\wedge\mathrm{d}p + \frac{\partial p^*}{\partial q}\frac{\partial q^*}{\partial q}\mathrm{d}q\wedge\mathrm{d}q.$$

The exterior product is skew symmetric. In particular, it holds that

$$\mathrm{d}p \wedge \mathrm{d}p = \mathrm{d}q \wedge \mathrm{d}q = 0, \qquad \mathrm{d}p \wedge \mathrm{d}q = -\mathrm{d}q \wedge \mathrm{d}p.$$

Thus

$$\mathrm{d}p^* \wedge \mathrm{d}q^* = \left(\frac{\partial p^*}{\partial p}\frac{\partial q^*}{\partial q} - \frac{\partial p^*}{\partial q}\frac{\partial q^*}{\partial p}\right)\mathrm{d}p \wedge \mathrm{d}q$$

and from (3.2) we see that conservation of area is equivalent to

$$\mathrm{d}p^* \wedge \mathrm{d}q^* = \mathrm{d}p \wedge \mathrm{d}q.$$

This usually provides a convenient way of checking preservation of area.

4. Symplectic transformations

4.1. Hamiltonian flows and symplectic transformations

It is now time to consider the case $d > 1$. Is there something analogous to the area that is being conserved by Hamiltonian flows? The $2d$-dimensional volume in Ω appears to be a natural candidate and indeed this volume *is* conserved. However this is not what we really want. What does the trick is to consider *two-dimensional* surfaces Σ in Ω, to find the projections Σ_i, $1 \leq i \leq d$ onto the d two-dimensional planes of the variables (p_i, q_i) and sum the two-dimensional oriented areas of these projections. This yields a number $m(\Sigma)$. It can be proved (see e.g. Arnold (1989) Section 44) that the flow of (2.1) preserves m: $m(\phi_{t,H}(\Sigma)) = m(\Sigma)$ whenever Σ is contained in the domain of $\phi_{t,H}$. Now transformations that have this preservation property are called *symplectic* or canonical, so that we have the theorem:

Theorem 4.1 For each t, the flow $\phi_{t,H}$ of a Hamiltonian system is a symplectic transformation.

Furthermore, if Ω is simply connected (i.e. each closed curved in Ω may be shrunk down to a single point without leaving Ω), then the converse is also true: an m-preserving differential system is a Hamiltonian system, see Arnold (1989, Section 40D) (once more, if Ω fails to be simply connected then preservation of m implies that the system is locally Hamiltonian). In this respect the symplecticness of the flow is the hallmark of Hamiltonian systems and once more the dynamical features that are specific to Hamiltonian problems can be traced back to the symplectic character of the flow.

4.2. Checking symplecticness

The condition (3.3) which we used to decide whether a transformation ψ in the plane was area-preserving or was otherwise generalized to read

$$\forall (\mathbf{p}, \mathbf{q}) \in \Omega, \quad \psi'^T J \psi' = \frac{\partial (\mathbf{p}^*, \mathbf{q}^*)}{\partial (\mathbf{p}, \mathbf{q})}^T J \frac{\partial (\mathbf{p}^*, \mathbf{q}^*)}{\partial (\mathbf{p}, \mathbf{q})} = J, \tag{4.1}$$

where now

$$J = \begin{bmatrix} 0_d & I_d \\ -I_d & 0_d \end{bmatrix}, \tag{4.2}$$

with I_d and 0_d denoting the unit and zero d-dimensional matrix. Note that the matrix J has the property that, for each pair $(\mathbf{v}, \mathbf{w})$ of vectors in $\mathbb{R}^{2d}$, $\mathbf{v}^T J \mathbf{w}$ represents the sum of the two-dimensional areas of the d parallelograms that result from projecting the parallelogram determined by $\mathbf{v}$ and $\mathbf{w}$ onto the planes of the variables (p_i, q_i).

Differential forms can also be used. In the present context, 1-differential forms are formal expressions of the form $P_1 dp_1 + \cdots + P_d dp_d + Q_1 dq_1 + \cdots + Q_d dq_d$, with P_i and Q_i smooth real-valued functions defined in Ω. Again, two 1-forms give rise to a 2-form via the exterior product. The transformation ψ is symplectic if and only if

$$dp_1^* \wedge dq_1^* + \cdots + dp_d^* \wedge dq_d^* = dp_1 \wedge dq_1 + \cdots + dp_d \wedge dq_d,$$

a relationship that we can rewrite more compactly as

$$d\mathbf{p}^* \wedge d\mathbf{q}^* = d\mathbf{p} \wedge d\mathbf{q}.$$

4.3. Conservation of volume

Let $\phi_{t,H}$ play the role of ψ in (4.1) and take determinants. The result is that $\det(\phi'_{t,H})$ is either $+1$ or -1. The value -1 is excluded since, by Liouville's theorem, the flow of any differential system has a Jacobian matrix with a positive determinant. Hence $\det(\phi'_{t,H}) \equiv 1$: Hamiltonian flows preserve the oriented volume in $\mathbb{R}^{2d}$ or, in other words, points in phase-space convected by a Hamiltonian flow behave like particles of an incompressible fluid flow. Note that preservation of volume $\det(\psi') \equiv 1$ is a direct generalization to $d > 1$ of the property (3.2). However, when going from $d = 1$ to $d > 1$, the right generalization of preservation of area is symplecticness rather than preservation of volume. Symplecticness *characterizes* Hamiltonian flows; conservation of volume is a much weaker property shared by some nonHamiltonian systems.

5. Symplectic integrators

5.1. Numerical methods

Even though some attention has been given in the literature to symplectic multistep methods (see Aizu, 1985; Feng and Qin, 1987; Eirola and Sanz-Serna, 1990; Sanz-Serna and Vadillo, 1986, 1987), in this paper we restrict our interest to *one-step* integrators. If h denotes the step-length and $(\mathbf{p}^n, \mathbf{q}^n)$ denotes the numerical approximation at time $t_n = nh$, n an integer, to the value $(\mathbf{p}(t_n), (t_n))$ of a solution of (2.1), then a one-step method is specified by a smooth mapping

$$(\mathbf{p}^{n+1}, \mathbf{q}^{n+1}) = \psi_{h,H}(\mathbf{p}^n, \mathbf{q}^n). \tag{5.1}$$

The transformation $\psi_{h,H}$ itself is assumed to depend smoothly on h and H. The domain Ω_h of $\psi_{h,H}$ need not be, for each h, the whole Ω. In fact for implicit methods, where the actual computation of $(\mathbf{p}^{n+1}, \mathbf{q}^{n+1})$ involves the solution of some system of equations, it is often the case that, for fixed $(\mathbf{p}^n, \mathbf{q}^n)$, the new approximation $\psi_{h,H}(\mathbf{p}^n, \mathbf{q}^n)$ is only defined if $|h|$ is suitably small.

The method (5.1) is of order r with r an integer, if, as $h \to 0$, $\psi_{h,H}$ differs from the flow $\phi_{h,H}$ by $\mathcal{O}(h^{r+1})$ terms whenever the Hamiltonian H is suitably smooth. Consistency means order ≥ 1.

Given an initial condition $(\mathbf{p}^0, \mathbf{q}^0)$, the numerical approximation at time t_n is found by iterating the mapping $\psi_{h,H}$ n times, i.e.

$$(\mathbf{p}^n, \mathbf{q}^n) = \psi_{h,H}^n(\mathbf{p}^0, \mathbf{q}^0),$$

whereas for the true solution

$$(\mathbf{p}(t_n), \mathbf{q}(t_n)) = \phi_{t_n,H}(\mathbf{p}^0, \mathbf{q}^0) = \phi_{h,H}^n(\mathbf{p}^0, \mathbf{q}^0).$$

5.2. Symplectic numerical methods

Is it possible to construct numerical methods (5.1) that take into account the Hamiltonian nature of the problem being integrated? In other words, is there such a thing as a *Hamiltonian* numerical method? Before answering this question, let us first note that the discrete equations (5.1) do not intend to mimic the differential system (2.1). On the contrary $\psi_{h,H}$ tries to mimic the *flow* $\phi_{h,H}$. Now, we saw in Section 4 that the Hamiltonian form of the differential equations corresponds, in terms of flows, to symplecticness. Hence the right question to ask is: are there numerical methods (5.1) for which $\psi_{h,H}$ is a symplectic transformation for all Hamiltonians H and all step-lengths h? Such methods, that do exist, are called *symplectic* (or canonical) and are the subject of this paper.

Roughly speaking, there are two main groups of symplectic methods. The

first group consists of formulae that belong to standard families of numerical methods, such as Runge–Kutta or Runge–Kutta–Nyström methods, and
just 'happen' to be symplectic (Sanz-Serna, 1991b). These symplectic methods can be applied to general (i.e. not necessarily Hamiltonian) systems of
differential equations and, when applied to a Hamiltonian system, achieve
symplecticness through a suitable balance between the formula coefficients.
The second main group of symplectic integrators consists of methods that
are derived via a so-called *generating function*. These methods cannot be
applied to general systems of differential equations, not even to small dissipative perturbations of Hamiltonian systems.

The presentation to a numerical analysis audience of the methods of the
first group is easier than the presentation of the second group. We therefore consider the first group in Sections 6–8 and postpone the study of the
methods of the second group until Section 10. This somehow goes against
the history of the field, where methods based on generating functions came
first.

5.3. Composing methods

Before we present particular examples of symplectic integrators, it is ex-
pedient to consider the issue of composition of methods, as this plays a role
in later developments. If $\psi_{h,H}^{[1]}$ and $\psi_{h,H}^{[2]}$ are consistent numerical methods,
then the mapping

$$\psi_{h,H} = \psi_{h/2,H}^{[2]}\psi_{h/2,H}^{[1]}$$

is clearly a new consistent numerical method. More general compositions of
the form

$$\psi_{\theta h,H}^{[2]}\psi_{(1-\theta)h,H}^{[1]},$$

θ a real constant, are also possible. Since it is obvious that the composition
of symplectic maps is a symplectic map, the composition of two symplectic
numerical methods gives rise to a new symplectic method.

On the other hand, along with each method (5.1) we consider its *adjoint*
$\widehat{\psi}_{h,H}$. By definition (see e.g. Hairer *et al.* (1987, Section II.8)), this is the
method such that $\widehat{\psi}_{-h,H}\psi_{h,H}$ is the identity map, i.e. stepping forward with
the given method is just stepping backward with its adjoint. The familiar
forward and backward Euler methods are mutually adjoint. The adjoint of
a symplectic method is itself a symplectic method, because the inverse of a
symplectic transformation is, clearly, a symplectic transformation.

Some methods, such as the implicit midpoint rule, happen to be their own
adjoints. These are called, unsurprisingly, *self-adjoint*. It is easy to see that
the order of consistency r of a self-adjoint method is necessarily even.

6. Runge–Kutta and related methods: conditions for symplecticness

6.1. Runge–Kutta methods

The application of the Runge–Kutta (RK) method to system (2.1) with tableau

$$
\begin{array}{|ccc|}
a_{11} & \cdots & a_{1s} \\
\vdots & \ddots & \vdots \\
a_{s1} & \cdots & a_{ss} \\
\hline
b_1 & \cdots & b_s
\end{array}
\tag{6.1}
$$

results in the relations

$$
\mathbf{P}_i = \mathbf{p}^n + h\sum_{j=1}^{s} a_{ij}\mathbf{f}(\mathbf{P}_j, \mathbf{Q}_j), \quad \mathbf{Q}_i = \mathbf{q}^n + h\sum_{j=1}^{s} a_{ij}\mathbf{g}(\mathbf{P}_j, \mathbf{Q}_j), \ 1 \le i \le s, \tag{6.2}
$$

$$
\mathbf{p}^{n+1} = \mathbf{p}^n + h\sum_{i=1}^{s} b_i \mathbf{f}(\mathbf{P}_i, \mathbf{Q}_i), \quad \mathbf{q}^{n+1} = \mathbf{q}^n + h\sum_{i=1}^{s} b_i \mathbf{g}(\mathbf{P}_i, \mathbf{Q}_i), \tag{6.3}
$$

where $\mathbf{f}$ and $\mathbf{g}$ respectively denote the d-vectors with components $-\partial H/\partial q_i$ and $\partial H/\partial p_i$ and $\mathbf{P}_i$ and $\mathbf{Q}_i$ are the internal stages corresponding to the $\mathbf{p}$ and $\mathbf{q}$ variables.

The following result was discovered independently by Lasagni (1988), Sanz-Serna (1988) and Suris (1989).

Theorem 6.1 Assume that the coefficients of the method (6.1) satisfy the relationships

$$
b_i a_{ij} + b_j a_{ji} - b_i b_j = 0, \quad 1 \le i, j \le s. \tag{6.4}
$$

Then the method is symplectic.

Proof. We follow the technique used by Sanz-Serna (1988). Suris (1989) resorts to Jacobians rather than to differential forms. No proof is presented in Lasagni (1988). We employ the notation

$$
\mathbf{k}_i = \mathbf{f}(\mathbf{P}_i, \mathbf{Q}_i), \quad \mathbf{l}_i = \mathbf{g}(\mathbf{P}_i, \mathbf{Q}_i)
$$

for the 'slopes' at the stages. Differentiate (6.3) and form the exterior product to arrive at

$$
d\mathbf{p}^{n+1} \wedge d\mathbf{q}^{n+1} = d\mathbf{p}^n \wedge d\mathbf{q}^n + h\sum_{i=1}^{s} b_i d\mathbf{k}_i \wedge d\mathbf{q}^n
$$

$$
+ h\sum_{j=1}^{s} b_j d\mathbf{p}^n \wedge d\mathbf{l}_j + h^2 \sum_{i,j=1}^{s} b_i b_j d\mathbf{k}_i \wedge d\mathbf{l}_j.
$$

Our next step is to eliminate $d\mathbf{k}_i \wedge d\mathbf{q}^n$ and $d\mathbf{p}^n \wedge d\mathbf{l}_j$ from this expression. This is easily achieved by differentiating (6.2) and taking the exterior product of the result with $d\mathbf{k}_i$, $d\mathbf{l}_j$. The outcome of the elimination is

$$d\mathbf{p}^{n+1} \wedge d\mathbf{q}^{n+1} - d\mathbf{p}^n \wedge d\mathbf{q}^n = h \sum_{i=1}^{s} b_i [d\mathbf{k}_i \wedge d\mathbf{Q}_i + d\mathbf{P}_i \wedge d\mathbf{l}_i]$$
$$- h^2 \sum_{i,j=1}^{s} (b_i a_{ij} + b_j a_{ji} - b_i b_j) \, d\mathbf{k}_i \wedge d\mathbf{l}_j.$$

The second term on the right-hand side vanishes in view of (6.4). To finish the proof is then sufficient to show that, for each i, $d\mathbf{k}_i \wedge d\mathbf{Q}_i + d\mathbf{P}_i \wedge d\mathbf{l}_i$ is 0. In fact, dropping the subscript i that numbers the stages, we can write

$$d\mathbf{k} \wedge d\mathbf{Q} + d\mathbf{P} \wedge d\mathbf{l} = \sum_{\mu=1}^{d} [dk_\mu \wedge dQ_\mu + dP_\mu \wedge dl_\mu]$$
$$= \sum_{\mu,\nu=1}^{d} \left[\frac{\partial f_\mu}{\partial p_\nu} dP_\nu \wedge dQ_\mu + \frac{\partial f_\mu}{\partial q_\nu} dQ_\nu \wedge dQ_\mu \right.$$
$$\left. + \frac{\partial g_\mu}{\partial p_\nu} dP_\mu \wedge dP_\nu + \frac{\partial g_\mu}{\partial q_\nu} dP_\mu \wedge dQ_\nu \right].$$

To see that this expression vanishes express f_μ and g_μ as derivatives of H and recall the skew-symmetry of the exterior product. $\square$

The symplecticness of the method must be understood in the following sense. Assume that, for a given h, $(\mathbf{p}^{n+1}, \mathbf{q}^{n+1}) = \psi_{h,H}(\mathbf{p}^n, \mathbf{q}^n)$ is a smooth function defined in a subdomain Ω_h of Ω and satisfying the RK equations (6.2)–(6.3), then $\psi_{h,H}$ is a symplectic transformation. In general, for a given h there can be several such functions (nonuniqueness of solutions of the RK scheme). Of course, for $h \to 0$ there is a *unique* RK solution that approximates the true solution and the corresponding domain of definition Ω_h tends to Ω. However spurious RK solutions may also exist and they are also symplectic. For material on the existence and uniqueness of RK solutions see e.g. Dekker and Verwer (1984, Chapter 5) and Sanz-Serna and Griffiths (1991). For spurious solutions see Iserles (1990a) and Hairer *et al.* (1990).

Lasagni (1988, 1990) has shown that, for RK methods without redundant stages, (6.4) is actually necessary for the method to be symplectic. A direct proof of this result is not available in the published literature. However the result is a corollary of Theorem 5.1 in Abia and Sanz-Serna (1990).

6.2. Partitioned Runge–Kutta methods

In the integration of systems of differential equations it is perfectly possible to integrate some components of the unknown vector with a numerical method and the remaining components with a different numerical method. For instance, one may wish to do this if the system includes both stiff and nonstiff components. In our setting, we may wish to integrate the **p** equations with an RK formula and the **q** equations with a different RK formula. The overall scheme is called a *partitioned Runge–Kutta* (PRK) scheme and is specified by two tableaux

$$
\begin{array}{c|ccc}
 & a_{11} & \cdots & a_{1s} \\
 & \vdots & \ddots & \vdots \\
 & a_{s1} & \cdots & a_{ss} \\
\hline
 & b_1 & \cdots & b_s
\end{array}
\qquad
\begin{array}{c|ccc}
 & A_{11} & \cdots & A_{1s} \\
 & \vdots & \ddots & \vdots \\
 & A_{s1} & \cdots & A_{ss} \\
\hline
 & B_1 & \cdots & B_s
\end{array}
\tag{6.5}
$$

The application of (6.5) to system (2.1) results in the relationships (cf. (6.2)–(6.3))

$$
\mathbf{P}_i = \mathbf{p}^n + h \sum_{j=1}^{s} a_{ij} \mathbf{f}(\mathbf{P}_j, \mathbf{Q}_j),
$$

$$
\mathbf{Q}_i = \mathbf{q}^n + h \sum_{j=1}^{s} A_{ij} \mathbf{g}(\mathbf{P}_j, \mathbf{Q}_j), \ 1 \le i \le s.
$$

$$
\mathbf{p}^{n+1} = \mathbf{p}^n + h \sum_{i=1}^{s} b_i \mathbf{f}(\mathbf{P}_i, \mathbf{Q}_i), \quad \mathbf{q}^{n+1} = \mathbf{q}^n + h \sum_{i=1}^{s} B_i \mathbf{g}(\mathbf{P}_i, \mathbf{Q}_i).
$$

Of course an RK method is a particular instance of (6.5) where both tableaux just happen to have the same entries. The following result was first given by the present author at the London 1989 ODE meeting (Sanz-Serna, 1989) and discovered independently by Suris (1990). The proof is analogous to that of Theorem 6.1.

Theorem 6.2 Assume that the coefficients of the method (6.5) satisfy the relationships

$$
b_i A_{ij} + B_j a_{ji} - b_i B_j = 0, \quad 1 \le i, j \le s. \tag{6.6}
$$

then the method is symplectic when applied to separable Hamiltonian problems (2.1), (2.2).

Symplecticness must again be understood as in Theorem 6.1 and, once more, (6.6) is necessary for symplecticness, provided that the method has no redundant stages, see Abia and Sanz-Serna (1990).

6.3. Runge–Kutta–Nyström methods

Systems of differential equations of the special form

$$\frac{d\mathbf{p}}{dt} = \mathbf{f}(\mathbf{q}), \qquad \frac{d\mathbf{q}}{dt} = \mathbf{p}, \tag{6.7}$$

or, equivalently, second-order systems $d^2\mathbf{q}/dt^2 = \mathbf{f}(\mathbf{q})$ can be efficiently integrated by means of Runge–Kutta–Nyström (RKN) methods (see e.g. Hairer *et al.* (1987, Section II.13)). For the RKN procedure with array

$$
\begin{array}{c|ccc}
\gamma_1 & \alpha_{11} & \cdots & \alpha_{1s} \\
\vdots & \vdots & \ddots & \vdots \\
\gamma_s & \alpha_{s1} & \cdots & \alpha_{ss} \\
\hline
 & \beta_1 & \cdots & \beta_s \\
\hline
 & b_1 & \cdots & b_s
\end{array} \tag{6.8}
$$

the intermediate stages $\mathbf{Q}_i$ are defined by

$$\mathbf{Q}_i = \mathbf{q}^n + h\gamma_i\mathbf{p}^n + h^2\sum_{j=1}^{s}\alpha_{ij}\mathbf{f}(\mathbf{Q}_j),$$

and the approximation at the next time level is

$$\mathbf{p}^{n+1} = \mathbf{p}^n + h\sum_{i=1}^{s}b_i\mathbf{f}(\mathbf{Q}_j), \quad \mathbf{q}^{n+1} = \mathbf{q}^n + h\mathbf{p}^n + h^2\sum_{i=1}^{s}\beta_i\mathbf{f}(\mathbf{Q}_i).$$

The system (6.7) is Hamiltonian if and only if $\mathbf{f}$ (the 'force') is the gradient of a scalar function $-V$. When this condition is satisfied the Hamiltonian is given by (2.3). The following result is due to Suris (1989), who used Jacobians in the proof. A proof based on differential forms, similar to that of Theorem 6.1 is easily given, and can be seen in Okunbor and Skeel (1990).

Theorem 6.3 Assume that the coefficients of the method (6.8) satisfy the conditions

$$
\begin{aligned}
\beta_i &= b_i(1 - \gamma_i), \quad 1 \le i \le s, & (6.9) \\
b_i(\beta_j - \alpha_{ij}) &= b_j(\beta_i - \alpha_{ji}), \quad 1 \le i, j \le s. & (6.10)
\end{aligned}
$$

Then the method is symplectic when applied to Hamiltonian problems (2.1), (2.3).

The conditions (6.9) and (6.10) are also necessary for methods without redundant stages to be symplectic, see Calvo (1991).

7. Runge–Kutta and related methods: order conditions for symplectic methods

Before we construct specific formulae within the classes of symplectic methods we have identified in the previous section, it is clearly appropriate to discuss the corresponding order conditions, i.e. the sets of relationships that the method coefficients must satisfy to ensure that a prescribed order of consistency r is reached.

7.1. Runge–Kutta methods

Since Hamiltonian problems are only a subclass of the family of all differential problems, it is *a priori* conceivable that the order of consistency r^* that an RK method achieves for Hamiltonian problems is higher than the classical order of consistency r, i.e. the order of consistency for the most general problem. This is not the case. By considering Hamiltonians of the form $H = \mathbf{p}^T \mathbf{g}(\mathbf{q})$, we see that any d-dimensional differential system $d\mathbf{q}/dt = \mathbf{g}(\mathbf{q})$ can be thought of as being the $\mathbf{q}$ equations of a Hamiltonian system with d-degrees of freedom. Hence $r = r^*$ and therefore the material in this subsection applies even if the system being integrated is not Hamiltonian.

The conditions that (6.1) should satisfy to achieve order $\geq r$ are well known (Butcher, 1987, Theorem 306A; Hairer *et al.*, 1987, Theorem 2.13). Each *rooted* tree $\rho\tau$ with r or fewer vertices gives rise to a condition

$$\Phi(\rho\tau) = 1/\gamma(\rho\tau). \tag{7.1}$$

Here the *density* $\gamma(\rho\tau)$ is an (easily computable) integer associated with $\rho\tau$ and the *elementary weight* $\Phi(\rho\tau)$ is a polynomial in the method coefficients a_{ij}, b_i. Figure 1 contains the rooted trees with four vertices or less; we have highlighted the roots by means of a cross.

As an illustration, let us recall that for consistency $r \geq 1$ we require, in connection with $\rho\tau_{1,1}$,

$$\sum_{i=1}^{s} b_i = 1.$$

For order $r \geq 2$ we further impose, in connection with $\rho\tau_{2,1}$,

$$\sum_{i,j=1}^{s} b_i a_{ij} = \tfrac{1}{2}.$$

For order $r \geq 3$ we add further, in view of $\rho\tau_{3,1}$,

$$\sum_{i,j,k=1}^{s} b_i a_{ij} a_{jk} = \tfrac{1}{6}, \tag{7.2}$$

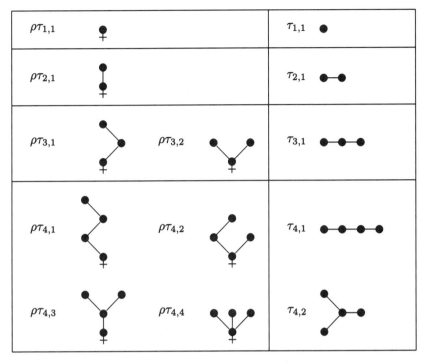

Fig. 1. Rooted n-trees and n-trees, $n = 1, 2, 3, 4$.

and, in view of $\rho\tau_{3,2}$,

$$\sum_{i,j,k=1}^{s} b_i a_{ij} a_{ik} = \tfrac{1}{3}. \tag{7.3}$$

Butcher (1987, Theorem 306A) proves that, if the number of stages s and the coefficients a_{ij}, b_i are regarded as free parameters, then each equation of the form (7.1) is independent of the others. However, when the symplecticness conditions (6.4) are imposed, the method coefficients are no longer free parameters and it turns out that some redundancies appear amongst the classical order conditions (7.1) arising from the various $\rho\tau$. As a result, in order to achieve order $\geq r$ it is not necessary to write down an equation for *every* rooted tree of order $\leq r$. This point has been studied by Sanz-Serna and Abia (1991), whose treatment we follow closely.

Assume that two rooted trees are identified if they only differ in the location of the root, but otherwise consist of the same vertices and edges. (In Figure 1, this is the case for the rooted trees $\rho_{3,1}$ and $\rho_{3,2}$, or for the rooted trees $\rho_{4,1}$ and $\rho_{4,2}$.) Each equivalence class under this equivalence relation is called a *tree*. Thus, in Figure 1, the eight rooted trees of order ≤ 4 give rise to only five trees.

Some trees are called *superfluous*. These are the trees that result when

Table 1. *Number of order conditions.*

Order	General RK	Symplectic RK	General PRK	Symplectic PRK	General RKN	Symplectic RKN
1	1	1	2	2	1	1
2	2	1	4	3	2	2
3	4	2	8	5	4	4
4	8	3	16	8	7	6
5	17	6	34	14	13	10
6	37	10	74	24	23	15
7	85	21	170	46	43	25
8	200	40	400	88	79	39

two copies of the same rooted tree with N vertices are joined by their roots to give rise to a graph with $2N$ vertices. For instance, in Figure 1, the tree $\tau_{2,1}$ is superfluous because it is the juxtaposition of two copies of $\rho\tau_{1,1}$. In a similar manner, $\tau_{4,1}$ is superfluous as it is the juxtaposition of two copies of $\rho\tau_{2,1}$. There is an alternative way of thinking of superfluous trees. Assume that trees are *coloured* in such a way that each vertex is painted either black or white with adjacent vertices receiving distinct colours. Most trees can be coloured in two different ways: in $\tau_{3,1}$ we could have either a black vertex between two white vertices or a white vertex between two black vertices. However some trees can be coloured in only one way: in $\tau_{2,1}$ we can only have a black vertex shaking hands with a white vertex. These trees are precisely the superfluous trees.

After these preliminaries we are ready for the main result.

Theorem 7.1 Assume that the RK method (6.1) satisfies the symplecticness requirement (6.4) and has order of consistency $\geq r \geq 1$. Then it has order of consistency $\geq r + 1$ if and only if for each *nonsuperfluous* tree τ with $r + 1$ or fewer vertices there is a rooted tree $\rho\tau \in \tau$ for which (7.1) holds.

For instance, since only the tree with two vertices is superfluous, each consistent symplectic RK method actually possesses an order of at least 2. To ensure order ≥ 3, it is sufficient to impose *either* (7.2) or (7.3). In other words (7.2) and (7.3) have become equivalent, as coming from the same tree. In general, for symplectic RK methods, the number of order conditions for order $\geq r$ equals the number of nonsuperfluous trees with r or fewer nodes, as distinct from the situation for general RK methods, where there is an order condition for each rooted tree with r or fewer vertices. The reduction in the number of order conditions is borne out in Table 1.

7.2. Partitioned Runge–Kutta methods

A similar theory exists for PRK methods (Abia and Sanz-Serna, 1990) applied to *separable* Hamiltonian systems. Again, there is no distinction between the order r of the method (6.5) for separable Hamiltonian systems and the classical order when applied to systems of the form

$$\frac{d\mathbf{p}}{dt} = \mathbf{f}(\mathbf{q}), \qquad \frac{d\mathbf{q}}{dt} = \mathbf{g}(\mathbf{p}),$$

where $\mathbf{f}$ and $\mathbf{g}$ are any smooth functions, rather than gradients of scalar functions $-V$ and T as they would be in the Hamiltonian case.

It is well known that graph theory can again be used to systematize the writing of the standard order conditions (e.g. Hairer *et al.* 1987, Section II.14). We now need *bicolour* rooted trees $\beta\rho\tau$, i.e. rooted trees with vertices coloured black or white as previously described. Clearly each rooted tree gives rise to two bicolour rooted trees: the root can be coloured either black or white and the colour of the root recursively determines the colour of all vertices (cf. Figure 2). There is an order condition for each $\beta\rho\tau$. The first of these are as follows. For the two bicolour rooted trees with one vertex we get

$$\sum_{i=1}^{s} b_i = 1, \qquad \sum_{i=1}^{s} B_i = 1.$$

Vertices of one colour bring in lower case letters and the vertices of the other colour bring in upper case letters. In connection with the two bicolour rooted trees with two vertices, we have

$$\sum_{i,j=1}^{s} b_i A_{ij} = \tfrac{1}{2}, \qquad \sum_{i,j=1}^{s} B_i a_{ij} = \tfrac{1}{2}, \tag{7.4}$$

etc. The symplecticness conditions (6.6) bring about some redundancies among the standard order conditions we have just presented (Abia and Sanz-Serna, 1990). Again the key point is to disregard the location of the root: bicolour rooted trees which only differ in the location of the root make an equivalence class called a bicolour tree $\beta\rho$ (see Figure 2). Then for symplectic methods, it is enough to consider an order condition for *a particular* bicolour rooted tree in each bicolour tree. For instance for a consistent symplectic method to have order 2 we impose *one* of the two conditions in (7.4). It should perhaps be emphasized that now bicolour trees arising from colouring a superfluous tree must also be considered. The difference between superfluous and nonsuperfluous trees is that a nonsuperfluous tree gives rise to *two* bicolour trees (two order conditions), while a superfluous tree only generates *one* bicolour tree (only one order condition).

The reduction in the number of order conditions is borne out in Table 1.

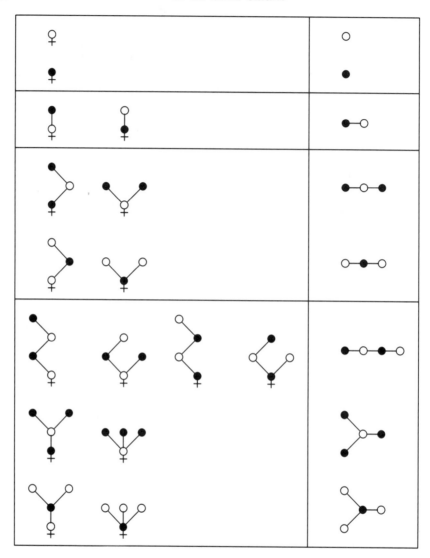

Fig. 2. Rooted bicolour n-trees and bicolour n-trees, $n = 1, 2, 3, 4$.

7.3. Runge–Kutta–Nyström methods

Similar considerations hold for RKN methods. In the interest of brevity we omit the corresponding results and the interested reader is referred to Calvo and Sanz-Serna (1991a). The reduction in the number of order conditions is apparent in Table 1. A word of warning: in the table, a general RKN method means a method satisfying (6.9); practical methods as a rule satisfy this condition.

7.4. The homogeneous form of the order conditions for symplectic methods

Let us return to the RK case. The fact that, in order to write the order conditions for symplectic methods, we are free to choose any rooted tree within each nonsuperfluous tree introduces some asymmetry among the various rooted trees. For instance, at the $r = 3$ stage, we are free to disregard $\rho\tau_{3,1}$ or $\rho\tau_{3,2}$, i.e. to omit (7.2) or (7.3). Sanz-Serna and Abia (1991) provide an alternative way of writing the order conditions, where all rooted trees belonging to a nonsuperfluous tree play a symmetric role. For a consistent method to have order $r \geq 4$, Sanz-Serna and Abia (1991) put

$$6 \sum_{ijk=1}^{s} b_i a_{ij} a_{jk} - 3 \sum_{ijk=1}^{s} b_i a_{ij} a_{ik} = 0, \tag{7.5}$$

$$12 \sum_{ijkl=1}^{s} b_i a_{ij} a_{jk} a_{jl} - 4 \sum_{ijkl=1}^{s} b_i a_{ij} a_{ik} a_{il} = 0. \tag{7.6}$$

It may be observed that in (7.5) we find the elemetary weights associated with (7.2) *and* (7.3), while in (7.6) we find the elementary weights arising from both rooted trees in the nonsuperfluous tree with four vertices $\tau_{4,2}$. This alternative form of the order conditions is called *homogeneous*. Full details concerning the systematic writing of the homogeneous order conditions and a proof of the equivalence between the homogeneous and standard forms can be found in Sanz-Serna and Abia (1991).

Homogeneous forms for PRK and RKN methods exist and can be seen in Abia and Sanz-Serna (1990) and Calvo (1991) respectively.

8. Runge–Kutta and related methods: available symplectic methods

8.1. Runge–Kutta methods

We start by noticing that, for methods satisfying the symplecticness condition (6.4), it may be assumed that all the weights b_i are not equal to 0. In fact, if $b_j = 0$, then (6.4) implies that $b_i a_{ij} = 0$ for all i and therefore neither does the jth stage, which does not contribute to the final quadrature (6.3), contribute to any other stage with nontrivial b_i: thus the method is equivalent to a method with fewer stages. Under the assumption of nonzero weights, (6.4) with $i = j$ reveals that a symplectic Runge–Kutta method *cannot be explicit*.

A second observation is that the left-hand side of (6.4) provides the entries of the M matrix that features in the definition of algebraic stability introduced by Burrage and Butcher (1979) and Crouzeix (1979) (see also Dekker and Verwer (1984)). The condition $M = 0$ was investigated by Cooper (1987) in a different context. It is well known that the Gauss–Legendre methods satisfy this condition (see e.g. Dekker and Verwer 1984, Theorem 4.6) and hence we have the following result (cf. Sanz-Serna, 1988).

Theorem 8.1   The Gauss–Legendre Runge–Kutta methods are symplectic.

We recall that with s stages the Gauss–Legendre method is the unique RK method that achieves order $2s$. It is also A- and B-stable. There is a price to be paid: the high computational cost deriving from implicitness. The efficient implementation of the Gauss–Legendre methods for Hamiltonian problems is an area where much work is needed. Current strategies for choosing the iterative method and initial guess for the solution of the nonlinear algebraic equations in RK processes are based on the assumption that the underlying system is stiff. This is reasonable: stiffness has been until now the main motivation for switching from explicit to implicit methods. However the problems with which we are concerned are not necessarily stiff and fresh ideas are required when dealing with the implementation. For references on the implementation of implicit RK methods see the references in Cooper and Vignesvaran (1990).

The two-stage, order 4 method has been successfully tested by Pullin and Saffman (1991) in a difficult Hamiltonian problem arising in fluid mechanics.

Of course, when the system being integrated is linear, the Gauss–Legendre methods generate diagonal Padé approximants to the exponential. The symplecticness of these rational approximants was shown by Feng (1986a), see also Feng *et al.* (1990).

A way of bringing down the implementation costs associated with implicitness is to resort to *diagonally implicit* methods. These satisfy (6.4) if and only if they have the tableau

$$
\begin{array}{c|ccccc}
b_1/2 & 0 & 0 & \cdots & 0 \\
b_1 & b_2/2 & 0 & \cdots & 0 \\
b_1 & b_2 & b_3/2 & \cdots & 0 \\
\vdots & \vdots & \vdots & \ddots & \vdots \\
b_1 & b_2 & b_3 & \cdots & b_s/2 \\
\hline
b_1 & b_2 & b_3 & \cdots & b_s.
\end{array}
\tag{8.1}
$$

A step of length h with the method (8.1) is just a concatenation of an implicit midpoint step of length $b_1 h$, an implicit midpoint step of length $b_2 h$, etc. Hence diagonally implicit symplectic methods are as easy to implement as the implicit midpoint rule. This sort of method is appealing when the number of degrees of freedom d is high, as would be the case if the system being integrated in time was the result of the space discretization of a partial differential equation.

Sanz-Serna and Abia (1991) show that the self-adjoint three-stage method (8.1) with

$$b_1 = b_3 = \frac{1 + \omega + \omega^{-1}}{3}, \quad \omega = 2^{1/3}, \quad b_2 = 1 - 2b_1$$

has order 4. This method has been applied to the time-integration of some partial differential equations by de Frutos and Sanz-Serna (1991). If $\psi_{h,H}^{[MP]}$ represents the midpoint rule, the fourth-order method is given by

$$\psi_{h,H}^{[4]} = \psi_{b_1 h,H}^{[MP]} \, \psi_{(b_2 h,H}^{[MP]} \, \psi_{b_1 h,H}^{[MP]}.$$

Following ideas in Yoshida (1990), this construction can be taken further. Consider the method

$$\psi_{\alpha h,H}^{[4]} \psi_{\beta h,H}^{[4]} \psi_{\alpha h,H}^{[4]}, \qquad (8.2)$$

with α and β chosen in such a way that $2\alpha + \beta = 1$ (consistency) and $2\alpha^5 + \beta^5 = 0$ (the leading h^5 term in the truncation error of the composition vanishes). Then (8.2) has order ≥ 5, but being self-adjoint the order must actually be ≥ 6. In turn, a sixth-order self-adjoint method can be composed to give rise to an eighth-order method etc. The conclusion is that there are diagonally implicit symplectic RK methods of arbitrarily high order. Of course it is an open question to decide whether high order methods constructed in this way have some *practical* interest.

Diagonally implicit methods are not the only 'easily implementable' implicit RK methods. It is well known that, following Butcher (1976), the RK matrix $A = (a_{ij})$ can be subjected to a transformation $A \mapsto T^{-1}AT$ with a view to simplifying the linear algebra. For symplectic methods this idea has been explored by Iserles (1990b) (see also Iserles and Nørsett, 1991, Section 3.7).

8.2. Partitioned Runge–Kutta methods

Unlike the class of RK methods, the class of PRK methods includes formulae that are both explicit and symplectic. However it should be emphasized that these properties only hold when dealing with separable Hamiltonians (2.2). In fact the methods of the form

$$
\begin{array}{|ccccc}
b_1 & 0 & 0 & \cdots & 0 \\
b_1 & b_2 & 0 & \cdots & 0 \\
b_1 & b_2 & b_3 & \cdots & 0 \\
\vdots & \vdots & \vdots & \ddots & \vdots \\
b_1 & b_2 & b_3 & \cdots & b_s \\
\hline
b_1 & b_2 & b_3 & \cdots & b_s
\end{array}
\qquad
\begin{array}{|ccccc}
0 & 0 & 0 & \cdots & 0 \\
B_1 & 0 & 0 & \cdots & 0 \\
B_1 & B_2 & 0 & \cdots & 0 \\
\vdots & \vdots & \vdots & \ddots & \vdots \\
B_1 & B_2 & B_3 & \cdots & 0 \\
\hline
B_1 & B_2 & B_3 & \cdots & B_s
\end{array}
\qquad (8.3)
$$

are explicit, symplectic and have the further favourable property that they can be implemented while only storing two d-dimensional vectors: $\mathbf{Q}_1$ is nothing but $\mathbf{q}^n$, $\mathbf{P}_1$ can be overwritten on $\mathbf{q}^n$, $\mathbf{Q}_2$ can be overwritten on $\mathbf{Q}_1$, etc.

The family of methods (8.3) was introduced by Ruth (1983) in one of the very first papers on symplectic integration. Ruth constructed methods with $s = r = 1, 2, 3$. In the case $s = r = 3$, there is a one-parameter family of methods and Ruth has chosen the parameter so as to obtain simple coefficients b_i, B_i. Sanz-Serna (1989) suggested a different choice for the parameter. Furthermore by composing the third-order, three-stage method $\psi_{h,H}$ with its adjoint $\widehat{\psi}_{h,H}$, he constructed an explicit symplectic method $\widehat{\psi}_{h/2,H}\psi_{h/2,H}$ that requires five evaluations per step, but produces fourth-order results both at the grid points $t_n = nh$ and at the points $t_{(n+1/2)} = (n+1/2)h$. A method with $s = r = 4$ has been constructed by Neri (1987), Forest and Ruth (1990) and Candy and Rozmus (1991). Yoshida (1990), by using a construction similar to that discussed earlier for diagonally implicit RK methods, has proved that there are methods of the form (8.3) possessing arbitrarily high orders. He furthermore derives sixth-order methods that use seven function evaluations per step and eighth-order methods requiring sixteen function evaluations per step.

8.3. Runge–Kutta–Nyström methods

There are explicit RKN methods that are symplectic. These have the tableau

$$
\begin{array}{c|cccc}
\gamma_1 & 0 & 0 & \cdots & 0 \\
\gamma_2 & b_1(\gamma_2 - \gamma_1) & 0 & \cdots & 0 \\
\vdots & \vdots & \vdots & \ddots & \vdots \\
\gamma_s & b_1(\gamma_s - \gamma_1) & b_2(\gamma_s - \gamma_2) & \cdots & 0 \\
\hline
 & b_1(1 - \gamma_1) & b_2(1 - \gamma_2) & \cdots & b_s(1 - \gamma_s) \\
\hline
 & b_1 & b_2 & \cdots & b_s
\end{array}
\qquad (8.4)
$$

and hence with s stages provide $2s$ free parameters. Okunbor and Skeel (1991) have pointed out that, for implementation purposes, (8.4) can be rewritten as an explicit PRK method, and hence only requires the storage of two d-dimensional vectors. Okunbor and Skeel (1990) prove that an explicit RKN method is symplectic if and only if its adjoint method is also explicit. This idea can be used to compose a method with its adjoint as shown earlier for the PRK case. Calvo and Sanz-Serna (1991b) have considered the family of fourth-order, five-stage methods of the form (8.4) that effectively require four function evaluations per step due to the fact that the last evaluation in the current step provides the first evaluation in the next step. An optimal method within this class has been obtained by minimizing the error constants. Similar work is under way for higher order methods.

Given a number of stages s and an order r, the tableaux (8.4) and (8.3) have the same number $2s$ of free parameters, while Table 1 makes it clear that the number of order conditions for the RKN case is substantially smaller than the number of order conditions for PRK methods. This is due to the fact that in the PRK case we are catering for all separable Hamiltonians (2.2) where RKN methods can only cope with the special case (2.3). However case (2.3) is very common in the applications and this should make the construction of explicit symplectic RKN methods an important practical issue.

9. Generating functions

We cannot make any further progress with the topic of symplectic integrators without first reviewing some basic facts about the generating functions of a symplectic transformation.

It is a remarkable feature of Hamiltonian problems that each system of the form (2.1) is fully determined by the choice of a *scalar* function H, whereas a general system $dy/dt = f(y)$ is determined by a *vector field* f. In a similar vein, a symplectic transformation $(p^*, q^*) = \psi(p, q)$ can be expressed in terms of a single real-valued function S, rather than in terms of the $2d$ components of ψ. The function S is called the *generating function* of ψ.

9.1. Generating functions of the first kind

Let $(p^*, q^*) = \psi(p, q)$ be a symplectic transformation defined in a simply connected domain Ω. For each closed path γ in Ω

$$\int_\gamma p \, dq - \int_\gamma p^* \, dq^* = 0, \tag{9.1}$$

where $p \, dq$ is the differential form $p_1 \, dq_1 + \cdots + p_d \, dq_d$, etc. In fact, by Stokes theorem, the first integral is the quantity $m(\Sigma)$, where m is the sum of two-dimensional areas considered in Section 4 and Σ is any two-dimensional surface bounded by γ. The second integral is $m(\psi(\Sigma))$ and hence (9.1) is just a way of saying that ψ is symplectic. The key observation is that (9.1) is the condition for $p \, dq - p^* \, dq^*$ to be the differential of a function S defined in Ω:

$$dS = p \, dq - p^* \, dq^*. \tag{9.2}$$

Now let us further assume that q and q^* are *independent* functions in Ω, i.e. each point in Ω may be uniquely specified by the corresponding values of q and q^*. Then we can express $S(p, q)$ in (9.2) as a function S^1 of q and

$\mathbf{q}^*$. It is evident from (9.2) that

$$\mathbf{p} = \frac{\partial S^1}{\partial \mathbf{q}}, \qquad \mathbf{p}^* = -\frac{\partial S^1}{\partial \mathbf{q}^*}. \tag{9.3}$$

These formulae implicitly define ψ by providing $2d$ relationships among the $4d$ components of $\mathbf{q}$, $\mathbf{p}$, $\mathbf{p}^*$, $\mathbf{q}^*$. The function $S^1(\mathbf{q}, \mathbf{q}^*)$ is called the generating function (of the first kind) of ψ. The reader may wish to check that for the rotation in (2.6)

$$S^1(q, q^*) = \frac{\cot t}{2} \left(q^2 + q^{*2} \right) - \operatorname{cosec} qq^*. \tag{9.4}$$

Conversely, if we choose any smooth function $S^1(\mathbf{q}, \mathbf{q}^*)$ satisfying the condition that the Hessian determinant $\det \partial^2 S^1 / \partial \mathbf{q} \partial \mathbf{q}^*$ does not vanish at a point $(\mathbf{q}_0, \mathbf{q}_0^*)$, then the formulae (9.3) implicitly define, in the neighbourhood of $(\mathbf{q}_0, \mathbf{q}_0^*)$, a *symplectic* transformation (see e.g. Arnold (1989, Section 47A)).

9.2. Generating functions of the third kind

For a symplectic transformation ψ to have a generating function of the first kind, it is clearly necessary that $\mathbf{q}$ and $\mathbf{q}^*$ are *independent*, a condition not fulfilled by the identity transformation. (Note that (9.4) has a singularity at $t = 0$, where the rotation (2.6) is just the identity.) Since we are interested in generating consistent numerical methods $\psi_{h,H}$, which, at $h = 0$, give the identity transformation, generating functions of the first kind are not really what we want.

Let us proceed as follows. Note that from (9.2)

$$d(\mathbf{p}^T \mathbf{q} - S) = \mathbf{q} \, d\mathbf{p} + \mathbf{p}^* \, d\mathbf{q}^* \tag{9.5}$$

and now assume that $\mathbf{p}$ and $\mathbf{q}^*$ are *independent* functions (which they are for the identity transformation). Then we can express the function in brackets in (9.5) in terms of the independent variables $\mathbf{p}$ and $\mathbf{q}^*$. The result $S^3(\mathbf{p}, \mathbf{q}^*)$ is called the generating function of the third kind of ψ and, from (9.5) we conclude that the formulae that now implicitly define ψ when S^3 is known are

$$\mathbf{p}^* = \frac{\partial S^3}{\partial \mathbf{q}^*}, \qquad \mathbf{q} = \frac{\partial S^3}{\partial \mathbf{p}}. \tag{9.6}$$

The generating function of the identity is $\mathbf{p}^T \mathbf{q}^*$. For the rotation (2.6) we find

$$S^3(p, q^*) = -\frac{\tan t}{2} \left(p^2 + q^{*2} \right) + \sec pq^*;$$

this is regular near $t = 0$, but breaks down when t approaches $\pm 2\pi$: at these values $p = q^*$ and p and q^* cannot be taken as independent coordinates.

Conversely given a function $S^3(\mathbf{p}, \mathbf{q}^*)$ with a locally nonvanishing Hessian $\det \partial^2 S^3 / \partial \mathbf{p} \partial \mathbf{q}^*$, the formulae (9.6) locally define a symplectic transformation.

9.3. Generating functions of all kinds

Some classical books on Hamiltonian mechanics considered four kinds of generating functions. Arnold (1989) has 2^n kinds. And in fact there are many more. According to Feng (1986a), Feng and Qin (1987), Feng et al. (1989), Wu (1988) the general idea is as follows. Collect in a ($2d$-dimensional) vector $\mathbf{y}$ the components of $(\mathbf{p}, \mathbf{q})$ and in a vector $\mathbf{y}^*$ the components of $(\mathbf{p}^*, \mathbf{q}^*)$. Introduce new $2d$-dimensional variables $\mathbf{w}$ and $\mathbf{w}^*$, such that $\mathbf{w}$ and $\mathbf{w}^*$ are linear functions of $\mathbf{y}$ and $\mathbf{y}^*$ (i.e. $\mathbf{w} = A\mathbf{y} + B\mathbf{y}^*$ and $\mathbf{w} = C\mathbf{y} + D\mathbf{y}^*$ for fixed $2d \times 2d$ matrices A, B, C, D). Under suitable hypotheses, the symplectic transformation $(\mathbf{p}^*, \mathbf{q}^*) = \psi(\mathbf{p}, \mathbf{q})$ reads, in terms of $\mathbf{w}$ and $\mathbf{w}^*$, $\mathbf{w}^* = \chi(\mathbf{w})$, where χ is the gradient of a scalar generating function σ. In the case of the generating functions of the first kind, A is the matrix that extracts the $\mathbf{q}$ variables of $\mathbf{y}$, B is the matrix that extracts the $\mathbf{q}^*$ variables of $\mathbf{y}^*$ etc.

A useful generating function is the so-called *Poincaré* generating function. Here $\mathbf{w}$ is taken to be the average of $\mathbf{y}$ and $\mathbf{y}^*$. The formulae for the transformation are (cf. MacKay, 1991)

$$\mathbf{p}^* = \mathbf{p} - \partial_2 S^P\left(\frac{\mathbf{p}^* + \mathbf{p}}{2}, \frac{\mathbf{q}^* + \mathbf{q}}{2}\right), \qquad \mathbf{q}^* = \mathbf{q} + \partial_1 S^P\left(\frac{\mathbf{p}^* + \mathbf{p}}{2}, \frac{\mathbf{q}^* + \mathbf{q}}{2}\right),$$

where ∂_1 and ∂_2 respectively represent differentiation with respect to the first and second groups of arguments in S^P. The Poincaré generating function of the identity is the 0 function.

9.4. Hamilton–Jacobi equations

Let us now complicate things and consider symplectic transformations ψ_t that depend on t. We assume that ψ_t has a generating function of the third kind S^3, which depends on t. Let us further consider a Hamiltonian system (2.1) in the variables $\mathbf{p}$, $\mathbf{q}$. If we change variables in this system we obtain a new differential system for the new unknowns $\mathbf{p}^*$, $\mathbf{q}^*$. Then the following holds true (Arnold, 1989, Section 45A).

Theorem 9.1 In the situation earlier, the transformed system is also a Hamiltonian system, with the nonautonomous Hamiltonian function

$$H^*(\mathbf{p}^*, \mathbf{q}^*; t) = H - \frac{\partial S^3}{\partial t}. \tag{9.7}$$

In (9.7) it is understood that once S^3 has been differentiated with respect to t with $\mathbf{p}$ and $\mathbf{q}^*$ constant, the formulae (9.6) that define the transformation are used to express the right-hand side in terms of the new variables $\mathbf{p}^*$ and $\mathbf{q}^*$.

A first corollary of this result refers to the case where the transformation is actually independent of t: then in the new variables the Hamiltonian system is still an autonomous Hamiltonian system and the new Hamiltonian is obtained by changing variables in the old Hamiltonian.

Another remarkable application arises when ψ_t is the t-flow of (2.1) and we see the old variables evolving under the Hamiltonian system with Hamiltonian $-H(\mathbf{p}, \mathbf{q})$, i.e. under the flow $\phi_{t,-H} = \phi_{t,H}^{-1}$. Then, the symplectic transformation ψ_t just undoes what the Hamiltonian evolution under $-H$ does; in the new variables, the solutions of the differential equations are $\mathbf{p}^* = constant$ and $\mathbf{q}^* = constant$ and the new Hamiltonian $H^* = -H - \partial S^3/\partial t$ must be 0 (or a constant: Hamiltonians are only defined up to an additive constant). We have proved that the generating function S^3 of the flow of the Hamiltonian system with Hamiltonian H satisfies

$$\frac{\partial S^3}{\partial t}(\mathbf{p}, \mathbf{q}^*; t) + H(\mathbf{p}, \mathbf{q}) = 0. \tag{9.8}$$

This is the celebrated *Hamilton–Jacobi* equation. Upon replacing $\mathbf{q}$ by $\partial S^3/\partial \mathbf{p}$ (cf. (9.6)), the relationship (9.8) is a partial differential equation of the first order for a function S^3 of the variables $\mathbf{p}$ and t (the $\mathbf{q}^*$ act just as parameters). If this equation can be solved explicitly, we find the generating function of the flow and hence the solution of the system (2.1). This is Jacobi's approach to the solution of Hamilton's equation. Jacobi and others used this technique explicitly to integrate problems of mechanics that had proved intractable by other techniques (see e.g. Arnold (1989, Section 47)). On the other hand, if we want to solve (9.8) by the method of characteristics, we find that the system of ordinary differential equations that defines the characteristics is none other than system (2.1)! The equivalence between the solution of a Hamiltonian ordinary differential system and the solution of a first-order partial differential equation with Hamilton–Jacobi structure is thus complete.

These ideas are not confined to generating functions of the third kind; they do work for all kinds of generating functions. The details of the construction of the new Hamiltonian H^* (and hence the form of the Hamilton–Jacobi equation) vary with the kind of generating function being used. The interested reader is referred to Feng (1986a), Feng and Qin (1987), Feng *et al.* (1989) and Wu (1988).

10. Symplectic integrators based on generating functions

Theorem 9.1 is the key to the construction of symplectic integrators via Hamiltonian functions (Channell, 1983; Menyuk, 1984; Feng, 1986a; Feng and Qin, 1987; Wu, 1988; Feng *et al.*, 1989; Channell and Scovel, 1990;

Miesbach and Pesch, 1990). Let $\psi_{t,H}$ be a symplectic numerical method consistent of order r with generating function S^3. An argument similar to that leading to the Hamilton–Jacobi equation, proves that $H^* = -H - \partial S^3/\partial t$ is $\mathcal{O}(t^r)$ as $t \to 0$; now the transformation $\psi_{t,H}$ undoes the effect of the evolution $\phi_{t,-H}$ except for terms of order $\mathcal{O}(t^r)$. Conversely, any function S^3 that makes $H^* = \mathcal{O}(t^r)$ generates a symplectic, rth-order numerical method, see Sanz-Serna and Abia (1991, Theorem 6.1).

Feng and his coworkers take the following approach (Feng, 1986a; Feng and Qin, 1987; Feng *et al.*, 1989; Wu, 1988). They begin by expanding S^3 in (9.8) in powers of t. On substituting this power series in (9.8), expanding H and collecting similar powers of t, the generating function S^3 can be expressed in terms of derivatives of H. When the series for S^3 is truncated, an approximate solution of the Hamilton–Jacobi equation is obtained, which is then used to generate the numerical method via (9.6).

Of course, similar approaches can be taken for generating functions other than generating functions of the third kind. The use of the Poincaré format is appealing, because it easily leads to self-adjoint schemes, with only odd powers of h in the Taylor expansion of the truncation error. The second-order method derived from the Poincaré generating funtion is none other than the familiar midpoint rule, with generating function $S^P = hH$.

The expression for the fourth-order method turns out to be

$$
\begin{aligned}
p_i^{n+1} = p_i^n - hH_{q_i} - \frac{h^3}{24} \Big[& H_{p_j p_k q_i} H_{q_j} H_{q_k} + 2H_{p_j p_k} H_{q_j q_i} H_{q_k} \\
& - 2H_{p_j q_k q_i} H_{p_j} H_{q_k} - 2H_{p_j q_k} H_{p_j q_i} H_{q_k} - 2H_{p_j q_k} H_{p_j} H_{q_k q_i} \\
& + 2H_{q_j q_k} H_{p_j q_i} H_{p_k} + H_{q_j q_k q_i} H_{p_j} H_{p_k} \Big],
\end{aligned}
$$

$$
\begin{aligned}
q_i^{n+1} = q_i^n + hH_{p_i} + \frac{h^3}{24} \Big[& H_{p_j p_k p_i} H_{q_j} H_{q_k} + 2H_{p_j p_k} H_{q_j p_i} H_{q_k} \\
& - 2H_{p_j q_k p_i} H_{p_j} H_{q_k} - 2H_{p_j q_k} H_{p_j p_i} H_{q_k} - 2H_{p_j q_k} H_{p_j} H_{q_k p_i} \\
& + 2H_{q_j q_k} H_{p_j p_i} H_{p_k} + H_{q_j q_k p_i} H_{p_j} H_{p_k} \Big].
\end{aligned}
$$

Here summation in repeated indices must be understood and the functions featuring in the right-hand sides are evaluated at the averages

$$
[\tfrac{1}{2}(\mathbf{p}^* + \mathbf{p}), \tfrac{1}{2}(\mathbf{q}^* + \mathbf{q})],
$$

so that the scheme is implicit. We have reported these formulae to emphasize the Taylor-series character of Feng's methods. As with any other Taylor-series method, these schemes would only be feasible if applied in conjunction with some automatic procedure for the computation of the higher derivatives.

Miesbach and Pesch (1990) note that, in Runge–Kutta methods for $\mathrm{d}y/\mathrm{d}t = \mathbf{f}(\mathbf{y})$, one obtains high-order schemes without resorting to higher derivatives of $\mathbf{f}$ by using as increment $\mathbf{y}^{n+1} - \mathbf{y}^n$ a weighted sum $h \sum b_i \mathbf{f}(\mathbf{Y}_i)$. Furthermore, the terms $h\mathbf{f}$ being weighted have the form of the increment in the Euler (i.e. simplest conceivable) method. In a similar vein, Miesbach and Pesch suggest methods where the Poincaré generating function is a weighted sum of terms, each term being the simplest generating function hH (i.e. that corresponding to the implicit midpoint rule) evaluated at some suitable inner stage. The resulting method is Runge–Kutta-like in that no higher derivatives of H are required; however it is not a symplectic Runge–Kutta method like those considered in Section 6.

11. Back to symplectic Runge–Kutta methods: the canonical theory of the order

The symplectic Runge–Kutta methods (6.1), (6.4) define a symplectic transformation which as $h \to 0$ approaches the identity. Hence they must have an S^3 generating function. Lasagni (1990) has found the corresponding expression

$$S^3(\mathbf{p}_n, \mathbf{q}_{n+1}; h) = \mathbf{p}_n^T \mathbf{q}_{n+1}$$
$$- h \sum_i b_i H(\mathbf{P}_i, \mathbf{Q}_i) - h^2 \sum_{ij} b_i a_{ij} H_{\mathbf{p}}(\mathbf{P}_i, \mathbf{Q}_i) H_{\mathbf{q}}(\mathbf{P}_j, \mathbf{Q}_j)^T.$$

Here $H_{\mathbf{p}}$ and H are row vectors of partial derivatives and the stages should be interpreted as functions of $\mathbf{p}_n$, $\mathbf{q}_{n+1}$ and h implicitly defined in (6.2), (6.3). (Actually, in Section 9 we showed that a generating function would exist if the domain Ω were simply connected. Lasagni's recipe for S^3 works for all domains. Symplectic RK have generating functions regardless of the geometry of Ω and therefore, in symplectic geometry jargon, they give rise to *exact symplectic* transformations, i.e. transformations for which (9.1) holds. Actually, the flow of a Hamiltonian system is also an exact symplectic transformation.)

In a manner similar to that used for symplectic PRK, Abia and Sanz-Serna (1990) find the generating function

$$S^3(\mathbf{p}_n, \mathbf{q}_{n+1}; h) = \mathbf{p}_n^T \mathbf{q}_{n+1}$$
$$- h \sum_i b_i V(\mathbf{Q}_i) - h \sum_i B_i T(\mathbf{P}_i) + h^2 \sum_{ij} B_i a_{ij} \mathbf{g}(\mathbf{P}_i)^T \mathbf{f}(\mathbf{Q}_j),$$

and for symplectic RKN schemes the generating function is given by (Calvo and Sanz-Serna, 1991a)

$$S^3(\mathbf{p}_n, \mathbf{q}_{n+1}; h) = \mathbf{p}_n^T \mathbf{q}_{n+1}$$
$$- h \sum_i b_i V(\mathbf{Q}_i) - \frac{h}{2} \mathbf{p}_n^T \mathbf{p}_n + \frac{h^3}{2} \sum_{ij} b_i (\beta_j - \alpha_{ij}) \, \mathbf{f}(\mathbf{Q}_i)^T \mathbf{f}(\mathbf{Q}_j).$$

We emphasize that, unlike the situation with the methods considered in

Section 10, these generating functions are not needed to derive or to implement the RK and related methods introduced in Section 6. However explicit knowledge of the generating function can be put to good use. In fact we can study the order of consistency by simply substituting the expression for S^3 with $h = t$ in $H + \partial S^3/\partial t$: an $\mathcal{O}(t^r)$ behaviour is, as we know, equivalent to order r. This is the methodology suggested by Sanz-Serna and Abia (1991). For the case of RK methods these authors give systematic rules, based on graph theory, to write the Taylor expansion of $H + \partial S^3/\partial t$ in powers of t. It turns out that the graphs to be used are *nonsuperfluous trees*: at the t^ρ level, $\rho > 1$, the Taylor expansion contains a term for each nonsuperfluous tree with ρ nodes. Hence the number of order conditions is the number of nonsuperfluous trees. Furthermore the coefficients that must be annihilated to impose $H + \partial S^3/\partial t = \mathcal{O}(t^r)$ are just the right-hand sides of the homogeneous order conditions we described at the end of Section 7, see e.g. (7.5)–(7.6). Thus the use of the Hamilton–Jacobi equation gives a very clear meaning to the results presented in Section 7.

12. Properties of symplectic integrators: backward error interpretation

Now that we have introduced the families of symplectic integrators available in the literature, it is time to investigate the general properties of symplectic integrators. The *a priori* motivation for resorting to symplectic methods was presented in Section 5: by making the integrator symplectic we reproduce an important property of the true flow. However there is a big gap in numerical analysis between a reasonably motivated method and a method that works well. It is therefore essential that theoretical analysis and numerical experiments are presented that show the advantages, if any, of symplectic integrators.

In our opinion, to the numerical analyst, the most appealing feature of symplectic integration is the possibility of *backward error interpretation*. This idea is very similar to the *method of modified equations*, see Warming and Hyett (1974) and, for a more rigorous treatment, Griffiths and Sanz-Serna (1986). Let us begin with an example. Consider the Hamiltonian $H = \frac{1}{2}p^2 + V(q)$, leading to the system

$$\mathrm{d}p/\mathrm{d}t = f(q), \qquad \mathrm{d}q/\mathrm{d}t = p, \tag{12.1}$$

where $f = -V'$. We assume that $f(0) = 0$ and $f'(0) < 0$; the first hypothesis implies that the origin is an equilibrium of (12.1), the second implies that this equilibrium is a stable centre (the origin is a minimum of the potential energy V). The system (12.1) is integrated by the following first-order, symplectic PRK method

$$p^{n+1} = p^n + hf(q^{n+1}), \qquad q^{n+1} = q^n + hp^n. \tag{12.2}$$

In order to describe the behaviour of the points (p^n, q^n) computed by (12.2), we could just say that they approximately behave like the solutions $(p(t_n), q(t_n))$ of (12.1). This would not be a very precise description, even for h small, because $\phi_{h,H}$ and $\psi_{h,H}$ differ in $\mathcal{O}(h^2)$ terms (first order of consistency). Can we find *another* differential system $\mathcal{S}_2$ so that (12.2) is consistent with the *second* order with $\mathcal{S}_2$? The points (p^n, q^n) would then be closer to the solutions of $\mathcal{S}_2$ than to the solutions of the system we want to integrate. To find the *modified* system $\mathcal{S}_2$ use an ansatz

$$dp/dt = f(q) + hF_1(p, q), \qquad dq/dt = p + hG_1(p, q)$$

(note that h features here as a parameter so that $\mathcal{S}_2 = \mathcal{S}_2(h)$), substitute the solutions of the modified system in the difference equations and ask for an $\mathcal{O}(h^3)$ residual. This leads to

$$\frac{dp}{dt} = f(q) + \frac{h}{2}pf'(q), \qquad \frac{dq}{dt} = p - \frac{h}{2}f(q), \qquad (12.3)$$

a Hamiltonian system, with Hamiltonian

$$H_2(h) = \tfrac{1}{2}p^2 + V(q) - (h/2)pf(q).$$

If we are not satisfied with $\mathcal{S}_2(h)$, we can find a differential system $\mathcal{S}_3(h)$ for which (12.2) is consistent with the third order. Again $\mathcal{S}_3(h)$ turns out to be a Hamiltonian problem; the expression for the Hamiltonian is

$$H_3(h) = (1/2)p^2 + V(q) - (h/2)pf(q) + (h^2/12)[f(q)^2 - p^2 f'(q)].$$

There is no limit: for any positive integer ρ a Hamiltonian system $\mathcal{S}_{H_\rho}(h)$ can be found such that the method $\psi_{h,H}$ differs from the flow $\phi_{h,H_\rho(h)}$ in $\mathcal{O}(h^{\rho+1})$ terms (see e.g. MacKay (1991)). By going from local to global errors, in any bounded time interval, the computed points are $\mathcal{O}(h^\rho)$ away from the solution of $\mathcal{S}_{H_\rho}(h)$.

What is the situation when using a nonsymplectic method? Take the standard forward Euler method as an illustration. Again a modified system $\mathcal{S}_2(h)$ can be found for which consistency is of the second order. This now reads

$$\frac{dp}{dt} = \left[f(q) + \frac{h}{2}pf'(q)\right] - hpf'(q), \qquad \frac{dq}{dt} = \left[p - \frac{h}{2}f(q)\right];$$

the terms in brackets replicate the Hamiltonian system (12.3), but there is an extra term $-hf'(q)p$. Since $f'(0) < 0$ this extra term introduces negative dissipation near the origin: in any bounded time interval, the computed points are $\mathcal{O}(h)$ away from the solutions of the Hamiltonian system we want to solve, but $\mathcal{O}(h^2)$ away from the solutions of a system where the Hamiltonian character has been lost and the origin is an unstable focus.

Even though these considerations have been presented by means of an example, they hold for all symplectic methods: provided that the system

(2.1) is smooth enough, for arbitrarily high ρ, a modified Hamiltonian system $\mathcal{S}_{H_\rho(h)}$ can be found such that the method $\psi_{h,H}$ differs from the flow $\phi_{h,H_\rho(h)}$ in $\mathcal{O}(h^{\rho+1})$ terms. The difference between the true Hamiltonian H and the modified Hamiltonian $H_\rho(h)$ is $\mathcal{O}(h^r)$, with r the order of the method.

As shown in this example, the functions $H_\rho(h)$, $\rho = 2, 3, \ldots$, are truncations of a power series in h. If this power series converges, its sum $H_\infty(h)$ gives rise to a modified Hamiltonian problem *that is integrated exactly by the symplectic numerical method*: $\psi_{h,H} \equiv \phi_{h,H_\infty(h)}$. In the previous example with $V(q) = \frac{1}{2}q^2$ (i.e. the harmonic oscillator (2.5)) this modified Hamiltonian problem is given by Beyn (1991, p. 221)

$$\frac{\mathrm{d}}{\mathrm{d}t} \left[\begin{array}{c} p \\ q \end{array} \right] = \left(h^{-1} \log \left[\begin{array}{cc} 1 - h^2 & -h \\ h & 1 \end{array} \right] \right) \left[\begin{array}{c} p \\ q \end{array} \right]; \qquad (12.4)$$

when solving this system analytically the matrix in brackets is exponentiated and the equations (12.2) of the numerical method are recovered.

In general, for nonlinear problems, the series does not converge: the computed points are not quite an exact solution of a differential problem (Sanz-Serna, 1991a, p. 168). However if H is very smooth, it can be shown (Neishtadt, 1984; cf. Lasagni, 1990; MacKay, 1991) that a Hamiltonian $H_\infty(h)$ can be constructed for which the corresponding h-flow differs from $\psi_{h,H}$ in terms that tend to 0 exponentially fast as $h \to 0$.

In any case the conclusion is the same: for a symplectic integrator applied to (2.1) modified Hamiltonian problems exist so that the computed points lie either exactly or 'very approximately' on the exact trajectories of the modified problem. This makes a backward error interpretation of the numerical results possible (cf. Sanz-Serna (1990)): the computed solutions are solving exactly (or 'very approximately') a nearby Hamiltonian problem. In a modelling situation where the exact form of the Hamiltonian H may be in doubt, or some coefficients in H may be the result of experimental measurements, the fact that integrating the model numerically introduces perturbations to H comparable with the uncertainty in H inherent in the model is the most one can hope for.

On the other hand, when a nonsymplectic formula is used the modified system is not Hamiltonian: the process of numerical integration perturbs the model in such a way as to take it out of the Hamiltonian class. The acceptability of such nonHamiltonian perturbations is a question that should be decided in each individual modelling problem.

12.1. An alternative approach

If

$$(\mathbf{p}, \mathbf{q}) = \psi_{h,H}(\mathbf{p}^0, \mathbf{q}^0) \qquad (12.5)$$

is a numerical method, it is a simple matter to find a differential equation satisfied by the functions $(\mathbf{p}(h), \mathbf{q}(h))$: we differentiate (12.5) with respect to h and eliminate $(\mathbf{p}^0, \mathbf{q}^0)$ in the result by using (12.5).

For the symplectic method (12.2) this procedure yields the system

$$\frac{dp}{dh} = f(q) + hpf'(q) - h^2 f(q) f'(q), \qquad \frac{dq}{dh} = p - hf(q),$$

which (this should not be surprising by now) is a Hamiltonian system, with Hamiltonian $\widetilde{H}(p, q; h) = \frac{1}{2}(p - hf(q))^2 + V(q)$. Since h is our 'time', the system is nonautonomous. Moving from $t = 0$ to $t = h$ with the method (12.2) is moving from $t = 0$ to $t = h$ with a nonautonomous system with Hamiltonian $\widetilde{H}$. The fact that H and $\widetilde{H}$ differ in $\mathcal{O}(h)$ terms is a reflection of the first-order accuracy of the method.

What is unsatisfactory with this approach is that taking two steps $0 \to h \to 2h$ with the numerical method is *not* going from $t = 0$ to $t = 2h$ with $S_{\widetilde{H}}$: given an initial condition, to move from 0 to $2h$ in a nonautonomous differential system is not quite the same as advancing the initial condition to $t = h$ and then using the result as new initial condition for another $0 \to h$ forward shift. There is a way around this problem: for $0 \le t < h$ we keep the Hamiltonian $\widetilde{H}(p, q; t)$ found earlier and for $h \le t < 2h$, $2h \le t < 3h$, ..., we repeat it periodically. The good news is that now the nonautonomous system is such that the transformation that moves the initial condition from $t = 0$ to $t = nh$ is the nth power of the transformation that advances the initial condition from $t = 0$ to $t = h$. Hence, the numerically computed points exactly lie on solutions of this nonautonomous system. The bad news is that the new Hamiltonian is not only nonautonomous, but also discontinuous as a function of t. Such a lack of smoothness is not very welcome.

The canonical formalism of generating functions provides a very clever way of finding $\widetilde{H}$ without having to differentiate $\psi_{h,H}$. The initial conditions $(\mathbf{p}^0, \mathbf{q}^0)$ do not vary with t: we could see them as solutions of the 0 Hamiltonian. By using Theorem 9.1, the functions $(\mathbf{p}(h), \mathbf{q}(h))$ then evolve with the Hamiltonian $\widetilde{H} = -\partial S^3/\partial h$. In the example, the generating function is $S^3 = p^0 q - (h/2)(p^0)^2 - hV(q)$ (now p and q play the role played by p^* and q^* in Section 9, while p^0 and q^0 play now the role of 'old' variables). Differentiation with respect to h in S^3 leads to $\widetilde{H} = \frac{1}{2}(p^0)^2 + V(q)$; in view of (12.2) this equals $\frac{1}{2}(p - hf(q))^2 + V(q)$, the same expression we found before.

McLachlan and Atela (1991) use the discrepancy between H and $\widetilde{H}$ as a measure of the accuracy of the method $\psi_{h,H}$. Since such a discrepancy equals $H + \partial S^3/\partial h$ this is just using the Hamilton–Jacobi methodology introduced by Sanz-Serna and Abia (1991).

13. Properties of symplectic integrators: conservation of energy

For the system (2.1) the Hamiltonian H is a conserved quantity: $H(\mathbf{p}(t),\mathbf{q}(t))$ does not vary with t if $(\mathbf{p}(t), \mathbf{q}(t))$ is a solution of (2.1). In applications in mechanics conservation of H usually corresponds to conservation of total mechanical energy. Do symplectic integrators possess the analogous property that, except for rounding errors, $H(\mathbf{p}^n, \mathbf{q}^n)$ does not vary with n along a numerically computed solution? Sometimes they do: if (2.1) is a *linear* system and the integrator is a symplectic RK method (6.1)–(6.4), then H is conserved along numerical trajectories. In fact in this case H is a quadratic function and symplectic RK methods conserve all quadratic functions that are conserved by the Hamiltonian system being integrated, Sanz-Serna (1988). However if we still assume linearity in the system and we use a PRK or a RKN method, conservation of H no longer holds. This is easily seen in the case of the harmonic oscillator (2.5) integrated by the method (12.2). We have noticed earlier that the computed points exactly lie on trajectories of the modified system (12.4) and hence on the lines $H_\infty(p, q; h) =$ constant in the (p, q) plane. But, for h small, these lines can be seen to be ellipses, while for conservation of energy we wanted the points to be on circles $p^2 + q^2 =$ constant. As $h \to 0$ the eccentricity of the ellipses decreases and they look more like circles: smaller values of h lead to smaller energy errors, as in the consistent method. Furthermore the fact that the computed points stay exactly on an ellipse near the theoretical circle implies that the error in energy remains bounded even if t gets very large.

The same ideas apply more generally. When problem (2.1) is very smooth but nonlinear, the computed points do not remain exactly on trajectories of the modified problem $\mathcal{S}_{H_\infty(h)}$. Nevertheless, the drift of the points away from the modified trajectories is very slow: the numerical scheme has exponentially small local truncation errors when seen as an approximation to the modified system. Therefore $H_\infty(h)$ is conserved by the numerical solutions, except for exponentially small errors, for long $(\mathcal{O}(h^{-1}))$ periods of time. This in turn implies that the errors in $H = H_\infty(h) + \mathcal{O}(h^r)$ possess an $\mathcal{O}(h^r)$ bound on time intervals of length $\mathcal{O}(h^{-1})$ (Lasagni, 1988).

For 'general' Hamiltonians, Ge and Marsden (1988) prove that a symplectic method $\psi_{h,H}$ cannot exactly conserve energy (except for the trivial cases where the function $\psi_{h,H}$ actually coincides with or is a time reparameterization of the true flow $\phi_{h,H}$). Hence conservation of the symplectic structure and conservation of energy are conflicting requirements that, in general, cannot be satisfied simultaneously by a numerical scheme. Since both the Hamiltonian and the symplectic structure are conserved by Hamiltonian systems, the question naturally arises of whether when constructing an integrator we should choose to conserve symplecticness and violate conservation of energy or *vice versa*. This is a question that should probably be answered differently for each specific application. However it should be pointed out that, as mentioned in Section 4, symplecticness is a property that fully characterizes Hamiltonian problems, while conservation of an energy-like function is a feature also present in many nonHamiltonian systems. Furthermore conservation of energy restricts the dynamics of the numerical solution by forcing the computed points to be on the correct $(2d - 1)$-dimensional manifold $H = $ constant, but otherwise poses no restriction to the dynamics: within the manifold the points are free to move anywhere and only motions orthogonal to the manifold are forbiden. When d is large this is clearly a rather weak restriction. On the other hand, symplecticness restricts the dynamics in a more global way: all directions in phase space are taken into account.

The literature has devoted a great deal of attention to the construction of numerical schemes that exactly conserve H (or more generally, to the construction of integrators for a system $d\mathbf{y}/dt = \mathbf{f}(\mathbf{y})$ that exactly conserve one or more invariants of motion). Several ideas have been suggested:

1   stepping from t_n to t_{n+1} with a standard method and then projecting the numerical result onto the correct energy surface;

2   adding the conservation constraints to the differential system to obtain a system of differential-algebraic equations; and

3   constructing *ad hoc* schemes. However conservation of energy is not the theme of this paper and we shall not attempt to review the relevant literature.

14. Properties of symplectic integrators: KAM theory

The Kolmogorov–Arnold–Moser (KAM) theory for Hamiltonian problems explains the behaviour of Hamiltonian systems that are perturbations of so-called integrable Hamiltonian systems (i.e. of Hamiltonian systems that can be explicitly solved in terms of quadratures). This material is covered in the books by Moser (1973), Arnold (1988, 1989) and MacKay and Meiss (1987). The theory also caters for the case of symplectic mappings that are perturbations of integrable symplectic mappings. Therefore KAM results can often be applied to the mappings $\psi_{h,H}$ associated with symplectic integrators.

To get the flavour of this sort of application, let us consider once more the method (12.2) applied to (12.1). Recall that the origin is a (stable) centre for the system (12.1). For the discrete equations (12.2), linearization around the origin leads to

$$p^{n+1} = p^n + hf(0)q^{n+1}, \qquad q^{n+1} = q^n + hp^n, \qquad (14.1)$$

a system that has, for h small, unit modulus eigenvalues. Thus the origin is also a centre for (14.1). However to go from (14.1) to the discretization (12.2) we must include the effects of the nonlinear terms that were discarded in the process of linearization. Since (14.1) is only neutrally stable, it may be feared that the nonlinear effects, small as they may be, will render the origin unstable for (12.2). The KAM theory can be used to show that the symplecticness of the method implies that such a destabilization does not occur. Full details of this example have been given in Sanz-Serna (1991a). Incidentally, we would like to point out that it is this mechanism that renders the standard explicit midpoint rule stable in many nonlinear problems, even though this rule is only neutrally stable in a linear analysis. The interested reader is referred to Sanz-Serna and Vadillo (1986, 1987).

15. Practical performance of symplectic integrators

Numerical tests provide the final verdict on the usefulness of any numerical method. For Hamiltonian problems, are symplectic methods more advantageous in practice than their nonsymplectic counterparts? Before we answer this question, let us observe that many symplectic methods are implicit. Even though explicit symplectic algorithms exist in the PRK and RKN families, they are only applicable to restricted classes of Hamiltonians. Furthermore, when deriving such explicit methods, free parameters are used to ensure symplecticness which could otherwise be directed at increasing accuracy. The result is that, to achieve a given order, a symplectic explicit PRK or RKN method usually needs more stages than a standard PRK or RKN method. All these considerations show that there is a price to pay for symplecticness. Symplecticness is expected to pay back when performing very long time integrations: then a symplectic scheme has some inbuilt features that may guarantee the right long-term qualitative behaviour and even result in a favourable error propagation mechanism. On the other hand for short-time integrations, where accuracy is of paramount importance, a good standard code is expected to outperform any symplectic method.

Menyuk (1984), Feng and Qin (1987), Sanz-Serna (1989), Channell and Scovel (1990), Miesbach and Pesch (1990), Candy and Rozmus (1991), McLachlan and Atela (1991), Okunbor and Skeel (1991) and Pullin and Saffman (1991) provide numerical experiments involving symplectic integrators. The sort of experiment performed often consists of the application of a symplectic method to the long-time integration of a Hamiltonian problem; some sort of graphic output is then examined. The conclusions appear to be that symplectic integrators are very successful in identifying most relevant qualitative features of Hamiltonian flows. In most of the papers cited here, the symplectic method is tested against a standard method of the same order of accuracy. The standard method is usually proved to require much smaller step-sizes to correctly identify the true dynamics.

This sort of experimentation, encouraging as it may be to the developer of symplectic methods, is open to criticism. To begin with, the reference standard method being used tends to be either the classical fourth-order RK method or a low accuracy RK formula like the modified Euler scheme. These reference methods are far away from state of the art numerical integrators. Furthermore, in the experiments we are discussing, both the symplectic integrator and the reference standard method are implemented with constant step-sizes, which again is far away from current numerical ODE practice. It is legitimate to ask what would happen if in the comparisons the nonsymplectic method would have been chosen to be a modern variable step-size code. On the other hand, this criticism may not be entirely fair: standard methods have been under development for several decades, while we are at the stone age of symplectic integration; it may then make sense to compare our symplectic integrators with stone-age standard methods.

A somewhat more severe test has been conducted by Calvo and Sanz-Serna (1991b,c). A fourth-order, explicit, symplectic RKN method is constructed which is optimal in the sense that the error constants have been minimized following a methodology due to Dormand *et al.* (1987). First, this symplectic integrator, implemented with *constant step-sizes*, is compared with a *variable step*-code based on an optimal fourth-order nonsymplectic formula of Dormand *et al.* (1987). The result of the comparison is that, in long time integrations, the symplectic method definitely needs less work to achieve a given accuracy. This holds even in cases where the solution possesses several time scales along the integration interval and the code is much benefiting from the step-changing facility. In the integration of Kepler's problem, it can be shown rigorously (Calvo and Sanz-Serna, 1991c) that for symplectic integrators the errors grow linearly with t, while for nonsymplectic methods grow like t^2. Hence the symplectic methods are guaranteed to win if t is large enough.

15.1. *Variable step-sizes*

Calvo and Sanz-Serna (1991b) then go on to compare the nonsymplectic code with a variable step-size implementation of the symplectic formula. For this implementation, due care was exercised in constructing the error estimator, etc. Before the experiments were conducted it was expected that the combination of the advantages of symplecticness with those of variable step-sizes would lead to a very efficient algorithm. The numerical results were very disappointing: in the variable step-size implementation, the symplectic formula does not show any advantage in the long-time error propagation mechanism. For instance, for Kepler's problem the error growth is quadratic, just as if a nonsymplectic formula were used. Since the cost per step of the simplectic algorithm is higher than that of the standard code (see earlier), the conclusion is that the variable step-size simplectic algorithm is not competitive with the standard code.

It thus appears that there is a future for the practical application of symplectic integration, especially if high order symplectic formulae are developed and if advances are made in efficiently implementing the implicit symplectic methods. However such a future seems to be limited to constant step-size implementations!

Before closing this section it is appropriate to say some words on the failure of variable step-size symplectic methods. In Section 12 we pointed out that a symplectic integrator $\psi_{h,H}$ 'almost' provides the exact flow of a Hamiltonian problem $\phi_{h,H_\infty(h)}$. If h is held constant during the integration, the initial condition is numerically advanced to $t = t_n$ by

$$\overbrace{\psi_{h,H}\psi_{h,H}\cdots\psi_{h,H}}^{n}$$

which for t_n in a compact time interval differs from

$$\overbrace{\phi_{h,H_\infty(h)}\phi_{h,H_\infty(h)}\cdots\phi_{h,H_\infty(h)}}^{n} = \phi_{t_n,H_\infty(h)}$$

in exponentially small terms: the computed points stay very close of a modified Hamiltonian trajectory. The situation is quite different for variable step-sizes. Now the initial condition is advanced by

$$\psi_{h_n,H}\psi_{h_{n-1},H}\cdots\psi_{h_1,H}, \tag{15.1}$$

an approximation to

$$\phi_{h_n,H_\infty(h_n)}\phi_{h_{n-1},H_\infty(h_{n-1})}\cdots\phi_{h_1,H_\infty(h_1)}.$$

The last expression cannot be interpreted as the t_n-flow of a Hamiltonian problem: the Hamiltonians being used at different time steps are different. This shows that the backward error interpretation of symplectic integration does not hold for variable step-sizes.

There is a difficulty here: in a variable step-size code the step points t_n are actually functions of the initial point $(\mathbf{p}^0, \mathbf{q}^0)$ (and also of the initial guess for the first step-size). Therefore the algorithm does not really effect a transformation mapping the phase space Ω at $t = 0$ into the phase space Ω at time t, rather $(\Omega \times (t = 0))$ is mapped into some curved $2d$-dimensional surface in the $(2d+1)$-dimensional spacetime. It is then possible to question the relevance of (15.1) to the analysis of the variable step implementation. However in the experiments reported by Calvo and Sanz-Serna (1991b) only one fixed initial condition was used so that, in a 'mental experiment', one could pretend that the sequence of step-sizes $h_1, h_2, \ldots$, actually used in the integration was recorded and would have been used to integrate neighbouring initial conditions. In this context, compatible with the numerical experiments, the initial condition is really advanced by the symplectic transformation (15.1).

The advantages of symplectic integration may well originate from the fact that one advances from $t = 0$ to time t_n by iterating n times a single symplectic mapping. Advancing by composing n *different* symplectic mappings does not appear to be as effective.

16. Concluding remarks

In the paper we have restricted ourselves to standard Hamiltonian problems (2.1) on a domain Ω in an even-dimensional oriented Euclidean space. One may also consider a so-called symplectic manifold, an even dimensional manifold endowed with a closed, nondegenerate differential 2-form that plays the role that was played here by $d\mathbf{p} \wedge d\mathbf{q}$ (Arnold, 1989; MacKay, 1991). In such a manifold to each scalar function H there corresponds a Hamiltonian-like system of differential equations. More generally one could consider a Poisson manifold. A reference where a symplectic integrator is derived for a Poisson system is de Frutos *et al.* (1990). Another area of active research in the physics literature is that of Lie–Poisson integrators, see e.g. Ge and Marsden (1988).

Many partial differential equations also possess a Hamiltonian structure. In connection with symplectic integration they pose two problems: how to discretize them in space to obtain a Hamiltonian semi-discretization and how to advance in time the semi-discrete solution to have an overall symplectic algorithm. Some references are Qin (1988), Li and Qin (1988), Qin and Zhang (1990), de Frutos *et al.* (1990).

Acknowledgments

This work has been supported by Junta de Castilla y León under project 1031-89 and by Dirección General de Investigación Científica y Técnica under project PB89-0351. I am very thankful to Miss M. P. Calvo for her help with the preparation of the manuscript.

REFERENCES

L. Abia and J.M. Sanz-Serna (1990), 'Partitioned Runge–Kutta methods for separable Hamiltonian problems', *Applied Mathematics and Computation Reports*, Report 1990/8, Universidad de Valladolid.

K. Aizu (1985), 'Canonical transformation invariance and linear multistep formula for integration of Hamiltonian systems', *J. Comput. Phys.* **58**, 270–274.

D.V. Anosov and V.I. Arnold (eds) (1988), *Dynamical Systems I*, Springer (Berlin).

V.I. Arnold (ed.) (1988), *Dynamical Systems III*, Springer (Berlin).

V.I. Arnold (1989), *Mathematical Methods of Classical Mechanics*, 2nd edition, Springer (New York).

V.I. Arnold and S.P. Novikov (1990), *Dynamical Systems IV*, Springer, Berlin.

W.-J. Beyn (1991), 'Numerical methods for dynamical systems', in *Advances in Numerical Analysis, Vol. I* (W. Light, ed.) Clarendon Press (Oxford) 175–236.

K. Burrage and J.C. Butcher (1979), 'Stability criteria for implicit Runge–Kutta methods', *SIAM J. Numer. Anal.* **16**, 46–57.

J.C. Butcher (1976), 'On the implementation of implicit Runge–Kutta methods', *BIT* **16**, 237–240.

J.C. Butcher (1987), *The Numerical Analysis of Ordinary Differential Equations*, John Wiley (Chichester).

M.P. Calvo (1991), Ph.D. Thesis, University of Valladolid (to appear).

M.P. Calvo and J.M. Sanz-Serna (1991a), 'Order conditions for canonical Runge–Kutta–Nyström methods', *BIT* to appear.

M.P. Calvo and J.M. Sanz-Serna (1991b), 'Variable steps for symplectic integrators', *Applied Mathematics and Computation Reports*, Report 1991/3, Universidad de Valladolid.

M.P. Calvo and J.M. Sanz-Serna (1991c), 'Reasons for a failure. The integration of the two-body problem with a symplectic Runge–Kutta–Nyström code with step-changing facilities', *Applied Mathematics and Computation Reports*, Report 1991/7, Universidad de Valladolid.

J. Candy and W. Rozmus (1991), 'A symplectic integration algorithm for separable Hamiltonian functions', *J. Comput. Phys.* **92**, 230–256.

P.J. Channell, 'Symplectic integration algorithms', *Los Alamos National Laboratory Report*, Report AT-6ATN 83-9.

P.J. Channell and C. Scovel (1990), 'Symplectic integration of Hamiltonian systems', *Nonlinearity* **3**, 231-259.

G.J. Cooper (1987), 'Stability of Runge–Kutta methods for trajectory problems', *IMA J. Numer. Anal.* **7**, 1–13.

G.J. Cooper and R. Vignesvaran (1990), 'A scheme for the implementation of implicit Runge–Kutta methods', *Computing* **45**, 321–332.

M. Crouzeix (1979), 'Sur la B-stabilité des méthodes de Runge–Kutta', *Numer. Math.* **32**, 75–82.

K. Dekker and J.G. Verwer (1984), *Stability of Runge–Kutta Methods for Stiff Nonlinear Differential Equations*, North Holland (Amsterdam).

J.R. Dormand, M.E.A. El-Mikkawy and P.J. Prince (1987), 'Families of Runge–Kutta–Nyström formulae', *IMA J. Numer. Anal.* **7**, 235–250.

T. Eirola and J.M. Sanz-Serna (1990), 'Conservation of integrals and symplectic structure in the integration of differential equations by multistep methods', *Applied Mathematics and Computation Reports*, Report 1990/9, Universidad de Valladolid.

Feng Kang (1985), 'On difference schemes and symplectic geometry', in *Proceedings of the 1984 Beijing Symposium on Differential Geometry and Differential Equations* (Feng Kang, ed.) Science Press (Beijing) 42–58.

Feng Kang (1986a), 'Difference schemes for Hamiltonian formalism and symplectic geometry', *J. Comput. Math.* **4**, 279–289.

Feng Kang (1986b), 'Symplectic geometry and numerical methods in fluid dynamics', in *Tenth International Conference on Numerical Methods in Fluid Dynamics (Lecture Notes in Physics 264)* (F. G. Zhuang and Y. L. Zhu, eds) Springer (Berlin) 1–7.

Feng Kang and Qin Meng-zhao (1987), 'The symplectic methods for the computation of Hamiltonian equations', in *Proceedings of the 1st Chinese Conference for Numerical Methods for PDE's, March 1986, Shanghai (Lecture Notes in Mathematics 1297)* (Zhu You-lan and Gou Ben-yu, eds) Springer (Berlin) 1–37.

Feng Kang, Wu Hua-mo and Qin Meng-zhao (1990), 'Symplectic difference schemes for linear Hamiltonian canonical systems', *J. Comput. Math.* **8**, 371–380.

Feng Kang, Wu Hua-mo, Qin Meng-zhao and Wang Dao-liu (1989), 'Construction of canonical difference schemes for Hamiltonian formalism via generating functions', *J. Comput. Math.* **7**, 71–96.

E. Forest and R. Ruth (1990), 'Fourth-order symplectic integration', *Physica D* **43**, 105–117.

J. de Frutos and J.M. Sanz-Serna (1991), 'An easily implementable fourth-order method for the time integration of wave problems', *Applied Mathematics and Computation Reports*, Report 1991/2, Universidad de Valladolid.

J. de Frutos, T. Ortega and J.M. Sanz-Serna (1990), 'A Hamiltonian, explicit algorithm with spectral accuracy for the "good" Boussinesq system', *Comput. Methods Appl. Mech. Engrg.* **80**, 417–423.

Ge Zhong and J.E. Marsden (1988), 'Lie–Poisson Hamilton–Jacobi theory and Lie-Poisson integrators', *Phys. Lett. A* **133**, 134–139.

D.F. Griffiths and J.M. Sanz-Serna (1986), 'On the scope of the method of modified equations ', *SIAM J. Sci. Stat. Comput.* **7**, 994–1008.

J. Guckenheimer and P. Holmes (1983), *Nonlinear Oscillations, Dynamical Systems and Bifurcations of Vector Fields*, Springer (New York).

E. Hairer, A. Iserles and J.M. Sanz-Serna (1990), 'Equilibria of Runge–Kutta methods', *Numer. Math.* **58**, 243–254.

E. Hairer, S.P. Nørsett and G. Wanner (1987), *Solving Ordinary Differential Equations I, Nonstiff Problems*, Springer (Berlin).

A. Iserles (1990a), 'Stability and dynamics of numerical methods for ordinary differential equations', *IMA J. Numer. Anal.* **10**, 1–30.

A. Iserles (1990b), 'Efficient Runge–Kutta methods for Hamiltonian equations', *Numerical Analysis Report DAMTP* , Report 1990/NA10, University of Cambridge.

A. Iserles and S.P. Nørsett (1991), *Order Stars*, Chapman and Hall (London).

F. Klein (1926), *Vorlesungen Über die Entwicklung der Mathematik im 19.Jahrhundert*, Teil I, Springer (Berlin).

F. Lasagni (1988), 'Canonical Runge–Kutta methods', *ZAMP* **39**, 952–953.

F. Lasagni (1990), 'Integration methods for Hamiltonian differential equations', unpublished manuscript.

Li Chun-wang and Qin Meng-zhao (1988), 'A symplectic difference scheme for the infinite dimensional Hamiltonian system', *J. Comput. Math.* **6**, 164–174.

R.S. MacKay (1991), 'Some aspects of the dynamics and numerics of Hamiltonian systems', in *Proceedings of the 'Dynamics of Numerics and Numerics of Dynamics' Conference* (D.S. Broomhead and A. Iserles, eds) (Oxford University Press (Oxford) to appear.

R.S. MacKay and J.D. Meiss (eds) (1987), *Hamiltonian Dynamical Systems*, Adam Hilger (Bristol).

R. McLachlan and P. Atela (1991), 'The accuracy of symplectic integrators', *Program in Applied Mathematics*, Report PAM 76, University of Colorado.

C.R. Menyuk (1984), 'Some properties of the discrete Hamiltonian method', *Physica D* **11**, 109–129.

S. Miesbach and H.J. Pesch (1990), 'Symplectic phase flow approximation for the numerical integration of canonical systems', *Schwerpunktprogramm der Deutschen Forschungsgemeinschaft, Anwendungsbezogene Optimierung und Steuerung*, Report 233, Technische Universität München.

J. Moser (1973), *Stable and Random Motions in Dynamical Systems*, Princeton University Press (Princeton).

A.I. Neishtadt (1984), 'The separation of motions in systems with rapidly rotating phase', *J. Appl. Math. Mech* **48**, 133-139.

F. Neri (1987), 'Lie algebras and canonical integration', University of Maryland Technical Report.

D. Okunbor and R.D. Skeel (1990), 'An explicit Runge–Kutta–Nyström method is canonical if and only if its adjoint is explicit', *Working Document* 90-3, Department of Computer Science, University of Illinois at Urbana-Champaign.

D. Okunbor and R.D. Skeel (1991), 'Explicit canonical methods for Hamiltonian systems ', *Working Document* 91-1, Department of Computer Science, University of Illinois at Urbana-Champaign.

D.I. Pullin and P.G. Saffman (1991), 'Long-time symplectic integration: the example of four-vortex motion', *Proc. Roy. Soc. Lond. A* **432**, 481-494.

Qin Meng-zhao (1988), 'Leap-frog schemes of two kinds of Hamiltonian systems for wave equations', *Math. Num. Sin.* **10**, 272–281.

Qin Meng-zhao and Zhang Mei-qing (1990), 'Multi-stage symplectic schemes of two kinds of Hamiltonian systems for wave equations', *Comput. Math. Appl.* **19**, 51–62.

R.D. Ruth (1983), 'A canonical integration techniqe', *IEEE Trans. Nucl. Sci.* **30**, 2669–2671.

J.M. Sanz-Serna (1988), 'Runge–Kutta schemes for Hamiltonian systems', *BIT* **28**, 877-883.

J.M. Sanz-Serna (1989) 'The numerical integration of Hamiltonian systems', in *Proc. 1989 London Numerical ODE Conference*, Oxford University Press (Oxford) to appear.

J.M. Sanz-Serna (1990), 'Numerical ordinary differential equations vs. dynamical systems', *Applied Mathematics and Computation Reports*, Report 1990/3, Universidad de Valladolid.

J.M. Sanz-Serna (1991a), 'Two topics in nonlinear stability', in *Advances in Numerical Analysis, Vol. I* (W. Light, ed.) Clarendon Press (Oxford) 147–174.

J.M. Sanz-Serna (1991b), 'Symplectic Runge–Kutta and related methods: recent results', *Applied Mathematics and Computation Reports*, Report 1991/5, Universidad de Valladolid.

J.M. Sanz-Serna and L. Abia (1991), 'Order conditions for canonical Runge–Kutta schemes', *SIAM J. Numer. Anal.* **28**, 1081–1096.

J.M. Sanz-Serna and D.F. Griffiths (1991), 'A new class of results for the algebraic equations of implicit Runge–Kutta processes', *IMA J. Numer. Anal.* to appear.

J.M. Sanz-Serna and F. Vadillo (1986), 'Nonlinear instability, the dynamic approach', in *Numerical Analysis* (D.F. Griffiths and G.A. Watson, eds) Longman (London) 187–199.

J.M. Sanz-Serna and F. Vadillo (1987), 'Studies in nonlinear instability III: augmented Hamiltonian problems', *SIAM J. Appl. Math.* **47**, 92–108.

Y.B. Suris (1989), 'Canonical transformations generated by methods of Runge–Kutta type for the numerical integration of the system $x'' = -\partial U \partial x$', *Zh. Vychisl. Mat. i Mat. Fiz.* **29**, 202-211 (in Russian). An English translation by D.V. Bobyshev, edited by R.D. Skeel, is available as Working Document 91-2, Numerical Computing Group, Department of Computer Science, University of Illinois at Urbana-Champaign.

Y.B. Suris (1990), 'Hamiltonian Runge–Kutta type methods and their variational formulation', *Math. Sim.* **2**, 78–87 (in Russian).

R.F. Warming and B.J. Hyett (1974), 'The modified equation approach to the stability and accuracy analysis of finite difference methods', *J. Comput. Phys.* **14**, 159–179.

H. Yoshida (1990), 'Construction of higher order symplectic integrators', *Phys. Lett. A* **150**, 262–268.

Wu Yuhua (1988), 'The generating function for the solution of ODE's and its discrete methods', *Comput. Math. Appl.* **15**, 1041–1050.

Acta Numerica (1991), *pp.* 287–339

Error analysis of boundary integral methods

Ian H. Sloan
School of Mathematics
University of New South Wales
Sydney, NSW 2033 Australia
E-mail: sloan@hydra.maths.unsw.oz.au

Many of the boundary value problems traditionally cast as partial differential equations can be reformulated as integral equations over the boundary. After an introduction to boundary integral equations, this review describes some of the methods which have been proposed for their approximate solution. It discusses, as simply as possible, some of the techniques used in their error analysis, and points to areas in which the theory is still unsatisfactory.

CONTENTS

1. Introduction

In the past decade there has been a dramatic growth of engineering interest in boundary integral or boundary element methods, witnessed by the large number of recent conference proceedings with these words in the title. At the same time, the former rivalry between advocates of BIE (boundary integral equation) and PDE (partial differential equation) approaches seems to have softened, as the relative strengths and weaknesses of each have become better understood.

Boundary integral methods may be used for interior and exterior problems, but have a special advantage for the latter. As a first introduction,

consider (without equations!) the problem of acoustic scattering from an
object. Under appropriate idealization, the pressure in the region exterior
to the object satisfies the wave equation, with an appropriate (typically
Neumann) condition on the scattering surface, and a radiation condition
at infinity. With the time variable separated, the equation becomes the
Helmholtz equation. Regarded as a PDE problem, the setting is an infinite
three-dimensional region. The boundary integral formulation of this problem, on the other hand, lives in a region that is only two-dimensional and
finite – namely the surface of the scatterer.

We leave until the next section any serious discussion of boundary integral
formulations (and refer to Colton and Kress (1983) for the specific matter of
the Helmholtz equation), but some readers may find the following thought
useful: If we knew the Green's function for this scattering problem, then
the pressure at any point could be found by quadrature over the surface.
But the true Green's function, incorporating the boundary condition on the
scatterer, is even harder to find than the solution itself. The next best thing
is to use the known fundamental solution, which is the Green's function for
the infinite region with *no* scatterer. That incorporates the boundary condition at infinity, and solves the differential equation, but takes no account
of the scatterer. To obtain a solution that satisfies the boundary conditions
on the scatterer we must therefore solve for an unknown function over the
surface of the scatterer. The equation to be solved is a (boundary) integral
equation.

Compared to PDE formulations, those involving BIEs are usually of lower
dimensionality (e.g. two-dimensional against three-dimensional in the earlier
example). On the other hand BIE methods almost invariably have dense
matrices, in contrast to the sparse matrices given by the PDE methods.
Moreover, the matrix elements are relatively hard to compute, involving
for example weakly or strongly singular kernels, perhaps (particularly in
the Galerkin method) several levels of integration, and difficult geometry.
Boundary integral equations rely fundamentally on the linear superposition
of solutions, and therefore are happiest when the underlying differential
equations are linear and homogeneous, and the material properties constant.
The PDE methods in contrast, being local in character, are not so fussy
about any of these matters.

Nevertheless, in the circumstances in which they are appropriate, boundary integral methods can be very useful. And their applicability can be
widened by the coupling of PDE and BIE methods, using each in the regions where they are appropriate. (See, for example, Zienkiewicz *et al.*
(1977), Johnson and Nedelec (1980), Costabel (1987) and recent reviews by
Hsiao (1990, 1991).)

In this review our concern is with the numerical analysis of boundary integral methods and, in particular, with certain recent developments. The

review could not hope to be exhaustive, given the range and the complexity of the subject. We are helped, however, by the existence of two other recent reviews, by Wendland (1990) and Atkinson (1990). The former gives a comprehensive overview of the recent theory of the Galerkin method for BIEs via the theory of strong ellipticity for pseudodifferential operators. It consists of lecture notes for an audience with a strong PDE background; those without such a background might find the present review a useful introduction. The review by Atkinson (1990) lays particular stress on the problems involved in the implementation of three-dimensional boundary integral equations, such as the problem of evaluating the (often weakly singular) integrals over boundary elements, and iterative methods for the solution of the dense linear systems that result. We shall not consider such questions in the present review.

Nor can we do justice to the large body of recent work on (Cauchy) singular integral equations in the plane (see, for example, Prössdorf and Silbermann (1977, 1991), Prössdorf (1989)). Mixed boundary value problems will be ignored (see Wendland *et al.* (1979), Lamp *et al.* (1984), Stephan and Wendland (1985), McLean (1990)). And we will have nothing to say about another topic currently attracting considerable interest, namely nonlinear aspects of BIE (see Ruotsalainen and Wendland (1988), Ruotsalainen and Saranen (1989), Atkinson and Chandler (1990), Eggermont and Saranen (1990), Ruotsalainen (1992); and, in connection with coupling of BIEs and PDEs, Gatica and Hsiao (1989)).

Many aspects of linear integral equations relevant to BIEs and their linear approximation are discussed carefully in the recent books by Kress (1989), Hackbusch (1989), and Prössdorf and Silbermann (1991).

In the later part of the review we will give particular attention to problems in the plane, because this has been an area which has seen considerable recent activity, with many new methods proposed, some new techniques of analysis, and some attempt to tackle the challenging problems posed by corners. Perhaps some of these methods will subsequently be extended to the even more challenging three-dimensional problems.

The structure of this review is as follows. In the next Section a simple introduction is given to the BIE formulation of the problem. Sobolev spaces and the mapping properties of boundary integral operators cannot be avoided in the modern numerical analysis of BIEs. They are introduced gently in Section 3, and then, based on this knowledge, existence and uniqueness questions are considered in Section 4. Aside from their importance for the basic theory, the Fourier series techniques introduced in Section 3 will play a major role later in the paper, for both the analysis and design of approximate methods.

Of all the methods mentioned in this review, the only one which is in a reasonably satisfactory condition for a wide class of BIEs is the Galerkin method. At heart, this is because it rests on a variational principle. A simple treatment is given in Section 5. Section 6 is devoted to the collocation method, and Section 7 to the so-called qualocation method and its discrete variants, for BIEs on plane curves. In general the analysis of these methods for smooth curves is reasonably satisfactory, but problems remain where there are corners. Section 8 summarizes some challenges for the future from corner and other problems. At the same time it discusses briefly an extreme case of a corner (the case of the logarithmic-kernel integral equation for a slit), for which a complete analysis is available. Perhaps this case may give some insight into the proper handling of general corners for this and other problems.

It might reasonably be said that in its theoretical analysis the boundary integral method is a decade or more behind the finite element method. A defence might be that the problem is genuinely harder, because of the non-local nature of integral operators. In any event, there can be no argument that there is still much to be done.

2. Boundary integral equations

The reformulation of elliptic boundary value problems as boundary integral equations has been discussed by many people, including Jaswon (1963) and Jaswon and Symm (1977) for potential theory and elastostatics, Kupradze (1965) for elasticity, and Colton and Kress (1983) for the Helmholtz equation. Clements (1981) considers general second-order elliptic problems, Hsiao and MacCamy (1973) and Hsiao (1989) concentrate on first-kind formulations, Ingham and Kelmanson (1984) consider biharmonic and singular problems, and Wendland (1990) discusses a range of examples. In addition there are many books and papers with an engineering flavour, among which we may mention Hess and Smith (1967), Brebbia et al. (1984), Banerjee and Watson (1986), and the introductory book by Hartmann (1989); for a more complete bibliography of engineering works see Atkinson (1990).

The classical mathematical formulations are discussed thoroughly by Mikhlin (1970). An excellent source for modern mathematical developments is the recent review of boundary integral equations by Maz'ya (1991).

Here our aim is merely to introduce some of the principal ideas in a simple setting, with no attempt at completeness or maximum generality.

2.1. Indirect methods

Consider the two-dimensional Laplace equation

$$\Delta \phi = 0, \quad t \in \Omega, \tag{2.1}$$

subject to the Dirichlet boundary conditions

$$\phi = g \quad \text{on} \quad \Gamma = \partial \Omega, \tag{2.2}$$

where Ω is a simply-connected open domain in the plane with a piecewise-smooth boundary Γ. The simplest boundary integral formulation of this problem is via the 'single-layer representation' of the potential ϕ; that is, one seeks a representation of ϕ in the form

$$\phi(t) = -\frac{1}{\pi}\int_\Gamma \log|t-s| z(s)\,dl_s, \quad t \in \Omega, \tag{2.3}$$

where $|t-s|$ is the Euclidean distance between t and s, dl is the element of arc length, and z is an unknown function, the 'single-layer density', or 'charge density'. The motivation is easily stated: because $(2\pi)^{-1}\log|t-s|$ is the fundamental solution of the Laplace equation, (2.3) yields a solution of the Laplace equation, no matter how z is chosen; thus all that remains is to satisfy the boundary condition (2.2). Letting t approach the boundary, and assuming that the right-hand side of (2.3) is continuous onto the boundary, we obtain

$$g(t) = -\frac{1}{\pi}\int_\Gamma \log|t-s| z(s)\,dl_s, \quad t \in \Gamma.$$

This is an integral equation of the first kind (which merely means that the unknown z occurs only under the integral sign). Introducing the single-layer integral operator V defined by

$$Vv(t) = -\frac{1}{\pi}\int_\Gamma \log|t-s| v(s)\,dl_s, \quad t \in \Gamma, \tag{2.4}$$

we may write the integral equation as

$$Vz = g. \tag{2.5}$$

That the integral on the right-hand side of (2.3) is continuous as $t \to \Gamma$ has been shown by Gaier (1976), under the assumption that $z \in L_p(\Gamma)$ for some $p > 1$, and that the curve is piecewise smooth and has no cusps.

Next, consider the exterior problem

$$\Delta\phi = 0, \quad t \in \Omega_e, \tag{2.6}$$

where $\Omega_e = \mathbb{R}^2\backslash\bar{\Omega}$ and Ω is defined as for (2.1). Again we assume the Dirichlet condition (2.2) on Γ, but this time we need also a regularity condition at infinity,

$$\phi \quad \text{bounded at infinity.} \tag{2.7}$$

Following Jaswon and Symm (1977), it is natural to seek a representation in the form

$$\phi(t) = -\frac{1}{\pi}\int_\Gamma \log|t-s| z(s)\,dl_s + \omega, \tag{2.8}$$

where ω is a constant, and where, in order to satisfy the condition at infinity, the side condition

$$\int_\Gamma z(s)\,\mathrm{d}l_s = 0$$

is imposed. In this case the corresponding boundary integral equation is the pair

$$Vz + \omega = g, \quad \int_\Gamma z = 0. \tag{2.9}$$

In the preceding paragraph we considered the exterior problem, but in fact there is nothing to stop us from using the same approach, of an additional unknown ω and a side condition on z, even for the interior problem (2.1), (2.2). That approach has been advocated by Hsiao and MacCamy (1973), in order to avoid the existence/uniqueness problems that can beset (2.5) (see Subsection 4.3). There is a close relationship between the two approaches: for example, if the pair ω, $z^{(1)}$ satisfies (2.9) and if $Vz^{(2)} = 1$ then it is obvious that $z = z^{(1)} + \omega z^{(2)}$ satisfies (2.5). For an elaboration of this relationship see Sloan and Spence (1988a, Section 3).

Three-dimensional interior and exterior problems for the Laplace equation with Dirichlet boundary condition (and with the regularity condition $\phi(t) \to 0$ as $|t| \to \infty$ in the exterior case) may be approached in a manner analogous to (2.3), the fundamental solution in this case being the Newtonian potential, the single-layer representation being

$$\phi(t) = \frac{1}{2\pi}\int_\Gamma \frac{1}{|t - s|} z(s)\,\mathrm{d}S_s, \quad t \in \Omega \text{ or } \Omega_e, \tag{2.10}$$

and the single-layer operator on $\Gamma = \partial\Omega$ being

$$Vz(t) = \frac{1}{2\pi}\int_\Gamma \frac{1}{|t - s|} z(s)\,\mathrm{d}S_s, \quad t \in \Gamma. \tag{2.11}$$

The resulting integral equation, for both interior and exterior problems, is (2.5).

2.2. The classical BIEs of potential theory

Returning to the two-dimensional case, the classical approach to the interior Dirichlet problem (2.1) and (2.2) is to seek a 'double-layer' representation for ϕ, i.e.

$$\begin{aligned}
\phi(t) &= \frac{1}{\pi}\int_\Gamma \left(\frac{\partial}{\partial n_s} \log|t - s|\right) z(s)\,\mathrm{d}l_s \\
&= \frac{1}{\pi}\int_\Gamma \frac{n(s)\cdot(s - t)}{|t - s|^2} z(s)\,\mathrm{d}l_s, \quad t \in \Omega. \tag{2.12}
\end{aligned}$$

Here the derivative is the normal derivative (with respect to s), in the direction of the outward unit normal n (i.e. the normal directed into the exterior region Ω_e). This approach leads to a quite different kind of equation, because the double-layer operator on the right of (2.12) is generally not continuous onto Γ. The following theorem is proved by Mikhlin (1970) for the case of a Lyapunov curve, and by Hackbusch (1989) and Wendland (1990) for a curve with corners.

Theorem 1 Let Γ be piecewise Lyapunov without cusps, and let $z \in C(\Gamma)$. Then the integral

$$\frac{1}{\pi} \int_\Gamma \left(\frac{\partial}{\partial n_s} \log |t - s| \right) z(s) \, dl_s, \quad t \in \mathbb{R}^2 \backslash \Gamma, \tag{2.13}$$

has limiting values as t approaches Γ from Ω and Ω_e separately. If $t' \in \Gamma$ is a point at which Γ has a tangent, the limiting values as $t \to t'$ are

$$\frac{1}{\pi} \int_\Gamma \left(\frac{\partial}{\partial n_s} \log |t' - s| \right) z(s) \, dl_s \pm z(t'), \tag{2.14}$$

where the upper and lower signs hold for $t \in \Omega$ and $t \in \Omega_e$ respectively.

The proof proceeds by representing $z(s)$ in (2.13) as $z(t') + (z(s) - z(t'))$, and showing that the integral corresponding to the second term is continuous onto Γ. For the first term, because $z(t')$ can be taken outside the integral, it is sufficient to prove the result for $z \equiv 1$. Briefly, for $s \in \Gamma$ and for t a fixed point in Ω, Ω_e or Γ (but not a corner point of Γ), let ρ, θ be polar coordinates of $s - t$, and let ψ be the angle between the outward normal $n(s)$ and the vector $s - t$. Then

$$\frac{\partial}{\partial n_s} \log |t - s| = \frac{n(s) \cdot (s - t)}{\rho^2} = \frac{\cos \psi}{\rho},$$

and

$$dl_s = \frac{\rho \, d\theta}{\cos \psi},$$

thus

$$\frac{1}{\pi} \int_\Gamma \left(\frac{\partial}{\partial n_s} \log |t - s| \right) dl_s = \frac{1}{\pi} \int d\theta = \frac{1}{\pi} \begin{cases} 2\pi & \text{if} \quad t \in \Omega, \\ \pi & \text{if} \quad t \in \Gamma, \\ 0 & \text{if} \quad t \in \Omega_e. \end{cases} \tag{2.15}$$

That is the integral in (2.13) has in this case the value 2 or 0 for t in Ω or Ω_e respectively, while (2.14) has the value 1 ± 1.

Let

$$\begin{aligned} Kz(t) &= \frac{1}{\pi} \int_\Gamma \left(\frac{\partial}{\partial n_s} \log |t - s| \right) z(s) \, dl_s \\ &= \frac{1}{\pi} \int_\Gamma \frac{n(s) \cdot (s - t)}{|t - s|^2} z(s) \, dl_s, \quad t \in \Gamma, \end{aligned} \tag{2.16}$$

be the double-layer operator on Γ. Then if t' is a point on Γ at which a tangent exists, the limiting values (2.14) become

$$Kz(t') \pm z(t'). \tag{2.17}$$

If we now return to the double-layer representation (2.12) of the interior Dirichlet problem (2.1), (2.2), we see, by taking the limit as t approaches a point on the boundary and using the theorem, that z satisfies

$$g(t) = Kz(t) + z(t), \quad t \in \Gamma, \quad t \text{ not a corner point.} \tag{2.18}$$

This is an equation of the *second* kind, in the nomenclature introduced by Fredholm. If Γ is a Lyapunov curve the kernel of the integral operator K when appropriately parametrized turns out to be weakly singular (see Mikhlin (1970)). Indeed, if Γ is a C^2 curve then the kernel is even continuous. In these cases the integral operator K is a compact operator on $C(\Gamma)$, and the classical Fredholm theory applies.

For a region with corners K is no longer compact on $C(\Gamma)$ (or indeed any other space), and the Fredholm theory is inapplicable. However, it is by now well understood that the double-layer equation (2.18) can still be a very effective tool (see, for example, Verchota (1984), Costabel (1988), Hackbusch (1989), Maz'ya (1991)). In particular, Verchota (1984) shows that the jump relations in Theorem 1 hold, almost everywhere on Γ, even for general Lipschitz curves (and hence for all piecewise Lyapunov curves without cusps), and with the density function z allowed to be merely in $L_2(\Gamma)$, provided that the double-layer operator K is defined with appropriate care. (Specifically, one need only replace 3 by 2 and 2π by π in the three-dimensional generalization (2.27) given below.) The precise nature of the operator K at a two-dimensional corner was first elucidated by Radon (1919), and further discussed by Cryer (1970); see also Atkinson and de Hoog (1984) for a study of the Dirichlet problem for a wedge.

In the same way the exterior Dirichlet problem (2.6), (2.2), (2.7) may be approached by the double-layer representation (2.12). In this case the jump relations lead to an operator equation on Γ with a different sign,

$$g(t) = Kz(t) - z(t), \quad t \in \Gamma, \quad t \text{ not a corner point.} \tag{2.19}$$

The classical approach to the Neumann problem

$$\Delta\phi = 0, \quad t \in \Omega, \quad \frac{\partial\phi}{\partial n} = h \text{ on } \Gamma, \tag{2.20}$$

or to the corresponding exterior problem satisfying also (2.7), is via the single-layer representation (2.3) or (2.10). It can be shown (Mikhlin, 1970) that for z integrable on Γ and $t \notin \Gamma$ the potential $\phi(t)$ can be differentiated under the integral sign, giving for the two-dimensional case

$$\nabla\phi = -\frac{1}{\pi}\int_\Gamma(\nabla_t\log|t-s|)z(s)\,dl_s$$

$$= -\frac{1}{\pi}\int_\Gamma\frac{t-s}{|t-s|^2}z(s)\,dl_s, \quad t\notin\Gamma.$$

Letting t' denote a point on Γ and $\partial\phi/\partial n$ the directional derivative in the direction of the (outward) normal at t', we have

$$\frac{\partial\phi}{\partial n} = -\frac{1}{\pi}\int_\Gamma\frac{n(t')\cdot(t-s)}{|t-s|^2}z(s)\,dl_s, \quad t\notin\Gamma. \tag{2.21}$$

The normal derivative has jump discontinuities analogous to those in Theorem 1 as $t\to t'$. For the case of a Lyapunov surface the limits as $t\to t'\in\Gamma$ are (Mikhlin, 1970)

$$-K^*z(t')\pm z(t'), \tag{2.22}$$

where again the upper and lower signs hold for $t\in\Omega$ and Ω_e respectively, and

$$K^*z(t) = \frac{1}{\pi}\int_\Gamma\left(\frac{\partial}{\partial n_t}\log|t-s|\right)z(s)\,dl_s,$$

$$= \frac{1}{\pi}\int_\Gamma\frac{n(t)\cdot(t-s)}{|t-s|^2}z(s)\,dl_s, \quad t\in\Gamma. \tag{2.23}$$

Note that the normal derivative in this case is with respect to t, whereas in the double-layer operator (2.16) it is with respect to s. In fact these operators are adjoints of each other.

It follows from this that the two-dimensional interior and exterior Neumann problems, at least for a Lyapunov curve, are characterized by the equation

$$h(t) = -K^*z(t)\pm z(t), \quad t\in\Gamma. \tag{2.24}$$

For a curve with corners comments similar to those made earlier for the double-layer equation are applicable: the same equation holds for t not a corner point (Hackbusch, 1989).

For the three-dimensional Laplace equation the double-layer representation is

$$\phi(t) = -\frac{1}{2\pi}\int_\Gamma\left(\frac{\partial}{\partial n_s}\frac{1}{|t-s|}\right)z(s)\,dS_s$$

$$= \frac{1}{2\pi}\int_\Gamma\frac{n(s)\cdot(s-t)}{|t-s|^3}z(s)\,dS_s, \quad t\notin\Gamma. \tag{2.25}$$

Corresponding to (2.15) are the Gauss laws as a result of which jump relations analogous to those of Theorem 1 hold. This time, however, we state the more powerful version due to Verchota (1984). (The analogous two-dimensional result also holds.) Here Ω is allowed to be an open, bounded, Lipschitz domain with connected boundary Γ, and $\Omega_e = \mathbb{R}^3 \backslash \bar{\Omega}$. Verchota (1984) shows, by making use of the celebrated Coifman *et al.* (1982) theorem, that for $z \in L_2(\Gamma)$ the limit of (2.25) as $t \to t' \in \Gamma$ exists almost everywhere on Γ, and has the value

$$Kz(t') \pm z(t'), \tag{2.26}$$

where the upper and lower signs are for the interior and exterior cases respectively, and

$$Kz(t) = \lim_{\epsilon \to 0} \frac{1}{2\pi} \int_{\Gamma, |t-s| > \epsilon} \frac{n(s) \cdot (s-t)}{|t-s|^3} z(s) \, dS_s, \quad t \in \Gamma. \tag{2.27}$$

Thus the BIEs for the interior and exterior Dirichlet problem become, as in the two-dimensional case,

$$g = Kz \pm z. \tag{2.28}$$

The operator K defined by (2.27) is a bounded operator on $L_2(\Gamma)$, about which we will have more to say when we turn to the question of existence and uniqueness. The jump relations for the normal derivatives of the single-layer potential extend to the three-dimensional situation in a similar way. Thus one obtains again BIEs of the form (2.24) for the interior and exterior Neumann problem, where for a general Lipschitz surface (Verchota, 1984)

$$K^*z(t) = \lim_{\epsilon \to 0} \frac{1}{2\pi} \int_{\Gamma, |t-s| > \epsilon} \frac{n(t) \cdot (t-s)}{|t-s|^3} z(s) \, dS_s. \tag{2.29}$$

2.3. Direct methods

The methods discussed so far are termed indirect methods, because they introduce quantities (namely, the single- or double-layer densities z on Γ) which are not part of the problem as originally formulated. Direct methods, in contrast, deal only with physically meaningful quantities, and for that reason are often favoured.

Direct methods are based on Green's theorem or its analogues. Suppose we are considering the Laplace equation (2.1) for an interior domain Ω having a smooth boundary, and suppose that $\phi \in C^2(\bar{\Omega})$ and ϕ satisfies (2.1). Then Green's theorem gives (Mikhlin, 1970, p. 224)

$$\phi(t) = \frac{1}{2\pi} \int_\Gamma \left[\left(\frac{\partial}{\partial n_s} \log |t-s| \right) \phi(s) - \log |t-s| \frac{\partial \phi(s)}{\partial n_s} \right] dl_s, \quad t \in \Omega.$$
$$\tag{2.30}$$

An equation on the boundary may now be obtained by letting $t \to \Gamma$ and using the continuity properties of the single-and double-layer potentials discussed earlier: we obtain

$$\phi(t) = \frac{1}{2}(K\phi(t) + \phi(t)) + \frac{1}{2}V\frac{\partial\phi}{\partial n}(t), \quad t \in \Gamma,$$

or

$$\phi = K\phi + V\frac{\partial\phi}{\partial n}. \tag{2.31}$$

This equation is an identity, which holds whenever ϕ satisfies the Laplace equation on Ω. Thus far we have assumed stringent conditions on Γ and ϕ, but these can be relaxed significantly (see, for example, Costabel (1988)), to allow curves that are merely Lipschitz, and hence may have corners.

Now let us introduce boundary conditions. Consider first the case of the Dirichlet boundary condition (2.2). Then (2.31) gives an integral equation of the first kind for $z \equiv \partial\phi/\partial n$,

$$Vz = g - Kg. \tag{2.32}$$

Now suppose instead that the boundary condition is

$$\frac{\partial\phi}{\partial n} = \kappa\phi + h, \tag{2.33}$$

with κ a constant. Then (2.31) becomes a second kind equation for ϕ,

$$\phi = (K + \kappa V)\phi + Vh. \tag{2.34}$$

The direct method is particularly attractive in the common situation in which the boundary conditions are mixed, for example with Dirichlet boundary conditions imposed on $\Gamma_1 \subset \Gamma$, and Neumann conditions on $\Gamma\backslash\Gamma_1$. This is because starting from the identity (2.31) (which is appropriate if we assume still that the equation is the Laplace equation), we may easily develop coupled boundary integral equations for $\partial\phi/\partial n$ on Γ_1 and ϕ on $\Gamma\backslash\Gamma_1$.

3. Sobolev spaces and mappings of operators

The modern study of boundary integral equations and their numerical approximation needs some acquaintance with the mapping properties of boundary integral operators in Sobolev spaces. In the case of the Galerkin method, discussed in Section 5, information of this kind is needed for the analysis. For some of the other methods discussed in later sections a precise understanding of the operators is even more critical, in that this understanding is built into the very design of the methods. For that reason we defer the discussion of numerical methods until we have more machinery available to us.

We shall concentrate here on the two-dimensional case, with a few remarks on the three-dimensional case at the end. We make every attempt to make the presentation as elementary as possible. A more detailed presentation from a similar point of view has been given by Kress (1989).

We shall assume for the present that Γ is a C^1 closed Jordan curve, parametrized by $t = \nu(x)$, where

$$\nu : [0,1] \to \Gamma, \quad \nu \text{ is 1-periodic}, \quad \nu \in C^1, \quad |\nu'(x)| \neq 0.$$

Any integrable function defined on Γ can be represented after this parametrization as a Fourier series,

$$v \sim \sum_{k \in \mathbb{Z}} \hat{v}(k) e^{2\pi i k x},$$

where

$$\hat{v}(k) = \int_0^1 e^{-2\pi i k x} v(x)\, dx, \quad k \in \mathbb{Z}.$$

For any real number s we define the Sobolev norm $\|v\|_s$ of v by

$$\|v\|_s = \left(|\hat{v}(0)|^2 + \sum_{k \neq 0} |k|^{2s} |\hat{v}(k)|^2 \right)^{1/2}. \tag{3.1}$$

When $s = 0$ the norm $\|v\|_0$ is just the L_2 norm. The norm $\|v\|_s$ also has a simple enough interpretation when s is a positive integer: if we recall that the sth derivative of v has the Fourier series

$$v^{(s)} \sim \sum_{k \in \mathbb{Z}} (2\pi i k)^s \hat{v}(k) e^{2\pi i k x},$$

we see that, apart from an unimportant constant factor, $\|v\|_s$ is essentially the L_2 norm of the sth derivative. (The term $|\hat{v}(0)|^2$ is included on the right of (3.1) to make this a norm, and not just a semi-norm.) Similarly, for negative integer values of s the norm is essentially the L_2 norm of the sth anti-derivative of v.

Corresponding to the norm $\| \cdot \|_s$, we introduce the Sobolev space H^s, which may be defined as the closure with respect to the norm $\| \cdot \|_s$ of the space of 1-periodic C^∞ functions. The elements of H^s are 1-periodic functions (or more generally distributions) with finite $\| \cdot \|_s$ norm. The space H^s is a Hilbert space with respect to the inner product

$$(v, w)_s = \hat{v}(0)\overline{\hat{w}(0)} + \sum_{k \neq 0} |k|^{2s} \hat{v}(k)\overline{\hat{w}(k)}. \tag{3.2}$$

An important inequality, holding for all real s and α, is

$$|(v, w)_s| \leq \|v\|_{s-\alpha} \|w\|_{s+\alpha}, \quad v \in H^{s-\alpha}, \quad w \in H^{s+\alpha}. \tag{3.3}$$

The proof is an easy application of the Cauchy–Schwarz inequality, starting from

$$(v, w)_s = \hat{v}(0)\overline{\hat{w}(0)} + \sum_{k \neq 0} |k|^{s-\alpha}\hat{v}(k)|k|^{s+\alpha}\overline{\hat{w}(k)}.$$

We also have the stronger result

$$\|v\|_{s-\alpha} = \sup_{w \in H^{s+\alpha}} \frac{(v, w)_s}{\|w\|_{s+\alpha}}, \quad v \in H^{s-\alpha}, \tag{3.4}$$

with the supremum achieved if $\hat{w}(0) = \hat{v}(0)$, $\hat{w}(k) = |k|^{-2\alpha}\hat{v}(k)$ for $k \neq 0$. In the jargon of the trade, $H^{s-\alpha}$ and $H^{s+\alpha}$ provide a 'duality pairing' with respect to the inner product $(\cdot, \cdot)_s$.

Now that the spaces are defined, we turn to the boundary integral operators, beginning with the single-layer operator V. Writing $t = \nu(x)$, we have, from (2.4),

$$\begin{aligned}
Vz(\nu(x)) &= -\frac{1}{\pi} \int_0^1 \log|\nu(x) - \nu(y)|\, z(\nu(y))\, |\nu'(y)|\, dy \\
&= -2 \int_0^1 \log|\nu(x) - \nu(y)|\, u(y)\, dy \\
&=: Lu(x), \tag{3.5}
\end{aligned}$$

where we have introduced a new unknown function

$$u(y) = \frac{1}{2\pi} z(\nu(y))\, |\nu'(y)| \tag{3.6}$$

which incorporates the Jacobian $|\nu'(y)|$ and also a convenient normalization factor.

If the curve Γ is smooth, or equivalently $\nu \in C^\infty$, the operator L defined by (3.5) behaves rather like the corresponding operator for a circle. Let A denote the operator L for the specific case of a circle of radius α. With the circle parametrized by $t = (t_1, t_2) = \alpha(\cos 2\pi x, \sin 2\pi x)$, we have explicitly

$$Au(x) = -2 \int_0^1 \log|2\alpha \sin \pi(x - y)|\, u(y)\, dy. \tag{3.7}$$

Then the operator L for the general curve Γ can be written as

$$L = A + B, \tag{3.8}$$

where

$$Bu(x) = -2 \int_0^1 \log\left|\frac{\nu(x) - \nu(y)}{2\alpha \sin \pi(x - y)}\right| u(y)\, dy. \tag{3.9}$$

For the case in which Γ is a C^∞ curve, whereas L and A have kernels which contain logarithmic singularities, the kernel of B is a C^∞ 1-periodic function of two variables. Thus for $v \in H^s$ with $s \in \mathbb{R}$ it follows that Bv is a C^∞ function, from which we see easily that

$$B : H^s \to H^t \quad \text{for all} \quad s, t \in \mathbb{R}. \tag{3.10}$$

This fact will often allow us to treat B as a compact perturbation. (Be warned, however, that this strategy fails if Γ has corners: for then B does *not* have a smooth kernel, and the compact perturbation approach fails.)

The operator A (i.e. the single-layer operator for the case of a circle of radius α) turns out to have the following extraordinarily simple Fourier representation.

Proposition 1

$$Av(x) \sim -2\log \alpha \hat{v}(0) + \sum_{k \neq 0} \frac{1}{|k|} \hat{v}(k) e^{2\pi i k x}. \tag{3.11}$$

This follows from the well known Fourier cosine series representation, valid for $x \neq 0$,

$$-\log\left(2 \left|\sin \pi x\right|\right) = \sum_{k=1}^{\infty} \frac{1}{k} \cos 2\pi k x = \frac{1}{2} \sum_{k \neq 0} \frac{1}{|k|} e^{2\pi i k x},$$

or equivalently

$$-2\log\left(2\alpha \left|\sin \pi x\right|\right) = -2\log \alpha + \sum_{k \neq 0} \frac{1}{|k|} e^{2\pi i k x}.$$

Equation (3.11) tells us that the effect of the operator A on the kth Fourier component of v, $k \neq 0$, is to multiply that component by $1/|k|$. Recalling the definition of the Sobolev norm $\| \cdot \|_s$, it follows immediately that, for $v \in H^s$,

$$\|Av\|_{s+1} \leq c\|v\|_s, \tag{3.12}$$

and hence

$$A : H^s \to H^{s+1}. \tag{3.13}$$

Assuming for the present that Γ is a C^∞ curve, it follows that

$$L : H^s \to H^{s+1}. \tag{3.14}$$

(Throughout the paper c denotes a constant which may take different values at its different occurrences.)

The mapping property (3.14) tells us, in effect, that L is a 'once-smoothing' operator, but conveys only limited information about L. A more precise statement is that L is a 'pseudo-differential operator of order -1 and principal symbol $|\xi|^{-1}$'. That means (following Agranovich (1979)) that L can be represented in the form

$$Lv(x) = \sum_{k \neq 0} \frac{1}{|k|} \hat{v}(k) \mathrm{e}^{2\pi \mathrm{i} k x} + \int_0^1 m(x, y) v(y) \, \mathrm{d}y, \qquad (3.15)$$

where m is a smooth kernel. This follows from (3.8), (3.9) and (3.11). Technically, the order is -1 because $|\xi|^{-1}$ is a positive-homogeneous function of degree -1. More general pseudo-differential operators exist, for example the principal symbol may change sign with ξ, or may depend on x. For the general form see Agranovich (1979) or Wendland (1990).

The identity operator I is a pseudo-differential operator of order 0 and principal symbol 1. So too is the operator $I + K$ arising in Section 2 from the double-layer approach to the Dirichlet problem (see (2.16)) for the case of a C^∞ curve Γ, since the double-layer operator K has in that case a C^∞ kernel.

Other pseudo-differential operators which arise in boundary integral methods are the Cauchy singular integral operator

$$Cv(t) = \frac{1}{\pi \mathrm{i}} \int_\Gamma \frac{v(s)}{s - t} \, \mathrm{d}s, \quad t \in \Gamma,$$

where s and t are taken to be complex numbers and the integral is to be understood in the principal-value sense, which is a pseudo-differential operator of order 0 and principal symbol sign ξ; and the normal derivative of the double-layer potential (or the 'hypersingular' operator), which is a pseudo-differential operator of order $+1$ and principal symbol $|\xi|$.

For three-dimensional surfaces $\Gamma = \partial\Omega$ the Sobolev spaces cannot be defined in quite such an elementary way, because there is no equivalent of the 1-periodic parametrization. Rather, one must use the machinery of local coordinate transformations, C^∞ cut-off functions, and Fourier transforms (see, for example, Wendland (1990)). Correspondingly, the definition of pseudo-differential operators needs to be based on Fourier transforms, rather than Fourier series. (In fact, strictly speaking this is true even in the two-dimensional case. However, the equivalence of the simpler Fourier series approach has been demonstrated by Agranovich (1979); see also Saranen and Wendland (1987) and McLean (1991).)

Nevertheless, the main results can be stated just as simply: for example, assuming that Γ is the smooth boundary of a simply connected open region, the single-layer operator defined by (2.11) is a pseudo-differential operator of order -1, while the operator $I + K$, with K the double-layer operator, is a pseudo-differential operator of order 0, and so on.

For regions with corners all of the considerations in this section require substantial modification. The Fourier series approach to the two-dimensional single-layer operator becomes less useful, because the kernel of the operator B is no longer smooth. For the case of a polygon, parametrized for example by arc length, the single-layer potential V may still be represented in the form (3.8) where B is given by (3.9), but now the kernel of B, far from being smooth, is discontinuous at the vertices of the polygon. This defect notwithstanding, Yan and Sloan (1988, Section 5) studied the single-layer equation for a polygon in the space $H^0 = L_2$ by using the fact that the nonsmooth part of B is, in a certain precise way, not too large. However, this kind of analysis has only limited applicability to the analysis of numerical methods, because the discretized operators typically have larger norms – for example Yan (1990) in using this approach to study a collocation method for this equation was forced to restrict attention to polygons with angles no smaller than a certain minimum. A related problem is that the boundary integral operators are no longer classical pseudo-differential operators. Considerable progress has been made in the study of these operators for regions with corners and edges, see Costabel and Stephan (1985) and Costabel (1988), with, for example, Mellin transforms replacing the Fourier transforms of the classical theory. However, this is a difficult subject, into which we will not venture further.

4. Existence and uniqueness questions

4.1. Introduction

Knowledge of existence and uniqueness of the exact solution is always a precondition for a satisfactory numerical analysis. In the present context we have an added interest, in that the methods used for the exact equation often have a parallel in the analysis of approximate methods.

The classical boundary integral formulations are equations of the second kind. The analysis of these, indicated in the next subsection, uses the classical Fredholm theory in the case of reasonably smooth curves or surfaces, and more sophisticated variants when corners or edges are present.

In more recent times there has been great interest in other formulations, particularly integral equations of the first kind such as those seen already in Section 2. The extension of these to more general differential equations has been considered by Fichera (1961), and more recently by Hsiao and MacCamy (1973); see also Giroire and Nedelec (1978) and, for a review, Hsiao (1989). We consider in detail the case of the logarithmic-kernel first-kind equation in the plane, which has some interesting features, and then consider briefly more general problems. In the analysis of these more general problems the notion of strong ellipticity has come to play an important role.

4.2. Equations with second kind structure

We begin with the classical case, treated for example by Mikhlin (1970), in which Γ is taken to be a connected C^2 curve or surface. In this situation the double-layer operator K defined by (2.16) or (2.27) is a compact operator on $L_2(\Gamma)$, as is its adjoint K^*. As the Fredholm theory is applicable, it is convenient to consider together the integral equations (2.18) for the interior Dirichlet problem,

$$z + Kz = g, \tag{4.1}$$

and (2.24) for the exterior Neumann problem

$$z + K^* z = -h, \tag{4.2}$$

since these are mutually adjoint. An argument of potential theory (see Mikhlin (1970, Chapter 18, Section 10 for the three-dimensional case, and Section 13 for the two-dimensional case)) shows that the homogeneous equation corresponding to (4.2) has only the trivial solution. The Fredholm theory (see, for example, Kress (1989)) then tells us that the same is true for the homogeneous equation corresponding to (4.1), and that both (4.1) and (4.2) have (unique) solutions $z \in L_2(\Gamma)$ for arbitrary $g \in L_2(\Gamma)$ or $h \in L_2(\Gamma)$ respectively. In other words, both $I + K$ and $I + K^*$ are boundedly invertible in $L_2(\Gamma)$.

Now consider the integral equation pair (2.19) for the exterior Dirichlet problem,

$$z - Kz = -g, \tag{4.3}$$

and (2.24) for the interior Neumann problem,

$$z - K^* z = h, \tag{4.4}$$

again mutually adjoint. This time the situation is slightly more interesting, since the Gauss laws (stated explicitly for the two-dimensional case as (2.15)) are equivalent to

$$1 - K1 = 0, \tag{4.5}$$

where 1 denotes the function on Γ whose values everywhere equal 1, so that the solution of (4.3) is not unique. It can be shown (Mikhlin 1970, Chapter 18, Sections 11 and 13) that the solution space of $z - Kz = 0$ is one-dimensional, thus from the Fredholm theory the same is true of the adjoint homogeneous equation. Let f_ϵ denote the unique solution of

$$f_\epsilon - K^* f_\epsilon = 0, \qquad \int_\Gamma f_\epsilon = 1. \tag{4.6}$$

Then by the Fredholm alternative (4.3) has a solution $z \in L_2(\Gamma)$ if and only if g is orthogonal to all solutions of the adjoint homogeneous equation, i.e. if and only if

$$\int_\Gamma g f_\epsilon = 0, \tag{4.7}$$

and similarly (4.4) has a solution if and only if

$$\int_\Gamma h = 0. \tag{4.8}$$

If h satisfies (4.8) and $z_0 \in L_2(\Gamma)$ is a particular solution of (4.4), it now follows that the general solution of (4.4) is $z_0 + \alpha f_\epsilon$, with α an arbitrary real number. A unique solution lying in the space

$$\overset{\circ}{L}_2(\Gamma) = \left\{ z \in L_2(\Gamma) : \int_\Gamma z = 0 \right\} \tag{4.9}$$

is then obtained by the choice $\alpha = -\int_\Gamma z_0$. In other words, $I - K^*$ is boundedly invertible in the space $\overset{\circ}{L}_2(\Gamma)$. Similarly, $I - K$ is boundedly invertible in the space

$$\tilde{L}_2(\Gamma) = \left\{ z \in L_2(\Gamma) : \int_\Gamma z f_\epsilon = 0 \right\}. \tag{4.10}$$

These arguments assume considerable regularity of Γ, but it is now known that these results hold in great generality. In particular, Verchota (1984) shows for general Lipschitz curves and surfaces that it is still true that $I + K$ is boundedly invertible in $L_2(\Gamma)$ and that $I - K^*$ is boundedly invertible in $\overset{\circ}{L}_2(\Gamma)$. It then follows by duality (Verchota, private communication) that $I + K^*$ and $I - K$ are boundedly invertible in $L_2(\Gamma)$ and $\tilde{L}_2(\Gamma)$ respectively.

The quantity f_ϵ introduced in (4.6) has an interesting interpretation. Defining a potential ψ in Ω by

$$\psi(t) = -\frac{1}{\pi} \int_\Gamma \log |t - s| f_\epsilon(s) \, dl_s, \quad t \in \Omega$$

in the two-dimensional case, or

$$\psi(t) = \frac{1}{2\pi} \int_\Gamma \frac{1}{|t - s|} f_\epsilon(s) \, dS_s, \quad t \in \Omega$$

in the three-dimensional case, it follows from the jump relation (2.22) for the normal derivative combined with (4.6) that

$$\frac{\partial \psi}{\partial n_-} = 0 \quad \text{on } \Gamma,$$

in which the normal derivative is the limit as Γ is approached from the interior Ω. (In the case of a Lyapunov surface this holds everywhere on Γ; for a Lipschitz surface it is valid almost everywhere – see Verchota (1984).) Since ψ is harmonic, it follows from the usual uniqueness theorem for the interior Neumann problem that ψ is constant in Ω and, hence, by the continuity of the single-layer potential,

$$V f_\epsilon = \text{constant} \quad \text{on } \Gamma \tag{4.11}$$

(almost everywhere, in the case of a Lipschitz curve). Thus f_ϵ is the 'equilibrium distribution'. In physical terms we may think of f_ϵ as the charge distribution on Γ (where the total charge is 1, since $\int_\Gamma f_\epsilon = 1$) that gives rise to a constant potential on Γ (and hence also in the interior region Ω).

We conclude this subsection with the observation that integral equation formulations are not always perfect reflections of the underlying boundary value problem. The condition (4.7), which we have seen is necessary and sufficient for the exterior Dirichlet integral equation (4.3) to have a solution, is by no means a necessary condition for the exterior Dirichlet problem itself. Mikhlin (1970, Chapter 18) discusses a modification of the equation which is solvable for every choice of the boundary-data function $g \in L_2(\Gamma)$. In a different direction, the exterior Neumann problem in two dimensions has a necessary condition, namely $\int_\Gamma h = 0$ (this is shown for example by Mikhlin (1970, Lemma 18.13.1)), which is not apparent in the integral equation formulation. Mikhlin shows (Lemma 18.13.2) that if this condition is satisfied then $\int_\Gamma z = 0$ (i.e. $I + K^*$ is boundedly invertible in $\overset{\circ}{L}_2(\Gamma)$, as well as in $L_2(\Gamma)$). The general solution of the exterior Neumann problem in two dimensions is then given by (2.8), where ω is an arbitrary constant. Because $\int_\Gamma z = 0$, this solution satisfies the boundary condition (2.7) at infinity.

4.3. The logarithmic-kernel BIE and the transfinite diameter

Before turning to more general equations, we consider the first-kind logarithmic-kernel integral equation in the plane

$$Vz(t) = -\frac{1}{\pi} \int_\Gamma \log|t - s| z(s) \, dl_s = g(t), \quad t \in \Gamma, \qquad (4.12)$$

which we have seen arising in Section 2 from both direct and indirect approaches to the Laplace equation with Dirichlet boundary conditions. It turns out that there is a genuine uniqueness/existence difficulty if the linear scale of the problem is inappropriate (Jaswon and Symm, 1977; Hsiao, 1986; Sloan and Spence, 1988a). Even for the case of a circle equation (4.12) may run into trouble: from (3.11), which gives the explicit Fourier representation of Vz for a circle of radius α, we see that if the radius α is 1 then $z = $ constant implies $Vz = 0$ (since then $\log \alpha = \log 1 = 0$). Thus the solution is not unique for the case of a circle of unit radius. Moreover, for a circle of this radius it is clear from (3.11) that there is *no* solution if $g \equiv 1$.

It is well known that a similar problem arises no matter what the geometry of Γ: there is always some linear scaling of Γ for which the solution is nonunique, and no solution exists for a constant right-hand side. (Jaswon and Symm (1977) refer to a contour with this bad scaling as a 'Γ-contour'.) The essential argument depends on nothing more than the properties of the logarithm: suppose that for a given contour Γ the equation

$$V f_\epsilon(t) = \frac{u}{\pi}, \quad t \in \Gamma, \quad \int_\Gamma f_\epsilon = 1, \tag{4.13}$$

has a solution $f_\epsilon \in L_1(\Gamma)$ for some real number u. Then for the re-scaled contour $\Gamma' = C^{-1}\Gamma$, where

$$C = \exp(-u) \tag{4.14}$$

we find, for $t' \in \Gamma'$ and $t = Ct'$, that

$$
\begin{aligned}
-\frac{1}{\pi} \int_{\Gamma'} \log|t' - s'| f_\epsilon(Cs') \, \mathrm{d}l_{s'} &= -\frac{C^{-1}}{\pi} \int_\Gamma \log(C^{-1}|t - s|) f_\epsilon(s) \, \mathrm{d}l_s \\
&= C^{-1}\left(-(\log C^{-1})\frac{1}{\pi} \int_\Gamma f_\epsilon(s) \, \mathrm{d}l_s + V f_\epsilon(t)\right) \\
&= C^{-1}\left(-\frac{u}{\pi} + \frac{u}{\pi}\right) = 0.
\end{aligned}
$$

Thus the logarithmic-kernel equation on the rescaled contour Γ' has a non-unique solution.

The number $C = C_\Gamma$ is a length associated with the contour Γ: it is easily seen that $C_{a\Gamma} = aC_\Gamma$. It is called the 'transfinite diameter' or 'logarithmic capacity' of Γ. (We prefer the former name, as it reminds us that C_Γ scales as a length.) In the preceding paragraph the rescaled curve Γ' has a transfinite diameter equal to 1. It also has a nontrivial solution of the homogeneous logarithmic-kernel equation. This observation should persuade us that for this equation contours of transfinite diameter 1 are to be avoided.

The argument in the preceding paragraphs depends on the existence of a solution of (4.13). Fortunately, it can be shown that a solution exists under very general conditions. For example, Hille (1962, p.280), assuming only that Γ is a closed bounded set in the plane, gives a variational definition of $u = u_\Gamma$ (the 'Robin constant'), as

$$u_\Gamma = \inf \left(-\int_\Gamma \int_\Gamma \log|t - s| \, \mathrm{d}\mu(t) \, \mathrm{d}\mu(s)\right), \tag{4.15}$$

where the infimum is over all normalized positive measures μ defined on Γ (i.e. $\mu \geq 0$, $\int_\Gamma \mathrm{d}\mu = 1$). The transfinite diameter C_Γ is then defined by (4.14). (Actually Hille gives independent definitions of transfinite diameter and logarithmic capacity, but shows them to be equivalent, in Theorem 16.4.4.) He shows moreover (in Theorem 16.4.3) that there exists a unique normalized positive measure μ_ϵ which achieves the infimum in (4.15), and that, except possibly for t in a set of transfinite diameter zero, one has (Hille, 1962, Theorem 16.4.8)

$$-\int_\Gamma \log|t - s| \, \mathrm{d}\mu_\epsilon(s) = u_\Gamma, \quad t \in \Gamma. \tag{4.16}$$

Thus a solution of (4.13) always exists in the sense of a measure.

Some useful properties of the transfinite diameter established in Hille (1962, Chapter 16) are:

1   the transfinite diameter of Γ does not exceed its Euclidean diameter;
2   if Γ lies inside Γ', then $C_\Gamma \leq C_{\Gamma'}$;
3   the transfinite diameter of a circle of radius α is α; and
4   the transfinite diameter of an interval of length l is $l/4$.

For our present purposes it is sufficient to restrict Γ to be the union of a finite number of C^2 arcs, having only a finite number of points of intersection. Note that this is both more restrictive and less restrictive than we have assumed in preceding sections: more restrictive because we do not allow general Lipschitz curves; less restrictive in that open arcs, cusps and multiple points of intersection are allowed. Under these conditions it follows from classical arguments that $d\mu_\epsilon$ has the form of a classical distribution $f_\epsilon \, dl$, where $f_\epsilon \in L_1(\Gamma)$. Indeed, much more can be said about f_ϵ. Let

$$
\begin{aligned}
\phi_\epsilon(t) &= -\frac{1}{\pi} \int_\Gamma \log|t - s| \, d\mu_\epsilon(s) \\
&= -\frac{1}{\pi} \int_\Gamma \log|t - s| f_\epsilon(s) \, dl_s, \quad t \in \mathbb{R}^2,
\end{aligned}
$$

be the potential corresponding to the equilibrium distribution. By standard arguments (e.g. Gaier (1976)) ϕ_ϵ is continuous on $\mathbb{R}^2$, except possibly at ends of arcs, cusps and points of intersection, so that

$$
\phi_\epsilon(t) \to V f_\epsilon(t) = \frac{u_\Gamma}{\pi} \quad \text{as } t \to \Gamma.
$$

With the same exceptions the jump relations (2.22) for the normal derivative hold in a pointwise sense on Γ, from which it follows that

$$
f_\epsilon = -\frac{1}{2} \left(\frac{\partial \phi_\epsilon}{\partial n_+} - \frac{\partial \phi_\epsilon}{\partial n_-} \right), \tag{4.17}
$$

where the normal derivatives are the limits as Γ is approached from the positive and negative sides (with respect to a normal with arbitrary but fixed sense). The known regularity property of the solutions of the Laplace equation now allows us to infer that f_ϵ is continuous on Γ except at points of intersection, cusps or ends. On the other hand f_ϵ is singular at a free end: for example, for an arc lying on the positive x-axis, with one end at the origin, in a neighbourhood of the origin we have

$$
\phi_\epsilon(r\cos\theta, r\sin\theta) = \frac{u_\Gamma}{\pi} + cr^{1/2}\sin\frac{\theta}{2} + c'r\sin\theta + \mathcal{O}(r^{3/2}), \quad 0 < \theta < 2\pi,
$$

from which it follows, using (4.17), that

$$
f_\epsilon(x) = cx^{-1/2} + \mathcal{O}(x^{1/2}) \tag{4.18}
$$

in a neighbourhood of the end. Similar but weaker singularities occur at corners; for details see Sloan and Spence (1988b). Note that $f_\epsilon \notin L_2(\Gamma)$ if there is a free end. On the other hand if Γ is the piecewise-smooth-without-cusps boundary of an open domain Ω, then $f_\epsilon \in L_2(\Gamma)$, and is just the function we met in the previous subsection as the solution of the second-kind integral equation (4.6) (since (4.11) is equivalent to (4.13)).

For $C_\Gamma \neq 1$ the solution of (4.12), if it exists, is unique. One way to show this is to decompose $z \in L_1(\Gamma)$ in the form

$$z = \alpha f_\epsilon + z_0, \tag{4.19}$$

with

$$z_0 \in \overset{\circ}{L}_1(\Gamma) = \left\{ w \in L_1(\Gamma) : \int_\Gamma w = 0 \right\}. \tag{4.20}$$

Since this decomposition is always possible with a uniquely determined α, namely

$$\alpha = \int_\Gamma z,$$

the representation (4.19) corresponds to a direct sum decomposition of $L_1(\Gamma)$,

$$L_1(\Gamma) = \{\alpha f_\epsilon : \alpha \in \mathbb{R}\} \oplus \overset{\circ}{L}_1(\Gamma). \tag{4.21}$$

Corresponding to the representation (4.19) we have

$$\begin{aligned} Vz &= \alpha V f_\epsilon + V z_0 \\ &= \frac{u}{\pi}\alpha 1 + V z_0. \end{aligned} \tag{4.22}$$

By a change in the order of integration (using Fubini's theorem), we see that

$$\int_\Gamma (V z_0) f_\epsilon = \int_\Gamma z_0 (V f_\epsilon) = \frac{u}{\pi} \int_\Gamma z_0 = 0, \tag{4.23}$$

thus $V : \overset{\circ}{L}_1 \to \tilde{L}_1$, where

$$\tilde{L}_1(\Gamma) = \left\{ w \in L_1(\Gamma) : \int_\Gamma w f_\epsilon = 0 \right\}. \tag{4.24}$$

Thus (4.19) and (4.22) correspond to a direct sum representation of V, namely

$$V = V_\epsilon \oplus \overset{\circ}{V}, \tag{4.25}$$

where

$$V_\epsilon : \{\alpha f_\epsilon : \alpha \in \mathbb{R}\} \to \{\alpha 1 : \alpha \in \mathbb{R}\} \tag{4.26}$$

with $V_\epsilon f_\epsilon = V f_\epsilon = (u/\pi)1$, and

$$\overset{\circ}{V} : \overset{\circ}{L}_1(\Gamma) \to \tilde{L}_1(\Gamma). \tag{4.27}$$

The operator V_ϵ, having one-dimensional domain and co-domain, is certainly one-to-one if $u \neq 0$ (and hence if $C_\Gamma \neq 1$); and Doob (1984) shows that the operator $\overset{\circ}{V}$ is positive definite (and hence one-to-one), in the sense that

$$-\int_\Gamma \int_\Gamma \log|t - s| z_0(s)\, dl_s z_0(t)\, dl_t \geq 0 \qquad (4.28)$$

for all $z_0 \in \overset{\circ}{L}_1$. (In fact Doob's result holds for the much larger class of signed measures with mean zero.) Thus uniqueness is proved.

When does a solution of (4.12) exist? If Γ is a C^∞ curve then the existence can be discussed in terms of the Sobolev spaces introduced in Section 3. As in (3.5) we define $V z(\nu(x)) = Lu(x)$, and as in (3.8) we write L as

$$L = A + B, \qquad (4.29)$$

where A is the single-layer operator for the case of a circle of radius α. It is convenient to choose $\alpha = e^{-1/2}$, because then we see from (3.11) that A has the especially simple Fourier series representation

$$Av(x) \sim \hat{v}(0) + \sum_{k \neq 0} \frac{1}{|k|} \hat{v}(k) e^{2\pi i k x}. \qquad (4.30)$$

From this it follows that A is an invertible operator from H^s onto H^{s+1} for arbitrary $s \in \mathcal{R}$, that is

$$A : H^s \to H^{s+1}, \qquad A^{-1} : H^{s+1} \to H^s, \qquad (4.31)$$

and, moreover, from the definition (3.1) of the Sobolev norms A is isometric:

$$\|Av\|_{s+1} = \|v\|_s. \qquad (4.32)$$

Since A is invertible we may write (4.29) as

$$L = A(I + K), \qquad (4.33)$$

where

$$K = A^{-1}B. \qquad (4.34)$$

Now from (3.10) it follows that

$$K : H^s \to H^t \quad \text{for all } s, t \in \mathbb{R},$$

thus K is a compact operator on H^s. If we assume that $C_\Gamma \neq 1$ then, as discussed earlier, L is a one-to-one operator, thus, from (4.33), so too is $I + K$. It now follows from the Fredholm alternative that $I + K$ is boundedly invertible on H^s or equivalently,

$$I + K : H^s \to H^s, \qquad 1\text{–}1 \text{ and onto}. \qquad (4.35)$$

The final conclusion is that if $C_\Gamma \neq 1$ the operator L behaves just like the operator for the case of a circle, in that

$$L : H^s \to H^{s+1}, \quad \text{1–1 and onto.} \qquad (4.36)$$

Thus the equation $Lu = f$ has a solution $u \in H^s$ for arbitrary $f \in H^{s+1}$.

For Γ a closed Lipschitz curve Verchota (1984) shows, for $C_\Gamma \neq 1$, that (4.12) has a solution $z \in L_2(\Gamma)$ for arbitrary $g \in L'_2(\Gamma) := \{ w \in L_2(\Gamma) : w' \in L_2(\Gamma) \}$, w' being the (tangential) derivative on Γ.

Before leaving the single-layer equation, it should be said that for the three-dimensional first-kind boundary integral equation $Vz = g$, with V given by (2.11), no uniqueness difficulty arises; there is no scaling for which the homogeneous equation has a nontrivial solution, and in fact V, for any scaling of Γ, is a positive definite operator. The difficulties that arise with (4.12) may be thought of as an idiosyncrasy of two dimensions.

4.4. More general equations – strong ellipticity

Many of the boundary integral equations that arise in practice are 'strongly elliptic' and hence 'coercive' with respect to an appropriate Hilbert space. We shall see that this not only provides a simple way of establishing the existence and uniqueness of the exact solution (in an appropriate weak sense), but also gives a very satisfactory framework for analysing the Galerkin method (see Section 5).

For simplicity we restrict ourselves to boundary integral equations which can be written as single equations of the form

$$Lu = f. \qquad (4.37)$$

In the two-dimensional case it is convenient to assume that the boundary curve Γ has already been parametrized in the manner of (3.5), so that u and f are 1-periodic functions. In the three-dimensional case u and f are functions on Γ. By restricting ourselves to equations of the form (4.37) we are excluding systems of equations, and also equations such as (2.9), in which there is a scalar unknown in addition to the unknown function u. For generalizations see Stephan and Wendland (1976) and Wendland (1983, 1985, 1987).

We shall say that (4.37) has a 'weak' solution u if, for all χ in an appropriate space,

$$(Lu, \chi)_0 = (f, \chi)_0, \qquad (4.38)$$

where

$$(v, w)_0 = \int_0^1 v\overline{w} \quad \text{or} \quad \int_\Gamma v\overline{w} \qquad (4.39)$$

in the two-dimensional or three-dimensional case respectively. In one important circumstance the existence of a weak solution is guaranteed: if for some

Hilbert space H the bilinear form $(L\phi, \psi)_0$ is both bounded and coercive, i.e. if for some positive constants D and ν

$$|(L\phi, \psi)_0| \le D\|\phi\|_H\|\psi\|_H \quad \forall \phi, \psi \in H, \tag{4.40}$$

and

$$\mathrm{Re}\,(L\phi, \phi)_0 \ge \nu\|\phi\|_H^2 \quad \forall \phi \in H, \tag{4.41}$$

then the Lax–Milgram theorem (Gilbarg and Trudinger, 1983, Theorem 5.8; Ciarlet, 1978) is applicable:

Theorem 2 (Lax–Milgram) Assume that $a(\phi, \psi)$ is a bounded, coercive bilinear form on a Hilbert space H, and that F is a bounded linear functional on H. Then there exists a unique $u \in H$ such that

$$a(u, \chi) = F(\chi) \quad \forall \chi \in H.$$

It follows that $Lu = f$ has a weak solution $u \in H$ for each f for which $(f, \cdot)_0$ is a bounded linear functional on H.

A simple example is provided by the single-layer equation for a circle of radius $\alpha < 1$, already discussed in the preceding subsection. From (3.11) and (3.1) we have in this case

$$\begin{aligned}
(L\phi, \psi)_0 &= (A\phi, \psi)_0 = -2\log\alpha\hat{\phi}(0)\overline{\hat{\psi}(0)} + \sum_{k\neq 0}\frac{1}{|k|}\hat{\phi}(k)\overline{\hat{\psi}(k)} \\
&\le \max(-2\log\alpha, 1)\|\phi\|_{-1/2}\|\psi\|_{-1/2}
\end{aligned} \tag{4.42}$$

and

$$(L\phi, \phi)_0 \ge \min(-2\log\alpha, 1)\|\phi\|_{-1/2}^2, \tag{4.43}$$

so that $(L\phi, \phi)_0$ is bounded and coercive with respect to the Sobolev space $H^{-1/2}$. Thus the logarithmic-kernel equation $Lu = f$ for a circle of radius $\alpha < 1$ has a weak solution $u \in H^{-1/2}$ for each $f \in H^{1/2}$. (Recall that H^s and H^{-s} are a dual pair with respect to the L_2 inner product $(\cdot, \cdot)_0$ – see (3.3), (3.4).) This is consistent with the previously established result (4.31).

How can we establish, for more general boundary integral operators, that the conditions (4.40) and (4.41) are satisfied for some Hilbert space H? For boundary integral operators on smooth closed curves or surfaces, the theory of pseudo-differential operators, already discussed briefly in Section 3, can be used to good effect. This theory, and its application to the Galerkin method, has been discussed with admirable thoroughness in a number of places, for example Stephan and Wendland (1976), Wendland (1983, 1987, 1990) and Hsiao and Wendland (1981). Here we content ourselves with a brief look at a two-dimensional case.

Suppose that L is an operator on 1-periodic functions defined by

$$Lv(x) = \sum_{k \neq 0} a(x, k)\hat{v}(k)e^{2\pi ikx} + \int_0^1 m(x, y)v(y)\,dy, \quad x \in [0, 1], \quad (4.44)$$

where $m \in C^\infty([0, 1] \times [0, 1])$, and $a(x, \xi)$ is a 1-periodic C^∞ function of x for each $\xi \neq 0$, and for some $\beta > 0$ and each $x \in \mathbb{R}$ is a positive-homogeneous function of degree β in ξ. Then L is a pseudo-differential operator of order β and principal symbol $a(x, \xi)$. (The logarithmic-kernel operator defined by (3.8)–(3.11) is a pseudo-differential operator of order -1 and principal symbol $|\xi|^{-1}$.) A pseudo-differential operator of order β is (Hörmander, 1965; Wendland, 1987) a continuous operator from H^σ to $H^{\sigma-\beta}$ for all $\sigma \in \mathbb{R}$. In particular, therefore, L is a continuous operator from $H^{\beta/2}$ to $H^{-\beta/2}$. From this and (3.3) it follows that

$$|(Lv, w)_0| \leq \|Lv\|_{-\beta/2}\|w\|_{\beta/2} \leq c\|v\|_{\beta/2}\|w\|_{\beta/2},$$

so that (4.40) is satisfied with $H = H^{\beta/2}$.

Now suppose, in addition, that the principal symbol $a(x, \xi)$ is 'strongly elliptic', that is to say that for some $\mu > 0$

$$\text{Re } a(x, \pm 1) \geq \mu \quad \forall x \in [0, 1]. \quad (4.45)$$

Then it is known (Kohn and Nirenberg, 1965, p.283) that, for any $\varepsilon > 0$, $L = L_0 + M(\varepsilon)$, where L_0 is coercive with respect to $H^{\beta/2}$,

$$\text{Re}(L_0\phi, \phi)_0 \geq (\mu - \varepsilon)\|\phi\|_{\beta/2}^2 \quad \forall \phi \in H, \quad (4.46)$$

and $M = M(\varepsilon)$ is a compact operator from $H^{\beta/2}$ to $H^{-\beta/2}$. The addition of the compact term M leaves the conclusion of the Lax–Milgram theorem unaltered, provided that L remains one-to-one (Hildebrandt and Wienholtz (1964), Remark 3). Thus it follows in the strongly elliptic case that the equation $Lu = f$ has a weak solution $u \in H^{\beta/2}$ for each $f \in H^{-\beta/2}$.

It should be noted that the pseudo-differential operator arguments need serious modification as soon as corners or edges appear (see, for example, Costabel and Stephan (1985)).

A different and in some ways more versatile approach is to found the strong ellipticity theory for boundary integral operators on the well studied strong ellipticity properties of the associated elliptic PDEs. For further details, see Costabel and Wendland (1986). This approach has the advantage that it remains available even when corners are present, and indeed even for general Lipschitz curves (Costabel, 1988).

5. The Galerkin method

Most theoretical treatments of the boundary element method give great attention to the Galerkin method, a method originated in the context of the differential equations of elasticity by a Russian engineer (Galerkin, 1915). The present treatment will be briefer, not least because of the very complete discussions that exist elsewhere (see, for example, Stephan and Wendland (1976), Hsiao and Wendland (1977, 1981), Wendland (1983, 1987, 1990), Rannacher and Wendland (1985, 1988)). However, the example of the two-dimensional logarithmic-kernel integral equation will be worked out in some detail.

Let us assume, as in (4.37), that the problem is expressible as a single equation of the form

$$Lu = f, \tag{5.1}$$

with u and f 1-periodic in the two-dimensional case, and functions on Γ in the three-dimensional case.

Let S_h be a finite-dimensional space within which the approximate solution is to be sought. Typically, S_h is defined by partitioning Γ into a finite number of pieces with simple geometry (e.g. plane or curved triangles) and maximum diameter h, on each of which the restriction of S_h is a piecewise polynomial space with respect to an appropriate local parametrization (e.g. one in which the element boundary is a triangle). Continuity conditions across elements may or may not be imposed, depending on the circumstances, one important constraint being $S_h \subset H$, where H is the space in (4.40–1). (For further details see, for example, Brebbia *et al.* (1984).) Then the Galerkin method is: find $u_h \in S_h$ such that

$$(Lu_h, \chi)_0 = (f, \chi)_0 \quad \forall \chi \in S_h, \tag{5.2}$$

where the inner product is defined by (4.39).

While the Galerkin method is the theorist's favourite, it is in truth not easy to implement. Let $\{\phi_1, \ldots, \phi_N\}$ be a basis for S_h. Then we may write

$$u_h = \sum_{j=1}^{N} a_j \phi_j, \tag{5.3}$$

and the equations to be solved in practice are

$$\sum_{j=1}^{N} (L\phi_j, \phi_k)_0 a_j = (f, \phi_k)_0, \quad k = 1, \ldots, N, \tag{5.4}$$

in which each matrix element on the left, even in the two-dimensional case, is a two-dimensional integral – one integral for the integral operator, and one for the inner product. In the three-dimensional case four levels of integration are needed for each matrix element. And the difficulty is compounded by the fact that the matrix in the boundary element method is invariably dense.

The error analysis for the Galerkin method rests on the variational formulation of the exact problem given in the preceding section. Suppose that

the bilinear form $(L\phi, \psi)_0$ satisfies the boundedness and coercivity properties (4.40) and (4.41) for some Hilbert space H, and that $u \in H$ is the weak
solution of the exact equation (5.1). Then, provided only that $S_h \subset H$,
Céa's lemma (Ciarlet, 1978) ensures that the Galerkin equation (5.2) has a
unique solution $u_h \in S_h$, whose error in the $\| \cdot \|_H$ norm is within a constant
factor of the error of best approximation by an element of S_h:

Theorem 3 (Céa's lemma) Assume that the bilinear form $(L\phi, \psi)_0$ satisfies (4.40) and (4.41), and that $Lu = f$ has the weak solution $u \in H$. Assume also that S_h is a finite-dimensional subspace of H. Then the Galerkin
approximation (5.2) has a unique solution $u_h \in S_h$, which satisfies

$$\|u_h - u\|_H \leq \frac{D}{\nu} \inf_{v_h \in S_h} \|v_h - u\|_H. \tag{5.5}$$

Proof. The existence and uniqueness of u_h follows from the Lax–Milgram
theorem applied to S_h as a subspace of H. Then (4.40), (4.41) and (5.2)
give, for arbitrary $v_h \in S_h$,

$$\begin{aligned}
\nu \|u_h - u\|_H^2 &\leq |(L(u_h - u), u_h - u)_0| = |(L(u_h - u), v_h - u)_0| \\
&\leq D\|u_h - u\|_H \|v_h - u\|_H,
\end{aligned}$$

from which the result follows. $\square$

The result (5.5) has the nice property of reducing the Galerkin error esti-
mation in the 'natural' or 'energy' norm $\| \cdot \|_H$ to a problem of approximation.

Now let us be more explicit, and assume that Γ is a smooth curve in the
two-dimensional case, or a smooth surface in the three-dimensional case, and
that L is a strongly elliptic pseudo-differential operator of order β. Then, as
noted in the preceding subsection, $L = L_0 + M$, where L_0 is bounded and
coercive with respect to the space $H^{\beta/2}$, and M is a compact operator from
$H^{\beta/2}$ to $H^{-\beta/2}$. It can be shown (Hildebrandt and Wienholtz, 1964) that the
addition of the compact term leaves the essential conclusion of Céa's lemma
unaltered, provided L remains one-to-one: specifically, it follows that $h_0 > 0$
exists such that $u_h \in S_h$ exists for $h < h_0$, and satisfies

$$\|u_h - u\|_{\beta/2} \leq c \inf_{v_h \in S_h} \|v_h - u\|_{\beta/2}, \tag{5.6}$$

for some constant $c > 0$.

We now specialize further to the case of the logarithmic-kernel integral
equation. In the following example we indicate 'power of h' results for the
Galerkin error not only in the energy norm, but also in a range of other
Sobolev norms. We shall also return to the same example later, in other
sections, to illustrate other numerical methods.

Example. Let Γ be a smooth curve with C_Γ (the transfinite diameter) not
equal to 1, and let L be the logarithmic-kernel operator on $[0,1]$ defined

by (3.5). Let S_h be the space of 1-periodic smoothest splines of order r (i.e. degree $\leq r - 1$), where $r \geq 1$, on a partition

$$\Pi_h : 0 = x_0 < x_1 < \cdots < x_{N-1} < x_N = 1, \tag{5.7}$$

with

$$h_k = x_{k+1} - x_k, \quad k = 0, \ldots, N - 1, \tag{5.8}$$

and

$$h = \max h_k.$$

That is, $v \in S_h$ satisfies $v \in C^{r-2}(\mathbb{R})$, and $v|_{(x_k, x_{k+1})} \in \mathbb{P}_{r-1}$. Assume, moreover, that $h \to 0$. Then it is known that for $-\infty < t \leq s \leq r$ and $t < r - \frac{1}{2}$, there exists a constant c depending only on t and s such that

$$\inf_{v_h \in S_h} \|v_h - u\|_t \leq ch^{s-t}\|u\|_s \quad \text{if } u \in H^s. \tag{5.9}$$

(For a discussion see, for example, Arnold and Wendland (1983).) In particular, therefore, it follows from (5.6) that the Galerkin error estimate in the natural norm is

$$\|u_h - u\|_{-1/2} \leq ch^{r+1/2}\|u\|_r \quad \text{if } u \in H^r. \tag{5.10}$$

Error estimates in 'lower' norms can now be deduced by a duality argument ('Nitsche's trick'), apparently first used for BIEs by Hsiao and Wendland (1981). We illustrate this for the most extreme case, namely the $\|\cdot\|_{-r-1}$ norm, in which the maximum order of convergence can be doubled over that obtained in (5.10) for the natural norm. For simplicity, we assume here that Γ is a circle of radius $e^{-1/2}$, so that $L = A$, in the language of (4.29) and (4.30). Using (3.4), (3.2) and (4.30), we have

$$\|u_h - u\|_{-r-1} = \sup_{v \in H^r} \frac{(u_h - u, v)_{-1/2}}{\|v\|_r} = \sup_{v \in H^r} \frac{(A(u_h - u), v)_0}{\|v\|_r}$$

$$= \sup_{v \in H^r} \frac{(A(u_h - u), v - v_h)_0}{\|v\|_r}$$

$$\leq \sup_{v \in H^r} \frac{\|u_h - u\|_{-1/2}\|v - v_h\|_{-1/2}}{\|v\|_r},$$

where v_h is an arbitrary element of S_h, which makes its appearance at the second-last step because we have used again the Galerkin equation (5.2). Thus, by (5.10) and another application of the approximation theory result (5.9) we obtain

$$\|u_h - u\|_{-r-1} \leq \|u_h - u\|_{-1/2} \sup_{v \in H^r} \frac{ch^{r+1/2}\|v\|_r}{\|v\|_r}$$

$$\leq ch^{2r+1}\|u\|_r. \tag{5.11}$$

Finally, error estimates in higher Sobolev norms than the $\|\cdot\|_{-1/2}$ norm may be established if the sequence of meshes is quasi-uniform; i.e. if there exists $c > 0$ such that

$$h_k \geq ch \qquad \forall\, k,$$

with c independent of h. From this follows the inverse estimate (see, for example, Arnold and Wendland (1983)), for $\tau \leq \sigma < r - \frac{1}{2}$,

$$\|v\|_\sigma \leq ch^{\tau-\sigma}\|v\|_\tau \quad \text{for } v \in S_h. \tag{5.12}$$

We also make use of the fact (again see Arnold and Wendland (1983) for a discussion) that for given $u \in H^s$ there exists $\psi_h \in S_h$, *independent of t,* such that for $t \leq s \leq r$ and $t < r - \frac{1}{2}$

$$\|u - \psi_h\|_t \leq ch^{s-t}\|u\|_s \quad \text{for } u \in H^s,$$

with c independent of u. Then for $-\frac{1}{2} \leq t < r - \frac{1}{2}$ it follows that

$$\begin{aligned}
\|u_h - u\|_t &\leq \|u_h - \psi_h\|_t + \|\psi_h - u\|_t \\
&\leq ch^{-r-1-t}\|u_h - \psi_h\|_{-r-1} + \|\psi_h - u\|_t \\
&\leq ch^{-r-1-t}\|u_h - u\|_{-r-1} + ch^{-r-1-t}\|\psi_h - u\|_{-r-1} + \|\psi_h - u\|_t \\
&\leq ch^{r-t}\|u\|_r. \tag{5.13}
\end{aligned}$$

Here we have assumed, for simplicity, that $u \in H^r$. Results for u with lesser smoothness, and correspondingly fewer powers of h, are easily written down.

The highest-order convergence in this example – of order $\mathcal{O}(h^{2r+1})$ – is obtained in the $\|\cdot\|_{-r-1}$ norm. At first sight it may not be clear why 'negative norm' results of this kind are of interest, given that they cannot be observed directly. The answer is that we do indeed see the benefit if we are interested finally not in u, but rather in an inner product $(u, w)_0$, where w is a reasonably smooth function. For from (3.3) we have

$$|(u_h, w)_0 - (u, w)_0| = |(u_h - u, w)_0| \leq \|u_h - u\|_{-r-1}\|w\|_{r+1}, \tag{5.14}$$

so that the $\mathcal{O}(h^{2r+1})$ order of convergence in the example is observable if $w \in H^{r+1}$. As a specific example of such an inner product, suppose that in the case of the logarithmic-kernel integral equation we are interested in computing, in the context of the indirect method for the interior Dirichlet problem for the Laplace equation, the potential $\phi(t)$ given by (2.3) at a point $t \notin \Gamma$. Then from (2.3) and (3.6)

$$\begin{aligned}
\phi(t) &= -\frac{1}{\pi}\int_\Gamma \log|t - s|z(s)\,\mathrm{d}l_s \\
&= -2\int_0^1 \log|t - \nu(y)|u(y)\,\mathrm{d}y \\
&= (u, w_t)_0, \tag{5.15}
\end{aligned}$$

where w_t is the C^∞ function defined by

$$w_t(x) = -2\log|t - \nu(x)|, \quad t \notin \Gamma, \quad x \in \mathbb{R}. \tag{5.16}$$

Thus if

$$\phi_h(t) = -2\int_0^1 \log|t - \nu(y)|u_h(y)\,\mathrm{d}y \tag{5.17}$$

then

$$\phi_h(t) - \phi(t) = (u_h, w_t)_0 - (u, w_t)_0, \tag{5.18}$$

and (5.14) applies.

Many modifications of the Galerkin method have been proposed for the case of smooth plane curves. Arnold (1983) has shown that the order of negative-norm convergence can be made arbitrarily large by the use of an unsymmetric (or Petrov–Galerkin) approximation, in which the trial space remains a spline space but the test space is a space of trigonometric polynomials. McLean (1986) obtains exponential rates of convergence in stronger norms by the use of trigonometric polynomials for both test and trial spaces, if the exact solution is smooth. Atkinson (1988) obtains a similar rate of convergence with a fully discrete version of the Galerkin method with trigonometric polynomials.

The Galerkin-collocation method (Hsiao *et al.* 1980, 1984), as the name suggests, has some relation to both the Galerkin and collocation methods. In this method the logarithmic-kernel integral equation for a smooth curve, in the modified form (2.9) as advocated by Hsiao and MacCamy (1973), is handled by decomposing the operator into two parts: a principal part, which is a convolution operator, treated by a Galerkin method; and a second part, which is an integral operator with a smooth kernel, treated by a discrete approximation. (The separation into the two terms is similar to that in (3.8) and (3.9), but is different in detail.) The analysis exploits the close relation to the Galerkin method, yet the implementation is much less laborious, since the matrix corresponding to the principal part is a Toeplitz matrix, and so is representable as a vector; and moreover it is independent of the particular curve, and so can be computed once and for all. An extensive discussion of applications is given in Hsiao *et al.* (1984).

Before leaving the Galerkin method, we may mention that Rannacher and Wendland (1985,1988), by the clever use of weighted Sobolev norms, have established *uniform* error estimates for the Galerkin approximation for the single-layer equation on closed curves and surfaces.

6. The collocation method

For the solution of boundary integral equations in practice the collocation method is generally the method of choice, because it is so much easier to

implement than the Galerkin method. It does, however, have disadvantages:
a theoretical analysis is available only in special cases (see later); the convergence rate in negative norms is often inferior; and the matrix is generally
not symmetric even if the operator is self-adjoint.

Let us assume, as in the discussion of the Galerkin method, that the
equation to be solved is a single equation of the form

$$Lu = f, \tag{6.1}$$

where u and f are either 1-periodic functions on $\mathbb{R}$, or are functions on Γ in
the three-dimensional case. Let S_h be the finite-dimensional space within
which the approximation is to be sought. In the collocation method one
chooses also a set of 'collocation points' $t_1, \ldots, t_N$, where $N = N_h$ is the
dimension of S_h. Then the collocation method is: find $u_h \in S_h$ such that

$$Lu_h(t_k) = f(t_k), \quad k = 1, \ldots, N. \tag{6.2}$$

Needless to say, the choice of the collocation points is a very important
question, one to which we shall return.

Letting $\{\phi_1, \ldots, \phi_N\}$ be a basis for S_h, and writing u_h in the form (5.3),
the equations to be solved in practice are

$$\sum_{j=1}^{N} L\phi_j(t_k)a_j = f(t_k), \quad k = 1, \ldots, N. \tag{6.3}$$

Clearly, the labour involved in setting up the matrix is much less than in the
Galerkin method: in the two-dimensional case each matrix element requires
just one integration.

Theoretical analyses of the collocation method are available in a variety
of situations. For the double-layer equation (2.18) on smooth curves or surfaces, the standard analysis for Fredholm integral equations of the second
kind (see, for example, Atkinson (1976) or Baker (1977)) is available. For
regions with corners we saw in Section 2 that the double-layer integral operator is no longer compact, so that the standard theory is not applicable, but
at least in the plane case considerable progress has nevertheless been made
(see, for example, Atkinson and de Hoog (1984), Chandler and Graham
(1988), Elschner (1988)). In the latter papers piecewise-polynomial collocation is shown to be stable and of optimal order, provided the approximating
space is suitably modified near the corner, and the mesh is appropriately
'graded'. (The mesh is graded at a corner z, with grading parameter $q > 0$,
if the points of the partition satisfy $|x_k - z| = ck^q, k = 0, 1, \ldots$, near z).

For more general boundary integral equations on smooth plane curves an
important contribution to the theoretical study of the collocation method
is that of Arnold and Wendland (1983). In that paper certain collocation
methods are shown to be equivalent, after integration by parts, to Galerkin

methods with nonstandard inner products. As a consequence stability and convergence are established, with almost no restrictions on the mesh. The principal limitation is that the analysis is restricted to smoothest splines of even order or odd degree (for example, continuous piecewise-linear functions), and to the particular case of collocation at the knots.

For the case of the logarithmic-kernel equation $Lu = f$ with L defined by (3.5), and for smoothest splines of (even) order r on the partition (5.7), the convergence result of highest order obtained by Arnold and Wendland (1983) is

$$\|u_h - u\|_{-1} \le ch^{r+1}\|u\|_r, \qquad (6.4)$$

if $u \in H^r$. That is, the highest order of convergence obtainable in any norm is $\mathcal{O}(h^{r+1})$, compared with $\mathcal{O}(h^{2r+1})$ for the standard Galerkin method (see (5.11)). The nonstandard Galerkin method to which the collocation method is equivalent has as inner product the Sobolev inner product $(\cdot, \cdot)_{r/2}$, defined by (3.2). The Arnold and Wendland (1983) analysis also handles more general pseudo-differential operators L, provided that the bilinear form $(L\phi, \psi)_{r/2}$ is coercive with respect to an appropriate norm. For pseudo-differential operators of order β the highest order convergence result they obtain is

$$\|u_h - u\|_\beta \le ch^{r-\beta}\|u\|_r, \qquad (6.5)$$

compared to

$$\|u_h - u\|_{-r+\beta} \le ch^{2r-\beta}\|u\|_r$$

for the standard Galerkin method.

A generalization this above approach to piecewise-linear collocation on the torus has been given by Hsiao and Prössdorf (1992).

The present theoretical situation for the collocation method is much less satisfactory for approximation by piecewise-constants, or other splines of even degree. For the very special case of smooth plane curves, smoothest splines and a uniform mesh, a satisfactory analysis has been developed (de Hoog, 1974; Saranen and Wendland, 1985; Arnold and Wendland, 1985; Saranen, 1988) by the use of Fourier series methods, combined with localization arguments. In this analysis the collocation points must always be chosen in an appropriate way; for a full discussion of the correct choice see Wendland (1990). For example, for the case of the logarithmic-kernel equation with a uniform mesh and smoothest splines of even degree, the collocation points should be taken to be the midpoints of each sub-interval. In this case it is known (Saranen and Wendland, 1985; Arnold and Wendland, 1985) that the result (6.4) holds. More surprisingly, in this even degree case an order of convergence one power of h higher can be obtained if u has the appropriate smoothness: Saranen (1988) shows that

$$\|u_h - u\|_{-2} \le ch^{r+2}\|u\|_{r+1}, \tag{6.6}$$

if $u \in H^{r+1}$.

In the following section we shall see that the collocation method is a special case of the so-called 'qualocation' method. At that point we will demonstrate the Fourier series method of analysis. A generalization of the Fourier series techniques to the case of equations on a torus has been given by Costabel and McLean (1992).

Another situation for which a reasonably satisfactory collocation analysis exists is that of singular integral operators, or systems of singular operators, on plane curves; see Prössdorf and Schmidt (1981), Prössdorf and Rathsfeld (1984), and, for an overview, Prössdorf (1989). Here the principal tool is a localization technique, combined with the observation that translationally invariant operators yield circulant matrices (in the case of closed smooth curves) or Toeplitz matrices (in the case of open arcs). The circulant matrix methods are closely related to the Fourier series techniques mentioned earlier. A generalization to multi-dimensional equations is given by Prössdorf and Schneider (1991).

In spite of the successes in the analysis of the collocation method, there remain some large gaps. Most strikingly, there is apparently no analysis as yet of piecewise-constant collocation for the single-layer equation (2.5) on a three-dimensional sphere. Fourier series methods have no obvious extension, because on a sphere there is no such thing as a uniform partition. In the two-dimensional analogue, however, significant progress with nonuniform partitions has recently been achieved: Chandler (1989, 1990, 1991) has shown for the logarithmic-kernel integral equation that piecewise-constant collocation at the midpoints is stable and convergent even for an essentially arbitrary mesh. Chandler's analysis exploits the specific structure of the collocation matrix for this problem.

7. The qualocation and related methods

In this section we consider the qualocation method (Sloan, 1988; Sloan and Wendland, 1989; Chandler and Sloan, 1990), and its fully discrete variants (Sloan and Burn, 1991; Saranen and Sloan, 1992). For a review with a more limited focus but some more details, see Sloan (1992). An earlier review, restricted to the qualocation method, is that of Wendland (1989).

7.1. The qualocation method

The qualocation method (or 'quadrature-modified collocation method') is an approximation which aims to achieve an order of convergence better than that of the collocation method, while not being too much more expensive to implement. Significant results are so far available only for boundary integral equations on smooth plane curves, thus we shall assume that the equation to be solved is a single equation of the form

$$Lu = f, \tag{7.1}$$

where u and f are 1-periodic functions on $\mathbb{R}$.

The qualocation method is characterized by three things: a trial space S_h, which is the finite-dimensional space within which the approximate solution is to be sought; a test space T_h of the same dimension as S_h; and a quadrature rule Q_h. Given these three ingredients, the method is: find $u_h \in S_h$ such that

$$(Lu_h, \chi)_h = (f, \chi)_h \qquad \forall \, \chi \in T_h, \tag{7.2}$$

where

$$(v, w)_h = Q_h(v\overline{w}). \tag{7.3}$$

Letting $\{\phi_1, \ldots, \phi_N\}$ be a basis for S_h and $\{\chi_1, \ldots, \chi_N\}$ a basis for T_h, the equations to be solved in practice are

$$\sum_{j=1}^{N} (L\phi_j, \chi_k)_h a_j = (f, \chi_k)_h, \qquad k = 1, \ldots, N. \tag{7.4}$$

The method is in effect a semi-discrete version of the Petrov–Galerkin method, i.e. the Galerkin method with different test and trial spaces. It reduces to the Petrov–Galerkin method if $(\cdot, \cdot)_h$ is replaced by the exact inner product $(\cdot, \cdot)$. The novel feature of the qualocation method lies in the discretization: for we shall see that the recommended quadrature rules can be curious indeed.

First, though, we note the important fact that the qualocation formalism includes the collocation method as a special case. For if the quadrature rule is

$$Q_h g = \sum_{\ell=1}^{N} w_\ell g(t_\ell), \tag{7.5}$$

an N-point quadrature rule with nonzero weights $w_1, \ldots, w_N$, then (7.2) is equivalent to

$$\sum_{\ell=1}^{N} w_\ell [Lu_h(t_\ell) - f(t_\ell)] \overline{\chi}_k(t_\ell) = 0, \quad k = 1, \ldots, N,$$

which is in turn equivalent to the collocation equations (6.2) if the $N \times N$ matrix $\{\overline{\chi}_k(t_\ell)\}$ is nonsingular. It is easy to see that quadrature rules with fewer than N points necessarily make the matrix in (7.4) singular, thus only quadrature rules with N or more points are of interest.

At the present time theoretical results are available only if S_h is a space of smoothest splines of order $r, r \geq 1$, and the partition Π_h is uniform, with $h = 1/N$. For the trial space T_h Sloan (1988) and Sloan and Wendland (1989) used a trigonometric polynomial space,

$$T_h = \mathrm{span}\left\{e^{2\pi ijx} : -\frac{N}{2} < j \leq \frac{N}{2}\right\}.$$

This choice was inspired by the Arnold and Wendland (1985) analysis of the collocation method, in which, effectively, the collocation method was treated as a qualocation method with an N-point rectangle rule for Q_h. Here, however, we shall follow Chandler and Sloan (1990) in taking $T_h = S'_h$, the space of smoothest splines of order $r', r' \geq 1$, on the partition Π_h. In practice a low-order spline test space is likely to be preferred over the trigonometric polynomial test space, because it admits a (B-spline) basis in which each element has small support. (If $r' = 1$ the value of the piecewise-constant test function at a point of discontinuity must be understood to be the mean of the left-hand and the right-hand limits. This becomes important if a quadrature point is a point of the partition.)

Following Chandler and Sloan (1990), the operator L in (7.1) is taken to be of the form

$$L = A + B, \tag{7.6}$$

where A has as its Fourier series representation either

$$Av(x) \sim \hat{v}(0) + \sum_{k \neq 0} |k|^\beta \hat{v}(k) e^{2\pi ikx} \tag{7.7}$$

or

$$Av(x) \sim \hat{v}(0) + \sum_{k \neq 0} \mathrm{sign}\, k |k|^\beta \hat{v}(k) e^{2\pi ikx}, \tag{7.8}$$

where $\beta \in \mathbb{R}$, and where

$$B : H^s \to H^t \quad \text{for all } s, t \in \mathbb{R}. \tag{7.9}$$

In the language of the paragraph containing (4.44), L is a pseudo-differential operator of order β and principal symbol either $|\xi|^\beta$ or $\mathrm{sign}\xi|\xi|^\beta$. Since the principal symbol is constant, i.e. independent of x, the operator A can be represented as a convolution. If (7.7) holds then the principal symbol is even, and A is said to be even. Similarly, if (7.8) holds then A is said to be odd. An important special case is that of the logarithmic-kernel integral operator: setting $\beta = -1$ and taking the even case, the operators L, A and B defined by (3.5), (3.7–11) are exactly of the prescribed form, if the free parameter α is set equal to $e^{-1/2}$.

The quadrature rule is taken to be a composite rule of the form

$$Q_h g = h \sum_{\ell=0}^{N-1} \sum_{j=1}^{J} w_j g(\frac{\ell + \xi_j}{N}), \tag{7.10}$$

where

$$0 \le \xi_1 < \xi_2 < \cdots < \xi_J < 1, \tag{7.11}$$

and

$$\sum_{j=1}^{J} w_j = 1, \qquad w_j > 0 \quad \text{for} \quad j = 1, \ldots, J. \tag{7.12}$$

Thus Q_h is the composition, onto each subinterval of the partition, of the J-point rule

$$Q g = \sum_{j=1}^{J} w_j g(\xi_j), \tag{7.13}$$

a quadrature rule defined in $[0, 1]$.

How should the rule Q be chosen? Since the choice $J = 1$ is equivalent to a collocation method, it is natural to consider $J = 2$. Chandler and Sloan (1990) restrict attention to $J = 2$ rules that are symmetric, i.e. having the property that if ξ is a quadrature point then either $\xi = 0$, or else $1 - \xi$ is also a quadrature point with the same associated weight as ξ. There are just two kinds of symmetric rule with $J = 2$, namely

$$Q g = w g(0) + (1 - w) g(\tfrac{1}{2}), \tag{7.14}$$

where $0 < w < 1$, and

$$Q g = \tfrac{1}{2} g(\xi) + \tfrac{1}{2} g(1 - \xi), \tag{7.15}$$

with $0 < \xi < \tfrac{1}{2}$. The first of these is analogous to Simpson's rule, and becomes Simpson's rule if $w = \tfrac{1}{3}$; and the second is analogous to 2-point Gauss quadrature, and becomes so if $\xi = 0.21132\ 48654\ldots$. We shall see, however, that these are usually *not* the recommended values of w or ξ. Rather, the value of w or ξ should be the unique value that will increase the maximum order of (negative-norm) convergence. In some circumstances, for example, the recommended value of w will turn out to be $w = \tfrac{3}{7}$ (giving the '$\tfrac{3}{7}, \tfrac{4}{7}$ rule').

The next two theorems give the highest-order results obtained by Chandler and Sloan (1990) for the two kinds of quadrature rule. First we collect the main assumptions.

Assumption Y: the equation to be solved is (7.1), with L given by (7.6), (7.9), and one of (7.7), (7.8); L is one-to-one; the partition Π_h is uniform; the test space is S'_h, the space of smoothest splines on Π_h of order $r' \ge 1$;

and r' has the same parity as r if A is even and the opposite parity if A is odd.

Theorem 4 Assume that Y holds, that Q is given by (7.14), with $0 < w < 1$, and that $r > \beta + 1$.

(i) The qualocation equation (7.2) has a unique solution $u_h \in S_h$ for all h sufficiently small.

(ii) If r and A are both even or both odd then u_h satisfies

$$\|u_h - u\|_\beta \le ch^{r-\beta}\|u\|_r. \tag{7.16}$$

It satisfies also

$$\|u_h - u\|_{\beta-2} \le ch^{r-\beta+2}\|u\|_{r+2} \tag{7.17}$$

if and only if, in addition,

$$w = \frac{2^{r-\beta-1} - 1}{2^{r-\beta} - 1}. \tag{7.18}$$

(iii) If r and A are of opposite parity then u_h satisfies

$$\|u_h - u\|_{\beta-1} \le ch^{r-\beta+1}\|u\|_{r+1}. \tag{7.19}$$

It satisfies also

$$\|u_h - u\|_{\beta-3} \le ch^{r-\beta+3}\|u\|_{r+3} \tag{7.20}$$

if and only if, in addition,

$$w = \frac{2^{r-\beta} - 1}{2^{r-\beta+1} - 1}. \tag{7.21}$$

A sketch of the proof of this theorem follows Theorem 5. A first observation about the content of the theorem is that (7.16) is the same as (6.5), the fastest convergence result obtained by Arnold and Wendland (1983, 1985) for the collocation method and L a pseudo-differential operator of order β; and the one higher order result (7.19) is the improved collocation result obtained by Saranen (1988) for the case of an even operator and odd r, already referred to in Section 6. More interestingly, we see in (7.17) or (7.20) that the maximum order of convergence jumps by yet another two if (and only if) w has the precise values specified in (7.18) or (7.21).

For the particular case of the logarithmic-kernel operator A is even and $\beta = -1$, so if r also is even then the special value of w is

$$w = \frac{2^r - 1}{2^{r+1} - 1}, \tag{7.22}$$

which yields

$$\|u_h - u\|_{-3} \le ch^{r+3}\|u\|_{r+2}, \tag{7.23}$$

whereas the best result available if w has any other value is the $\mathcal{O}(h^{r+1})$

result (6.4). For example, in the piecewise-linear case (i.e. $r = 2$) the value $w = 3/7$ yields an $\mathcal{O}(h^5)$ result, compared with $\mathcal{O}(h^3)$ for the collocation method. If r is odd then the choice

$$w = \frac{2^{r+1} - 1}{2^{r+2} - 1} \tag{7.24}$$

yields

$$\|u_h - u\|_{-4} \le ch^{r+4}\|u\|_{r+3}, \tag{7.25}$$

whereas the best result if w has any other value is the $\mathcal{O}(h^{r+2})$ result (6.6). Thus in the piecewise-constant case (i.e. $r = 1$) the value $w = 3/7$ again yields an $\mathcal{O}(h^5)$ result, compared with $\mathcal{O}(h^3)$ for the midpoint collocation method.

We should note, though, that the higher order of convergence apparent in (7.17) or (7.20) require both higher regularity of u and a more negative norm in which to observe the error. It follows that in some applications the maximum order of convergence will not be achieved.

In the next theorem the recommended quadrature points for the rule (7.15) of 2-point Gauss type are the zeros of the function

$$G_\alpha(x) = \sum_{n=1}^{\infty} \frac{1}{n^\alpha} \cos 2\pi n x, \tag{7.26}$$

for appropriate values of $\alpha \ge 1$. It is known that G_α has exactly two zeros on $(0, 1)$, located symmetrically with respect to the midpoint. Some values of the first zero (taken from Sloan and Wendland (1989)) are given in Table 1.

Table 1. *The unique zero of G_α in $(0, \frac{1}{2})$*

α	Zero of G_α
1	1/6
2	0.21132 48654
3	0.23082 96503
4	0.24033 51888
5	0.24511 88417
∞	1/4

Theorem 5   Assume that Y holds, that Q is given by (7.15) with $0 < \xi < \frac{1}{2}$, and that $r > \beta + \frac{1}{2}$.

(i)   The qualocation equation (7.2) has a unique solution $u_h \in S_h$ for all h sufficiently small.

(ii) If r and A are both even or both odd then u_h satisfies (7.16). It satisfies also (7.17) if, in addition, ξ is the unique zero in $(0, \frac{1}{2})$ of $G_{r-\beta}$.

(iii) If r and A are of opposite parity then u_h satisfies (7.19). It satisfies also (7.20) if, in addition, $r' \geq 3$ and ξ is the unique zero in $(0, \frac{1}{2})$ of $G_{r-\beta+1}$.

The complete proofs of Theorems 4 and 5 are lengthy. Here we indicate only the outline, with main emphasis on the argument that determines the special values of w or ξ. (For a more complete sketch of that part of the argument, see Sloan (1992).)

The main task is to prove the theorems for the special case $L = A$, the result then being extended to the full operator $L = A + B$ by a standard perturbation argument, given, for example, by Arnold and Wendland (1985). Because A is a convolution operator, and so invariant under translation, and because also the partition is uniform, the qualocation matrix in (7.4) can be made diagonal if the basis functions of S_h and S'_h are chosen so as to behave in an appropriate way under translation by h. An appropriate basis for S_h is $\{\psi_\mu : \mu \in \Lambda_N\}$, where

$$\Lambda_N = \left\{\mu \in \mathbb{Z} : -\frac{N}{2} < \mu \leq \frac{N}{2}\right\}, \tag{7.27}$$

and

$$\psi_\mu(x) = \begin{cases} 1, & \mu = 0, \\ \sum_{k \equiv \mu} (\mu/k)^r e^{2\pi i k x}, & \mu \in \Lambda_N^*. \end{cases} \tag{7.28}$$

Here $\Lambda_N^* = \Lambda_N \backslash \{0\}$, and $k \equiv \mu$ means that $k - \mu$ is a multiple of N. (If $r = 1$ the Fourier series, which is then not absolutely convergent, is to be understood as the limit of the symmetric partial sums.) That ψ_μ really is a spline of order r on the uniform partition Π_h follows from the fact that the Fourier coefficients satisfy the appropriate recurrence relation for a function $v \in S_h$, namely (Arnold (1983), extending Quade and Collatz (1938))

$$k^r \hat{v}(k) = \mu^r \hat{v}(\mu) \qquad \text{if } k \equiv \mu.$$

Since $\hat{\psi}_\mu(\nu) = \delta_{\mu\nu}$ for $\mu, \nu \in \Lambda_N$, the expansion coefficients of $v \in S_h$ in terms of $\{\psi_\mu\}$ are just the Fourier coefficients; that is

$$v = \sum_{\mu \in \Lambda_N} \hat{v}(\mu)\psi_\mu \quad \text{for } v \in S_h. \tag{7.29}$$

The function ψ_μ is, in essence, the spline equivalent of the trigonometric polynomial $e^{2\pi i \mu x}$. In particular, the two functions behave in exactly the same way under translation by h.

With a basis $\{\psi'_\mu : \mu \in \Lambda_N\}$ for S'_h defined in a similar way, and with the aid of the expression (7.7) or (7.8) for A, it is a straightforward if tedious matter to evaluate the matrix element $(A\psi_\nu, \psi'_\mu)_h$ explicitly, and to verify that it vanishes for $\mu \neq \nu$. In detail, we find (Chandler and Sloan, 1990, Lemma 1)

$$(A\psi_\nu, \psi'_\mu)_h = \begin{cases} 1 & \text{if } \mu = \nu = 0, \\ (\text{sign } \mu)|\mu|^\beta D(\mu h) & \text{if } \mu = \nu \in \Lambda_N^*, \\ 0 & \text{if } \mu \neq \nu, \end{cases} \tag{7.30}$$

where the factor $(\text{sign } \mu)$ in this equation is present only if A is odd, and where

$$D(y) = \sum_{j=1}^{J} w_j [1 + \Omega(\xi_j, y)] \left[1 + \overline{\Delta'(\xi_j, y)}\right], \tag{7.31}$$

with

$$\Delta'(\xi, y) = y^{r'} \sum_{\ell \neq 0} \frac{1}{(\ell + y)^{r'}} e^{2\pi i \ell \xi}, \tag{7.32}$$

and with

$$\Omega(\xi, y) = |y|^{r-\beta} \sum_{\ell \neq 0} \frac{1}{|\ell + y|^{r-\beta}} e^{2\pi i \ell \xi} \tag{7.33}$$

if r and A are both even or both odd, or

$$\Omega(\xi, y) = \text{sign } y |y|^{r-\beta} \sum_{\ell \neq 0} \frac{\text{sign } \ell}{|\ell + y|^{r-\beta}} e^{2\pi i \ell \xi} \tag{7.34}$$

if r and A are of opposite parity.

Since we will have to divide by $D(\mu h)$, for stability of the method it is essential that $D(y)$ be bounded away from zero for $y \in [-\frac{1}{2}, \frac{1}{2}]$ – a property that is not quite trivial, since it is well known that some collocation methods (e.g. midpoint collocation if r and A are both even) are unstable. Nevertheless it is shown in Chandler and Sloan (1990), by appeal to known properties of trigonometric sums, that under the conditions of the theorem there exists $d > 0$ such that

$$|D(y)| \geq d \qquad \text{for all } y \in [-\frac{1}{2}, \frac{1}{2}].$$

It now follows from the qualocation equation (7.4) and from (7.29) and (7.30) that

$$\hat{u}_h(\mu) = \begin{cases} (Au, \psi'_0)_h & \text{if } \mu = 0, \\ \dfrac{(\text{sign } \mu)}{|\mu|^\beta D(\mu h)} (Au, \psi'_\mu)_h & \text{if } \mu \in \Lambda_N^*. \end{cases} \tag{7.35}$$

After evaluating the right-hand side (using e.g. Chandler and Sloan (1990,

Lemma 1)), we find

$$\hat{u}_h(\mu) - \hat{u}(\mu) = \begin{cases} P_h, & \mu = 0, \\ -[E(\mu h)/D(\mu h)]\hat{u}(\mu) + R_h(\mu), & \mu \in \Lambda_N^*, \end{cases} \qquad (7.36)$$

where

$$E(y) = \sum_{j=1}^{J} w_j \Omega(\xi_j, y) \left[1 + \overline{\Delta'(\xi_j, y)}\right], \qquad (7.37)$$

$$|P_h| \leq {\sum_{n \equiv 0}}' |n|^\beta |\hat{u}(n)|, \qquad (7.38)$$

$$|R_h(\mu)| \leq \frac{c}{|\mu|^\beta} {\sum_{n \equiv \mu}}' |n|^\beta |\hat{u}(n)|, \qquad (7.39)$$

and where

$${\sum_{n \equiv \mu}}' = \sum_{\substack{n \equiv \mu \\ n \neq \mu}}.$$

The error expression (7.36) is the key to the theorems. The quantities P_h and $R_h(\mu)$ in the expression depend only on the Fourier coefficients $\hat{u}(n)$ for which $|n| \geq N/2$, and so can be made to decay as rapidly as desired by requiring u to be in a sufficiently high Sobolev space. The first term for $\mu \in \Lambda_N^*$ is in a quite different category, because it is this that imposes an absolute restriction on the maximum order of convergence that can be achieved: if $E(y) = \mathcal{O}(|y|^\rho)$ as $y \to 0$ then the best order of convergence we can hope for, given (7.36), is $\mathcal{O}(h^\rho)$. For this reason ρ is called by Chandler and Sloan (1990) the 'order' of the particular qualocation method.

If r and A are both even or both odd then it follows from (7.32), (7.33) and (7.37) that, for any symmetric rule Q,

$$E(y) = |y|^{r-\beta} 2 \sum_{j=1}^{J} w_j \sum_{\ell=1}^{\infty} \frac{\cos 2\pi \ell \xi_j}{\ell^{r-\beta}} + \mathcal{O}(|y|^{r-\beta+2}). \qquad (7.40)$$

(Note that $r' \geq 2$, since under the present assumptions r' is even.) Thus the qualocation method is of order $r - \beta$ – unless, that is

$$\sum_{j=1}^{J} w_j \sum_{\ell=1}^{\infty} \frac{\cos 2\pi \ell \xi_j}{\ell^{r-\beta}} = 0, \qquad (7.41)$$

in which case the order jumps to $r - \beta + 2$. For a rule Q of the form (7.14) the latter equation becomes

$$w \sum_{\ell=1}^{\infty} \frac{1}{\ell^{r-\beta}} + (1-w) \sum_{\ell=1}^{\infty} \frac{(-1)^\ell}{\ell^{r-\beta}} = 0,$$

or by a standard zeta function trick

$$\left[w - (1-w)(1 - \frac{1}{2^{r-\beta-1}})\right] \sum_{\ell=1}^{\infty} \frac{1}{\ell^{r-\beta}} = 0,$$

which is satisfied if and only if w has the value given by (7.18). And for a rule Q of the form (7.15) it is immediately obvious that (7.41) is satisfied if and only if ξ is the zero in $(0, \frac{1}{2})$ of the function $G_{r-\beta}$ defined by (7.26).

Similarly, if r and A are of opposite parity (and hence r' is odd) then it follows from (7.32), (7.34) and (7.37) that, for Q a symmetric rule,

$$E(y) = -|y|^{r-\beta+1} 2(r-\beta) \sum_{j=1}^{J} w_j \sum_{\ell=1}^{\infty} \frac{\cos 2\pi\ell\xi_j}{\ell^{r-\beta+1}} + \mathcal{O}\left(|y|^{r-\beta+\min(r',3)}\right).$$

Thus the method is of order $r - \beta + 1$, unless

$$\sum_{j=1}^{J} w_j \sum_{\ell=1}^{\infty} \frac{\cos 2\pi\ell\xi_j}{\ell^{r-\beta+1}} = 0, \qquad (7.42)$$

in which case it is of order $r - \beta + 3$, provided $r' \geq 3$. The special values of w or ξ in rules of the form (7.14), (7.15) are as before, but with r replaced by $r + 1$.

The orders of convergence in every case are now seen to correspond exactly to the maximum orders of convergence in Theorems 4 and 5. For the remainder of the proof we refer to Chandler and Sloan (1990). In particular, Theorem 2 of that paper shows that all the results follow once the order and stability of the method has been established.

The qualocation analysis indicated above conforms to one of the great paradigms of numerical analysis: first identify the form of the leading term of the errors, then adjust the method so as to eliminate that leading term. Looked at that way, the 3/7, 4/7 qualocation rule is no stranger than, say, the formulas of Romberg integration.

7.2. Fully discrete variants

We have seen that the Galerkin method for (7.1) requires two levels of integration for each matrix element, whereas the collocation and qualocation methods need only one level. But the following variant of the qualocation method proposed by Sloan and Burn (1991) for the logarithmic-kernel integral equation on a smooth curve requires no exact integrals at all.

In this method, the exact integral (3.5) is first replaced by its rectangle rule approximation

$$L_h u(x) = -2h \sum_{k=0}^{N-1} \log |\nu(x) - \nu(kh)| u(kh). \qquad (7.43)$$

Then one proceeds as in the qualocation method: find $u_h \in S_h$ such that

$$(L_h u_h, \chi)_h = (f, \chi)_h \qquad \forall \chi \in T_h. \qquad (7.44)$$

Here $(\cdot, \cdot)_h$ is defined again by (7.3) and (7.10–13), but now the parameters in the rule Q must be chosen differently, as the quadrature rule has the added burden of compensating for the damage caused by replacing L by L_h.

In working out this method u_h is evaluated only at the points of the rectangle rule (7.43), thus the trial space S_h becomes significant only if one wants to interpolate between the points. For the analysis, however, the choice of trial space is important. In Sloan and Burn (1991) a trigonometric trial space

$$S_h = \{e^{2\pi i \mu x} : \mu \in \Lambda_N\} \qquad (7.45)$$

was assumed.

The following result was established in Sloan and Burn (1991) for the case of a circle by Fourier methods similar to those used above, but for general smooth curves was proved only under additional restrictions. The result for general curves was proved without the extra restrictions by Saranen and Sloan (1992).

Theorem 6 Assume that the equation to be solved is $Lu = f$, where L is the logarithmic-kernel integral operator in (3.5); that the transfinite diameter is different from 1, so that L is one-to-one; that the partition Π_h is uniform; that the trial space is given by (7.45); that the test space is S_h', the space of smoothest splines of order r'; that r' is even; and that

$$Qg = \tfrac{1}{2}g(\xi) + \tfrac{1}{2}g(1 - \xi),$$

with $0 < \xi < \tfrac{1}{2}$. Then

(i) Equation (7.44) has a solution $u_h \in S_h$ for all h sufficiently small.
(ii) For $s \geq -1$, u_h satisfies

$$\|u_h - u\|_s \leq ch^1 \|u\|_{s+1}. \qquad (7.46)$$

It satisfies also

$$\|u_h - u\|_s \leq ch^3 \|u\|_{s+3} \qquad (7.47)$$

if and only if $\xi = \tfrac{1}{6}$.

Versions with maximum order higher than $\mathcal{O}(h^3)$ have been foreshadowed in Sloan (1992).

An alternative version proposed by Saranen and Sloan (1992) replaces the right-hand side of (7.44) by the exact inner product. Thus the method becomes: find $u_h \in S_h$ such that

$$(L_h u_h, \chi)_h = (f, \chi) \qquad \forall \chi \in T_h. \qquad (7.48)$$

This has an advantage if u is of low regularity, in that the condition $s \geq -1$ in Theorem 6 is replaced by $s \geq -r' - 1$. On the other hand the estimate (7.47) is replaced by the more restrictive estimate

$$\|u_h - u\|_s \leq ch^{\min(r',3)} \|u\|_{s+\min(r',3)}. \tag{7.49}$$

8. Corners, cracks and challenges

At many points we have mentioned difficulties caused by corners. Part of the problem is that the techniques of analysis (e.g. pseudo-differential operator arguments, compactness of operators, Fourier series methods) break down when corners are present. Part of it is that corners force us to consider modifications (such as mesh grading near the corner) which further complicate the analysis. Sometimes (as in the case of the double-layer equations) the presence of corners forces changes in a method (such as modifications in the trial space for the collocation method near a corner), even though there is little or no evidence that such changes are needed other than to make the proofs go through. It is fair to say that even for plane problems corners still present many theoretical challenges.

Consider, for a moment, the qualocation method and its discrete variants, described in Section 7. Once corners are present the Fourier series arguments outlined there break down, because it is no longer possible to write the boundary integral operator in the form $L = A + B$ with A given by (7.7) or (7.8) and B a smoothing operator as in (7.9). Yet numerical experiments (Chandler and Sloan, 1990; Sloan and Burn, 1991) suggest very strongly that the methods can remain useful, and even yield orders of convergence similar to those predicted for smooth curves, if the mesh is suitably graded in a neighbourhood of the corner.

Curiously, there is one extreme case of a corner, namely the exterior Dirichlet problem for a slit or crack, for which the theoretical understanding is reasonably complete. Taking for simplicity a straight slit of length 2γ, and forgetting the boundary condition (2.7) at infinity, the single-layer equation (2.5) becomes

$$-\frac{1}{\pi} \int_{-\gamma}^{\gamma} \log |t - s| z(s) \mathrm{d}s = g(t), \quad t \in (-\gamma, \gamma). \tag{8.1}$$

Applying the transformation (Yan and Sloan, 1988)

$$t = \gamma \cos 2\pi x, \qquad s = \gamma \cos 2\pi y, \tag{8.2}$$

$$f(x) = g(\gamma \cos 2\pi x), \tag{8.3}$$

$$u(x) = \gamma z(\gamma \cos 2\pi x)|\sin 2\pi x|, \tag{8.4}$$

we obtain (on noting that f and u are even and 1-periodic)

$$- \int_0^1 \log |\gamma(\cos 2\pi x - \cos 2\pi y)| u(y) \mathrm{d}y = f(x), \quad x \in \mathbb{R}. \qquad (8.5)$$

The cosine transformation used here has a long history, particularly in connection with the airfoil equation (Multhopp, 1938; Weissinger, 1950; Schleiff, 1968a,b). Its particular advantage in the present context is that it reduces the problem to one we have met already, namely the logarithmic-kernel equation for a circle: using only trigonometric and logarithmic identities and the fact that u is even, it can easily be shown (see Prössdorf et al. (1992), Lemma 2.1) that (8.5) is equivalent to

$$- 2 \int_0^1 \log |(2\gamma)^{1/2} \sin \pi(x - y)| u(y) \mathrm{d}y = f(x), \quad x \in \mathbb{R}. \qquad (8.6)$$

But from (3.7) this is just the single-layer integral equation for a circle of radius $(\gamma/2)^{1/2}$.

The transformations leading to (8.6) tell us that the solution z of (8.1) usually has singularities of the form $(\gamma \mp t)^{-1/2}$ at the two ends (this follows from (8.4)). It also tells us, since a circle has a transfinite diameter equal to its radius (see Subsection 4.3), that equation (8.1) for z is singular when $(\gamma/2)^{1/2} = 1$, or $\gamma = 2$. (This corresponds to the fact, mentioned in Subsection 4.3, that the transfinite diameter of an interval is one quarter of its length.)

More importantly for our present purposes, this transformation lies at the heart of several theoretical analyses of numerical methods for the logarithmic-kernel integral equation on open arcs. These include Atkinson and Sloan (1991), which gives an analysis of a discrete Galerkin method; Sloan and Stephan (1992), analysing a collocation method with Chebyshev polynomials; Prössdorf et al. (1992), adapting to an open arc the discrete method of Sloan and Burn (1991) discussed in Subsection 7.2; and Joe and Yan (1991, 1992), analysing a piecewise-constant collocation method on a graded mesh, with the collocation points taken to be the midpoints with respect to the transformed variable x in (8.2) rather than with respect to the original variable t.

That work of Joe and Yan (1991, 1992) establishes the (surprising) conclusion that the order of convergence can be increased by a seemingly insignificant shift in the collocation points. Indeed, the order of convergence established by Joe and Yan is even higher than the apparent order of convergence of the Galerkin method for the same piecewise-constant basis and the same graded mesh (Yan and Sloan, 1989). A lesson for the future seems to be that in both the theory and the practice of mesh grading we need to take more seriously than in the past the transformed independent variable implicit in the mesh grading: if the partition is uniform with respect to the transformed variable x, then perhaps we should specify all collocation points and quadrature rules with respect to that variable.

A serious challenge for many boundary element methods is the extension of the analysis to irregular meshes on plane curves, and thence to three-dimensional surfaces, which is after all where the main game is. This certainly poses a problem for methods such as the qualocation method which rely on Fourier series methods for their analysis. It is even a problem, as we have remarked before, for a method as simple as piecewise-constant collocation for the single-layer equation on a sphere.

There is still much to be done. Those of us who enjoy the field are happy that this is so.

Acknowledgment

The writer was supported in part by the US Army Research Office through the Mathematical Sciences Institute at Cornell University, and by the Australian Research Council. He is also indebted to Professor H.J. Stetter and to the Technical University of Vienna for their hospitality at the time when much of this review was written.

REFERENCES

M.S. Agranovich (1979), 'Spectral properties of elliptic pseudodifferential operators on a closed curve', *Funct. Anal. Appl.* **13**, 279–281.

D.N. Arnold (1983), 'A spline-trigonometric Galerkin method and an exponentially convergent boundary integral method', *Math. Comput.* **41**, 383–397.

D.N. Arnold and W.L. Wendland (1983), 'On the asymptotic convergence of collocation methods', *Math. Comput.* **41**, 349–381.

D.N. Arnold and W.L. Wendland (1985), 'The convergence of spline collocation for strongly elliptic equations on curves', *Numer. Math.* **47**, 317–341.

K.E. Atkinson (1976), *A Survey of Numerical Methods for the Solution of Fredholm Integral Equations of the Second Kind*, SIAM (Philadelphia).

K.E. Atkinson (1988), 'A discrete Galerkin method for first kind integral equations with a logarithmic kernel', *J. Int. Eqns Appl.* **1**, 343–363.

K.E. Atkinson (1990), 'A survey of boundary integral equation methods for the numerical solution of Laplace's equation in three dimensions', in *The Numerical Solution of Integral Equations* (M. Golberg, ed.), Plenum Press (New York), 1–34.

K.E. Atkinson and G.A. Chandler (1990), 'Boundary integral equation methods for solving Laplace's equation with nonlinear boundary conditions: the smooth boundary case', *Math. Comput.* **55**, 451–472.

K.E. Atkinson and F.R. de Hoog (1984), 'The numerical solution of Laplace's equation on a wedge', *IMA J. Numer. Anal.* **4**, 19–41.

K.E. Atkinson and I.H. Sloan (1991), 'The numerical solution of first-kind logarithmic-kernel integral equations on smooth open arcs', *Math. Comput.* **56**, 119–139.

C.T.H. Baker (1977), *The Numerical Treatment of Integral Equations*, Clarendon Press (Oxford).

P. Banerjee and J. Watson (eds) (1986), *Developments in Boundary Element Methods – 4*, Elsevier (New York).

C.A. Brebbia, J.C.F. Telles and L.C. Wrobel (1984), *Boundary Element Techniques*, Springer-Verlag (Berlin).

G.A. Chandler (1989), 'Mid-point collocation for Cauchy singular integral equations', submitted.

G.A. Chandler (1990), 'Optimal order convergence of midpoint collocation for first kind integral equations', submitted.

G.A. Chandler (1991), 'Discrete norms for the convergence of boundary element methods', in *Workshop on Theoretical and Numerical Aspects of Geometric Variational Inequalities*, (G. Dzuik, G. Huisken and J. Hutchinson, eds), Centre for Mathematics and its Applications (Canberra).

G.A. Chandler and I.G. Graham (1988), 'Product integration-collocation methods for noncompact integral operators', *Math. Comput.* **50**, 125–138.

G.A. Chandler and I.H. Sloan (1990), 'Spline qualocation methods for boundary integral equations', *Numer. Math.* **58**, 537–567.

P.G. Ciarlet (1978), *The Finite Element Method for Elliptic Problems*, North-Holland (Amsterdam).

D.L. Clements (1981), *Boundary Value Problems Governed by Second Order Elliptic Systems*, Pitman (Boston).

R.R. Coifman, A. McIntosh and Y. Meyer (1982), 'L'intégrale de Cauchy définit un opérateur borné sur L_2 pour les courbes Lipschitziennes', *Ann. Math. (II Ser.)* **116**, 361–387.

D. Colton and P. Kress (1983), *Integral Equation Methods in Scattering Theory*, J. Wiley & Sons (New York).

M. Costabel (1987), 'Symmetric methods for the coupling of finite elements and boundary elements', in *Boundary Elements XI*, vol. **1**, (C.A. Brebbia, W.L. Wendland and G. Kuhn, eds) Springer-Verlag (Berlin), 411–420.

M. Costabel (1988), 'Boundary integral operators on Lipschitz domains: elementary results', *SIAM J. Math. Anal.* **19**, 613–626.

M. Costabel and W. McLean (1992) 'Spline collocation for strongly elliptic equations on the torus', submitted.

M. Costabel and E.P. Stephan (1985), 'Boundary integral equations for mixed boundary value problems in polygonal domains and Galerkin approximation', in *Mathematical Models and Methods in Mechanics 1981* (W. Fiszdon and K. Wilmanski, eds), Banach Center Publications **15**, PN-Polish Scientific Publications (Warsaw), 175–251.

M. Costabel and W.L. Wendland (1986), 'Strong ellipticity of boundary integral operators', *J. Reine Angew. Math.* **372**, 34–63.

C. Cryer (1970), 'The solution of the Dirichlet problem for Laplace's equation when the boundary data is discontinuous and the domain has a boundary which is of bounded rotation by means of the Lebesgue–Stieltjes integral equation for the double layer potential', *Computer Science Technical Report* 99, University of Wisconsin (Madison).

J.L. Doob (1984), *Classical Potential Theory and its Probabilistic Counterpart*, Springer-Verlag (New York).

P.P.B. Eggermont and J. Saranen (1990), 'L^p estimates of boundary integral equations for some nonlinear boundary value problems', *Numer. Math.* **58**, 465–478.

J. Elschner (1988), 'On spline approximation for a class of integral equations I: Galerkin and collocation methods with piecewise polynomials' *Math. Meth. Appl. Sci.* **10**, 543–559.

G. Fichera (1961), 'Linear elliptic equations of higher order in two independent variables and singular integral equations', in *Partial Differential Equations and Continuum Mechanics*, (R.E. Langer, ed.), University of Wisconsin Press (Madison), 55–80.

D. Gaier (1976), 'Integralgleichungen erster Art und konforme Abbildung', *Math. Z.* **147**, 113–129.

B.G. Galerkin (1915), 'Expansions in stability problems for elastic rods and plates', *Vestnik inzkenorov* **19**, 897–908 (in Russian).

G.N. Gatica and G.C. Hsiao (1989), 'The coupling of boundary element and finite element methods for a nonlinear exterior boundary value problem', *Z. Anal. Anw.* **8**, 377–387.

D. Gilbarg and N.S. Trudinger (1983), *Elliptic Partial Differential Equations of Second Order*, 2nd edition, Springer-Verlag (Berlin).

J. Giroire and J.C. Nedelec (1978), 'Numerical solution of an exterior Neumann problem using a double layer potential', *Math. Comput.* **32**, 973–990.

W. Hackbusch (1989), *Integralgleichungen: Theorie und Numerik*, Teubner Studienbücher (Stuttgart).

F. Hartmann (1989), *Introduction to Boundary Elements*, Springer-Verlag (Berlin).

J.L. Hess and A.M.O. Smith (1967), 'Calculation of potential flows about arbitrary bodies', in *Progress in Aeronautical Sciences* vol. 8 (D. Küchemann, ed.), Pergamon Press (London).

S. Hildebrandt and E. Wienholtz (1964), 'Constructive proofs of representation theorems in separable Hilbert space', *Comm. Pure Appl. Math.* **17**, 369–373.

E. Hille (1962), *Analytic Function Theory*, Vol. II, Ginn and Company (Boston).

F.R de Hoog (1974), 'Product integration techniques for the numerical solution of integral equations', Ph.D. Thesis, Australian National University (Canberra).

L. Hörmander (1965), 'Pseudo-differential operators', *Comm. Pure Appl. Math.* **18**, 501–507.

G.C. Hsiao (1986), 'On the stability of integral equations of the first kind with logarithmic kernels', *Arch. Rat. Mech. Anal.* **94**, 179–192.

G.C. Hsiao (1989), 'On boundary integral equations of the first kind', *J. Comput. Math. (Beijing)* **7**, 121–131.

G.C. Hsiao (1990), 'The coupling of boundary element and finite element methods', *ZAMM* **70**, T493–T503.

G.C. Hsiao (1991), 'Some recent developments on the coupling of finite element and boundary element methods', in *Proc. Int. Conf. on Numerical Methods in Applied Science and Industry, Torino 1990*, to appear.

G.C. Hsiao, P. Kopp and W.L. Wendland (1980), 'A Galerkin collocation method for some integral equations of the first kind', *Computing* **25**, 89–130.

G.C. Hsiao, P. Kopp and W.L. Wendland (1984), 'Some applications of a Galerkin-collocation method for boundary integral equations of the first kind', *Math. Meth. Appl. Sci.* **6**, 280–325.

G.C. Hsiao and R.C. MacCamy (1973), 'Solution of boundary value problems by integral equations of the first kind', *SIAM Review* **15**, 687–705.

G.C. Hsiao and S. Prössdorf (1992), 'A generalization of the Arnold–Wendland lemma to collocation methods for boundary integral equations in $\mathbb{R}^n$', *Math. Meth. Appl. Sci.*, to appear.

G.C. Hsiao and W.L. Wendland (1977), 'A finite element method for some integral equations of the first kind', *J. Math. Anal. Appl.* **58**, 449–481.

G.C. Hsiao and W.L. Wendland (1981), 'The Aubin–Nitsche lemma for integral equations', *J. Int. Eqns* **3**, 299–315.

D.B. Ingham and M.A. Kelmanson (1984), *Boundary Integral Equation Analyses of Singular, Potential and Biharmonic Problems*, Springer-Verlag (Berlin).

M.A. Jaswon (1963), 'Integral equation methods in potential theory, I', *Proc. Roy. Soc. Ser. A* **275**, 23–32.

M.A. Jaswon and G. Symm (1977), *Integral Equation Methods in Potential Theory and Elastostatics*, Academic Press (London).

S. Joe and Y. Yan (1991), 'Numerical solution of a first kind integral equation on [-1,1]', Proc. Centre for Mathematical Analysis, Canberra, to appear.

S. Joe and Y. Yan (1992), 'A collocation method using cosine mesh grading for Symm's equation on the interval [-1,1]', submitted.

C. Johnson and J.C. Nedelec (1980), 'On the coupling of boundary integral and finite element methods', *Math. Comput.* **35**, 1063–1079.

J.J. Kohn and L. Nirenberg (1965), 'An algebra of pseudodifferential operators', *Comm. Pure Appl. Math.* **18**, 269–305.

R. Kress (1989), *Linear Integral Equations*, Springer-Verlag (Berlin).

V.D. Kupradze (1965), *Potential Methods in the Theory of Elasticity*, Israel Programme of Scientific Translations (Jerusalem).

U. Lamp, T. Schleicher, E.P. Stephan and W.L. Wendland (1984), 'Galerkin collocation for an improved boundary element method for a plane mixed boundary value problem', *Computing* **33**, 269–296.

V.G. Maz'ya (1991), 'Boundary integral equations', in *Analysis IV (Springer-Verlag, Encyclopaedia of Mathematical Sciences, vol. 27)* (V.G. Maz'ya and S.M. Nikolskii, eds), (Berlin), 127–222.

W. McLean (1986), 'A spectral Galerkin method for a boundary integral equation', *Math. Comput.* **47**, 597–607.

W. McLean (1990), 'An integral equation method for a problem with mixed boundary conditions', *SIAM J. Math. Anal.* **21**, 917–934.

W. McLean (1991), 'Local and global descriptions of periodic pseudodifferential operators', *Math. Nachr.* **150**, 151–161.

S.G. Mikhlin (1970), *Mathematical Physics, an Advanced Course*, North-Holland (Amsterdam).

H. Multhopp (1938), 'Die Berechnung der Auftriebsverteilung von Tragflügeln', *Luftfahrt-Forschung (Berlin)* **15**, 153–169.

S. Prössdorf (1989), 'Numerische Behandlung singulärer Integralgleichungen', *ZAMM* **69** (4), T5–T13.

S. Prössdorf and A. Rathsfeld (1984), 'A spline collocation method for singular integral equations with piecewise continuous coefficients', *Int. Eqns Oper. Theory* **7**, 537–560.

S. Prössdorf, J. Saranen and I.H. Sloan (1992), 'A discrete method for the logarithmic-kernel integral equation on an arc', *J. Austral. Math. Soc. (Ser. B)*, to appear.

S. Prössdorf and G. Schmidt (1981), 'A finite element collocation method for singular integral equations', *Math. Nachr.* **100**, 33–60.

S. Prössdorf and R. Schneider (1991), ' A spline collocation method for multidimensional strongly elliptic pseudodifferential operators of order zero', *Int. Eqns Oper. Theory* **14**, 399–435.

S. Prössdorf and B. Silbermann (1977), *Projektionsverfahren und die näherungsweise Lösung singulärer Gleichungen*, Teubner (Leipzig).

S. Prössdorf and B. Silbermann (1991), *Numerical Analysis for Integral and Related Operator Equations*, Birkhäuser (Basel).

W. Quade and L. Collatz (1938), 'Zur Interpolationstheorie der reellen periodischen Funktionen', *Sonderausgabe d. Sitzungsber. d. Preußischen Akad. d. Wiss. Phys.-math. Kl. Verlag d. Wiss.* (Berlin), 1–49.

J. Radon (1919), 'Über die Randwertaufgaben beim logarithmischen Potential', *Sitzungsber. Akad. Wiss. Wien* **128**, Abt. IIa, 1123–1167.

R. Rannacher and W.L. Wendland (1985), 'On the order of pointwise convergence of some boundary element methods. Part I: Operators of negative and zero order', *RAIRO Modél. Math. Anal. Numer.* **19**, 65–88.

R. Rannacher and W.L. Wendland (1988), 'On the order of pointwise convergence of some boundary element methods. Part II: Operators of positive order', *Math. Modelling Numer. Anal.* **22**, 343–362.

K. Ruotsalainen (1992), 'Remarks on the boundary element method for strongly nonlinear problems', *J. Austral. Math. Soc. (Ser. B)*, to appear.

K. Ruotsalainen and J. Saranen (1989), 'On the collocation method for a nonlinear boundary integral equation', *J. Comput. Appl. Math.* **28**, 339–348.

K. Ruotsalainen and W.L. Wendland (1988), 'On the boundary element method for some nonlinear boundary value problems', *Numer. Math.* **53**, 299–314.

J. Saranen (1988), 'The convergence of even degree spline collocation solution for potential problems in smooth domains of the plane', *Numer. Math.* **53**, 499–512.

J. Saranen and I.H. Sloan (1992), 'Quadrature methods for logarithmic kernel integral equations on closed curves', *IMA J. Numer. Anal.*, to appear.

J. Saranen and W.L. Wendland (1985), 'On the asymptotic convergence of collocation methods with spline functions of even degree', *Math. Comput.* **45**, 91–108.

J. Saranen and W.L. Wendland (1987), 'The Fourier series representation of pseudodifferential operators on closed curves', *Complex Variables Theory Appl.* **8**, 55–64.

M. Schleiff (1968a), 'Untersuchung einer linearen singulären Integrodifferentialgleichung der Tragflügeltheorie', *Wiss. Z. der Univ Halle XVII* **68** M(6), 981–1000.

M. Schleiff (1968b), 'Über Näherungsverfahren zur Lösung einer singulären linearen Integrodifferentialgleichung', *ZAMM* **48**, 477–483.

I.H. Sloan (1988), 'A quadrature-based approach to improving the collocation method', *Numer. Math.* **54**, 41–56.

I.H. Sloan (1992), 'Unconventional methods for boundary integral equations in the plane' in *Proc. 1991 Dundee Conference on Numerical Analysis* (D.F. Griffiths and G.A. Watson, eds), Longman Scientific and Technical (Harlow), to appear.

I.H. Sloan and B.J. Burn (1991), 'An unconventional quadrature method for logarithmic-kernel integral equations on closed curves', *J. Int. Eqns Appl.*, to appear.

I.H. Sloan and A. Spence (1988a), 'The Galerkin method for integral equations of the first kind with logarithmic kernel: theory', *IMA J. Numer. Anal.* **8**, 105–122.

I.H. Sloan and A. Spence (1988b), 'The Galerkin method for integral equations of the first kind with logarithmic kernel: applications', *IMA J. Numer. Anal.* **8**, 123–140.

I.H. Sloan and E.P. Stephan (1992), 'Collocation with Chebyshev polynomials for Symm's integral equation on an interval', *J. Austral. Math. Soc. (Ser. B)*, to appear.

I.H. Sloan and W.L. Wendland (1989), 'A quadrature based approach to improving the collocation method for splines of even degree', *Z. Anal. Anw.* **8**, 361–376.

E.P. Stephan and W.L. Wendland (1976), 'Remarks to Galerkin and least squares methods with finite elements for general elliptic problems', in *Partial Differential Equations (Lecture Notes in Mathematics 564)* Springer-Verlag (Berlin) 461–471, and in *Manuscripta Geodaetica* **1**, 93–123.

E.P. Stephan and W.L. Wendland (1985), 'An augmented Galerkin procedure for the boundary integral method applied to mixed boundary value problems', *Appl. Numer. Math.* **1**, 121–143.

G. Verchota (1984), 'Layer potentials and regularity for the Dirichlet problem for Laplace's equation in Lipschitz domains', *J. Funct. Anal.* **59**, 572–611.

J. Weissinger (1950), 'Über Integrodifferentialgleichungen vom Typ der Prandtlschen Tragflügelgleichung', *Math. Nachr.* **3**, 316–326.

W.L. Wendland (1983), 'Boundary element methods and their asymptotic convergence', in *Theoretical Acoustics and Numerical Techniques, CISM Courses* vol. 277 (P. Filippi, ed.), Springer (Berlin, Heidelberg, New York), 135–216.

W.L. Wendland (1985), 'Asymptotic accuracy and convergence for point collocation methods', in *Topics in Boundary Element Research* vol. 3 (C.A. Brebbia, ed.), Springer-Verlag (Berlin), 230–257.

W.L. Wendland (1987), 'Strongly elliptic boundary integral equations', in *The State of the Art in Numerical Analysis* (A. Iserles and M.J.D. Powell, eds), Clarendon Press (Oxford), 511–561.

W.L. Wendland (1989), 'Qualocation, the new variety of boundary element methods', *Wiss. Z. d TU Karl–Marx–Stadt* **31**(2) 276–284.

W.L. Wendland (1990), 'Boundary element methods for elliptic problems', in *Mathematical Theory of Finite and Boundary Element Methods* (A.H. Schatz, V. Thomée and W.L. Wendland, eds), Birkhäuser (Basel), 219–276.

W.L. Wendland, E.P. Stephan and G.C. Hsiao (1979), 'On the integral equation method for the plane mixed boundary value problem for the Laplacian', *Math. Meth. Appl. Sci.* **1**, 265–321.

Y. Yan (1990), 'The collocation method for first-kind boundary integral equations on polygonal domains', *Math. Comput.* **54**, 139–154.

Y. Yan and I.H. Sloan (1988), 'On integral equations of the first kind with logarithmic kernels', *J. Int. Eqns Applic.* **1**, 517–548.

Y. Yan and I.H. Sloan (1989), 'Mesh grading for integral equations of the first kind with logarithmic kernel', *SIAM J. Numer. Anal.* **26**, 574–587.

O.C. Zienkiewicz, D.W. Kelly and P. Bettess (1977), 'The coupling of the finite element method and boundary solution procedures', *Int. J. Num. Meth. Eng.* **11**, 335–375.

Acta Numerica (1991), *pp.* 341–407

Interior methods
for constrained optimization

Margaret H. Wright

AT&T Bell Laboratories
Murray Hill, New Jersey 07974 USA
E-mail: mhw@research.att.com

Interior methods for optimization were widely used in the 1960s, primarily in the form of barrier methods. However, they were not seriously applied to linear programming because of the dominance of the simplex method. Barrier methods fell from favour during the 1970s for a variety of reasons, including their apparent inefficiency compared with the best available alternatives. In 1984, Karmarkar's announcement of a fast polynomial-time interior method for linear programming caused tremendous excitement in the field of optimization. A formal connection can be shown between his method and classical barrier methods, which have consequently undergone a renaissance in interest and popularity. Most papers published since 1984 have concentrated on issues of computational complexity in interior methods for linear programming. During the same period, implementations of interior methods have displayed great efficiency in solving many large linear programs of ever-increasing size. Interior methods have also been applied with notable success to nonlinear and combinatorial problems. This paper presents a self-contained survey of major themes in both classical material and recent developments related to the theory and practice of interior methods.

CONTENTS

1. Introduction to interior methods

1.1. The way we were

Before 1984, the question 'How should I solve a linear program?' would have been answered almost without exception by 'Use the simplex method'. In fact, it would have been extremely difficult to find serious discussion of any method for linear programming (LP) other than the famous simplex method developed by George B. Dantzig in 1947.

As most readers already know, the simplex method is an iterative procedure derived from a fundamental property of essentially all linear programs: an optimal solution lies at a vertex of the feasible region. Beginning with a vertex, the simplex method moves between adjacent vertices, decreasing the objective as it goes, until an optimal vertex is found.

Although nonsimplex strategies for LP were suggested and tried from time to time, such techniques had never approached the simplex method in overall speed and reliability. Hence the simplex method retained unquestioned preeminence as the linear programming method of choice for nearly 40 years. (We describe later the persistent unhappiness with the simplex method on grounds of its theoretical complexity.)

Such an exclusive focus on the simplex method had several effects on the field of optimization. Largely for historical reasons, the simplex method is surrounded by a bevy of highly specialized terminology ('basic feasible solution') and pedagogical constructs (the tableau) with little apparent connection to other continuous optimization problems. Many researchers and practitioners consequently viewed linear programming as philosophically distinct from nonlinear programming. This conceptual gap reinforced a tendency to develop 'new' linear programming methods only as variations on the simplex method.

In marked contrast, the field of *nonlinear* optimization was characterized not only by the constant development of new methods with differing flavours, but also by a shift over time in the preferred solution techniques. Since the late 1970s, for example, nonlinearly constrained optimization problems have been solved with sequential quadratic programming (SQP) methods, which involve a sequence of constrained subproblems based on the Lagrangian function. In the 1960s, however, constrained problems were most often converted to *unconstrained* subproblems. Penalty and barrier methods were especially popular, both motivated by minimizing a composite function that reflects the original objective function as well as the influence of the constraints. Classical barrier methods, intended for inequality constraints, include a composite function containing an impassable positive singularity ('barrier') at the boundary of the feasible region, and thereby maintain strict feasibility while approaching the solution.

Although barrier methods were widely used and thoroughly analysed during the 1960s (see Section 3 for details and references), they nonetheless suffered a severe decline in popularity in the 1970s for various reasons, including inherent ill-conditioning as well as perceived inefficiency compared to alternative strategies. By the late 1970s, barrier methods were considered for the most part an interesting but passé solution technique.

As we shall see, the situation today (1991) in both linear and nonlinear programming has altered dramatically since 1984, primarily as a result of dissatisfaction with the theoretical computational complexity of the simplex method.

1.2. Concerns about the simplex method

On 'real-world' problems, the simplex method is invariably extremely efficient, and consistently requires a number of iterations that is a small multiple (2–3) of the problem dimension. Since the number of vertices associated with any LP is finite, the simplex method is also guaranteed under quite mild conditions to converge to the optimal solution. The number of vertices, however, can be exponentially large. The well known 'twisted cube' example of Klee and Minty (1972) is a linear program with n variables and $2n$ inequality constraints for which the simplex method with the standard pivot-selection rule visits each of the 2^n vertices. The *worst-case* complexity of the simplex method (the number of arithmetic operations required to solve a general LP) is consequently *exponential* in the problem dimension. The gigantic gap between the observed and worst-case performance of the simplex method is still puzzling; the issue of whether an (undiscovered) simplex pivot rule could improve its complexity is also unresolved.

As the formal study of computational complexity increased in importance during the 1960s and 1970s, it became a strongly held article of faith among computer scientists that a 'fast' algorithm must be *polynomial-time*, meaning that the number of operations required to solve the problem should be bounded above by a polynomial in the problem size. The simplex method clearly does not satisfy this property. Although practitioners routinely and happily solved large linear programs with the simplex method, the existence of a provably polynomial algorithm remained a major open question.

In 1979, to the accompaniment of worldwide publicity, Leonid Khachian published the first polynomial algorithm for LP. The *ellipsoid method* of Khachian is based on earlier techniques for nonlinear programming developed by other mathematicians, notably Shor, Yudin and Nemirovsky. An interesting feature of Khachian's approach is that it does not rely on combinatorial features of the LP problem. Rather, it constructs a sequence of ellipsoids such that each successive ellipsoid both encloses the optimal solution and undergoes a strict reduction in volume. The ellipsoid method generates improving iterates in the sense that the region of uncertainty surrounding the solution is monotonically 'squeezed'. (Simplex iterates are also improving in the sense that the objective value is decreasing, but they provide no information about the closeness of the current iterate to the solution.)

The crucial elements in polynomiality of the ellipsoid method are what might be termed outer and inner bounds for the solution. The outer bound guarantees an initial enclosing ellipsoid, and the inner bound specifies the size of the final ellipsoid needed to ensure sufficient closeness to the exact solution. Similar features also figure prominently in the complexity analysis of interior methods, and are discussed in Section 6.

Despite its polynomial complexity, the ellipsoid method's performance was extremely disappointing. In practice, the number of iterations tended to be almost as large as the worst-case upper bound, which, although polynomial, is *very* large. The simplex method accordingly retained its position as the clear winner in any comparison of actual solution times. Creation of the ellipsoid method led to an unexpected anomaly in which an algorithm with the desirable theoretical property of polynomiality compared unfavourably in speed to an algorithm with worst-case exponential complexity. The quest therefore continued for an LP algorithm that was not only polynomial, but also efficient in practice.

This search ended in 1984, when Narendra Karmarkar presented a novel interior method of polynomial complexity for which he reported solution times 50 times faster than the simplex method. Once again, international coverage in the popular press surrounded the event, which has had remarkable and lasting scientific consequences.

Karmarkar's announcement led to an explosion of interest among researchers and practitioners, with substantial progress in several directions. Interior methods are indeed 'fast'; extensive numerical trials have shown conclusively that a variety of interior methods can solve many very large linear programs substantially faster than the simplex method. After a formal relationship was shown between Karmarkar's method and classical barrier methods (Gill *et al.*, 1986), much research has concentrated on the common theoretical foundations of linear and nonlinear programming.

Unlike the simplex method, interior techniques can obviously be applied to nonlinear optimization problems. (In fact, they were devised more than 30 years ago for this purpose!) Interior methods have already been developed for quadratic and nonlinear programming, and extensions of the interior approach to difficult combinatorial problems have also been proposed; see Karmarkar (1990).

A fundamental theme permeating the motivation for interior methods is the creation of continuously parametrized families of approximate solutions that asymptotically converge to the exact solution. As the parameter approaches its limit, the paths to the solution trace smooth trajectories whose geometric properties can be analysed. Each iteration of a 'path-following' method constructs a step intended to follow one of these trajectories, moving both 'toward' and 'along' the path. In the first heyday of barrier methods, these ideas led to great interest in extrapolation. Today, they are being generalized and extended to new problem areas; for a discussion of such ideas in linear programming, see Megiddo (1987), Bayer and Lagarias (1989, 1991), and Karmarkar (1990). The field of interior methods seems to offer the continuing promise of original theory and efficient methods.

1.3. Overview

This article covers only a small part of the large and rapidly expanding number of topics related to interior methods. Although the term 'interior methods' is not precisely defined, several themes perceived as disparate before 1984 can now be placed in a unified framework. For reasons of space, we motivate interior methods only through a 'classical' barrier function. Karmarkar's original 1984 algorithm was based on nonlinear projection, a perspective that provides interesting geometric insights. See Gonzaga (1992), Nesterov and Nemirovsky (1989), and Powell (1990) for further interpretations.

Work in interior methods today is a melange of rediscovered as well as new methods, complexity analysis, and sparse linear algebra. The approach taken in this article is to present some initial background on optimization (Section 2), followed by a detailed treatment of the theory of classical barrier methods (Section 3). After reviewing Newton's method (Section 4), we turn in Section 5 to the special case of linear programming, and describe the structure of several interior methods. A particular interior LP method and its complexity analysis are given in detail (Section 6) to give the flavour of such proofs. The practical success of interior methods is dependent on efficient linear algebra; the relevant techniques for linear and nonlinear problems are described in Section 7. Finally, we close by mentioning selected directions for future research.

2. Background in optimization

2.1. Definitions and notation

Optimization problems, broadly speaking, involve finding the 'best' value of some function. A continuous optimization problem has three ingredients: a set of variables, usually denoted by the real n-vector x; an *objective* function $f(x)$ to be optimized (minimized or maximized); and *constraints* (equality and/or inequality) that restrict acceptable values of the variables.

Except for the linear programming case, our main interest is in inequality constraints. We consider a generic optimization problem of the form

$$\underset{x \in \mathbb{R}^n}{\text{minimize}} \ f(x) \quad \text{subject to} \quad c_i(x) \geq 0, \quad i = 1, \ldots, m. \qquad (2.1)$$

It is assumed throughout that all functions of interest are smooth. This assumption is stronger than necessary, and is imposed mainly to simplify the discussion.

The set of points satisfying the constraints of (2.1) is denoted by

$$\mathcal{F} = \{x \mid c_i(x) \geq 0, \ i = 1, \ldots, m\}, \qquad (2.2)$$

and is called the *feasible region*. If x is in $\mathcal{F}$, x is said to be *feasible*.

A *linear programming* problem is an optimization problem in which the objective function and all the constraint functions, both equalities and inequalities, are linear. An optimization problem is called a *nonlinear program* if the objective or any constraint function is nonlinear. A *quadratic program* has a quadratic objective and linear constraints.

Several definitions involving sets will be important in our discussion. All sets are in $\mathbb{R}^n$ unless stated otherwise.

Definition 1 (Interior of a set.) Given a set S, a point x is an *interior point* of S if $x \in S$ and there exists a neighbourhood of x that is entirely contained in S. The *interior* of S, denoted by int(S), is the collection of all interior points of S.

Definition 2 (Boundary of a set.) Given a set S, a point x is a *boundary point* of S if every neighbourhood of x contains at least one point in S and at least one point not in S. The *boundary* of S is the collection of all boundary points of S.

It is straightforward to show that a closed set contains all its boundary points.

For the feasible region $\mathcal{F}$ (2.2) associated with our generic optimization problem, the subset of points in $\mathcal{F}$ for which all the constraint functions are *strictly positive* is denoted by strict($\mathcal{F}$) and defined as

$$\text{strict}(\mathcal{F}) = \{x \mid c_i(x) > 0, \quad i = 1, \ldots, m\}. \qquad (2.3)$$

A point x in strict($\mathcal{F}$) is said to be *strictly feasible*.

Although the sets strict($\mathcal{F}$) and int($\mathcal{F}$) are identical in many instances, they can be different. For example, consider the single constraint $x_1^2 + x_2^2 \geq 0$ in $\mathbb{R}^2$. The corresponding feasible region $\mathcal{F}$ includes all of $\mathbb{R}^2$; consequently, every point in $\mathbb{R}^2$ is an interior point, and int($\mathcal{F}$) $= \mathbb{R}^2$. In contrast, the set strict($\mathcal{F}$) includes all points in $\mathbb{R}^2$ *except* the origin.

The idea of a level set will be used in several proofs.

Definition 3 (Level set.) For x in a set S, the *level set* of the function

$f(x)$ corresponding to the constant τ is the set of points in S for which the value of f is less than or equal to τ:

$$\{x \in S \mid f(x) \le \tau\}.$$

For reference, we state formal definitions of local, global and isolated minimizers for the generic problem (2.1). The definitions given here, taken from Fiacco and McCormick (1968), are tailored to our treatment of interior methods, and are slightly different from those in standard textbooks. They can be specialized in an obvious way to include additional restrictions on x.

Definition 4 (Local constrained minimizer.) The point x^* is a *local (constrained) minimizer* of problem (2.1) if there exists a compact set S such that

$$x^* \in \text{int}(S) \cap \mathcal{F} \quad \text{and} \quad f(x^*) = \min\{f(x) \mid x \in S \cap \mathcal{F}\}.$$

Definition 5 (Global constrained minimizer.) The point x^* is a *global (constrained) minimizer* of problem (2.1) if

$$x^* \in \mathcal{F} \quad \text{and} \quad f(x^*) = \min\{f(x) \mid x \in \mathcal{F}\}.$$

Definition 6 (Isolated constrained minimizer.) A constrained minimizer x^* is *isolated* if there is a neighbourhood of x^* in which x^* is the only constrained minimizer.

For the nonlinear function $f(x)$, the n-vector $g(x)$ denotes the gradient (vector of first partial derivatives) of f, and the $n \times n$ symmetric matrix $H(x)$ denotes the Hessian (matrix of second partial derivatives) of f. Given a nonlinear constraint function $c_i(x)$, its gradient will be denoted by $a_i(x)$, and its Hessian by $H_i(x)$. For an m-vector $c(x)$ of constraint functions, the $m \times n$ Jacobian matrix of c is denoted by $A(x)$, whose ith row (the transposed gradient of c_i) is $a_i(x)^T$.

2.2. Optimality conditions

We now state optimality conditions for three varieties of nonlinear optimization problems, without any explanation of the origin of these conditions. (Optimality conditions for linear programming are given in Section 5.) Detailed derivations of optimality conditions are given in, for example, Avriel (1976); Fiacco and McCormick (1968); Fletcher (1987); Gill *et al.* (1981); and Luenberger (1984). Optimality conditions are extremely important because they not only allow us to recognize that a solution has been found, but also suggest algorithms for finding a solution.

Unconstrained optimization. The definition of a local unconstrained minimizer will be important in our discussion of barrier functions, where 'unconstrained' implies that no constraints are *locally* relevant.

Definition 7 (Local unconstrained minimizer.) The point x^* is a local unconstrained minimizer of $f(x)$ if there exists a compact set S such that

$$x^* \in \text{int}(S) \quad \text{and} \quad f(x^*) = \{\min f(x) \mid x \in S\}.$$

The following conditions are well known to be necessary for x^* to be an unconstrained minimizer of $f(x)$:

$$g(x^*) = 0 \quad \text{and} \quad H(x^*) \geq 0, \tag{2.4}$$

where the notation '$M \geq 0$' means that the matrix M is positive semi-definite. (Similarly, '$M > 0$' means that M is positive definite.)

Sufficient conditions for x^* to be an isolated unconstrained minimizer of $f(x)$ are

$$g(x^*) = 0 \quad \text{and} \quad H(x^*) > 0. \tag{2.5}$$

The 'order' of an optimality condition refers to the highest order of the derivatives that it contains. For example, the requirement that $g(x^*) = 0$ is a first-order optimality condition.

Linear equality constraints. Consider the problem of minimizing $f(x)$ subject to linear equality constraints:

$$\underset{x \in \mathbb{R}^n}{\text{minimize}} \quad f(x) \quad \text{subject to} \quad Ax = b, \tag{2.6}$$

where A is a constant $m \times n$ matrix. (Note that A is the Jacobian of the linear constraints $Ax - b = 0$.) Let N denote any matrix whose columns form a basis for the null space of A, i.e. for the subspace of vectors p such that $Ap = 0$. Although the null space itself is unique, in general there are an infinite number of associated bases.

The following conditions are necessary for the point x^* to be a local solution of (2.6):

$$Ax^* = b; \tag{2.7a}$$
$$g(x^*) = A^T \lambda^* \quad \text{for some } \lambda^*; \tag{2.7b}$$
$$N^T H(x^*) N \geq 0. \tag{2.7c}$$

Sufficient conditions for x^* to be an isolated solution of (2.6) are that (2.7a–b) hold and that $N^T H(x^*) N$ is positive definite.

The first-order condition $g(x^*) = A^T \lambda^*$ of (2.7b) means that the gradient of f at an optimal point can be expressed as a linear combination of the columns of A^T, and hence lies in the range space of A^T. The *Lagrange multiplier* λ^* represents the set of coefficients in this linear combination, and is unique if A has full row rank.

The *Lagrangian function* for problem (2.6) is

$$L(x, \lambda) = f(x) - \lambda^T (Ax - b), \tag{2.8}$$

where λ is an m-vector. Condition (2.7b) can be interpreted as a statement that the gradient of the Lagrangian function with respect to x vanishes when $\lambda = \lambda^*$.

The relation $g(x^*) = A^T\lambda^*$ is also equivalent to the condition $N^Tg(x^*) = 0$, namely, the projection of $g(x^*)$ into the null space of A vanishes. (The vector $N^Tg(x)$ is called the *reduced gradient* of f at x.) Condition (2.7b) is therefore analogous to the requirement in the unconstrained case that the gradient itself must be zero.

The matrix $N^TH(x^*)N$ appearing in the second-order optimality condition (2.7c) is the Hessian of f projected into the null space of A, and is called the *reduced Hessian* of f. For linear equality constraints, the reduced Hessian plays the same role in optimality conditions as the full Hessian in the unconstrained case.

The feasibility and first-order optimality conditions (2.7a–b) satisfied by x^* and λ^* can conveniently be summarized as a system of $(n+m)$ nonlinear equations in the variables (x, λ):

$$\Phi(x, \lambda) = \begin{pmatrix} g(x) - A^T\lambda \\ Ax - b \end{pmatrix} = \begin{pmatrix} 0 \\ 0 \end{pmatrix}. \tag{2.9}$$

These equations state that the gradient of the Lagrangian function (2.8) and the constraint vector $Ax - b$ should both be zero.

Nonlinear inequality constraints. The final problem category to be discussed is the generic problem with nonlinear inequality constraints:

$$\underset{x \in \mathbb{R}^n}{\text{minimize}} \ f(x) \quad \text{subject to} \quad c(x) \geq 0, \tag{2.10}$$

where $c(x)$ consists of m component functions. The constraint $c_i(x) \geq 0$ is said to be *active* at $\bar{x}$ if $c_i(\bar{x}) = 0$ and *inactive* if $c_i(\bar{x}) > 0$. Let $\hat{A}(x)$ denote the Jacobian of the *active* constraints at x, and let $N(x)$ denote a matrix whose columns form a basis for the null space of $\hat{A}$.

Nonlinear constraints can be extremely complicated, and necessary optimality conditions can be stated only after making assumptions about the constraints (called regularity assumptions or constraint qualifications); see, for example, Avriel (1976), Fiacco and McCormick (1968) or Fletcher (1987). The most common form of constraint qualification is an assumption that the gradients of the active constraints are linearly independent (or that the constraints are linear).

The Lagrangian function for problem (2.10) is defined as

$$L(x, \lambda) = f(x) - \lambda^Tc(x) \tag{2.11}$$

(see (2.8)). For future reference, we note that the Hessian of the Lagrangian with respect to x, denoted by W, is given by

$$W(x, \lambda) = \nabla^2 L(x, \lambda) = H(x) - \sum_{i=1}^{m} \lambda_i H_i(x). \qquad (2.12)$$

If a suitable constraint qualification holds at x^*, the following conditions can be shown to be necessary for the point x^* to be a constrained minimizer of (2.10):

$$c(x^*) \geq 0; \qquad (2.13a)$$
$$g(x^*) = A(x^*)^T \lambda^* \quad \text{for some } \lambda^*; \qquad (2.13b)$$
$$\lambda_i^* c_i(x^*) = 0, \quad i = 1, \dots, m; \qquad (2.13c)$$
$$\lambda_i^* \geq 0, \quad i = 1, \dots, m; \qquad (2.13d)$$
$$N(x^*)^T W(x^*, \lambda^*) N(x^*) \geq 0. \qquad (2.13e)$$

Condition (2.13c), which forces at least one of $c_i(x^*)$ and λ_i^* to be zero for every i, is called a *complementarity* condition. In particular, it means that if $c_i(x^*) > 0$, i.e. constraint i is *inactive* at x^*, then λ_i^* must be zero.

Because the multipliers for inactive constraints are zero, the first-order condition (2.13b) states that the gradient of f at x^* is a linear combination of the *active* constraint gradients, so that $N(x^*)^T g(x^*) = 0$. Trivial rearrangement of (2.13b) also reveals that the gradient of the Lagrangian function with respect to x vanishes at x^* when $\lambda = \lambda^*$, i.e. x^* is a stationary point of the Lagrangian function when $\lambda = \lambda^*$. However, x^* is not necessarily a minimizer of the Lagrangian function.

A crucial distinction arising from constraint nonlinearities can be seen in the second-order condition (2.13e), which involves the reduced Hessian of the Lagrangian, rather than the reduced Hessian of f alone. The inclusion of constraint curvature is an essential feature of efficient algorithms for nonlinearly constrained problems.

Sufficient conditions for x^* to be an isolated constrained minimizer of (2.10) are: (i) a suitable constraint qualification applies at x^* (for example, the gradients of the active constraints at x^* are linearly independent); (ii) conditions (2.13a–c) are satisfied; and (iii) the following strengthened versions of (2.13d–e) hold:

$$\lambda_i^* > 0 \quad \text{if} \quad c_i(x^*) = 0; \qquad (2.14)$$
$$N(x^*)^T W(x^*, \lambda^*) N(x^*) > 0. \qquad (2.15)$$

Inequality (2.14) is called *strict complementarity*, and holds when all Lagrange multipliers associated with active constraints are positive. The property of strict complementarity is often assumed because the presence of a zero multiplier for an active constraint creates complications.

Condition (2.15) is equivalent to the existence of $\alpha > 0$ such that

$$p^T W(x^*, \lambda^*)p \geq \alpha \|p\|^2 \quad \text{for all } p \text{ such that } \hat{A}(x^*)p = 0. \qquad (2.16)$$

2.3. Convexity

Most work on interior methods to date has focused on convex optimization problems, of which linear programming is the most obvious instance. As we shall see, many complications that arise for general optimization problems disappear in the presence of convexity. See Rockafellar (1970) for a complete treatment of convex analysis.

Definition 8 (Convex set.) The set S is convex if, for every x_1 and x_2 in S, and for all θ satisfying $0 \leq \theta \leq 1$, the point $z = (1 - \theta)x_1 + \theta x_2$ is also in S.

Definition 9 (Convex and concave functions.) The function $f(x)$, defined for x in a nonempty open convex set S, is convex if, for every two points x_1 and x_2 in S, and for all θ satisfying $0 \leq \theta \leq 1$,

$$f\big((1 - \theta)x_1 + \theta x_2\big) \leq (1 - \theta)f(x_1) + \theta f(x_2). \qquad (2.17)$$

(If the set S is not specified, it is assumed to be $\mathbb{R}^n$.) The function f is concave if $-f$ is convex. The function f is strictly convex if the inequality in (2.17) is strict when $x_1 \neq x_2$ and $0 < \theta < 1$.

Several useful results associated with convexity are:

1   the intersection of a finite number of convex sets is convex;
2   all level sets of a convex function are convex;
3   given a set of convex functions $\{\varphi_i(x)\}$, $i = 1, \ldots, m$, the set of points satisfying $\varphi_i(x) \leq 0$ is convex;
4   the smooth function $f(x)$, defined for x in an open convex set S, is convex if its Hessian matrix $H(x)$ is positive semi-definite for all $x \in S$, and strictly convex if $H(x)$ is positive definite for all $x \in S$.

Convex programs are constrained optimization problems with important special properties. It should be stressed that the only *equality* constraints permitted in a convex program are linear constraints.

Definition 10 (Convex program.) The problem of minimizing $f(x)$ subject to the linear equality constraints $Ax = b$ and the inequality constraints $c_i(x) \geq 0$, $i = 1, \ldots, m$, is a convex program if $f(x)$ is convex and $-c_i(x)$ is convex for $i = 1, \ldots, m$.

A slight irritation is that our generic form for inequality constraints involves a 'greater than' relation ($c_i(x) \geq 0$) for expositional convenience. Unfortunately, an optimization problem with constraints in this form is a convex program only if each *negative* constraint function $-c_i(x)$ is convex, i.e. if $c_i(x)$ itself is concave. Hence minus signs appear throughout our discussion of the constraints in convex programs.

Using these definitions, it is easy to see that a linear function is convex (and also concave), and that a linear programming problem is a convex program. Two properties that are important in interior methods for linear programming are stated formally in the following theorems; see Fiacco and McCormick (1968) or Fletcher (1987) for details.

Theorem 1 If x^* is a local constrained minimizer of a convex programming problem, it is also a global constrained minimizer. Further, the set of minimizers of a convex program is convex.

Theorem 2 If the optimization problem (2.10) is a convex program, and if x^* satisfies the feasibility and first-order necessary conditions (2.13a–d), then x^* is a global constrained minimizer of (2.10).

3. Barrier methods

3.1. Intuition and motivation

Suppose that we wish to minimize $f(x)$ subject to a set of inequality constraints $c_i(x) \geq 0$, $i = 1, \ldots, m$. If the constraints affect the solution, either an unconstrained minimizer of $f(x)$ is infeasible (for example, when minimizing x^2 subject to $x \geq 1$), or else $f(x)$ is unbounded below when the constraints are removed (for example, when minimizing x^3 subject to $x \geq 1$). Consequently, if an optimization method tries to achieve a 'large' reduction in the objective function from its value at a feasible point, the iterates tend to move *outside* the feasible region. In fact, many popular algorithms for nonlinearly constrained optimization (such as SQP methods; see, for example, Fletcher (1987), and Gill *et al.* (1981)) typically produce infeasible iterates that approach feasibility only in the limit.

When feasibility at intermediate points is essential – for example, in practical problems where the objective function is meaningless unless the constraints are satisfied – it seems desirable for iterates to approach the constrained solution from the *interior* of the feasible region. *Barrier methods* constitute a well known class of methods with this property.

Barrier methods may be applied only to inequality constraints for which strictly feasible points exist. This property does not hold for all inequality constraints, even if the feasible region is nonempty; for example, consider the constraints $x_1 + x_2 \geq 0$ and $-x_1 - x_2 \geq 0$, for which the feasible region consists of the line $\{x_1 + x_2 = 0\}$.

Given an initial strictly feasible point and mild assumptions about the feasible region, strict feasibility can be retained by minimizing a composite function consisting of the original objective $f(x)$ plus a positive multiple of an infinite 'barrier' at the boundary of strict($\mathcal{F}$). The most effective methods for unconstrained optimization (such as Newton's method; see Section 4) require differentiability. A suitable barrier term is therefore composed of functions that are smooth at strictly feasible points, but contain a positive singularity if any constraint is zero. Under these conditions, a minimizer of the composite function must occur at a strictly feasible point.

When the barrier term is heavily weighted, a minimizer of the composite function will lie, informally speaking, 'far away' from the boundary. If the coefficient of the barrier term is reduced, the singularity becomes less influential, except at points near the boundary; minimizers of the composite function can then move closer (but not 'too close') to the boundary. The weight on the barrier term thus tends to regulate the distance from the iterates to the boundary. In the parlance of modern interior methods, the barrier term forces the iterates to remain *centred* in the strictly feasible region.

As the factor multiplying the barrier term decreases to zero, intuition suggests that minimizers of the composite function will converge to a constrained solution x^* that lies on the boundary of strict($\mathcal{F}$). We shall see later (Sections 3.3 and 3.4) that this intuition can be verified rigorously under reasonably mild conditions.

We stress that there is ample room for many formulations of a 'barrier function', as indicated by the range of definitions in Fiacco and McCormick (1968) and in Nesterov and Nemirovsky (1989). Other varieties of composite functions – called 'potential' and 'centering' functions – have also been proposed for use in interior methods; see, for example, Sonnevend (1986) and Gonzaga (1992). Karmarkar's original (1984) LP algorithm included a logarithmic potential function. The method of centres of Huard (1967) imposes an additional constraint at each iteration based on the current value of the objective function; see Renegar (1988) for an LP method based on this idea.

In all cases, the composite functions display a common motivation of simultaneously reflecting the objective function (thereby encouraging its reduction) as well as forcing iterates to stay 'nicely centred' in the feasible region. They differ, however, in the balance of these sometimes conflicting aims.

3.2. The logarithmic barrier function

For simplicity, we discuss only the simplest barrier function based on a logarithmic singularity, which was not only the most popular in the 1960s, but also has received substantial attention since 1984. The logarithmic barrier function was first defined by Frisch in 1955, and was extensively studied and analysed during the 1960s. Detailed theoretical discussions of classical barrier methods, along with historical background, are given in Fiacco and McCormick (1968) and Fiacco (1979).

The *logarithmic barrier function* associated with minimizing $f(x)$ subject to $c(x) \geq 0$ is

$$B(x, \mu) = f(x) - \mu \sum_{i=1}^{m} \ln c_i(x), \qquad (3.1)$$

where the *barrier parameter* μ is strictly positive. (When the meaning is clear, we may write B with a single argument μ or without arguments.) Since the logarithm is undefined for nonpositive arguments, the logarithmic barrier function is defined only in strict($\mathcal{F}$).

Simply stating the definition (3.1) does not give an adequate impression of the dramatic effects of the imposed barrier. Figure 1 depicts the one-dimensional variation of a barrier function for two values of μ. Even for the modest value $\mu = 0.1$, the (visually) extreme steepness of the singularity is evident.

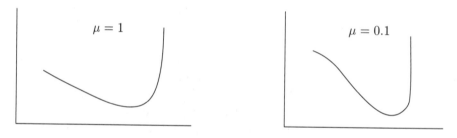

Fig. 1. The one-dimensional behaviour of a barrier function.

The intuitive motivation for a barrier method is that we seek unconstrained minimizers of $B(x, \mu)$ for values of μ decreasing to zero. If the solution x^* of the constrained problem lies on the boundary and exact arithmetic is used, a barrier method can never produce the exact solution. Barrier methods consequently terminate when the current iterate satisfies some approximation to the desired optimality conditions. 'Classical' barrier algorithms as well as many recent interior methods have the following form:

Generic Barrier Algorithm

0. Set x_0 to a strictly feasible point, so that $c(x_0) > 0$, and set μ_0 to a positive value; $k \leftarrow 0$.
1. Check whether x_k qualifies as an approximate local constrained minimizer for the original problem (2.10). If so, stop with x_k as the solution.
2. Compute an unconstrained minimizer $x(\mu_k)$ of $B(x, \mu_k)$.
3. $x_{k+1} \leftarrow x(\mu_k)$; choose $\mu_{k+1} < \mu_k$; $k \leftarrow k+1$; return to Step 1.

In practice, the calculation of $x(\mu_k)$ in Step 2 is carried out approximately, and only a few iterations of an unconstrained method may be performed before the barrier parameter is updated. In the theoretical results given here, we assume that $x(\mu_k)$ is an exact unconstrained minimizer.

We illustrate the behaviour of the generic algorithm on a simple two-variable example:

$$\begin{aligned}
\text{minimize} \quad & x_1 x_2 - \tfrac{1}{2}x_1^2 - x_2 \\
\text{subject to} \quad & x_1^2 + x_2^2 \le 2 \\
& x_1^2 x_2^2 \le 10.
\end{aligned}$$

The first constraint is satisfied inside the circle of radius $\sqrt{2}$ centred at the origin; although the second constraint is redundant, it nonetheless affects each minimizer of the barrier function. The point $x^* = (-1, 1)^T$ is an isolated local constrained minimizer at which only the first constraint is active. Figure 2 depicts selected barrier minimizers converging to x^*, which lies on the boundary of the feasible region (depicted as a dashed curve).

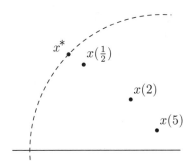

Fig. 2. Convergence of barrier minimizers to $x^* = (-1, 1)^T$.

The next two sections provide a rigorous foundation for the generic approach, including the assumptions necessary to make it succeed in converging to a solution x^* of the original constrained problem. After establishing local convergence properties, we return in Section 3.5 to a more detailed analysis of the sequence of barrier minimizers.

3.3. Theoretical results for convex programs

Pre-1984 presentations of barrier methods for nonlinear problems typically begin with general results, which are then specialized to convex programs. We have chosen instead to give a self-contained presentation of the convex results first. Readers whose primary interest is in interior methods for linear and convex programming can read this section only and skip to Section 3.5.

Consider the convex programming problem

$$\underset{x \in \mathbb{R}^n}{\text{minimize}} \ f(x) \quad \text{subject to} \quad c_i(x) \geq 0, \quad i = 1, \ldots, m, \qquad (3.2)$$

where f and $\{-c_i\}$ are convex. In this section, $\mathcal{F}$ denotes the feasible region for the constraints of (3.2). Recall from Theorem 1 that every local minimizer of a convex program is a global minimizer; hence, if any minimizer exists, the optimal value of f in $\mathcal{F}$ is unique.

An obvious fundamental question involves the conditions under which a solution x^* of (3.2) is the limit of a sequence of unconstrained minimizers of the barrier function. The main assumption needed to prove convergence results is that the set $\mathcal{M}$ of minimizers of (3.2) is *bounded*. (We know already from Theorem 1 that $\mathcal{M}$ is convex.) Boundedness of the set of minimizers holds automatically under the much stronger assumption that the feasible region itself is bounded.

The major results of this section are given in Theorem 5. Two other theorems serve as a prelude.

Theorem 3 (a version of Theorem 24 in Fiacco and McCormick (1968)) shows that, if a set of convex functions defines a *bounded* feasible region, then suitably perturbed versions of the same functions also define a bounded feasible region. The application of this theorem in the proof of Theorem 4 involves a level set derived from the objective function.

Theorem 3 (Boundedness of perturbed convex sets.) Let $-\varphi_i(x)$ be a convex function for $i = 1, \ldots, m$, and assume that the convex set

$$\mathcal{N} = \{x \mid \varphi_i(x) \geq 0, \quad i = 1, \ldots, m\}$$

is nonempty and bounded. Then for any set of values $\{\Delta_i\}$, where $\Delta_i \geq 0$, $i = 1, \ldots, m$, the set

$$\{x \mid \varphi_i(x) \geq -\Delta_i, \quad i = 1, \ldots, m\}$$

is bounded.

Proof. The result will follow in an obvious way if verified for $\Delta_1 > 0$ and $\Delta_i = 0$, $i \neq 1$. Given $\Delta_1 > 0$, let $\mathcal{N}_1$ denote the set

$$\mathcal{N}_1 = \{x \mid \varphi_1(x) \geq -\Delta_1 \text{ and } \varphi_i \geq 0, \quad i = 2, \ldots, m\}.$$

Because $\mathcal{N}_1$ is the intersection of a finite number of convex sets, $\mathcal{N}_1$ is convex.

To prove by contradiction that $\mathcal{N}_1$ is bounded, we assume the contrary: for any point $x_1 \in \mathcal{N}$, there exists a ray emanating from x_1 that does not intersect the boundary of $\mathcal{N}_1$, so that $x_1 + \alpha p$ lies in $\mathcal{N}_1$ for some direction p and any $\alpha \geq 0$. (The fact that any unbounded convex set must contain a ray is standard; see, for example, Grünbaum (1967).)

Because $\mathcal{N}$ is bounded by assumption, there must be a point x_2 on this ray that does *not* lie in $\mathcal{N}$. Let x_2 be such a point, given by $x_2 = x_1 + \alpha_2 p$ for some $\alpha_2 > 0$, for which φ_1 assumes a *negative* value, say $\varphi_1(x_2) = -\delta < 0$, where $\delta < \Delta_1$.

Let x_3 denote a point on the ray that lies *beyond* x_2, i.e. $x_3 = x_1 + \alpha_3 p$, where $\alpha_3 > \alpha_2$. The point x_2 can then be written as

$$x_2 = x_1 + \theta(x_3 - x_1) = (1 - \theta)x_1 + \theta x_3, \qquad (3.3)$$

where $0 < \theta < 1$.

Applying Definition 9 of a convex function to the expression (3.3) for x_2, we obtain

$$(1 - \theta)\varphi_1(x_1) + \theta\varphi_1(x_3) \leq \varphi_1(x_2) = -\delta,$$

which gives

$$\theta\varphi_1(x_3) \leq -\delta - (1 - \theta)\varphi_1(x_1).$$

Because $\varphi_1(x_1) \geq 0$ and $0 < \theta < 1$, it follows that

$$\varphi_1(x_3) \leq \frac{-\delta}{\theta}.$$

If θ is sufficiently small, namely $\theta < \delta/\Delta_1$, the value of $\varphi_1(x_3)$ must be strictly less than $-\Delta_1$, which shows that x_3 cannot lie in $\mathcal{N}_1$. This gives the desired contradiction, and shows that $\mathcal{N}_1$ must be bounded. $\square$

The next result is related to Lemma 12 in Fiacco and McCormick (1968), which applies to a general barrier function. Given a convex program with a nonempty strict interior and a bounded set of minimizers, the theorem states that any particular level set of the logarithmic barrier function is bounded and closed. The boundedness property is important because it implies that the set of *minimizers* of the barrier function is bounded.

Theorem 4 (Compactness of barrier function level sets.) Consider the convex program of minimizing $f(x)$ subject to $c_i(x) \geq 0$, $i = 1, \ldots, m$. Let $\mathcal{F}$ denote the (convex) feasible region. Assume that strict($\mathcal{F}$) is nonempty and that the set of minimizers $\mathcal{M}$ for the convex program is nonempty and bounded. Then for any $\mu_k > 0$ and any constant τ, the level set

$$S(\tau) = \{x \in \text{strict}(\mathcal{F}) \mid B(x, \mu_k) \leq \tau\}$$

is bounded and closed, where $B(x, \mu_k)$ is the logarithmic barrier function.

Proof. Boundedness of $S(\tau)$ will be established by showing that, under the stated assumptions, the barrier function cannot remain bounded above while its argument becomes unbounded.

Let $\hat{x}$ denote any point in strict($\mathcal{F}$) (which is assumed to be nonempty). Given any $\epsilon > 0$, let $\hat{D}$ denote the level set defined by the values of $f(\hat{x})$ and ϵ:

$$\hat{D} = \{x \in \mathcal{F} \mid f(x) \leq f(\hat{x}) + \epsilon\}. \qquad (3.4)$$

Convexity of f implies that $\hat{D}$ is convex (see Section 2.3). The functions f and $\{c_i\}$ are smooth, so that $\hat{D}$ is closed. The first step in proving the theorem is to show that $\hat{D}$ is bounded, from which it will follow that $\hat{D}$ is compact.

To show that the set $\hat{D}$ is bounded, we invoke Theorem 3. Let f^* denote the minimum value of $f(x)$ for $x \in \mathcal{F}$. Because every local minimizer of a convex program is a global minimizer (see Theorem 1), the quantity $\Delta = f(\hat{x}) + \epsilon - f^*$ must be positive. By assumption, the set $\mathcal{M}$, which may be written as $\{x \in \mathcal{F} \mid f(x) \leq f^*\}$, is nonempty and bounded; further, $\mathcal{M}$ is convex because f is convex. We now define the function $\phi(x)$ as $f^* - f(x)$, and observe that $-\phi$ is convex. Theorem 3 then applies to ϕ and the positive perturbation Δ, and implies *boundedness* of the set

$$\{x \in \mathcal{F} \mid \phi(x) \geq -\Delta\} = \{x \in \mathcal{F} \mid f^* - f(x) \geq f^* - f(\hat{x}) - \epsilon\},$$

which is simply a rearranged definition of $\hat{D}$. Consequently, $\hat{D}$ is compact. It is straightforward to see that its boundary, $\mathrm{bnd}(\hat{D})$, is also compact. The definition (3.4) of $\hat{D}$ shows that $\hat{x}$ does not lie on the boundary of $\hat{D}$.

Having established the compactness of $\hat{D}$ and its boundary, we can now prove boundedness of $S(\tau)$ by contradiction. Assume the contrary of the desired result, namely that for some $\mu_k > 0$, there is an *unbounded* sequence $\{y_j\}$ of points in strict$(\mathcal{F})$ for which the barrier function values $B(y_j, \mu_k)$ remain bounded above.

For such a sequence, let j be sufficiently large so that y_j lies outside $\hat{D}$. By definition of $\hat{D}$, it must hold that

$$f(y_j) > f(\hat{x}) + \epsilon.$$

Let z_j be the point on the boundary of $\hat{D}$ where the line connecting $\hat{x}$ and y_j intersects the boundary. (Because $\hat{D}$ is convex, z_j is unique.) Let λ_j be the scalar satisfying $0 < \lambda_j < 1$ such that

$$z_j = (1 - \lambda_j)\hat{x} + \lambda_j y_j. \tag{3.5}$$

We have assumed that $\|y_j\|$ is unbounded for sufficiently large j. Since $\|z_j\|$ is finite, (3.5) shows that

$$\lambda_j \to 0 \quad \text{as} \quad j \to \infty. \tag{3.6}$$

Because $\hat{x}$ and y_j are both in strict$(\mathcal{F})$, we know that $c_i(\hat{x}) > 0$ and $c_i(y_j) > 0$ for $i = 1, \ldots, m$. Convexity of $-c_i(x)$ combined with (3.5) gives

$$c_i(z_j) \geq (1 - \lambda_j)c_i(\hat{x}) + \lambda_j c_i(y_j) > 0, \tag{3.7}$$

which shows that $z_j \in \text{strict}(\mathcal{F})$. Since z_j is by definition in $\mathrm{bnd}(\hat{D})$, we conclude from (3.4) that $f(z_j) = f(\hat{x}) + \epsilon$. Because f is convex (see Definition 9), (3.5) implies

$$f(z_j) \leq (1 - \lambda_j)f(\hat{x}) + \lambda_j f(y_j).$$

Dividing by λ_j and substituting $f(z_j) = f(\hat{x}) + \epsilon$, we obtain a *lower bound* on $f(y_j)$:

$$f(y_j) \geq f(\hat{x}) + \frac{\epsilon}{\lambda_j}. \tag{3.8}$$

It then follows from (3.6) that

$$f(y_j) \to \infty \quad \text{as} \quad j \to \infty,$$

so that the objective function values at $\{y_j\}$ become unbounded.

Turning back to the constraint functions, positivity of λ_j means that the first inequality in (3.7) can be rewritten as

$$c_i(y_j) \leq c_i(\hat{x}) + \frac{c_i(z_j) - c_i(\hat{x})}{\lambda_j}. \tag{3.9}$$

Since the set $\text{bnd}(\hat{D})$ is compact, the function $c_i(x) - c_i(\hat{x})$ achieves its maximum for some $x \in \text{bnd}(\hat{D})$. Let d_i denote

$$d_i = \max\{c_i(x) - c_i(\hat{x}) \mid x \in \text{bnd}(\hat{D})\}.$$

We now wish to demonstrate that $d_i \geq 0$. Because $z_j \in \text{bnd}(\hat{D})$ and $c_i(y_j) > 0$, we apply the definition of d_i and relation (3.9) to show that

$$c_i(\hat{x}) + \frac{d_i}{\lambda_j} \geq c_i(y_j) > 0, \quad i = 1, \dots, m. \tag{3.10}$$

If d_i were negative, the first expression in (3.10) would eventually become negative as $\lambda_j \to 0$, which is impossible. It follows that $d_i \geq 0$ for $i = 1, \dots, m$.

Finally, the barrier function $B(y_j, \mu_k)$ is formed. Using (3.8), (3.10), monotonicity of the logarithm function, and positivity of μ_k, we have:

$$
\begin{aligned}
B(y_j, \mu_k) &= f(y_j) - \mu_k \sum \ln c_i(y_j) \\
&\geq f(\hat{x}) + \frac{\epsilon}{\lambda_j} - \mu_k \sum \ln(c_i(\hat{x}) + (d_i/\lambda_j)) \\
&= f(\hat{x}) + \frac{\epsilon - \mu_k \lambda_j \sum \ln(c_i(\hat{x}) + (d_i/\lambda_j))}{\lambda_j}. \tag{3.11}
\end{aligned}
$$

The logarithm function has the property that, for a positive constant ν and $\delta \geq 0$,

$$\lim_{\lambda \to 0_+} \lambda \ln\left(\nu + \frac{\delta}{\lambda}\right) = 0.$$

Thus the limit of the numerator in (3.11) is ϵ, and the quotient in (3.11) is unbounded above as $\lambda_j \to 0$. It follows that $B(y_j, \mu_k)$ is unbounded above as $j \to \infty$, thereby contradicting our assumption that the barrier function values $\{B(y_j, \mu_k)\}$ are bounded above for an unbounded sequence $\{y_j\}$. This proves that $S(\tau)$ is bounded.

To show that $S(\tau)$ is closed, we prove that it contains all its accumulation points. Let $\{x_j\}$ be a convergent sequence in $S(\tau)$, with limit point $\bar{x}$. It follows from the continuity of f and $\{c_i\}$ in strict$(\mathcal{F})$ that $\bar{x}$ must satisfy $B(\bar{x}, \mu_k) \leq \tau$. Further, $\bar{x}$ must either be in strict$(\mathcal{F})$ or else have the property that $c_i(\bar{x}) = 0$ for at least one index i.

If $\bar{x}$ is in strict$(\mathcal{F})$, by definition $\bar{x}$ is in $S(\tau)$. Suppose that $\bar{x}$ is not in strict$(\mathcal{F})$. Then, because $c_i(\bar{x}) = 0$ for some index i, unboundedness of the logarithm for a zero argument and convergence of $\{x_j\}$ to $\bar{x}$ together imply that, for sufficiently large j, the barrier term $-\sum_{i=1}^m \ln c_i(x_j)$ cannot be bounded above. In particular, for any constant γ and sufficiently large j,

$$-\sum_{i=1}^m \ln c_i(x_j) > \gamma. \tag{3.12}$$

We now define γ as $\gamma = (\tau - f^*)/\mu_k$; the value of γ is finite because f^* is finite. Since x_j lies in strict$(\mathcal{F})$, we know from the convexity of f that $f(x_j) \geq f^*$, which means that $-f^* \geq -f(x_j)$. Applying this inequality and the definition of γ in (3.12), we obtain

$$-\sum_{i=1}^m \ln c_i(x_j) > \frac{\tau - f(x_j)}{\mu_k}.$$

After rearrangement, this relation implies that $B(x_j, \mu_k) > \tau$, i.e., that $x_j \notin S(\tau)$, a contradiction. We conclude that any accumulation point of a sequence in $S(\tau)$ must lie in $S(\tau)$, which means that $S(\tau)$ is closed.

We have shown that $S(\tau)$ is both bounded and closed; its compactness is immediate. $\square$

We are now ready to give the main theorem concerning barrier methods for convex programs. The most important result is (vi), which shows that limit points of a minimizing sequence for the barrier function converge to constrained minimizers of the convex program.

Theorem 5 (Convergence of barrier methods on convex programs.) Consider the convex program of minimizing $f(x)$ subject to $c_i(x) \geq 0$, $i = 1$, $\ldots, m$. Let $\mathcal{F}$ denote the feasible region for this problem, and assume that strict$(\mathcal{F})$ is nonempty. Let $\{\mu_k\}$ be a decreasing sequence of positive barrier parameters such that $\lim_{k\to\infty} \mu_k = 0$. Assume that the set $\mathcal{M}$ of constrained local minimizers of the convex program is nonempty and bounded, and let f^* denote the optimal value of f. Then

(i)   the logarithmic barrier function $B(x, \mu_k)$ is convex in strict$(\mathcal{F})$;

(ii) $B(x, \mu_k)$ has a finite unconstrained minimizer in strict($\mathcal{F}$) for *every*
     $\mu_k > 0$, and the set $\mathcal{M}_k$ of unconstrained minimizers of $B(x, \mu_k)$ in
     strict($\mathcal{F}$) is convex and compact for every k;

(iii) any unconstrained local minimizer of $B(x, \mu_k)$ in strict($\mathcal{F}$) is also a
     global unconstrained minimizer of $B(x, \mu_k)$;

(iv) let y_k denote an unconstrained minimizer of $B(x, \mu_k)$ in strict($\mathcal{F}$); then,
     for all k,

$$f(y_{k+1}) \leq f(y_k) \quad \text{and} \quad -\sum_{i=1}^{m} \ln c_i(y_k) \leq -\sum_{i=1}^{m} \ln c_i(y_{k+1});$$

(v)   there exists a compact set S such that, for all k, every minimizing point
     y_k of $B(x, \mu_k)$ lies in $S \cap \text{strict}(\mathcal{F})$;

(vi) any sequence $\{y_k\}$ of unconstrained minimizers of $B(x, \mu_k)$ has at least
     one convergent subsequence, and every limit point of $\{y_k\}$ is a local
     constrained minimizer of the convex program;

(vii) let $\{x_k\}$ denote a convergent subsequence of unconstrained minimizers
     of $B(x, \mu_k)$; then $\lim_{k \to \infty} f(x_k) = f^*$;

(viii) $\lim_{k \to \infty} B_k = f^*$, where B_k denotes $B(x_k, \mu_k)$.

Proof. It is straightforward to prove convexity of $B(x, \mu_k)$ using the convexity of f and $\{-c_i\}$, monotonicity of the logarithm function and Definition 9
of a convex function. Thus (i) is established.

The assumptions of this theorem are the same as those of Theorem 4.
Let x_0 denote the strictly feasible point at which the barrier iterations are
initiated. For the barrier parameter μ_k and some $\epsilon > 0$, we define the set
S_0 as:

$$S_0 = \{x \in \text{strict}(\mathcal{F}) \mid B(x, \mu_k) \leq B(x_0, \mu_k) + \epsilon\}.$$

Theorem 4 implies that S_0 is *compact* for all $\mu_k > 0$. It follows that the
smooth function $B(x, \mu_k)$ assumes its minimum in S_0, necessarily at an
interior point of S_0. We then apply Definition 7 and conclude that $B(x, \mu_k)$
has at least one finite *unconstrained* minimizer.

Because $B(x, \mu_k)$ is convex, any local minimizer is also a global minimizer,
so that *every* unconstrained minimizer of $B(x, \mu_k)$ must be in the set S_0.
Thus the set $\mathcal{M}_k$ of unconstrained minimizers of $B(x, \mu_k)$ is bounded. The
set $\mathcal{M}_k$ is closed because the minimum value of $B(x, \mu_k)$ is unique, and it
follows that $\mathcal{M}_k$ is compact. Convexity of $\mathcal{M}_k$ follows from Theorem 1, and
result (ii) has been verified.

Result (iii) follows from Theorem 1, and results (i) and (ii).

To show result (iv), let y_k and y_{k+1} denote global minimizers of the barrier
function for the barrier parameters μ_k and μ_{k+1}. By definition of y_k and

y_{k+1} as minimizers, we have

$$f(y_k) - \mu_k \sum_{i=1}^{m} \ln c_i(y_k) \leq f(y_{k+1}) - \mu_k \sum_{i=1}^{m} \ln c_i(y_{k+1}); \quad (3.13)$$

$$f(y_{k+1}) - \mu_{k+1} \sum_{i=1}^{m} \ln c_i(y_{k+1}) \leq f(y_k) - \mu_{k+1} \sum_{i=1}^{m} \ln c_i(y_k).$$

We multiply the first of these inequalities by the ratio μ_{k+1}/μ_k, which lies strictly between 0 and 1, add the resulting inequality to the second inequality, cancel the terms involving logarithms and obtain

$$f(y_{k+1})\left(1 - \frac{\mu_{k+1}}{\mu_k}\right) \leq f(y_k)\left(1 - \frac{\mu_{k+1}}{\mu_k}\right).$$

Since $0 < \mu_{k+1} < \mu_k$, it follows that $f(y_{k+1}) \leq f(y_k)$. Applying this result in (3.13) and dividing by the positive number μ_k, we obtain

$$-\sum_{i=1}^{m} \ln c_i(y_k) \leq -\sum_{i=1}^{m} \ln c_i(y_{k+1}), \quad (3.14)$$

as required for the second part of (iv).

To verify existence of the set S in (v), we use result (iv). Let f_k denote $f(y_k)$. Since $f_{k+1} \leq f_k$ for each k, the compact convex level set $\{x \in \mathcal{F} \mid f(x) \leq f_k\}$ not only contains all minimizers of $B(x, \mu_k)$, but also contains all minimizers of $B(x, \mu_{k+1})$. The compact convex level set S defined by the strictly feasible point x_0,

$$S = \{x \in \mathcal{F} \mid f(x) \leq f(x_0)\}, \quad (3.15)$$

accordingly contains $\mathcal{M}$ as well as all minimizers of $B(x, \mu_k)$ for all k.

Now we show (vi). It follows from the last statement of the preceding paragraph that every minimizer y_k must lie in the compact set S defined by (3.15). We conclude that the sequence $\{y_k\}$ is uniformly bounded, and hence contains at least one convergent subsequence, say with limit point $\hat{x}$. Because y_k lies in S for all k, $\hat{x}$ must be feasible.

To prove that $\hat{x}$ is a local constrained minimizer of the convex program, we assume otherwise, that $\hat{x} \notin \mathcal{M}$. Since every local solution of a convex program is a global solution, this would imply that $f(\hat{x}) > f^*$. A contradiction is now established from this inequality and the definition of $\hat{x}$ as a limit point of a convergent subsequence of minimizers of $B(x, \mu_k)$.

Let $\{x_k\}$ denote a subsequence of $\{y_k\}$ converging to $\hat{x}$. Continuity of f and the relation $f_k \geq f_{k+1}$ imply that, for all k,

$$f(x_k) \geq f(\hat{x}). \quad (3.16)$$

We next show that there must exist a strictly feasible point x_{int} such that

$$f(\hat{x}) > f(x_{\text{int}}).$$

Let x^* denote any point in the set $\mathcal{M}$ of constrained minimizers, so that $f(x^*) = f^*$ and x^* is in the set S defined by (3.15). If x^* itself is strictly feasible, we simply take $x_{\text{int}} = x^*$, since our initial assumption was that $f(\hat{x}) > f(x^*)$.

If x^* is not strictly feasible, x_{int} is found as follows. By assumption, $\text{strict}(\mathcal{F})$ is nonempty, and hence contains at least one point, say z; the definition and uniqueness of f^* guarantee that $f(z) \geq f(x^*)$. If $f(z) < f(\hat{x})$, z may be taken as x_{int}. If $f(z) \geq f(\hat{x})$, consider a generic point $\bar{x}$ on the line segment joining x^* and z, defined by $\bar{x} = (1 - \lambda)x^* + \lambda z$ for λ satisfying $0 < \lambda < 1$. Because z is strictly feasible and $-c_i$ is convex for $i = 1, \ldots, m$, we have

$$c_i(\bar{x}) \geq (1 - \lambda)c_i(x^*) + \lambda c_i(z) > 0,$$

so that $\bar{x}$ is strictly feasible.

Convexity of f implies that

$$f(\bar{x}) \leq (1 - \lambda)f(x^*) + \lambda f(z),$$

where $f(z) \geq f(\hat{x}) > f(x^*)$. Using continuity of f, we see that $f(\bar{x}) < f(\hat{x})$ for some suitably small λ, namely λ such that

$$\lambda < \frac{f(\hat{x}) - f(x^*)}{f(z) - f(x^*)} \leq 1. \tag{3.17}$$

For any λ satisfying (3.17), $\bar{x}$ may be taken as x_{int}.

Thus far we have shown that, if $\hat{x}$ is not in the minimizing set $\mathcal{M}$, then a strictly feasible point x_{int} exists such that

$$f(x_k) \geq f(\hat{x}) > f(x_{\text{int}}). \tag{3.18}$$

Since x_k is a global minimizer of $B(x, \mu_k)$,

$$f(x_k) - \mu_k \sum_{i=1}^{m} \ln c_i(x_k) \leq f(x_{\text{int}}) - \mu_k \sum_{i=1}^{m} \ln c_i(x_{\text{int}}). \tag{3.19}$$

The barrier term involving x_{int} in (3.19) is finite, and

$$\lim_{k \to \infty} B(x_{\text{int}}, \mu_k) = f(x_{\text{int}}).$$

If the limit point $\hat{x}$ of $\{x_k\}$ is also strictly feasible, the barrier term involving x_k in (3.19) is similarly finite as $k \to \infty$, and

$$\lim_{k \to \infty} B(x_k, \mu_k) = f(\hat{x}).$$

Letting $k \to \infty$ in (3.19), we obtain the inequality $f(\hat{x}) \leq f(x_{\text{int}})$, which contradicts (3.18).

Suppose, on the other hand, that $\hat{x}$ is not in $\text{strict}(\mathcal{F})$, so that $c_i(\hat{x}) = 0$ for at least one index i. Adding a barrier term involving x_{int} to both sides of the inequality $f(x_{\text{int}}) < f(x_k)$, we have

$$f(x_{\text{int}}) - \mu_k \sum_{i=1}^{m} \ln c_i(x_{\text{int}}) < f(x_k) - \mu_k \sum_{k=1}^{m} \ln c_i(x_{\text{int}}).$$

Combining this inequality with (3.19), rearranging and dividing by μ_k, we obtain

$$f(x_k) - \mu_k \sum_{k=1}^{m} \ln c_i(x_k) < f(x_k) - \mu_k \sum_{k=1}^{m} \ln c_i(x_{\text{int}}).$$

Cancelling $f(x_k)$ from both sides then gives

$$-\sum_{k=1}^{m} \ln c_i(x_k) < -\sum_{k=1}^{m} \ln c_i(x_{\text{int}}).$$

The sum on the right-hand side involving x_{int} is finite. However, since $\hat{x}$ is not strictly feasible, $-\ln c_i(x_k)$ approaches infinity for at least one i. The left-hand side is therefore unbounded above, and we again have a contradiction.

The conclusion is that $\hat{x}$ lies in $\mathcal{M}$, the set of minimizers. Because $\hat{x}$ is the limit point of $\{x_k\}$, we have obtained the crucial result (vi). For the remainder of the proof, x^* will denote the limit point of $\{x_k\}$.

Part (vii) follows immediately from the fact that $\lim_{k \to \infty} x_k = x^*$.

To show (viii), note first that the optimal value of $B(x, \mu_k)$ is unique, and is equal to $B(x_k, \mu_k)$. We distinguish two cases, depending on whether or not x^* (the limit point of $\{x_k\}$) is strictly feasible.

If x^* is strictly feasible, the sum of logarithms of the constraints at x_k remains finite as $k \to \infty$. It is easy to see that in this case $\lim_{k \to \infty} B(x_k, \mu_k) = f^*$.

Consider the other possibility, that x^* is not strictly feasible. Since at least one constraint is converging to zero, the barrier term of $B(x_k, \mu_k)$ must be *positive* for all sufficiently large k. Combining this property with (3.14), we have

$$0 < -\sum_{i=1}^{m} \ln c_i(x_k) \leq -\sum_{i=1}^{m} \ln c_i(x_{k+1}). \tag{3.20}$$

One implication of this result is that, for sufficiently large k,

$$B(x_k, \mu_k) > f(x_k). \tag{3.21}$$

In addition, the minimizing property of x_{k+1}, the first inequality in (3.20), and the relation $\mu_{k+1} \leq \mu_k$ together give:

$$f(x_{k+1}) - \mu_{k+1} \sum_{i=1}^{m} \ln c_i(x_{k+1}) \leq f(x_k) - \mu_{k+1} \sum_{i=1}^{m} \ln c_i(x_k)$$

$$f(x_k) - \mu_{k+1} \sum_{i=1}^{m} \ln c_i(x_k) \leq f(x_k) - \mu_k \sum_{i=1}^{m} \ln c_i(x_k),$$

which shows that, for sufficiently large k,

$$B(x_{k+1}, \mu_{k+1}) \leq B(x_k, \mu_k). \tag{3.22}$$

It follows from (3.21) and (3.22) that, for sufficiently large k,

$$f^* \leq \cdots \leq B_{k+1} \leq B_k, \tag{3.23}$$

where B_k denotes $B(x_k, \mu_k)$. The sequence $\{B_k\}$ of barrier function values is consequently nonincreasing and bounded from below, and must converge monotonically from above to a limit, say B^*, where $B^* \geq f^*$.

Suppose that $B^* > f^*$. In this case, we define δ as the positive number $\frac{1}{2}(B^* - f^*)$. It follows from continuity of f and the relation $f^* < B^*$ that there must be a neighbourhood of x^* in which

$$f(x) \leq B^* - \delta \tag{3.24}$$

for all x in the neighbourhood. Consider a particular strictly feasible point $\bar{x}$ in this neighbourhood. (Such a point must exist because $\mathcal{F}$ is convex and strict($\mathcal{F}$) is nonempty.) Strict feasibility of $\bar{x}$ implies that the quantity $\sum_{i=1}^{m} \ln c_i(\bar{x})$ is finite. Because $\mu_k > 0$ and $\mu_k \to 0$, there must be an integer K such that, for $k \geq K$,

$$- \mu_k \sum_{i=1}^{m} \ln c_i(\bar{x}) < \tfrac{1}{2}\delta. \tag{3.25}$$

Since x_k is a global minimizer of $B(x, \mu_k)$, we know that

$$B(x_k, \mu_k) \leq B(\bar{x}, \mu_k) = f(\bar{x}) - \mu_k \sum_{i=1}^{m} \ln c_i(\bar{x}).$$

If we apply (3.24) and (3.25), the result is

$$B(x_k, \mu_k) < B^* - \delta + \tfrac{1}{2}\delta = B^* - \tfrac{1}{2}\delta,$$

which contradicts the monotonic convergence of $\{B_k\}$ to B^* from above. We conclude that $B^* = f^*$, which gives result (viii). $\square$

The implications of this theorem are remarkably strong. For any convex program with a bounded set of minimizers, the barrier function has a finite unconstrained minimizer for *every* value of the barrier parameter, and every limit point of a minimizing sequence for the barrier function is a constrained minimizer. It is not necessarily true in general, however, that every minimizing sequence converges.

To every convex program, there corresponds a related *dual* convex program. For reasons of space, general results from duality theory will not be considered here, except for the following important result: the objective function value at any unconstrained minimizer x_k of $B(x, \mu_k)$ satisfies the inequality $f(x_k) - f^* \leq m\mu_k$, where m is the number of constraints. We know from results (iv) and (vii) of Theorem 5 that $f^* \leq f(x_k)$. Combining these bounds, we have

$$0 \leq f(x_k) - f^* \leq m\mu_k. \tag{3.26}$$

This somewhat surprising property implies that, when a barrier method is applied to a convex program, the deviation of $f(x_k)$ from optimality is always bounded by $m\mu_k$, independently of the particular problem functions. For comments about duality in linear programming, see Section 5.1.

3.4. Results for general nonlinear programs

Once we move from a convex program to a general nonlinear program, matters become far more complicated. In particular, certain topological assumptions are required to avoid pathological cases. Furthermore, the results apply only in a neighbourhood of a constrained minimizer, and involve convergence of subsequences of *global* minimizers of the barrier function. The general approach in this section follows that in Fiacco and McCormick (1968).

At the most basic level, the nice property given by Theorem 4 that the level sets of the barrier function are bounded if the set of constrained minimizers is bounded does not hold for the nonconvex case. If the feasible region is bounded, the barrier function is obviously bounded below. The following example of Powell (1972), however, shows that difficulties may arise when the feasible region is unbounded:

$$\text{minimize} \quad \frac{-1}{x^2 + 1} \quad \text{subject to} \quad x \geq 1. \tag{3.27}$$

The objective function is bounded below in the feasible region, and the unique solution is $x^* = 1$. In contrast, the barrier function

$$B(x, \mu) = \frac{-1}{x^2 + 1} - \mu \ln(x - 1)$$

is *unbounded below* in the feasible region, although it has a local minimizer that approaches x^* as $\mu \to 0$.

The major local convergence results will be given in Theorem 7. To build up to the statement of this theorem, several preliminary results are required.

The following lemma, an adaptation of Corollary 8 from Fiacco and McCormick (1968), plays the role of Theorem 4 for the convex case. The general result is that, if a continuous function is unbounded above for all sequences of points in strict($\mathcal{F}$) and converging to its boundary, then the function must achieve its minimum value at a *strictly interior* point. The obvious application of Lemma 1 is when $B(x, \mu)$ plays the role of φ.

Lemma 1   Given a set of m smooth constraint functions $\{c_i(x)\}$, $i = 1$, ..., m, let strict$(\mathcal{F})$ denote the set defined by (2.3). Let S be a compact set, and assume that the set strict$(\mathcal{F}) \cap S$ is nonempty. Consider any convergent sequence $\{y_k\} \in$ strict$(\mathcal{F}) \cap S$ whose limit point $\bar{y}$ lies on the boundary of strict$(\mathcal{F})$, i.e. such that

$$\lim_{k \to \infty} y_k = \bar{y}, \quad \text{where} \quad \bar{y} \in \text{bnd}(\text{strict}(\mathcal{F})) \cap S. \tag{3.28}$$

Suppose that φ is a continuous function on strict$(\mathcal{F}) \cap S$ with the property that $\varphi(y_k)$ is unbounded above as $k \to \infty$ for every sequence $\{y_k\}$ satisfying (3.28). Then the global minimum value of φ in strict$(\mathcal{F}) \cap S$, denoted by φ^*, is finite, and is achieved at some point x^* in strict$(\mathcal{F}) \cap S$:

$$\min\{\varphi(x) \mid x \in \text{strict}(\mathcal{F}) \cap S\} = \varphi(x^*) = \varphi^*.$$

Proof.   Given any point $\hat{x}$ in strict$(\mathcal{F}) \cap S$, define the associated level set $\hat{W}$ as

$$\hat{W} = \{x \in \text{strict}(\mathcal{F}) \cap S \mid \varphi(x) \leq \hat{\varphi}\},$$

where $\hat{\varphi} = \varphi(\hat{x})$. Because S is compact, $\hat{W}$ is bounded. Compactness of $\hat{W}$ will follow if we show that $\hat{W}$ is closed, i.e., contains all its accumulation points.

Let R denote the closed set

$$R = \text{strict}(\mathcal{F}) \cup \text{bnd}(\text{strict}(\mathcal{F})).$$

Because S is compact and R is closed, the set $R \cap S$ is compact. Consider any convergent sequence $\{x_k\}$ such that $x_k \in \hat{W}$ for all k, with limit point $\bar{x}$. Since $x_k \in$ strict$(\mathcal{F}) \cap S$, $\bar{x}$ must lie in $R \cap S$. Hence $\bar{x}$ must lie in either strict$(\mathcal{F}) \cap S$ or bnd$(\text{strict}(\mathcal{F})) \cap S$.

If $\bar{x}$ is in bnd$(\text{strict}(\mathcal{F})) \cap S$, then $\{x_k\}$ is a sequence satisfying (3.28), which means that $\varphi(x_k) \to \infty$. Since $\hat{\varphi}$ is an upper bound on the value of φ at any point in $\hat{W}$, we conclude that $x_k \notin \hat{W}$ for sufficiently large k, which is a contradiction. Any limit point $\bar{x}$ of a sequence in $\hat{W}$ therefore cannot be in bnd$(\text{strict}(\mathcal{F})) \cap S$, and must lie in strict$(\mathcal{F}) \cap S$.

Because x_k is in $\hat{W}$, the relation $\varphi(x_k) \leq \hat{\varphi}$ holds for all k. Continuity of φ in strict$(\mathcal{F}) \cap S$ then implies that the limit point $\bar{x}$ satisfies $\varphi(\bar{x}) \leq \hat{\varphi}$, so that $\bar{x}$ possesses both properties required for membership in $\hat{W}$. Since $\{x_k\}$ is an arbitrary convergent sequence in $\hat{W}$, it follows that $\hat{W}$ contains all its accumulation points and is closed.

We know already that $\hat{W}$ is bounded, so that $\hat{W}$ is compact. Because φ is continuous in the compact set $\hat{W}$, it attains its global minimum in $\hat{W}$ at some point x^*. By definition of $\hat{W}$, the value of φ at any point in strict$(\mathcal{F}) \cap S$ but outside $\hat{W}$ must be strictly larger than the smallest value of φ at any point in $\hat{W}$. Hence x^* is the global minimizer of φ in the entire set strict$(\mathcal{F}) \cap S$, which is the desired conclusion. $\square$

A property needed for local convergence is that a particular subset of local constrained minimizers is 'isolated' within the full set of local constrained minimizers. Such a definition is unnecessary for the convex case, since the set of minimizers is convex.

Definition 11 (Isolated subset.) Let $\mathcal{M}$ and $\mathcal{M}^*$ be sets in $\mathbb{R}^n$ such that $\mathcal{M}^* \subseteq \mathcal{M}$. The set $\mathcal{M}^*$ is called an *isolated* subset of $\mathcal{M}$ if there exists a closed set E such that $\mathcal{M}^* \subset \text{int}(E)$ and $E \cap \mathcal{M} = \mathcal{M}^*$.

Broadly speaking, $\mathcal{M}^*$ is 'separated' by E from any other points of $\mathcal{M}$. The definition is satisfied if $\mathcal{M}^* = \mathcal{M}$, or if $\mathcal{M}^*$ is an isolated point in $\mathcal{M}$.

The next theorem (a version of Theorem 7 of Fiacco and McCormick (1968)) shows that, if a set of constrained minimizing points is *compact* and *isolated*, there is a compact set S, strictly enclosing the set of minimizers, within which the minimizers are *global*. The role of the set S is critical: if we can restrict attention to points in S, the value of f at any minimizing point in S is a strict lower bound on the value of f at any other feasible (nonoptimal) point in S. For the convex case, a suitable set S is provided 'automatically' by the level set for f at any strictly feasible point; see (3.15).

Theorem 6 (Existence of compact enclosing set.) Consider the problem of minimizing $f(x)$ subject to $c_i(x) \geq 0$, $i = 1, \ldots, m$. Let $\mathcal{M}$ denote the set of all local constrained minimizers with objective function value f^*, and assume that $\mathcal{M}$ is nonempty. Assume further that the set $\mathcal{M}^* \subseteq \mathcal{M}$ is a nonempty compact isolated subset of $\mathcal{M}$. Then there exists a compact set S such that $\mathcal{M}^*$ lies in $\text{int}(S) \cap \mathcal{F}$, with the property that for any feasible point y in S but not in $\mathcal{M}^*$, $f(y) > f^*$. The points in $\mathcal{M}^*$ are thus global minimizers of the nonlinear program for $x \in S \cap \mathcal{F}$.

Proof. Applying Definition 11, the assumption that $\mathcal{M}^*$ is an isolated subset of $\mathcal{M}$ implies existence of a closed set E strictly containing $\mathcal{M}^*$ such that $\text{int}(E) \cap \mathcal{M} = \mathcal{M}^*$.

The assumption that $\mathcal{M}^*$ is compact means that we can construct a sequence of strictly nested compact sets $\{S_j\}$ converging to $\mathcal{M}^*$, each strictly containing $\mathcal{M}^*$, namely such that $\mathcal{M}^* \subset \text{int}(S_j) \subset \text{int}(E)$,

$$S_{j+1} \subset S_j, \quad \text{and} \quad \lim_{j \to \infty} S_j = \mathcal{M}^*. \tag{3.29}$$

The proof will show by contradiction that the desired compact set S may be taken as S_j for some finite j.

If this is impossible, then for every j we can find a feasible point x_j with the following properties:

$$x_j \in \mathcal{F} \cap S_j, \quad x_j \notin \mathcal{M}^* \quad \text{and} \quad f(x_j) \leq f^*. \tag{3.30}$$

Consider this hypothetical sequence $\{x_j\}$. The nested structure of $\{S_j\}$ means that $\{x_j\}$ is bounded and hence has at least one limit point, say $\bar{x}$. It follows from (3.29) and the fact that $\mathcal{M}^*$ is closed that $\bar{x} \in \mathcal{M}^*$, so that $f(\bar{x}) = f^*$ and $\bar{x}$ is a constrained minimizer. Let $\{y_k\}$ denote a subsequence of $\{x_j\}$ converging to $\bar{x}$, where $y_k = x_{j_k}$.

If $f(y_k)$ is strictly less than $f(\bar{x})$ for an infinite number of indices k, then every neighbourhood of $\bar{x}$ contains feasible points with a strictly smaller value of f. This means that $\bar{x}$ cannot be a constrained minimizer (see Definition 4) and gives a contradiction.

We conclude that eventually, for some $k = \bar{k}$, any point in $S_{j_{\bar{k}}}$ that qualifies as $y_{\bar{k}}$ (i.e. any feasible point in $S_{j_{\bar{k}}}$ for which $f(y_{\bar{k}}) \leq f^*$) must have an objective function value *equal to* f^*. This result implies that f^* is the *smallest* value of f achieved at feasible points in $S_{j_{\bar{k}}}$.

Let $\bar{S}$ denote $S_{j_{\bar{k}}}$. Since all subsequent sets S_{j_k} for $k \geq \bar{k}$ are subsets of $\bar{S}$, as is the minimizing set $\mathcal{M}^*$, it follows that f^* is the smallest value of f assumed at any feasible point in S_{j_k} for all $k \geq \bar{k}$.

The strictly nested property of the sets $\{S_j\}$ means that, for sufficiently large j, *any* point x_j satisfying (3.30), not necessarily a member of the subsequence converging to $\bar{x}$, must lie in the interior of the compact set $\bar{S}$. Because x_j is feasible and satisfies $f(x_j) \leq f^*$, and the smallest value of f for any feasible point in $\bar{S}$ is equal to f^*, it must be true that $f(x_j) = f^*$. Since x_j lies in the interior of the compact set $\bar{S}$ and $f(x_j) = f^*$, x_j satisfies Definition 4 of a local constrained minimizer with function value f^*, and hence x_j is in $\mathcal{M}$, the set of such minimizers. However, x_j is by definition in S_j, which is contained in the interior of E. Because $\text{int}(E) \cap \mathcal{M} = \mathcal{M}^*$, x_j must be in $\mathcal{M}^*$. It follows that, for sufficiently large j, no points in S_j satisfy (3.30).

We have constructed a compact set $\bar{S}$ that strictly contains $\mathcal{M}^*$. Further, $\bar{S}$ contains no feasible points with objective function values less than f^*, and every feasible point in $\bar{S}$ with objective function value equal to f^* lies in $\mathcal{M}^*$. It follows that any feasible point y in $\bar{S}$ but not in $\mathcal{M}^*$ must satisfy $f(y) > f^*$. The set $\bar{S}$ thus satisfies all the criteria specified for S, and the theorem is proved. □

We now give a fundamental theorem, analogous to Theorems 8 and 10 in Fiacco and McCormick (1968), about local convergence of logarithmic barrier methods. This theorem assumes two important properties: (a) a compactness requirement that the relevant set $\mathcal{M}^*$ of local minimizers is nonempty and compact (in the simplest case, $\mathcal{M}^*$ is a single point); and (b) a topological restriction that at least one of the points in $\mathcal{M}^*$ lies in the closure of strict($\mathcal{F}$), i.e. is either strictly feasible or else an accumulation point of strict($\mathcal{F}$).

Assumption (b) disallows minimizers that occur at isolated feasible points (points in a neighbourhood containing no other feasible points). For example, consider the constraints $x \geq 1$ and $x^2 - 5x + 4 \geq 0$. The function $x^2 - 5x + 4$ is nonnegative if $x \leq 1$ and if $x \geq 4$, so that the feasible points lie in two separated regions. The constraint $x \geq 1$ eliminates all of the region $\{x \leq 1\}$ except the single point $x = 1$. The feasible region for *both* constraints therefore consists of the isolated point $\{x = 1\}$ and the set of points $\{x \geq 4\}$. Hence strict($\mathcal{F}$) is the set $\{x > 4\}$, and the point $x = 1$ does not lie in the closure of strict($\mathcal{F}$).

Barrier methods can be viewed as finding the infimum of f subject to $c(x) > 0$, and consequently cannot converge to minimizers occurring at isolated points. Isolated minimizers do not arise in the convex case because a convex set with a nonempty interior cannot contain an isolated point.

Theorem 7 (Local convergence for barrier methods.) Consider the problem of minimizing $f(x)$ subject to $c_i(x) \geq 0$, $i = 1, \ldots, m$. Let $\mathcal{F}$ denote the feasible region, and let $\mathcal{M}$ denote the set of minimizers corresponding to the objective function value f^*. Let $\{\mu_k\}$ be a decreasing sequence of positive barrier parameters such that $\lim_{k \to \infty} \mu_k = 0$. Assume that

(a) there exists a nonempty compact set $\mathcal{M}^*$ of local minimizers that is an isolated subset of $\mathcal{M}$;

(b) at least one point in $\mathcal{M}^*$ is in the closure of strict($\mathcal{F}$).

Then the following results hold:

(i) there exists a compact set S strictly containing $\mathcal{M}^*$ such that for any feasible point $\bar{x}$ in S but not in $\mathcal{M}^*$, $f(\bar{x}) > f^*$;

(ii) for all sufficiently small μ_k, $B(x, \mu_k)$ has at least one unconstrained minimizer in strict($\mathcal{F}$) $\cap$ int(S), and any sequence of global unconstrained minimizers of $B(x, \mu_k)$ in strict($\mathcal{F}$) $\cap$ int(S) has at least one convergent subsequence;

(iii) let $\{x_k\}$ denote any convergent subsequence of global unconstrained minimizers of $B(x, \mu_k)$ in strict($\mathcal{F}$) $\cap$ int(S); then the limit point of $\{x_k\}$ is in $\mathcal{M}^*$;

(iv) $\lim_{k \to \infty} f(x_k) = f^* = \lim_{k \to \infty} B(x_k, \mu_k)$.

Proof. Result (i) follows immediately from Theorem 6, which implies the existence of a strictly enclosing compact set S within which all points in $\mathcal{M}^*$ are global constrained minimizers.

Consider the behaviour of the barrier function $B(x, \mu_k)$ in the bounded set strict($\mathcal{F}$) $\cap S$. Continuity of f and $\{c_i\}$ in $\mathcal{F}$ implies that $B(x, \mu_k)$ is continuous in strict($\mathcal{F}$) $\cap S$. The barrier function possesses the properties of φ in Lemma 1, which then implies that $B(x, \mu_k)$ achieves a finite global minimum value at some point in strict$(\mathcal{F}) \cap S$. (This result is close but not equivalent to (ii), which states that the minimizing point lies in int(S).) Let y_k be any point in strict$(\mathcal{F}) \cap S$ for which the minimum value is achieved.

The sequence $\{y_k\}$ is bounded and hence has at least one limit point. Let $\hat{x}$ denote a limit point of $\{y_k\}$. Because y_k is strictly feasible for all k and the set S is compact, it follows that $\hat{x} \in \mathcal{F} \cap S$, so that $\hat{x}$ is feasible.

We wish to show that $\hat{x}$ lies in the set $\mathcal{M}^*$ of constrained minimizers, with $f(\hat{x}) = f^*$. The result will be proved by contradiction, and we accordingly assume the contrary, that $\hat{x} \notin \mathcal{M}^*$.

Since $\hat{x}$ is feasible and in S, result (i) implies that $f(\hat{x}) > f^*$. We next prove that this inequality implies the existence of a strictly feasible point x_{int} in S such that

$$f(\hat{x}) > f(x_{\text{int}}). \tag{3.31}$$

The point x_{int} can be found as follows. We know from assumption (b) that at least one point in $\mathcal{M}^*$ is in the closure of strict$(\mathcal{F})$. Let x^* denote such a point, which must either lie in strict$(\mathcal{F})$ or else be an accumulation point of strict$(\mathcal{F})$. Because $\mathcal{M}^*$ is contained in int(S), x^* is also in the interior of S.

If x^* itself is strictly feasible, x_{int} may be taken as x^*. If x^* is not strictly feasible, x^* is an accumulation point of strict$(\mathcal{F})$, which means that every neighbourhood of x^* contains strictly feasible points. Further, every neighbourhood of x^* contains points in S. We know that: f is continuous; $\hat{x}$ is feasible and lies in S; $f(\hat{x}) > f(x^*)$; and x^* is a global constrained minimizer of f for all feasible points in S. Hence there must be a strictly feasible point x_{int} in a neighbourhood of x^* for which $f(x_{\text{int}}) < f(\hat{x})$.

Let $\{x_k\}$ denote a convergent subsequence of $\{y_k\}$ with limit $\hat{x}$. The relation $f(\hat{x}) > f(x_{\text{int}})$ then implies that, for sufficiently large k,

$$f(x_k) > f(x_{\text{int}}). \tag{3.32}$$

Since x_{int} is in strict$(\mathcal{F}) \cap S$, our definition of x_k as a global minimizer of $B(x, \mu_k)$ in strict$(\mathcal{F}) \cap S$ implies the inequality

$$f(x_k) - \mu_k \sum_{i=1}^{m} \ln c_i(x_k) \le f(x_{\text{int}}) - \mu_k \sum_{i=1}^{m} \ln c_i(x_{\text{int}}). \tag{3.33}$$

Strict feasibility of x_{int} means that the barrier term involving x_{int} in (3.33) is *finite*, and

$$\lim_{k \to \infty} B(x_{\text{int}}, \mu_k) = f(x_{\text{int}}).$$

Suppose that the limit point $\hat{x}$ of $\{x_k\}$ is also strictly feasible, namely $\hat{x} \in \text{strict}(\mathcal{F}) \cap S$. Then the barrier term involving x_k is finite as $k \to \infty$

and

$$\lim_{k \to \infty} B(x_k, \mu_k) = f(\hat{x}).$$

Letting $k \to \infty$ in (3.33), we obtain the inequality $f(x_k) \leq f(x_{\text{int}})$, which contradicts the relation $f(x_k) > f(x_{\text{int}})$ of (3.32).

Suppose on the other hand that $\hat{x}$ is not strictly feasible. Adding a barrier term to both sides of the inequality $f(x_k) > f(x_{\text{int}})$ gives

$$f(x_{\text{int}}) - \mu_k \sum_{i=1}^{m} \ln c_i(x_{\text{int}}) < f(x_k) - \mu_k \sum_{k=1}^{m} \ln c_i(x_{\text{int}}).$$

Combining this inequality with (3.33), rearranging and then dividing by μ_k, we obtain

$$f(x_k) - \mu_k \sum_{k=1}^{m} \ln c_i(x_k) < f(x_k) - \mu_k \sum_{k=1}^{m} \ln c_i(x_{\text{int}}).$$

Cancelling $f(x_k)$ from both sides, the result is

$$-\sum_{k=1}^{m} \ln c_i(x_k) < -\sum_{k=1}^{m} \ln c_i(x_{\text{int}}). \tag{3.34}$$

As before, strict feasibility of x_{int} guarantees that the sum on the right-hand side is fixed and finite. However, since $\hat{x}$ is not strictly feasible, $-\ln c_i(x_k)$ approaches infinity for at least one i. The left-hand side of (3.34) is therefore unbounded above, which again gives a contradiction.

In either case, we have shown that $f(\hat{x}) = f^*$ and hence that $\hat{x} \in \mathcal{M}^*$. Since $\hat{x}$ was taken as any limit point of $\{y_k\}$, we conclude that every limit point of a convergent subsequence of global barrier minimizers lying in strict$(\mathcal{F}) \cap S$ must be a constrained minimizer with objective value f^*.

Result (ii) is proved by noting that the relation $\hat{x} \in \mathcal{M}^*$ means that $\hat{x} \in \text{int}(S)$. Since $\hat{x}$ is the limit point of $\{x_k\}$, it must hold that x_k is also in $\text{int}(S)$ for sufficiently large k. By definition, x_k is strictly feasible. Hence the global minimum of $B(x, \mu_k)$ in strict$(\mathcal{F}) \cap S$ is achieved at some point x_k lying strictly inside both $\mathcal{F}$ and S. Applying Definition 7, the global minimizer x_k of $B(x, \mu_k)$ in S is an *unconstrained* minimizer. Results (ii) and (iii) are thus proved.

The first relation in (iv), that $\lim_{k \to \infty} f(x_k) = f^*$, follows because $f(\hat{x}) = f^*$. The second, $\lim_{k \to \infty} B(x_k, \mu_k) = f^*$, follows from the same arguments used in proving (iv) and (viii) of Theorem 5, with the additional restriction here that all points must lie in S. $\square$

At this point we should emphasize what has *not* been proved. Even within the set S, a general global minimizing sequence $\{x_k\}$ of the barrier function is not guaranteed to converge. The properties of local minimizing sequences are even less secure. In particular, it is not true that every limit point of a local minimizing sequence is a constrained minimizer.

For example, consider the problem

$$\text{minimize } x \quad \text{subject to} \quad x^2 \geq 0, \quad x \geq -\gamma,$$

where $\gamma > 0$ (Moré and Wright, 1990); a similar example is given in Fiacco (1979). The unique solution is obviously the point $x^* = -\gamma$. The barrier function $B(x, \mu)$ has two feasible minimizers:

$$x(\mu) = \frac{3\mu - \gamma \pm \left((3\mu - \gamma)^2 + 8\gamma\mu\right)^{1/2}}{2}.$$

For $\mu \to 0$, the *global* minimizing sequence corresponds to the negative square root and converges to $-\gamma$, the unique solution of the constrained problem. However, the *nonglobal* minimizing sequence of $B(x, \mu)$, corresponding to the positive square root, converges to the origin, which is *not* a constrained minimizer.

Despite these cautions, the bright side is that barrier methods will converge to the solutions of constrained problems for which the usual sufficient conditions do not hold. Barrier methods can converge, for example, when the constrained minimizer is not locally unique. Barrier methods can succeed even when a local constrained minimizer does not satisfy a constraint qualification.

3.5. The barrier trajectory

In this section, we describe conditions under which a sequence $x(\mu)$ of barrier minimizers not only converges to x^*, but also defines a smooth path (the 'barrier trajectory') that is nontangential to the active constraint gradients.

Discussions of the logarithmic barrier function involve special diagonal matrices related to vectors, for which the following notation has become popular. When a lower-case letter refers to a vector, its upper-case version means the diagonal matrix of comparable dimension whose (i, i) element is the ith component of the vector. For example, C denotes the $m \times m$ diagonal matrix of constraint values $\{c_i\}$:

$$C = \text{diag}(c_i) = \begin{pmatrix} c_1 & & & \\ & c_2 & & \\ & & \ddots & \\ & & & c_m \end{pmatrix},$$

and C^{-1} is the diagonal matrix whose ith element is $1/c_i$. Using this convention, we have the general relation $Ce = c$, where e denotes the vector of appropriate dimension whose components are all equal to one, $e = (1, 1, \ldots, 1)^T$.

The gradient of $B(x, \mu)$ with respect to x is

$$\nabla B(x, \mu) = g - \sum_{i=1}^{m} \frac{\mu}{c_i} a_i = g - \mu A^T C^{-1} e, \qquad (3.35)$$

where all functions are evaluated at x, a_i is the gradient of $c_i(x)$ and A is the Jacobian of $c(x)$. (Recall that the constraint gradients are the transposed rows of A and hence the *columns* of A^T.) The Hessian of $B(x, \mu)$ is

$$\begin{aligned} \nabla^2 B(x, \mu) &= H - \sum_{i=1}^{m} \frac{\mu}{c_i} H_i + \sum_{i=1}^{m} \frac{\mu}{c_i^2} a_i a_i^T \\ &= H - \sum_{i=1}^{m} \frac{\mu}{c_i} H_i + \mu A^T C^{-2} A. \end{aligned}$$

The point $x(\mu)$ is an unconstrained minimizer of $B(x, \mu)$ only if the gradient vanishes at $x(\mu)$. Substituting from (3.35), the following relation must hold at $x(\mu)$:

$$g(\mu) = \sum_{i=1}^{m} a_i(\mu) \frac{\mu}{c_i(\mu)}, \qquad (3.36)$$

where the argument μ denotes evaluation at $x(\mu)$. Since $\mu > 0$ and $c_i(\mu) > 0$, it follows that the gradient of f at $x(\mu)$ is a *positive* linear combination of the gradients of *all* the constraints.

The first-order optimality conditions (2.13b–d) for nonlinear constraints are

$$g(x^*) = A(x^*)^T \lambda^* = \sum_{i=1}^{m} a_i(x^*) \lambda_i^*, \qquad (3.37)$$

where $\lambda_i^* \geq 0$ and $\lambda_i^* c_i(x^*) = 0$, $i = 1, \ldots, m$. At x^*, the gradient of f is thus a nonnegative linear combination of all the constraint gradients, where inactive constraints have zero multipliers.

The similar forms of (3.36) and (3.37) reveal that the ith coefficient $\mu/c_i(\mu)$ in the linear combination (3.36) is directly analogous to the ith Lagrange multiplier λ_i^*. When standard sufficient optimality conditions hold at x^* and the gradients of the active constraints are linearly independent, the multiplier estimates $\mu/c_i(\mu)$ do indeed converge to λ_i^*. In fact, under these conditions a differentiable curve $x(\mu)$ of barrier minimizers, parametrized by μ, exists near $\mu = 0$ and converges to x^*. This curve of minimizers is called the *barrier trajectory*; in linear programming, it is usually known as the *central path*. Its existence and properties define the broad class of 'path-following' algorithms that attempt to follow the trajectory to the solution; see Sections 5.2 and 6.2.

The results of the following theorem are essentially those of Theorem 12 of Fiacco and McCormick (1968).

Theorem 8   Consider the problem of minimizing $f(x)$ subject to $c_i(x) \geq 0$, $i = 1, \ldots, m$, where $m \geq 1$. Let $\mathcal{F}$ denote the feasible region, and assume that strict($\mathcal{F}$) is nonempty. Assume further that x^* is a local constrained minimizer at which

(a) $g(x^*) = A(x^*)^T \lambda^*$, with $\lambda_i^* c_i(x^*) = 0$;

(b) $\lambda_i^* > 0$ if $c_i(x^*) = 0$;

(c)   there exists $\alpha > 0$ such that $p^T W(x^*, \lambda^*) p \geq \alpha \|p\|^2$ for all p satisfying $\hat{A} p = 0$, where $\hat{A}$ denotes the Jacobian of the active constraints at x^* and W is the Hessian of the Lagrangian function (see (2.12));

(d)   the gradients of the active constraints at x^* are linearly independent.

Consider a logarithmic barrier method in which $B(x, \mu_k)$ is minimized for a sequence of positive values $\{\mu_k\}$ converging monotonically to zero as $k \to \infty$. Then

(i)   there is at least one subsequence of unconstrained minimizers of the barrier function $B(x, \mu_k)$ converging to x^*;

(ii)   For such a convergent subsequence $\{x_k\}$,

$$\lim_{k \to \infty} \mu_k / c_i^k = \lambda_i^*, \quad \text{where } c_i^k \text{ denotes } c_i(x_k);$$

(iii) for sufficiently large k, the Hessian matrix $\nabla^2 B(x_k, \mu_k)$ is positive definite;

(iv)   a unique, continuously differentiable vector function $x(\mu)$ of unconstrained minimizers of $B(x, \mu)$ exists in a neighbourhood of $\mu = 0$;

(v) $\lim_{\mu \to 0} x(\mu) = x^*$.

Proof.   The properties assumed about x^* ensure that it is an isolated constrained minimizer. Two implications follow from the linear independence of the active constraint gradients: the Lagrange multipliers λ^* are unique; and every neighbourhood of x^* contains points in strict($\mathcal{F}$), so that x^* is in the closure of the interior of the feasible region. Theorem 7 consequently applies to x^*, and implies that there is at least one subsequence of unconstrained minimizers of $B(x, \mu_k)$ converging to x^*. This proves (i).

Let $\{x_k\}$ denote such a convergent sequence, with redefinition of k as necessary, so that

$$\lim_{k \to \infty} x_k = x^*. \tag{3.38}$$

As convenient, we denote quantities associated with x_k by a subscript or superscript k; the subscript i always denotes the ith component of a vector.

For sufficiently large k, x_k is an unconstrained minimizer of $B(x, \mu_k)$, which means that the gradient (3.35) of the barrier function vanishes at x_k:

$$g^k = \sum_{i=1}^{m} a_i^k \lambda_i^k, \quad \text{where} \quad \lambda_i^k = \frac{\mu_k}{c_i^k}. \tag{3.39}$$

The quantity λ_i^k is strictly positive for any $\mu_k > 0$.

Suppose that constraint i is *inactive* at x^*. Then, from (3.38),

$$\lim_{k \to \infty} c_i^k = c_i(x^*) > 0, \quad \text{and hence} \quad \lim_{k \to \infty} \lambda_i^k = \lambda_i^* = 0. \tag{3.40}$$

If no constraints are active, we have verified (ii).

Otherwise, let $\mathcal{A}$ denote the set of indices of constraints active at x^*, so that $c_i(x^*) = 0$ for $i \in \mathcal{A}$. Let the positive numbers s_k and v_i^k be defined as

$$s_k = \sum_{i=1}^{m} \lambda_i^k \quad \text{and} \quad v_i^k = \frac{\lambda_i^k}{s_k}.$$

Note that $v_i^k > 0$ and $\sum_{i=1}^{m} v_i^k = 1$, so that $v_i^k \leq 1$. Since $s_k > 0$ and (3.39) holds at x_k, we have

$$\frac{1}{s_k} g^k - \sum_{i=1}^{m} a_i^k v_i^k = 0. \tag{3.41}$$

As $k \to \infty$, the sequence $\{v_i^k\}$ is bounded for $i = 1, \ldots, m$, and accordingly contains a convergent subsequence.

The value of $\liminf_{k \to \infty} s_k$, denoted by $\hat{s}$, must be finite. If not, consider (3.41) as $k \to \infty$. Because of (3.38), a_i^k converges to a_i^*, where the superscript $*$ denotes evaluation at x^*. The following relation must hold for any set $\{\hat{v}_i\}$ of limit points of $\{v_i^k\}$:

$$\sum_{i=1}^{m} a_i^* \hat{v}_i = 0, \quad \text{where} \quad \hat{v}_i \geq 0 \quad \text{and} \quad \sum_{i=1}^{m} \hat{v}_i = 1. \tag{3.42}$$

Because $\hat{v}_i = 0$ if constraint i is inactive at x^*, relation (3.42) states that a nontrivial linear combination of the active constraint gradients at x^* is zero, which contradicts our assumption of their linear independence.

Finiteness of $\hat{s}$ implies that each component λ_i^k is bounded for all k, and consequently the sequence $\{\lambda_i^k\}$ has at least one accumulation point, say $\bar{\lambda}_i$. It follows from (3.39) and (3.40) that

$$g^* = \hat{A}^T \bar{\lambda}.$$

Because the rows of $\hat{A}$ are linearly independent, the values satisfying this equation are unique, and we conclude that $\bar{\lambda}_i = \lambda_i^*$ for $i \in \mathcal{A}$, which completes the proof of (ii).

We now wish to demonstrate positive-definiteness of the barrier Hessian at x_k. This property will be verified using the asymptotic structure of the Hessian of the barrier function, which approaches the sum of the Hessian of the Lagrangian function and a 'large' matrix in the range space of the active constraint gradients.

Consider the ratio $\mu_k/(c_i^k)^2$, which we denote by d_i^k:

$$d_i^k = \frac{\mu_k}{(c_i^k)^2} = \frac{\lambda_i^k}{c_i^k}. \tag{3.43}$$

When constraint i is active at x^*, i.e. $i \in \mathcal{A}$, result (ii) and assumption (b) of strict complementarity imply that λ_i^k converges to a strictly positive constant. Since c_i^k converges to zero for $i \in \mathcal{A}$, the final ratio in (3.43) is clearly *unbounded*, and

$$\liminf_{k\to\infty} d_i^k = \liminf_{k\to\infty} \frac{\mu_k}{(c_i^k)^2} = \infty \quad \text{for } i \in \mathcal{A}. \tag{3.44}$$

Recall that the Hessian of the barrier function is given by

$$\nabla^2 B(x, \mu) = H - \sum_{i=1}^m \frac{\mu}{c_i} H_i + \sum_{i=1}^m \frac{\mu}{c_i^2} a_i a_i^T.$$

Let H_B^k denote $\nabla^2 B(x_k, \mu_k)$. The limiting properties of this matrix are revealed by expressing it in the following form:

$$H_B^k = W^* + M^* + M_1^k + M_2^k + M_3^k.$$

The first two matrices on the right-hand side depend on x^* and a bounded positive constant γ:

$$W^* = H^* - \sum_{i=1}^m \lambda_i^* H_i^*$$

$$M^* = \gamma \sum_{i \in \mathcal{A}} a_i^* (a_i^*)^T = \gamma \hat{A}^T \hat{A}.$$

The remaining three matrices are expressed as perturbations involving x_k, x^* and γ:

$$M_1^k = H^k - H^* - \left(\sum_{i=1}^m \lambda_i^k H_i^k - \sum_{i=1}^m \lambda_i^* H_i^* \right)$$

$$M_2^k = \gamma \sum_{i \in \mathcal{A}} \left(a_i^k (a_i^k)^T - a_i^* (a_i^*)^T \right)$$

$$M_3^k = \sum_{i \in \mathcal{A}} (d_i^k - \gamma) a_i^k (a_i^k)^T + \sum_{i \notin \mathcal{A}} d_i^k a_i^k (a_i^k)^T.$$

The matrix W^* is the Hessian of the Lagrangian function at x^*. For sufficiently large k, the matrices M_1^k and M_2^k can be made arbitrarily small in norm; this statement follows from continuity of the problem functions, convergence of x_k to x^*, convergence of λ_k to λ^*, and boundedness of γ. Because d_i^k is unbounded above for $i \in \mathcal{A}$ (see (3.44)), the quantity $(d_i^k - \gamma)$ is positive for sufficiently large k; hence the matrix M_3^k is the sum of two positive semi-definite matrices and must itself be positive semi-definite.

Positive-definiteness of H_B^k will follow if the matrix $W^* + \gamma \hat{A}^T \hat{A}$ is guaranteed to be positive definite for some constant γ. This property is shown by examining the effect of the matrix in two orthogonal subspaces: the range space of $\hat{A}^T$ and the null space of $\hat{A}$.

It is well known that any n-vector p may be uniquely expressed as the sum of two orthogonal components,

$$p = p_R + p_N,$$

where p_R lies in the range space of $\hat{A}^T$ and p_N lies in the null space of $\hat{A}$. Using this form, the product $p^T(W^* + \gamma \hat{A}^T \hat{A})p$ can be written as

$$p_N^T W^* p_N + 2p_N^T W^* p_R + p_R^T W^* p_R + \gamma p_R^T \hat{A}^T \hat{A} p_R. \tag{3.45}$$

To ensure positive-definiteness, this quantity must be bounded below by a positive number when $p \neq 0$. To develop the bound, we use a relation that holds for any matrix C and vectors x and y:

$$x^T C y \geq -\|C\| \, \|x\| \, \|y\|.$$

Assumption (c) guarantees the existence of $\alpha > 0$ (the smallest eigenvalue of the reduced Hessian of the Lagrangian; see (2.16)) such that

$$p_N^T W^* p_N \geq \alpha \|p_N\|^2.$$

By definition, p_R is in the range of $\hat{A}^T$. Hence, if $p_R \neq 0$, it holds that

$$\hat{A} p_R \neq 0 \quad \text{and} \quad p_R^T \hat{A}^T \hat{A} p_R > \beta \|p_R\|^2$$

for some positive β (the square of the smallest nonzero singular value of $\hat{A}$). Let ω denote $\|W^*\|$.

Applying these inequalities to (3.45), we obtain

$$p^T(W^* + \gamma \hat{A}^T \hat{A})p \geq \alpha \|p_N\|^2 + \gamma \beta \|p_R\|^2 - 2\omega \|p_R\| \, \|p_N\| - \omega \|p_R\|^2.$$

If $\|p_R\| = 0$, so that p lies entirely in the null space of $\hat{A}$, the expression on the right-hand side is simply $\alpha \|p_N\|^2$, which must be positive. Otherwise, if $\|p_R\| \neq 0$, the right-hand side is guaranteed to be positive if γ is bounded below as follows:

$$\gamma > \frac{\omega^2 + \alpha \omega}{\alpha \beta}. \tag{3.46}$$

We have shown that the Hessian of the barrier function at x_k must be positive definite for sufficiently large k, which is result (iii). The point x_k

is consequently an *isolated* unconstrained minimizer of $B(x, \mu_k)$ (see (2.5)), and is locally unique.

To verify the existence of a unique, differentiable function $x(\mu)$ in a neighbourhood of $x(\mu_k)$, we apply the *implicit function theorem* (see Ortega and Rheinboldt (1970)) to the $n+1$ variables (x, μ). At (x_k, μ_k), we know from (3.36) that the following system of nonlinear equations has a solution:

$$\Phi(x, \mu) = g(x) - \mu \sum_{i=1}^{m} \frac{1}{c_i(x)} a_i(x).$$

The Jacobian of Φ with respect to x is the barrier Hessian H_B^k, which was just shown to be positive definite at (x_k, μ_k). The implicit function theorem then implies that there is a locally unique, differentiable function $x(\mu)$ passing through $x(\mu_k)$ such that $\Phi(x, \mu) = 0$ for all μ in a neighbourhood of μ_k.

Using continuation arguments, it is straightforward to show that the function $x(\mu)$ exists for all $0 < \mu \le \mu_k$ for sufficiently large k, which gives result (iv).

The final result is immediate from the local uniqueness of $x(\mu)$ and result (i). $\square$

We have now verified the existence of both the barrier trajectory $x(\mu)$ and the associated multiplier estimate $\lambda(\mu)$. A remaining question involves existence and differentiability of the trajectory at x^* itself. For sufficiently small μ, the following $n+m$ equations are satisfied identically at every pair $(x(\mu), \lambda(\mu))$ on the trajectory:

$$g(x) - A(x)^T \lambda = 0; \tag{3.47a}$$
$$\lambda_i c_i(x) = \mu, \quad i = 1, \ldots, m. \tag{3.47b}$$

If we treat the multipliers $\lambda(\mu)$ as *separate variables*, (3.47) can be viewed as a system of nonlinear equations in the $n+m$ variables (x, λ). The Jacobian matrix of this system is given by

$$J(\mu) = \begin{pmatrix} H(\mu) - \sum \lambda_i(\mu) H_i(\mu) & -A(\mu)^T \\ \Lambda(\mu) A(\mu) & C(\mu) \end{pmatrix}, \tag{3.48}$$

where Λ and C are diagonal matrices corresponding to λ and c.

We can again apply the implicit function theorem to deduce the existence of a differentiable trajectory $(x(\mu), \lambda(\mu))$ at (x^*, λ^*) if the matrix (3.48) is nonsingular at $\mu = 0$. Let J^* denote the limiting version of (3.48):

$$J^* = \begin{pmatrix} W^* & -A^{*T} \\ \Lambda^* A^* & C^* \end{pmatrix}.$$

Nonsingularity of J^* will follow if there is no nontrivial solution z to the system $J^*z = 0$. Partitioning z into an n-vector u and an m-vector v and using the form of J^*, the condition $J^*z = 0$ implies that

$$W^*u - A^{*T}v = 0 \quad \text{and} \quad \Lambda^*A^*u + C^*v = 0. \tag{3.49}$$

If $c_i^* > 0$, we know that $\lambda_i^* = 0$, and the second equation in (3.49) then implies that $v_i = 0$ for all inactive constraints. If, on the other hand, $c_i^* = 0$, the same relation implies that $\lambda_i^*(a_i^*)^T u = 0$ for $i \in \mathcal{A}$. It follows that u must lie in the null space of $\hat{A}$, the Jacobian of the active constraint gradients. Combining these properties, we see that

$$v_i = 0 \text{ for } i \notin \mathcal{A} \quad \text{and} \quad (a_i^*)^T u = 0 \text{ for } i \in \mathcal{A}. \tag{3.50}$$

If $J^*z = 0$, the scalar $z^T J^* z$ must also be zero. Writing out $z^T J^* z$ in terms of u and v, we have

$$z^T J^* z = u^T W^* u - v^T A^* u + v^T \Lambda^* A^* u + v^T C^* v = 0.$$

It follows from (3.50) that $v^T A^* u = 0$, $v^T \Lambda^* A^* u = 0$, and $v^T C^* v = 0$. Therefore, $u^T W^* u = 0$. But by assumption (c), this can be true only if $u = 0$.

If $u = 0$, the first equation in (3.49) implies that $A^{*T}v = \sum_{i=1}^{m} a_i^* v_i = 0$. Because the components of v corresponding to inactive constraints are zero, it follows that that $\sum_{i \in \mathcal{A}} a_i^* v_i = 0$. Under assumption (d) that the active constraint gradients are linearly independent, this can be true only if $v_i = 0$ for all $i \in \mathcal{A}$. But in this case, $u = 0$ and $v = 0$, which means that $z = 0$. Since $J^*z = 0$ only for a zero vector z, J^* is nonsingular. The implicit function theorem applied to (3.47) thus implies that the trajectory $x(\mu)$ exists and is differentiable at x^*.

The approach of the trajectory $x(\mu)$ to x^* can be analysed as follows. Let $\dot{x}$ denote $dx(\mu)/d\mu$, with a similar meaning for $\dot{\lambda}$. Differentiating (3.47) with respect to μ, we see that, for sufficiently small μ, $x(\mu)$ and $\lambda(\mu)$ satisfy the system of differential equations

$$\begin{pmatrix} H - \sum \lambda_i H_i & -A^T \\ \Lambda A & C \end{pmatrix} \begin{pmatrix} \dot{x} \\ \dot{\lambda} \end{pmatrix} = \begin{pmatrix} 0 \\ e \end{pmatrix}, \tag{3.51}$$

with initial conditions $x(0) = x^*$ and $\lambda(0) = \lambda^*$.

Let y denote the vector $\dot{x}$ evaluated at $\mu = 0$, i.e., y is the tangent to the barrier trajectory at x^*. For an *active* constraint i, the second set of equations in (3.51) reveals that

$$\lambda_i^*(a_i^*)^T y = 1, \quad \text{so that} \quad (a_i^*)^T y = \frac{1}{\lambda_i^*} \quad \text{and} \quad \hat{A}y = (\hat{\Lambda}^*)^{-1}e. \tag{3.52}$$

The assumption of strict complementarity means that $\lambda_i^* \neq 0$ for any active constraint. Hence relation (3.52) shows that the barrier trajectory approaches x^* *nontangentially* with respect to the active constraints, i.e. the iterates do not converge 'along' the boundary.

Assume that constraint i is active, and let θ_i denote the angle between y and the normal to constraint i at x^*. It follows from (3.52) that

$$\cos \theta_i \sim \frac{1}{\|a_i^*\|\lambda_i^*}. \tag{3.53}$$

If all active constraint gradients are approximately equal in norm, relation (3.53) shows that the approach of barrier trajectory to x^* is 'closer to tangential' for active constraints with larger multipliers.

These properties are illustrated graphically with a two-variable example:

$$\begin{aligned}
\text{minimize} \quad & 2x_1x_2 - x_1^2 - x_2 \\
\text{subject to} \quad & x_1^2 + x_2^2 \leq 2 \\
& x_1^2 x_2^2 \leq 10 \\
& (x_1 - \tfrac{1}{2})^2 + (x_2 - 1)^2 \leq \tfrac{9}{4}.
\end{aligned}$$

The first and third constraints (shown as dashed curves in Figure 3) intersect at $x^* = (-1, 1)^T$, which is an isolated local minimizer with Lagrange multipliers $\lambda_1^* = \frac{3}{2}$ and $\lambda_3^* = \frac{1}{3}$. The trajectory of barrier minimizers is depicted as a solid line converging to x^*. As expected, the trajectory approaches both active constraints along a nontangential path. The figure also confirms the prediction of (3.53) concerning the relative angles of approach to these constraints.

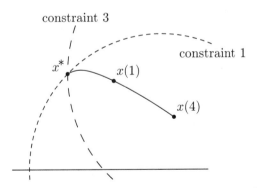

Fig. 3. The nontangential approach of $x(\mu)$ to x^*.

The nontangential property fails to hold without strict complementarity, even if the active constraint gradients are linearly independent. A complete analysis of this case is given in Jittorntrum (1978).

3.6. Properties of the barrier Hessian

The Hessian matrices H_B of the barrier function display a special structure as $\mu \to 0$. The barrier Hessian is given by

$$H_B(\mu) = H - \sum_{i=1}^{m} \frac{\mu}{c_i} H_i + \mu A^T C^{-2} A.$$

As $x(\mu)$ converges to x^*, the first two terms of the Hessian approach the Hessian of the Lagrangian function at x^*; the third matrix is given by

$$\mu A^T C^{-2} A = A^T D^2 A, \quad \text{where} \quad D^2 = \mu C^{-2} = \Lambda C^{-1} = \frac{1}{\mu} \Lambda^2. \qquad (3.54)$$

We have already shown that the elements of D corresponding to inactive constraints are converging to zero, and the elements corresponding to active constraints are becoming unbounded (see (3.44)).

Let $\hat{m}$ denote the number of active constraints whose gradients are linearly independent. The Hessian of the barrier function can be characterized in three ways, depending on $\hat{m}$.

If no constraints are active at x^*, then the Hessian of the barrier function converges (as we would expect) to the Hessian of f itself.

At the other extreme, suppose that $\hat{m} = n$, so that the Jacobian of the active constraints has rank n. If $\hat{\lambda}^* > 0$, the barrier Hessian along the trajectory approaches a large multiple of the *nonsingular* matrix $\hat{A}^T (\hat{\Lambda}^*)^2 \hat{A}$. In this case, the condition of the limiting Hessian depends on the condition of $\hat{A}$ and the condition of $\hat{\Lambda}^*$, but is not necessarily large. (This situation holds for linear programs in which there are no zero Lagrange multipliers.)

Finally, assume that $0 < \hat{m} < n$. The limiting matrix D^2 then contains $\hat{m}$ unbounded elements and $n - \hat{m}$ zero elements, which means that asymptotically $A^T D^2 A$ of (3.54) becomes not only unbounded, but also *rank-deficient*. Murray (1971) showed that $H_B(\mu)$ has $\hat{m}$ unbounded eigenvalues, corresponding to eigenvectors in the range space of $\hat{A}^T$, and $n - \hat{m}$ bounded eigenvalues, corresponding to eigenvectors in the null space of $\hat{A}$. The barrier Hessian accordingly becomes increasingly ill-conditioned for 'small' μ, and is *singular* in the limit. This property is one of the reasons that barrier function methods fell into disfavour in the 1970s, since standard unconstrained methods (such as Newton-based or quasi-Newton methods) tend to experience numerical difficulties when the Hessian is ill-conditioned. Various linear algebraic approaches have been proposed for dealing with this inherent ill-conditioning of the Hessian, and will be discussed in Section 7.2.

Our general analysis of barrier methods will be applied in Section 5 to the special case of linear programming. As background, we briefly summarize the relevant features of Newton's method in Section 4.

4. Newton's method

The application of interior methods to linear programming is heavily based on Newton's method, which we review in this section. Newton's method enters because satisfaction of certain nonlinear equations is definitive in optimality conditions (see Section 2.2), and the most popular technique for solving nonlinear equations is Newton's method.

4.1. Nonlinear equations and unconstrained minimization

We consider two forms for Newton's method. First, let $\Phi(x)$ denote an n-vector of smooth scalar functions $\varphi_i(x)$, $i = 1, \ldots, n$, and let $J(x)$ denote the Jacobian matrix of Φ. We seek a point x^* where $\Phi(x^*) = 0$. If x_k is the current point and $J(x_k)$ is nonsingular, the *Newton step* p_k is the step from x_k to the zero of the local affine model of Φ, and is the unique solution of the linear system

$$J_k p_k = -\Phi_k, \qquad (4.1)$$

where J_k denotes $J(x_k)$ and Φ_k denotes $\Phi(x_k)$.

The second form of Newton's method is designed for unconstrained minimization of $f(x)$. Here, a quadratic model of the local variation of f is obtained from the Taylor-series expansion about x_k:

$$f(x_k + p) - f(x_k) \approx g_k^T p + \tfrac{1}{2} p^T H_k p,$$

where $g_k = g(x_k)$ and $H_k = H(x_k)$. If H_k is positive definite, the Newton step p_k is the step from x_k to the minimizer of this model, and satisfies the nonsingular linear system

$$H_k p = -g_k. \qquad (4.2)$$

The direction p_k of (4.2), derived for minimization, is equivalent to the Newton step for solving the n-dimensional nonlinear system $g(x) = 0$.

4.2. Local convergence

A pure Newton method for either zero-finding or minimization begins with an initial point x_0, and generates a sequence of Newton iterates $\{x_k\}$, where

$$x_{k+1} = x_k + p_k \qquad (4.3)$$

and p_k is defined by (4.1) or (4.2). Newton's method is often regarded as an 'ideal', in large part because of its fast *quadratic* convergence. When x_0 is sufficiently close to the solution and the relevant matrix (Jacobian or Hessian) is nonsingular, the error after each pure Newton step is effectively *squared*:

$$\|x_{k+1} - x^*\| = O(\|x_k - x^*\|^2).$$

Far from the solution, a pure Newton method is unreliable and may fail to converge. The standard way to encourage convergence from a general starting point is to require a reduction at each iteration in some *merit function* that measures progress. The new iterate is then defined by

$$x_{k+1} = x_k + \alpha_k p_k, \tag{4.4}$$

where α_k is a positive scalar called the *step length*.

To guarantee convergence, the step α_k must satisfy conditions known as *sufficient decrease* in $\|\Phi\|$ or f, as appropriate. The process of choosing α_k to produce a sufficient decrease in F is called a *line search*. Standard sufficient decrease conditions are discussed in detail in, for example, Ortega and Rheinboldt (1970). For well-behaved problems, the ultimate quadratic convergence of Newton's method can be retained because the 'natural' step of unity ($\alpha_k = 1$ in (4.4)) asymptotically produces a sufficient decrease.

4.3. Linear equality constraints

Newton's method for minimizing $f(x)$ subject to linear equality constraints can be derived in two ways: solving the nonlinear equations associated with optimality, or solving the constrained minimization subproblem derived from the local quadratic model.

The optimal x^* and multiplier λ^* can be viewed as the solution of the system (2.9) of $n + m$ nonlinear equations in the variables (x, λ). Given x_k and λ_k, we substitute the Jacobian from (2.9) into the generic Newton equation (4.1), which leads to two equivalent linear systems satisfied by the Newton step (p_k, δ_k):

$$\begin{pmatrix} H_k & -A^T \\ A & 0 \end{pmatrix} \begin{pmatrix} p_k \\ \delta_k \end{pmatrix} = \begin{pmatrix} H_k & A^T \\ A & 0 \end{pmatrix} \begin{pmatrix} p_k \\ -\delta_k \end{pmatrix} = \begin{pmatrix} -g_k + A^T \lambda_k \\ b - A x_k \end{pmatrix}. \tag{4.5}$$

(The second form has been rewritten with $-\delta_k$ as an unknown so that the matrix is symmetric.) The matrices in (4.5) are nonsingular whenever A has full rank and the reduced Hessian $N^T H_k N$ is positive definite. Another option is to treat the 'new' Lagrange multiplier $\lambda_{k+1} = \lambda_k + \delta_k$ as an unknown, producing the linear system

$$\begin{pmatrix} H_k & A^T \\ A & 0 \end{pmatrix} \begin{pmatrix} p_k \\ -\lambda_{k+1} \end{pmatrix} = \begin{pmatrix} -g_k \\ b - A x_k \end{pmatrix}. \tag{4.6}$$

The form (4.6) is often called the *augmented system*; the symmetric indefinite matrix in (4.6) is sometimes called the *KKT matrix*.

Viewed from a minimization perspective, the Newton step p_k is chosen to minimize the Taylor-series quadratic model of f subject to satisfying the constraints $A x_{k+1} = b$. With this formulation, p_k solves the quadratic program

$$\underset{p\in\mathbb{R}^n}{\text{minimize}} \quad \tfrac{1}{2}p^T H_k p + g_k^T p \quad \text{subject to} \quad Ap = b - Ax_k. \qquad (4.7)$$

This subproblem is itself an optimization problem subject to linear equality constraints. The Newton iterate $x_k + p_k$ satisfies the conditions (2.7a–b) for optimality of (4.7) if

$$Ap_k = b - Ax_k \quad \text{and} \quad g_k + H_k p_k = A^T \lambda_{k+1}. \qquad (4.8)$$

After rearrangement, we again obtain the same linear system (4.6).

If x_k already satisfies the linear constraints, so that $Ax_k = b$, the Newton step is constrained to lie in the null space of A. In this case, the first equation in (4.8) becomes $Ap_k = 0$. If H_k is nonsingular, we may multiply the second equation by AH_k^{-1} and use the fact that $Ap_k = 0$, yielding the equations

$$AH_k^{-1}A^T\lambda_{k+1} = AH_k^{-1}g_k;$$
$$p_k = H_k^{-1}A^T\lambda_{k+1} - H_k^{-1}g_k.$$

5. Linear programming

The main focus of interior methods since 1984 has been on linear programming. Much of the work on interior methods for LP can be viewed as an application of the general theory for barrier functions (Section 3), with enormous simplifications arising from the special properties of linear programs. Before describing specific interior methods, we give the relevant background on LP, emphasizing the special structure that is relevant to barrier methods.

It should be stressed in advance that hundreds of papers have been and continue to be written about interior LP methods, so that preparation of a complete list of references would be a daunting task. Fortunately, the excellent survey article of Gonzaga (1992) contains an extensive bibliography covering most aspects of the subject through mid-1991. A general bibliography on interior methods has been compiled by Kranich (1991), and can be accessed via electronic mail.

5.1. Background

For various historical and computational reasons, linear programs are widely stated in so-called *standard form*:

$$\underset{x}{\text{minimize}} \quad c^T x \quad \text{subject to} \quad Ax = b, \quad x \geq 0, \qquad (5.1)$$

where A is $m \times n$. The nonnegativity bound constraints $x \geq 0$ are the only inequalities in a standard-form problem. It is customary to assume that the rows of A are linearly independent. The point x is called *strictly feasible* for the linear program (5.1) if $Ax = b$ and $x > 0$.

A standard-form LP is a convex program (see Definition 10), and first-order conditions are sufficient for optimality. Combining results for linear equalities and general inequalities, we see that the feasible point x^* is a minimizer of the standard-form LP (5.1) if and only if, for some m-vector y^* and n-vector z^*,

$$c = A^T y^* + z^*, \quad z^* \geq 0, \quad \text{and} \quad z_i^* x_i^* = 0 \quad \text{for } i = 1, \ldots, n.$$

The vector z^* is the Lagrange multiplier for the inequality (simple bound) constraints, and y^* is the Lagrange multiplier for the equality constraints. We note that x^* and z^* satisfy a complementarity relation (see (2.13c)); because of the special nature of standard form, the values of the variables $\{x_i\}$ are also the values of the inequality constraints.

A well known property of linear programs is that, if the optimal objective value is finite, a *vertex* minimizer must exist. For a standard-form LP, a point $x^* \geq 0$ satisfying $Ax^* = b$ is a vertex if at least $n - m$ components of x^* are zero. At a *nondegenerate* vertex x^*, exactly $n - m$ components are zero. For details about linear programming and its terminology, see, e.g., Chvátal (1983) or Goldfarb and Todd (1989).

The LP (5.1) is traditionally called the *primal* problem. Its *dual* may be written in the inequality form

$$\underset{y}{\text{maximize}} \ b^T y \quad \text{subject to} \quad A^T y \leq c, \tag{5.2}$$

or in standard form:

$$\underset{y,z}{\text{maximize}} \ b^T y \quad \text{subject to} \quad A^T y + z = c, \quad z \geq 0. \tag{5.3}$$

The vector z in (5.3) is called the *dual slack*. The solution y^* of the dual is the Lagrange multiplier for the m general equality constraints in the primal, and the primal solution x^* is the Lagrange multiplier for the n equality constraints of the standard-form dual problem (5.3).

The termination criteria in many interior LP methods are based on an important relationship between the primal and dual objective functions. Let x be any primal-feasible point (satisfying $Ax = b$, $x \geq 0$) and y any dual-feasible point (satisfying $A^T y \leq c$), with z the dual slack vector $c - A^T y$. It is straightforward to show that

$$c^T x - b^T y = x^T z \geq 0. \tag{5.4}$$

The necessarily nonnegative quantity $c^T x - b^T y$ is called the *duality gap*, and is zero if and only if x and y are *optimal* for the primal and dual.

Given a primal-feasible x and a dual-feasible y, the duality gap also provides a computable bound on the closeness of $c^T x$ to the optimal value $c^T x^*$. Assume that $c^T x - b^T y = \beta$. Since $c^T x^* = b^T y^*$,

$$c^T x - c^T x^* = \beta + b^T y - b^T y^*.$$

Because the dual objective is maximized by y^* for all dual-feasible y, we know that $b^T y^* \geq b^T y$, which means that $b^T y - b^T y^* \leq 0$. Similarly, because the primal objective is minimized among all primal-feasible x, $c^T x - c^T x^* \geq 0$. Hence a duality gap of β at (x, y) implies

$$0 \leq c^T x - c^T x^* \leq \beta. \tag{5.5}$$

5.2. The central path

Suppose that we wish to apply a barrier method to a standard-form LP for which the following assumptions are satisfied:

(a) the set of x satisfying $Ax = b$, $x > 0$, is nonempty;
(b) the set (y, z) satisfying $A^T y + z = c$, $z > 0$, is nonempty;
(c) $\text{rank}(A) = m$.

Because the inequality constraints in a standard-form problem are exclusively simple bounds, the corresponding logarithmic barrier function is $B(x, \mu) = c^T x - \mu \sum_{i=1}^n \ln x_i$. The barrier subproblem involves minimizing $B(x, \mu)$ subject to satisfying the linear equality constraints:

$$\text{minimize} \quad c^T x - \mu \sum_{i=1}^n \ln x_i \quad \text{subject to} \quad Ax = b. \tag{5.6}$$

The gradient and Hessian of $B(x, \mu)$ for this case have particularly simple forms:

$$\nabla B(x, \mu) = c - \mu X^{-1} e, \quad \nabla^2 B(x, \mu) = \mu X^{-2}, \tag{5.7}$$

where $X = \text{diag}(x_i)$.

The barrier subproblem (5.6) has a unique minimizer if assumption (b) is satisfied, i.e. there exist points that are strictly feasible for the dual problem. (This result can be deduced from the special nature of linear programs and Theorem 4.) The optimality conditions (2.7) for linear equality constraints imply the existence of y such that the solution of (5.6) satisfies

$$c - \mu X^{-1} e = A^T y \quad \text{or} \quad c = A^T y + \mu X^{-1} e.$$

Defining $z = \mu X^{-1} e$, we may write

$$c = A^T y + z \quad \text{and} \quad Xz = \mu e.$$

These equations are reminiscent of the equations (3.47) that hold along the barrier trajectory, since c is the objective gradient and the variables x are also the inequality constraints. The *central path* for a standard-form LP is defined by the vectors $x(\mu)$, $y(\mu)$ and $z(\mu)$ satisfying

$$\begin{align}
Ax &= b, \quad x > 0 \tag{5.8a} \\
A^T y + z &= c, \quad z > 0 \tag{5.8b} \\
Xz &= \mu e. \tag{5.8c}
\end{align}$$

The central path plays a crucial role in many interior LP methods; see Gonzaga (1992) for a detailed survey of methods based on the central path. We stress that relation (5.8c) formally defines the concept of 'centering' x and z, namely using the barrier parameter to control the distance of both vectors from zero (the boundary). Furthermore, the objective value along the path provides an estimate of the deviation from optimality; see (3.26).

5.3. The primal Newton barrier method

Assume that we are given a point x satisfying $Ax = b$ and $x > 0$, and that we wish to apply a barrier method to solve the standard-form LP (5.1). Using the forms (5.7) for the barrier gradient and Hessian, the Newton subproblem (4.7) for (5.6) is

$$\text{minimize}_p \ \tfrac{1}{2}\mu p^T X^{-2} p + c^T p - \mu p^T X^{-1} e \quad \text{subject to} \quad Ap = 0. \qquad (5.9)$$

The first-order optimality criteria of (2.7) applied to (5.9) show that the Newton direction p must satisfy

$$\mu X^{-2} p + c - \mu X^{-1} e = A^T y \qquad (5.10)$$

for some Lagrange multiplier vector y. (We use y rather than λ for the Lagrange multiplier to retain consistency with LP notation.) Multiplying through by X^2 and noting that $Xe = x$, we obtain two expressions for p:

$$p = \frac{1}{\mu} X^2 (A^T y - c + \mu X^{-1} e) \qquad (5.11a)$$

$$p = x + \frac{1}{\mu} X^2 (A^T y - c). \qquad (5.11b)$$

An expression for the Lagrange multiplier y is derived by multiplying (5.10) by AX^2 and using the relation $Ap = 0$ to eliminate p:

$$AX^2 A^T y = AX^2 c - \mu AXe = AX(Xc - \mu e). \qquad (5.12)$$

Because A has full rank and $x \neq 0$, the matrix $AX^2 A^T$ is positive definite.

Equation (5.12) has the familiar form of the *normal equations* for a linear least-squares problem with coefficient matrix XA^T. The vector y can therefore equivalently be represented as the solution of

$$\text{minimize}_y \ \|XA^T y - (Xc - \mu e)\|_2^2. \qquad (5.13)$$

The residual vector corresponding to (5.13) is given by $r = XA^T y - (Xc - \mu e)$. Applying (5.11a), we see that the Newton direction p satisfies

$$p = \frac{1}{\mu} Xr,$$

and is a diagonally scaled multiple of the least-squares residual.

Alternatively, we can think of p and y as forming a combined vector of dimension $n + m$ that solves the linear equations corresponding to the augmented system (4.6) for (5.9):

$$\begin{pmatrix} \mu X^{-2} & A^T \\ A & 0 \end{pmatrix} \begin{pmatrix} p \\ -y \end{pmatrix} = \begin{pmatrix} -c + \mu X^{-1} e \\ 0 \end{pmatrix}. \tag{5.14}$$

Since by construction $Ap = 0$, it follows that $A(x + \alpha p) = b$ for any step α along p whenever $Ax = b$. The pure Newton iterate for this problem (see (4.3)) is $\bar{x} = x + p$; however, a step of unity may violate strict feasibility. Because the constraints are linear, strict feasibility is retained if the step taken along p is less than the step $\hat{\alpha}$ to the boundary of the feasible region. With simple bound constraints, $\hat{\alpha}$ can be calculated directly: for all indices i such that $p_i < 0$, $\hat{\alpha}$ is the smallest value of $-x_i/p_i$.

A model primal Newton barrier algorithm includes both 'outer' and 'inner' iterations. The outer iterations reduce the barrier parameter, and the inner iterations apply Newton's method to solve the current barrier subproblem (5.6).

Primal Newton Barrier Algorithm
    $k \leftarrow 0$; $\mu_0 > 0$; x_0 satisfies $Ax_0 = b$ and $x_0 > 0$;
    while x_k is not sufficiently close to optimal for the LP **do**
        $x_k^0 \leftarrow x_k$; $i \leftarrow 0$;
        while x_k^i is not sufficiently close to optimal for (5.6) **do**
            Calculate the Newton direction p^i at x_k^i;
            $x_k^{i+1} \leftarrow x_k^i + \alpha^i p^i$, where $\alpha^i < \hat{\alpha}^i$ and $B(x_k^{i+1}, \mu_k) < B(x_k^i, \mu_k)$;
            $i \leftarrow i + 1$;
        end while
        $x_{k+1} \leftarrow x_k^i$;
        Choose $\mu_{k+1} < \mu_k$; $k \leftarrow k + 1$;
    end while

The major computational effort associated with a primal barrier algorithm is the calculation of the Newton direction. The equations satisfied by the Newton direction can be written in a variety of theoretically equivalent forms, each of which suggests different linear algebraic techniques; the linear algebra issues will be discussed in Section 7.1.

The unspecified, implementation-dependent aspects of this algorithm include the selection of x_0 and μ_0, the strategy for altering the barrier parameter, and the choice of termination criteria for the inner and outer iterations. With suitable modification, this algorithm does not necessarily require a strictly feasible point. See Gonzaga (1992) for references on approaches that allow a general starting point.

With respect to the choice of the step α, it has been universally observed in practice that a very simple line search strategy of choosing α as a fixed fraction close to one (say, 0.95 or 0.99) of the step $\hat{\alpha}$ to the boundary almost always produces an adequate sufficient decrease (see Section 4.2) in the barrier function. In the extremely rare cases when this step is inadequate, a standard backtracking line search may be used. (For *nonlinear* problems, the issue of the line search becomes more complicated; see Murray and Wright (1991).)

Affine scaling interior methods were originally derived in terms of scaled steepest descent, and at first sight appear unrelated to barrier functions. However, the primal affine scaling method corresponds to defining y from (5.12) with $\mu = 0$. In general, an affine scaling method may be viewed as the limiting case of Newton's method applied to a barrier function. Gonzaga (1992) discusses the history of affine scaling methods.

In the first few years following 1984, affine scaling techniques experienced considerable popularity, in large part because of their simplicity, and were among the most effective in practice. At present, *primal-dual* methods, to be discussed in the next section, are accepted as the most efficient interior methods for LP. Certainly they are the most widely implemented in major software packages.

5.4. Primal-dual barrier methods

The primal Newton barrier algorithm just described finds a Newton step in only the primal variables x; the Lagrange multiplier y arises from the equality-constrained Newton subproblem (5.9). An alternative approach is motivated by finding primal and dual variables x, y, and z that satisfy the (rearranged) nonlinear equations from (5.8) that define the central path:

$$\Phi(x, y, z) = \begin{pmatrix} XZe - \mu e \\ Ax - b \\ A^T y + z - c \end{pmatrix} = 0. \tag{5.15}$$

Note that the second and third equations are linear; all the nonlinearity occurs in the first equation.

Applying Newton's method (4.1) to this system, we obtain

$$J \begin{pmatrix} p_x \\ p_y \\ p_z \end{pmatrix} = \begin{pmatrix} Z & 0 & X \\ A & 0 & 0 \\ 0 & A^T & I \end{pmatrix} \begin{pmatrix} p_x \\ p_y \\ p_z \end{pmatrix} = \begin{pmatrix} \mu e - XZe \\ b - Ax \\ c - A^T y - z \end{pmatrix}, \tag{5.16}$$

where J is the Jacobian of Φ in (5.15) and p_x, p_y and p_z are the Newton directions for x, y and z. Despite the difference in derivation, the linear systems associated with the Newton step in a primal-dual method have the same character as those in a primal method.

The third equation in (5.16) gives the expression

$$p_z = -A^T p_y + c - A^T y - z.$$

Substituting in the first equation to eliminate p_z, we obtain an augmented system involving p_x and p_y:

$$\begin{pmatrix} X^{-1}Z & A^T \\ A & 0 \end{pmatrix} \begin{pmatrix} p_x \\ -p_y \end{pmatrix} = \begin{pmatrix} \mu X^{-1} e - c + A^T y \\ b - Ax \end{pmatrix} \tag{5.17}$$

(see (4.6)). Since both x and z are strictly positive, the matrix $X^{-1}Z$ may be written as a positive diagonal matrix D^2, with $d_i^2 = z_i/x_i$.

Using $X^{-1}Z$ as a block pivot to eliminate p_x from the second equation of (5.17), the result is

$$AZ^{-1}XA^T p_y = AZ^{-1}X(c - \mu X^{-1} e - A^T y) + b - Ax. \tag{5.18}$$

As in (5.12), the matrix is symmetric and positive definite, with the form AD^2A^T. Once p_y is known, p_z and p_x may be calculated without solving further equations.

Finally, if $b - Ax$ may be written as ADv for some vector v (for example, if $Ax = b$), the equations (5.18) are the normal equations for a linear least-squares problem with matrix DA^T; see (5.13).

Primal-dual algorithms typically have a form similar to the primal algorithm given in Section 5.3. Most implementations choose separate steps for the primal and dual variables, in each case to ensure a sufficient decrease in some suitable merit function. When x and z are respectively primal- and dual-feasible, the easily-computable duality gap provides a guaranteed measure of the deviation from the optimal objective value; see (5.5).

Primal-dual methods of several varieties have been implemented, many with great practical success. An important feature not discussed here is the use of a 'predictor–corrector' technique closely related to extrapolation along the trajectory. For detailed discussion of primal-dual methods, see, for example, Lustig *et al.* (1990) and Mehrotra (1990).

6. Complexity issues

It is interesting as well as ironic that interior methods possess the same property as the simplex method: they are much faster in practice than indicated by complexity analysis. As we shall see, the typical upper bound on the number of iterations required by an interior method is *extremely* large for a problem of even moderate size. However, a lighthearted 'rule of thumb' articulated by several implementors is that interior LP methods tend to converge in an effectively *constant* number of iterations for many problems. This discrepancy has yet to be explained rigorously.

A major reason for the widespread interest in interior methods has been

their provable polynomial complexity. Although formal complexity proofs
are not typical in the numerical analysis literature, we have included one to
indicate its flavour.

6.1. The role of problem size

In complexity proofs, it is standard to assume that *exact arithmetic* is used,
and that all data of the problem (i.e. the entries of A, b and c) are *integers*.
This is equivalent to assuming that the problem data are rational, since
rational values can be rescaled to become integers. We therefore assume, in
discussions of complexity only, that the entries of A, b and c are integers.

For a standard-form LP with n variables and m general constraints, the
worst-case complexity of the simplex method depends on the number of vertices of the feasible region (which provides an upper bound on the number
of iterations) and on the number of arithmetic operations required to perform an iteration. Both of these numbers can be bounded by expressions
involving only the dimensions n and m.

When analysing the complexity of interior methods, however, a 'new' integer L makes an appearance. The usual interpretation is that L measures the
'size' of a linear program, and indicates the amount of information needed
to represent an encoding of the problem.

The exact definition of L varies somewhat in the literature. For example,
in Goldfarb and Todd (1989), the value of L for a standard-form LP with n
variables and m equality constraints is defined as

$$L = \sum_{i=0}^{m} \sum_{j=0}^{n} \lceil \log(|a_{ij}| + 1) + 1 \rceil, \tag{6.1}$$

where $a_{i0} = b_i$ and $a_{0j} = c_j$. It should be stressed that L can be enormous
for problems of even moderate dimension.

The value of L enters the complexity analysis at both lower (termination)
and upper (initialization) extremes. The role of L in the initialization of
interior methods will be discussed following Theorem 9.

With respect to termination, Khachian (1979) showed that the smallest
possible nonzero variation in the objective function between any two distinct
vertices is expressible in terms of L. In particular, if x is any vertex, then $c^T x$
is either equal to the optimal value $c^T x^*$ or must exceed the optimal value
by at least 2^{-2L}. This bound depends on the fact that an optimal vertex
is the solution of a linear system involving b and a nonsingular submatrix
of A. Under an integrality assumption on the entries of A and b, Cramer's
rule shows that the exact solution of such a system is a vector of rational
numbers, such that the absolute value and denominator of each component
are bounded by $2^{O(L)}$.

A stopping rule that defines acceptable closeness to optimality is needed for interior methods because the exact solution of an LP (a vertex) cannot be produced in a finite number of iterations by a method that generates strictly feasible iterates. Given any feasible point x, a formal 'rounding' procedure requiring $O(n^3)$ operations is known that will produce a vertex $\bar{x}$ for which $c^T\bar{x} \leq c^Tx$. If the objective value at an interior point is known to be within 2^{-2L} of the optimal value, a single application of the rounding procedure will produce an optimal vertex. This result follows from the property stated above concerning the minimum nonzero variation in objective values between vertices; see Papadimitriou and Steiglitz (1982) and Gonzaga (1992) for details.

Although the optimal objective value is in general unknown, an interior method that constructs primal- and dual-feasible points can use the duality gap to provide a computable upper bound on the difference between the current and optimal objective values; see (5.5).

6.2. A polynomial-time path-following algorithm

The material in this section closely follows Roos and Vial (1988); similar proofs are given in Monteiro and Adler (1989a). See Gonzaga (1992) for a survey of path-following strategies and the associated complexity bounds.

The argument typifies complexity proofs involving Newton steps and the central path. The fundamental ideas are: first, defining a computable measure of closeness to the central path; second, showing that a Newton step retains a sufficient degree of closeness to the path; and finally, decreasing the barrier parameter at a rate that allows a polynomial upper bound on the number of iterations required to reduce the duality gap to less than $2^{-O(L)}$.

An important element of the proof is a suitable definition of a *proximity measure* $\delta(x,\mu)$, which measures closeness to the central path. This quantity is defined for a strictly feasible x and positive barrier parameter μ. Let $y(x,\mu)$ and $z(x,\mu)$ denote the vectors satisfying $A^Ty + z = c$ for which $\|Xz - \mu e\|$ is minimized. This requirement means that $y(x,\mu)$ and $z(x,\mu)$ solve an optimization problem with a quadratic objective function and linear equality constraints:

$$\underset{y,z}{\text{minimize}} \quad \tfrac{1}{2}z^TX^2z - \mu x^Tz \quad \text{such that} \quad A^Ty + z = c. \tag{6.2}$$

Problem (6.2) is merely a conceptual formalism; the required vectors are implicit in the calculations (5.11) and (5.12) associated with the primal Newton direction:

$$y(x,\mu) = y \quad \text{and} \quad z(x,\mu) = c - A^Ty.$$

The projected Newton direction p (5.11) may consequently be written as

$$p = x - \frac{1}{\mu}X^2z, \tag{6.3}$$

where z is the optimal $z(\mu)$ from (6.2).

The proximity measure $\delta(x, \mu)$ is defined as

$$\delta(x, \mu) = \|X^{-1}p\| = \left\|\frac{Xz}{\mu} - e\right\|, \quad \text{where} \quad z = z(x, \mu). \qquad (6.4)$$

If x solves (5.6), i.e. x lies on the central path, then $\delta(x, \mu) = 0$. The value of δ is thus a scaled measure of the Newton step, and indicates the distance of x from the central path.

It is convenient to use the n-vector s whose ith component is $s_i = x_i z_i / \mu$, so that

$$s = \frac{Xz}{\mu} \quad \text{and} \quad z = \mu X^{-1}s. \qquad (6.5)$$

The definitions of δ and s imply the relations

$$\delta^2 = \sum_{i=1}^{n}\left(\frac{x_i z_i}{\mu} - 1\right)^2 = \|s - e\|^2 = \sum_{i=1}^{n}(s_i - 1)^2. \qquad (6.6)$$

It follows from (6.3) and (6.5) that the next iterate $\bar{x}$ of a pure Newton method is

$$\bar{x} = x + p = 2x - \frac{X^2 z}{\mu} = 2x - Xs. \qquad (6.7)$$

Component-wise, the new iterate satisfies

$$\bar{x}_i = 2x_i - x_i s_i = (2 - s_i)x_i. \qquad (6.8)$$

Before treating the algorithm itself, we show that the duality gap is bounded if the proximity measure is sufficiently small.

Lemma 2 (Bounds on the duality gap.) If x is strictly feasible, $\delta(x, \mu) < 1$ and the vectors y and z solve (6.2), then y is dual-feasible (i.e., $A^T y \leq c$) and

$$\mu\left(n - \delta(x, \mu)\sqrt{n}\right) \leq c^T x - b^T y \leq \mu\left(n + \delta(x, \mu)\sqrt{n}\right).$$

Proof. Since $x > 0$ and $\delta(x, \mu) < 1$, it follows from the first equation in (6.6) that $x_i z_i \geq 0$. Hence $z \geq 0$, which means that $A^T y \leq c$ and y is dual-feasible.

Because x and y are primal- and dual-feasible, we know from (5.4) that the duality gap is given by $x^T z$, with $z = c - A^T y$. By definition of δ and e,

$$\delta(x, \mu)\sqrt{n} = \left\|\frac{Xz}{\mu} - e\right\|\|e\|.$$

Applying the Cauchy-Schwarz inequality, we obtain

$$\delta(x, \mu)\sqrt{n} = \left\|\frac{Xz}{\mu} - e\right\|\|e\| \geq \left|e^T\left(\frac{Xz}{\mu} - e\right)\right| = \left|\frac{x^T z}{\mu} - n\right|,$$

which leads to

$$n - \delta(x, \mu)\sqrt{n} \leq \frac{x^T z}{\mu} \leq n + \delta(x, \mu)\sqrt{n}.$$

Multiplying by μ gives the desired result. □

We now begin a sequence of lemmas that prove crucial relationships about Newton steps and the proximity measure. First, we show that, if the proximity measure is sufficiently small for given x and μ, the proximity measure for the same value of μ is squared at the next Newton iterate.

Lemma 3 (Quadratic convergence of proximity measure.) Let x satisfy $Ax = b$, $x > 0$, and assume that $\delta(x, \mu) < 1$. Then the next Newton iterate $\bar{x}$ (6.7) also satisfies $A\bar{x} = b$ and $\bar{x} > 0$. Further, $\delta(\bar{x}, \mu) \leq \delta(x, \mu)^2$.

Proof. Because $\delta(x, \mu) < 1$, it follows from (6.6) that $|s_i - 1| < 1$, so that $0 < s_i < 2$ for $1 \leq i \leq n$. Relation (6.8) then implies that $\bar{x} > 0$. The fact that $A\bar{x} = b$ is immediate from the construction (5.9) of p to satisfy $Ap = 0$.

Because $\delta(\bar{x}, \mu)$ is the smallest value of $\|\bar{X}z/\mu - e\|$ for all vectors y and z satisfying $A^T y + z = c$, we have

$$\delta(\bar{x}, \mu) \leq \left\|\frac{\bar{X}z}{\mu} - e\right\|.$$

Using the relations $z = \mu X^{-1}s$ and $\bar{x}_i = 2x_i - x_i s_i$ gives

$$\frac{\bar{X}z}{\mu} = \bar{X}X^{-1}s = (2X - XS)X^{-1}s = 2s - S^2 e.$$

Therefore, $\delta(\bar{x}, \mu) \leq \|2s - S^2 e - e\|$, which means that

$$\delta(\bar{x}, \mu)^2 \leq \sum_{i=1}^{n}(2s_i - s_i^2 - 1)^2 = \sum_{i=1}^{n}(s_i - 1)^4 \leq \left(\sum_{i=1}^{n}(s_i - 1)^2\right)^2 = \delta(x, \mu)^4.$$

The condition $\delta(x, \mu) < 1$ thus ensures that the pure Newton iterates converge quadratically to the point $x(\mu)$ on the central path. □

The next lemma develops a bound on the proximity criterion corresponding to a reduced value of the barrier parameter.

Lemma 4 (Effect of a reduction in μ.) If θ satisfies $0 < \theta < 1$ and $\bar{\mu}$ is defined as $(1 - \theta)\mu$, then

$$\delta(x, \bar{\mu}) \leq \frac{\delta(x, \mu) + \theta\sqrt{n}}{1 - \theta}.$$

Proof. It follows from the definition of δ that, for $z = z(x, \mu)$,

$$\delta(x, \bar{\mu}) \leq \left\|\frac{Xz}{\bar{\mu}} - e\right\|.$$

Let

$$\nu = \frac{1}{1-\theta} = \frac{\mu}{\bar{\mu}},$$

so that $\nu > 1$ and $\nu - 1 = \theta/(1-\theta)$. Then

$$\delta(x, \bar{\mu}) \leq \left\| \frac{Xz}{\bar{\mu}} - e \right\| = \left\| \nu \frac{Xz}{\mu} - e \right\| = \| \nu s - e \|.$$

Applying the triangle inequality gives

$$
\begin{aligned}
\delta(x, \bar{\mu}) &\leq \| \nu(s - e) + (\nu - 1)e \| \\
&\leq \nu \| s - e \| + (\nu - 1) \| e \| \\
&= \frac{\delta(x, \mu) + \theta \sqrt{n}}{1 - \theta},
\end{aligned}
$$

which is the desired result. $\square$

We now combine the preceding two lemmas to obtain a bound on the proximity measure for the Newton iterate with a barrier parameter that has been reduced by a factor related to the problem dimension.

Lemma 5 (Bounds on the proximity measure.) Assume that $\delta(x, \mu) \leq \frac{1}{2}$, and let

$$\theta = \frac{1}{6\sqrt{n}}, \quad \text{so that} \quad \theta \sqrt{n} = \frac{1}{6}.$$

When $\bar{x}$ is the Newton iterate (6.7) and $\bar{\mu} = (1 - \theta)\mu$, then $\delta(\bar{x}, \bar{\mu}) \leq \frac{1}{2}$.

Proof. Applying first Lemma 4 and then Lemma 3, we have

$$
\begin{aligned}
\delta(\bar{x}, \bar{\mu}) &\leq \frac{\delta(\bar{x}, \mu) + \theta \sqrt{n}}{1 - \theta} \leq \frac{\delta(x, \mu)^2 + \theta \sqrt{n}}{1 - \theta} \\
&\leq \frac{\frac{1}{4} + \frac{1}{6}}{1 - \theta} = \frac{\frac{5}{12}}{1 - \theta} \\
&\leq \frac{1}{2},
\end{aligned}
$$

where the last inequality holds because $1/(1-\theta) \leq \frac{6}{5}$. $\square$

An approximate path-following method based on reducing the barrier parameter and taking a single Newton step is obviously suggested by these results. Assume that we are given a strictly feasible x_0 and barrier parameter μ_0 such that $\delta(x_0, \mu_0) \leq \frac{1}{2}$; the latter condition can always be satisfied, as we shall discuss after the proof of Theorem 9. The following algorithm constructs a sequence of pairs (x_k, μ_k) such that every x_k is strictly feasible, $\mu_k > 0$, $\delta(x_k, \mu_k) \leq \frac{1}{2}$, and $\mu_k \to 0$ as $k \to \infty$.

Let q be an accuracy parameter, to be described later, and define θ as $1/(6\sqrt{n})$.

Algorithm I

   $k \leftarrow 0$;

   while $n\mu_k > \mathrm{e}^{-q}$

      $\mu_{k+1} \leftarrow (1-\theta)\mu_k$;

      $x_{k+1} \leftarrow 2x_k - X_k^2 z_k/\mu_k$, where $z_k = z(x_k, \mu_k)$ of (6.2);

      $k \leftarrow k+1$;

   end while

An upper bound for the number of iterations required by Algorithm I is given in the following theorem.

Theorem 9 (Worst-case number of iterations.) Define $q_0 = \lceil \ln(n\mu_0) \rceil$. Then Algorithm I will terminate after at most $6(q + q_0)\sqrt{n}$ steps, and the final iterate x and the corresponding y obtained from (6.2) satisfy

$$c^T x - b^T y \le \frac{3}{2}\,\mathrm{e}^{-q}.$$

Proof. We know from Lemma 3 that every iterate x_k is strictly feasible, and from Lemma 5 that $\delta(x_k, \mu_k) \le \frac{1}{2}$. The algorithm terminates when k satisfies $n\mu_k \le \mathrm{e}^{-q}$, where by construction $\mu_k = (1-\theta)^k \mu_0$. Applying the definition of q_0, termination will occur when

$$n\mu_k = n(1-\theta)^k \mu_0 \le (1-\theta)^k \mathrm{e}^{q_0} \le \mathrm{e}^{-q}.$$

Taking logarithms, the termination condition is

$$-k\,\ln(1-\theta) > q + q_0. \qquad (6.9)$$

Since $-\ln(1-\theta) > \theta$ for all $\theta < 1$, the inequality (6.9) holds if $k\theta > q + q_0$. Using the definition of θ, we see that the algorithm terminates if k satisfies

$$k > 6\sqrt{n}(q + q_0),$$

which gives the first desired result.

For the final iterate x_k, let $y_k = y(x_k, \mu_k)$. Lemma 2 implies that y_k is dual feasible and that

$$c^T x_k - b^T y_k \le \mu_k\left(n + \delta(x_k, \mu_k)\sqrt{n}\right).$$

Since $n\mu_k \le \mathrm{e}^{-q}$ and $\delta(x_k, \mu_k) \le \frac{1}{2}$, rearrangement gives

$$c^T x_k - b^T y_k \le \mathrm{e}^{-q}\left(1 + \frac{\delta(x_k, \mu_k)}{\sqrt{n}}\right) \le \frac{3}{2}\,\mathrm{e}^{-q},$$

which completes the proof. $\square$

To conclude that this bound is polynomial, we need to connect L (6.1) to both the initial barrier parameter μ_0 (which defines q_0) and to the accuracy parameter q.

The value of μ_0 is related to L in detailed proofs by Monteiro and Adler

(1989a, b), who show that any LP is polynomially equivalent (i.e. can be transformed in a polynomial number of operations) to an LP of similar size. For this related LP, a strictly feasible x_0 and an initial $\mu_0 = 2^{O(L)}$ are known such that x_0 lies on the central path. With these choices of μ_0 and x_0, the value of q_0 is $O(L)$, and $\delta(x_0, \mu_0) = 0$. It should be emphasized that this value of μ_0 is enormous, and would never be used in a practical algorithm.

Turning now to the accuracy parameter q, we know from our initial discussion of L that the algorithm should terminate when the duality gap is less than $2^{-O(L)}$. In Algorithm I, the duality gap is tested against e^{-q}. Consequently, q should be chosen as $O(L)$ to ensure that the rounding procedure will produce an optimal vertex from the final iterate.

With both q and q_0 taken as $O(L)$, the bound of Theorem 9 is indeed $O(\sqrt{n}L)$ iterations. Finally, each iteration of the algorithm requires $O(n^3)$ operations, to calculate p, y and z from (5.11) and (5.12). The total computational effort for Algorithm I is therefore $O(n^{3.5}L)$, which is (as promised) a polynomial in the problem size.

Polynomiality has been proved for a wide variety of interior methods for linear and quadratic programming. See, for example, Monteiro and Adler (1989b), in which the nature of the 'rounding' required for QP is described in detail. Various authors have proposed interior path-following methods for convex nonlinear problems satisfying certain assumptions. Recent discussion of these approaches is given in, for example, Nesterov and Nemirovsky (1989), den Hertog et al. (1990), Mehrotra and Sun (1990), and Jarre (1991). With these methods, polynomial bounds can be proved only on the number of iterations, since no rounding procedure exists for general nonlinear problems.

7. Linear algebraic issues

A persuasive argument can be made that the *practical* success of interior methods depends on numerical linear algebra. For very large problems, even (say) 40 iterations of an interior method would be inordinately time-consuming if the associated linear systems could not be solved efficiently and reliably.

7.1. Linear algebra in interior LP methods

The linear systems in interior methods for linear programming have a strikingly different nature from those associated with the simplex method. At each simplex iteration, two (transposed) square $m \times m$ systems are solved, and the matrix changes by only a single column per iteration. Typical implementations of the simplex method perform an initial sparse LU factorization of the basis, followed by Forrest–Tomlin or Bartels–Golub updates. As an aside, we stress that linear algebra in the 'real' simplex method bears almost no resemblance to a typical textbook tableau. Recent discussions of selected linear algebraic issues in the simplex method may be found in Bixby (1990), Duff *et al.* (1986) and Forrest and Tomlin (1990, 1991).

Interior LP methods have been of practical interest mainly for large problems, and we henceforth assume that the matrix A is large and sparse. In most implementations to date, the Newton direction is calculated from equations arising in two theoretically equivalent formulations:

(i) *Normal-equation* form, involving an $m \times m$ symmetric positive-definite matrix AD^2A^T (see (5.12) and (5.18));

(ii) *Augmented system* form, containing an $(n + m)$-dimensional specially-structured symmetric indefinite matrix (see (5.14) and (5.17)).

Least-squares problems such as (5.13) have primarily been solved by conversion to (i) or (ii), although some interest remains in application of sparse QR factorizations. A complete discussion of the relevant linear algebraic issues for all these approaches is given in Björck (1991), along with an extensive bibliography.

With either (i) or (ii), the following features are important:

• The $n \times n$ matrix D changes completely at every iteration, but its elements are converging to quantities associated with x^*;

• The Newton direction need not necessarily be computed with high accuracy, since it is only a means to follow the path. Unless there is a complete breakdown in accuracy, the line search ensures progress for any direction of descent with respect to the particular merit function.

The simplest and by far the most popular linear algebraic technique for the normal-equation approach is direct solution: we explicitly form AD^2A^T and compute its Cholesky factorization,

$$AD^2A^T = R^TR,$$

where R is upper triangular. Sparse Cholesky factorizations have been widely studied and carefully implemented in several sparse matrix packages. Comprehensive discussions are given in, for example, George and Liu (1981) and Duff *et al.* (1986).

Most implementations of a sparse Cholesky factorization perform an initial symbolic *analyse phase* that constructs a pivot ordering intended to produce a sparse factor R. When AD^2A^T is sufficiently positive definite, all pivoting orders are numerically stable, so that the ordering need not be altered later. Because only the diagonal scaling D changes at each iteration of an interior LP method, a single analyse phase suffices for all iterations. After a suitable ordering is determined, the triangular matrix R is calculated using the numerical values in AD^2A^T.

Standard ordering heuristics, most commonly minimum degree and minimum local fill, have been very effective in interior methods. The calculation of R has been organized in both 'left-looking' and 'right-looking' versions. The best choice of ordering and organization has been found, not surprisingly, to depend on details of the hardware such as vectorization and memory hierarchy.

Although the barrier Hessian is nonsingular at strictly interior iterates, it becomes asymptotically singular when the linear program is dual degenerate. Many (some would say most) real linear programs display a high degree of dual degeneracy, leading to obvious ill-conditioning in AD^2A^T. The small number of observed numerical difficulties with the normal-equation approach has therefore been a continuing surprise. A careful error analysis is likely to explain this phenomenon, but it remains slightly mysterious at this time.

The major practical difficulty with forming the Cholesky factorization of AD^2A^T is known as the 'dense column' problem. If any columns of A contain a relatively large number of nonzeros, the matrix AD^2A^T is much denser than A. (If A has even one entirely dense column, AD^2A^T fills in completely.) To retain efficiency, some strategy must be developed to detect and treat dense columns separately.

Suppose that A is partitioned into two subsets of columns, with a similar partition of D:

$$A = (A_1 \ A_2), \quad \text{so that} \quad AD^2A^T = A_1D_1^2A_1^T + A_2D_2^2A_2^T,$$

where A_2 contains the dense columns. The hope is to solve systems involving AD^2A^T without forming the matrix explicitly, using a Cholesky factorization of the 'sparse part':

$$A_1D_1^2A_1^T = R_1^TR_1.$$

A direct strategy can be devised by observing that the solution p of $AD^2A^Tp = d$ also satisfies

$$\begin{pmatrix} A_1D_1^2A_1^T & A_2 \\ A_2^T & -D_2^{-2} \end{pmatrix} \begin{pmatrix} p \\ z \end{pmatrix} = \begin{pmatrix} d \\ 0 \end{pmatrix}. \tag{7.1}$$

It is well known that the extended system (7.1) can be solved if we can solve linear systems involving $A_1D_1^2A_1^T$ and the (negative) Schur complement

$$C = D_2^{-2} + A_2^T(A_1D_1^2A_1^T)^{-1}A_2.$$

The matrix C can be expressed in terms of R_1 as

$$C = D_2^{-2} + A_2^T(R_1^TR_1)^{-1}A_2 = D_2^{-2} + U^TU,$$

where $U = R_1^{-T}A_2$. The desired vector p is found by solving (in order)

$$R_1^Tv = d, \quad Cz = U^Tv, \quad R_1p = v - Uz.$$

If the column dimension of A_2 is small, the positive-definite matrix C can be formed and factorized without undue effort. This technique for dealing with dense columns is discussed in, for example, Marxen (1989) and Choi *et al.* (1990).

A second approach involves applying an *iterative* technique such as the conjugate gradient method. Given that the equations need not be solved exactly, there is some hope that the required number of iterative steps will not be too large on average. Because AD^2A^T is often ill-conditioned, preconditioning is essential. An obvious source for the preconditioner is a 'sparsified' Cholesky factorization of AD^2A^T, such as the factorization of $A_1D_1^2A_1^T$; see, for example, Gill *et al.* (1986). Other strategies combining direct and iterative techniques have also been devised; see Lustig *et al.* (1990).

A drawback with either strategy is that the matrix $A_1D_1^2A_1^T$ remaining after removal of dense columns has frequently been found to be extremely ill-conditioned or even numerically singular. A second problematic aspect is that heuristic criteria must be developed to identify which columns qualify as dense.

We now turn to formulation (ii) – solving an augmented system in which the matrix has one of the forms

$$ K = \begin{pmatrix} D^{-2} & A^T \\ A & 0 \end{pmatrix} \quad \text{or} \quad M = \begin{pmatrix} \beta I & \tilde{D}A^T \\ A\tilde{D} & 0 \end{pmatrix}. \tag{7.2}$$

The second matrix arises from a least-squares formulation, and the scalar β is a scaling factor included to improve stability. Its selection is a compromise between preserving sparsity and maintaining stability; see Arioli *et al.* (1989) and Björck (1991).

Both K and M are symmetric but obviously indefinite. (We shall refer to K in the discussion, but most comments apply also to M.) The standard direct method for solving systems of this form involves calculation of the symmetric indefinite factorization

$$ P^T K P = LBL^T, $$

where P is a permutation matrix, L is unit lower-triangular, and B is block-diagonal, with 1×1 or 2×2 blocks. For dense problems, P is chosen using a stability criterion that determines whether to use a 1×1 or 2×2 pivot; see Bunch and Kaufman (1977).

In contrast to the positive-definite case, it cannot be guaranteed that all pivoting orders for a symmetric indefinite matrix are numerically stable. The analyse phase for the symmetric indefinite factorization thus attempts to choose a pivot ordering based solely on sparsity that will lead to low fill-in in L. When the factorization itself is computed with the actual numerical values, interchanges that alter the predicted pivot sequence may be required to retain numerical stability.

The augmented system approach involves an increase in dimension compared to the normal equations, as well as a more complicated factorization. Nonetheless, solving the augmented system should be more reliable numerically, particularly in avoiding instabilities attributable to dense columns. Very promising results have been reported by Fourer and Mehrotra (1990).

As a compromise between approaches (i) and (ii), some suggestions have been made for working with 'partially reduced' augmented systems of the form K in (7.2). The idea is to perform a block pivot in K with the 'good' part of D, simultaneously producing a smaller system and retaining numerical stability. In any such approach, the dense columns of A are placed in the portion of K that is not factorized; see, for example, Vanderbei (1991) and Gill *et al.* (1991). Alternatively, the indefinite system can be solved using an iterative method with a sparse preconditioner; see Gill *et al.* (1990).

Although taking advantage of symmetry often leads to savings in storage and computation, some linear algebra issues are simplified by ignoring symmetry. An approach that deserves exploration is the use of unsymmetric but highly structured systems, such as (5.16).

Much opportunity clearly remains for improvements and refinements in the linear algebraic aspects of interior LP methods.

7.2. Linear algebra for nonlinear problems

For nondegenerate linear programs, the results of Section 3.6 show that the barrier Hessian is asymptotically nonsingular, since $\hat{A}$ (the Jacobian matrix of the active constraints) has rank n. As soon as we consider *nonlinear* problems (including quadratic programming), however, in general the Hessian of the barrier function becomes increasingly ill-conditioned as the solution is approached along the trajectory Since the exact solution of an ill-conditioned problem is by definition extremely sensitive to small changes in the data, interior methods might appear to be fundamentally unsound.

Fortunately, a more optimistic view is justified by several observations. Inherent ill-conditioning afflicts the barrier Hessian only 'near' the solution, which is precisely where asymptotic properties of the Lagrangian function and the barrier trajectory apply. In particular, a 'good' step toward x^* from a point sufficiently near x^* is *not* poorly determined. The ill-conditioning is consequently an artifact of the barrier transformation rather than inherent to the constrained problem. In effect, the ill-conditioning gradually and implicitly reveals subspace information whose asymptotic nature is known.

If the correct active set is identified, a highly accurate approximation to the Newton step can be calculated in two orthogonal 'pieces' lying in the range of $\hat{A}^T$ and the null space of $\hat{A}$, where the condition of the relevant equations reflects that of the original problem; see Wright (1976). But since a definitive property of interior methods is that they do *not* make an explicit identification of the active set, it is arguably inappropriate to make such a prediction. More recent work on this issue has several flavours.

The Newton equations can be solved using a rank-revealing Cholesky (or modified Cholesky) factorization with symmetric interchanges (Higham, 1990), where linear algebraic criteria are invoked to define numerical rank-deficiency. When the condition of the Hessian becomes excessively large, its Cholesky factors lead to bases for the required range and null spaces (Wright, 1991).

If the nonlinear constraints are formulated in 'standard form', namely as $c(x) = 0$, $x \geq 0$, the barrier transformation applies only to the simple bounds. (Inequality constraints can always be converted to standard form by adding nonnegative slack variables.) The resulting Hessian of the barrier function asymptotically approaches the Hessian of the Lagrangian plus a diagonal matrix, some of whose entries are becoming unbounded. In this form, the ill-conditioning is concentrated entirely in large diagonal elements of the Hessian, and does not affect the sensitivity of the solution of the associated KKT system (Ponceleón, 1990).

Finally, in the spirit of seeking nonsymmetric matrices that may avoid difficulties with symmetric forms, we recall from Section 3.5 that the matrix

$$\begin{pmatrix} H - \sum \lambda_i H_i & -A^T \\ \Lambda A & C \end{pmatrix},$$

which arises in a primal-dual characterization of the barrier trajectory, is *nonsingular* at x^*, and does not suffer inevitable ill-conditioning. A nonlinear primal-dual algorithm can thus be developed in which the linear systems are unsymmetric but well-conditioned; see McCormick (1991).

It is still unknown which, if any, of these strategies will be most successful in overcoming the difficulties with conditioning that plagued barrier methods for nonlinear problems in the 1960s and 1970s.

8. Future directions

Many issues remain to be resolved for interior methods, even for linear programming. At the most basic level, the problem categories for which simplex and interior methods are best suited are not well understood. In addition, the gap between worst-case and average-case performance has not been satisfactorily explained.

One great strength of the simplex method is its efficient 'warm start' capability. Many large linear programs do not arise only once, in isolation, but are modified versions of an underlying model. After each change in the model, the resulting LP is re-solved. Because the simplex method can make effective use of *a priori* information, it is not uncommon for the solution to be found in a very small number of simplex iterations – many fewer than if the problem were solved from scratch. In contrast, the very nature of interior methods is to move away from the boundary and then approach the solution along a central path. No effective strategy has yet been devised for allowing interior methods to exploit 'strong hints' about the constraints active at the solution.

For nonlinear problems, researchers are returning with fresh enthusiasm to old topics, such as the treatment of ill-conditioning, the choice of merit function, and termination of the solution of each barrier subproblem. The work of Nesterov and Nemirovsky (1989) suggests new, previously unconsidered, barrier functions, which may be of practical as well as theoretical significance.

It seems safe to predict that the field of interior methods will continue to produce interesting research to suit every taste.

Acknowledgments

The figures in this paper were produced with the MetaPost system of John Hobby; his patient help with MetaPost and fine points of LaTeX is gratefully acknowledged. I thank Ken Clarkson, David Gay, Arieh Iserles, Jeff Lagarias, Mike Powell and Norm Schryer for helpful comments on content and exposition. Special thanks go to Philip Gill for his detailed reading of the manuscript and numerous suggestions for improvements.

REFERENCES

M. Arioli, I.S. Duff and P.P.M. de Rijk (1989), 'On the augmented system approach to sparse least-squares problems', *Numer. Math.* **55**, 667–684.

M. Avriel (1976), *Nonlinear Programming: Analysis and Methods*, Prentice-Hall (Englewood Cliffs, NJ).

D.A. Bayer and J.C. Lagarias (1989), 'The nonlinear geometry of linear programming, I: Affine and projective scaling trajectories', *Trans. Am. Math. Soc.* **314**, 499–526.

D.A. Bayer and J.C. Lagarias (1991), 'Karmarkar's linear programming algorithm and Newton's method', *Math. Program.* **50**, 291–330.

R. Bixby (1990), 'Implementing the simplex method: the initial basis', Report 90-32, Department of Mathematical Sciences, Rice University, Houston, Texas.

A. Björck (1991), 'Least squares methods', Institute of Technology, Linköping University, Linköping, Sweden.

J.R. Bunch and L.C. Kaufman (1977), 'Some stable methods for calculating inertia and solving symmetric linear systems', *Math. Comput.* **31**, 162–179.

I.C. Choi, C.L. Monma and D.F. Shanno (1990), 'Further development of a primal-dual interior point method', *ORSA J. Comput.* **2**, 304–311.

V. Chvátal (1983), *Linear Programming*, W. H. Freeman (New York).

D. den Hertog, C. Roos and T. Terlaky (1990), 'On the classical logarithmic barrier function method for a class of smooth convex programming problems', Report

90-01, Faculty of Technical Mathematics and Informatics, Delft University of Technology, Delft, Holland.

I.S. Duff, A.M. Erisman and J.K. Reid (1986), *Direct Methods for Sparse Matrices*, Oxford University Press (London).

A.V. Fiacco (1979), 'Barrier methods for nonlinear programming', in *Operations Research Support Methodology* (A.G. Holzman, ed.), Marcel Dekker (New York), 377–440.

A.V. Fiacco and G.P. McCormick (1968). *Nonlinear Programming: Sequential Unconstrained Minimization Techniques*, John Wiley and Sons (New York). Republished by SIAM (Philadelphia), 1990.

R. Fletcher (1987), *Practical Methods of Optimization* (second edition), John Wiley and Sons (Chichester).

J.J.H. Forrest and J.A. Tomlin (1990), 'Vector processing in simplex and interior methods for linear programming', *Annals of Operations Research* **22**, 71–100.

J.J.H. Forrest and J.A. Tomlin (1991), 'Implementing the simplex method for the Optimization Subroutine Library', Research Report RJ 8174, IBM Almaden Research Centre, San Jose, California.

R. Fourer and S. Mehrotra (1990), 'Performance of an augmented system approach for solving least-squares problems in an interior-point method for linear programming', Report, Department of Industrial Engineering and Management Sciences, Northwestern University, Evanston, Illinois.

K.R. Frisch (1955), 'The logarithmic potential method of convex programming', Report, University Institute of Economics, Oslo, Norway.

J.A. George and J.W.H. Liu (1981), *Computer Solution of Large Sparse Positive Definite Systems*, Prentice-Hall (Englewood Cliffs, NJ).

P.E. Gill, W. Murray, D.B. Ponceleón, and M.A. Saunders (1990), 'Preconditioners for indefinite systems arising in optimization', Report SOL 90-8, Department of Operations Research, Stanford University, Stanford, California.

P.E. Gill, W. Murray, D.B. Ponceleón, and M.A. Saunders (1991), 'Solving reduced KKT systems in barrier methods for linear and quadratic programming', Report SOL 91-7, Department of Operations Research, Stanford University, Stanford, California.

P.E. Gill, W. Murray, M.A. Saunders, J.A. Tomlin and M.H. Wright (1986), 'On projected Newton barrier methods for linear programming and an equivalence to Karmarkar's projective method', *Math. Program.* **36**, 183–209.

P.E. Gill, W. Murray and M.H. Wright (1981), *Practical Optimization*, Academic Press (London and New York).

D. Goldfarb and M.J. Todd (1989), 'Linear programming', in *Optimization* (G.L. Nemhauser, A.H.G. Rinnooy Kan and M.J. Todd, eds), North Holland (Amsterdam and New York), 73–170.

C.C. Gonzaga (1992), 'Path following methods for linear programming', *SIAM Review* **34**, to appear.

B. Grünbaum (1967), *Convex Polytopes*, John Wiley and Sons (London).

N.J. Higham (1990), 'Analysis of the Cholesky decomposition of a semi-definite matrix', in *Reliable Numerical Computation* (M.G. Cox and S. Hammarling, eds), Clarendon Press (Oxford), 161–185.

P. Huard (1967), 'Resolution of mathematical programming with nonlinear constraints by the method of centres', in *Nonlinear Programming* (J. Abadie, ed.), North Holland (Amsterdam and New York), 207–219.

F. Jarre (1991), 'Interior-point methods for convex programming', Report SOL 90-16, Department of Operations Research, Stanford University, Stanford, California.

K. Jittorntrum (1978), *Sequential Algorithms in Nonlinear Programming*, Ph.D. thesis, Australian National University.

N.K. Karmarkar (1984), 'A new polynomial time algorithm for linear programming', *Combinatorica* **4**, 373–395.

N.K. Karmarkar (1990), 'Riemannian geometry underlying interior-point methods for linear programming', in *Mathematical Developments Arising from Linear Programming* (J.C. Lagarias and M.J. Todd, eds), American Mathematical Society (Providence, RI), 51–75.

L.G. Khachian (1979), 'A polynomial algorithm in linear programming', *Doklady Akademiia Nauk SSSR* **244**, 1093–1096 (in Russian); English translation in *Sov. Math. Dokl.* **20**, 191–194.

V. Klee and G.J. Minty (1972), 'How good is the simplex algorithm?', in *Inequalities III* (O. Shisha, ed.), Academic Press (New York), 159–175.

E. Kranich (1991), 'Interior point methods for mathematical programming: a bibliography', Report 171, Universität Hagen, Hagen, Germany. This bibliography can be accessed electronically by sending email to 'netlib@research.att.com' with message 'send intbib.tex from bib' and/or 'send intbib.bbl from bib'.

D.G. Luenberger (1984), *Introduction to Linear and Nonlinear Programming*, Addison-Wesley (Menlo Park, CA).

I. Lustig, R.E. Marsten and D.F. Shanno (1990), 'On implementing Mehrotra's predictor–corrector interior point method for linear programming', Report SOR 90-03, Department of Civil Engineering and Operations Research, Princeton University, Princeton, New Jersey.

A. Marxen (1989), 'Primal barrier methods for linear programming', Ph.D. thesis, Stanford University, Stanford, California.

G.P. McCormick (1991), 'The superlinear convergence of a nonlinear primal-dual algorithm', Report T-550/91, Department of Operations Research, George Washington University, Washington, DC.

N. Megiddo (1987), 'Pathways to the optimal set in linear programming', in *Progress in Mathematical Programming* (N. Megiddo, ed.), Springer-Verlag (New York), 131–158.

S. Mehrotra (1990), 'On the implementation of a (primal-dual) interior method', Report 90-03, Department of Industrial Engineering and Management Sciences, Northwestern University, Evanston, Illinois.

S. Mehrotra and J. Sun (1990), 'An interior point algorithm for solving smooth convex programs based on Newton's method', in *Mathematical Developments Arising from Linear Programming* (J.C. Lagarias and M.J. Todd, eds), American Mathematical Society (Providence, RI), 265–284.

R.D.C. Monteiro and I. Adler (1989a), 'Interior path following primal-dual algorithms, Part I: Linear programming', *Math. Program.* **44**, 27–41.

R.D.C. Monteiro and I. Adler (1989b), 'Interior path following primal-dual algorithms, Part II: Convex quadratic programming', *Math. Program.* **44**, 43–66.

J.J. Moré and S.J. Wright (1990), private communication.

W. Murray (1971), Analytical expressions for the eigenvalues and eigenvectors of the Hessian matrices of barrier and penalty functions, *J. Optim. Theory Appl.* **7**, 189–196.

W. Murray and M.H. Wright (1991), 'Line search procedures for the logarithmic barrier function', Manuscript, AT&T Bell Laboratories, Murray Hill, New Jersey.

Y. Nesterov and A. Nemirovsky (1989), *Self-Concordant Functions and Polynomial-Time Methods in Convex Programming*, USSR Academy of Science (Moscow).

J.M. Ortega and W.C. Rheinboldt (1970), *Iterative Solution of Nonlinear Equations in Several Variables*, Academic Press (London and New York).

C.R. Papadimitriou and K. Steiglitz (1982), *Combinatorial Optimization: Algorithms and Complexity*, Prentice-Hall (Englewood Cliffs, NJ).

D.B. Ponceleón (1990), *Barrier Methods for Large-scale Quadratic Programming*, Ph.D. thesis, Stanford University, Stanford, California.

M.J.D. Powell (1972), 'Problems related to unconstrained optimization', in *Numerical Methods for Unconstrained Optimization* (W. Murray, ed.), Academic Press (London and New York), 29–55.

M.J.D. Powell (1990), 'Karmarkar's algorithm: a view from nonlinear programming', *IMA Bulletin* **26**, 165–181.

J. Renegar (1988), 'A polynomial-time algorithm based on Newton's method for linear programming', *Math. Program.* **40**, 59–94.

R. T. Rockafellar (1970), *Convex Analysis*, Princeton University Press (Princeton, NJ).

C. Roos and J.-Ph. Vial (1988), 'A polynomial method of approximate centres for linear programming', Report 88-68, Faculty of Mathematics and Informatics, Delft University of Technology, Delft, Holland. To appear in *Math. Program.*

G. Sonnevend (1986), 'An analytic centre for polyhedrons and new classes of global algorithms for linear (smooth, convex) programming', in *Lecture Notes in Control and Information Science*, Vol. 84, Springer-Verlag (New York), 866–876.

R.J. Vanderbei (1991), 'Symmetric quasi-definite matrices', Report 91-10, Department of Civil Engineering and Operations Research, Princeton University, Princeton, New Jersey.

M.H. Wright (1976), *Numerical Methods for Nonlinearly Constrained Optimization*, Ph.D. thesis, Stanford University, California.

M.H. Wright (1991), 'Determining subspace information from the Hessian of a barrier function', Manuscript, AT&T Bell Laboratories, Murray Hill, New Jersey.